Thomas Kieselbach · Simo Mannila (Eds.)

Unemployment, Precarious Work and Health

Psychologie sozialer Ungleichheit

Herausgegeben von
Prof. Dr. Thomas Kieselbach
Mitglied des Vorstands der International Commission on Occupational Health (ICOH)
Institut für Psychologie der Arbeit, Arbeitslosigkeit und Gesundheit (IPG)
Universität Bremen / Förderwerk Bremen

Ziel und Inhalt der Reihe „Psychologie sozialer Ungleichheit"

Die Entwicklung der Arbeitsmärkte in den hochindustrialisierten Ländern hat für viele Menschen in den vergangenen Jahrzehnten die Erfahrung von Arbeitsplatzverlust und Arbeitslosigkeit mit sich gebracht. Auch wenn die Bewältigung beruflicher Brüche nicht zwangsläufig zu persönlichen Krisen mit psychosozialen Schädigungen führen muss, ist dennoch zu betonen, dass besonders bei verletzlichen Gruppen eine solche Erfahrung den Weg in Langzeitarbeitslosigkeit und damit oft verknüpft soziale Exklusion begünstigt.
In der Reihe werden Themen behandelt, die sich mit den individuellen, organisationsbezogenen und sozialpsychologischen Folgen beruflicher Umbrüche sowie daraus folgenden Risiken sozialer Ausschließung befassen. Der Schwerpunkt liegt im Bereich der psychologischen Arbeitslosenforschung, welche die individuellen und gesellschaftlichen Kosten einer inzwischen weitgehend akzeptierten Massenarbeitslosigkeit aufzeigen will. Darüber hinaus wird der Blick auch auf jene indirekten Folgen der Arbeitsmarktkrise gelenkt, welche sich in Arbeitsplatzunsicherheit oder prekären Arbeitsverhältnissen zeigen und sich als eine verstärkte Einbeziehung von Merkmalen der Arbeitslosigkeit in Beschäftigungsverhältnisse charakterisieren lassen.
Ein wichtiger Ausgangspunkt der Reihe lag in dem von der Weltgesundheitsorganisation (WHO) in den 80er Jahren konzipierten Programm „Soziale Gerechtigkeit und Gesundheit". Dort wurden erstmalig umfassend Zusammenhänge zwischen Arbeitslosigkeit und gesundheitlichen Folgewirkungen aufgezeigt, Möglichkeiten der Begrenzung des schädigenden Einflusses durch Interventionsansätze diskutiert sowie Rückwirkungen von Massenarbeitslosigkeit auf Beschäftigte thematisiert.
Die Reihe versucht in einer unübersichtlicher gewordenen Berufswelt Perspektiven aufzuzeigen, welche die individuelle Bewältigung von erzwungenen Berufswechseln erleichtern und damit krisenhafte Verläufe begrenzen helfen. Dies erfolgt durch die Betonung sowohl der gesellschaftlichen wie auch der persönlichen Verantwortung für berufliche Neuorientierungen bei industriellen Restrukturierungen.
Die Reihe „Psychologie sozialer Ungleichheit" wendet sich an PsychologInnen, SoziologInnen, EpidemiologInnen, ÖkonomInnen, ArbeitswissenschaftlerInnen, PädagogInnen und PraktikerInnen im sozialen Bereich, die in ihrer täglichen Arbeit häufig mit den psychischen Folgen von beruflichen Umbrüchen und sozialer Ungleichheit konfrontiert sind. Einbezogen werden eigenständige empirische Arbeiten sowie Literaturüberblicke und Tagungsberichte. Neben theoretischen Erörterungen werden auch Praxisevaluationen veröffentlicht, welche die Möglichkeiten und Grenzen von Interventionsansätzen im Bereich von Arbeitslosigkeit und prekären Arbeitsbedingungen untersuchen.

Thomas Kieselbach
Simo Mannila (Eds.)

Unemployment, Precarious Work and Health

Research and Policy Issues

VS VERLAG

Bibliographic information published by the Deutsche Nationalbibliothek
The Deutsche Nationalbibliothek lists this publication in the Deutsche Nationalbibliografie; detailed bibliographic data are available in the Internet at http://dnb.d-nb.de.

1st Edition 2012

Editorial Office: Dorothee Koch | Marianne Schultheis

VS Verlag für Sozialwissenschaften is a brand of Springer Fachmedien.
Springer Fachmedien is part of Springer Science+Business Media.
www.vs-verlag.de

Cover design: KünkelLopka Medienentwicklung, Heidelberg
Printed on acid-free paper

ISBN 978-3-531-18509-5

Contents

Preface

Thomas Kieselbach & Simo Mannila[1]

The roots of unemployment research date back to the early 1930s, but it was not until the 1980s as the research into unemployment and health developed into a major field of research bringing forth high-level research in many Western countries. The development of the field was then also politically motivated: there was a general concern of unemployment as a social problem, and a keen interest on the impact of unemployment on health. This concern was also supported by a network of the WHO Regional Office for Europe (Copenhagen) addressing social inequity and health and promoting international co-operation (under the responsibility of Herbert Zöllner and Per-Gunnar Svensson).

The traditional research into links between unemployment and health had its focus on the health effects of job loss and long-term unemployment and later also on the positive impact of various interventions to limit the negative health effects. A policy-relevant issue was whether and how an integration of the experience from these interventions could be used in the development of occupational health (e.g. to integrate occupational health services (OHS) in enterprise restructuring; to monitor the health of unemployed people through regular health checks from side of OHS). Another policy concern was whether the interventions might reduce the unemployment spells and be, thus, socially cost-effective. Successful interventions would also combat against social exclusion reducing the hysteresis effect of unemployment, i.e., reducing the psychosocial and social barriers to re-employment as a consequence of long-term unemployment.

The increased interest in unemployment research, the changing nature of work life and the increase of psychosocial stressors and morbidity led to the establishment of two new Scientific Committees of the International Commission on Occupational Health (ICOH) at the end of the 1990´s to address these new challenges of occupational health: the SC Work Organisation and Psychosocial Factors and the SC Unemployment and Health (in 2006 renamed as SC Unemployment, Job Insecurity and Health). The idea of the creation of the SC Unemployment, Job Insecurity and Health was to bridge the areas of psychological, sociological and socio-medical unemployment research and integrate this research with considerations of social policy and human resource management, which then were mainly separated from the field of occupational health.

1 The preface is partly based on a paper that was published together with Jukka Vuori, Institute of Occupational Health, Helsinki, Finland (elected as the new chairperson of the SC at the ICOH World Congress in Cape Town in March 2009) in the ICOH Newsletter, Dec. 2008.

The predecessor of the SC Unemployment, Job Insecurity and Health was the ICOH Working Group on Unemployment and Health established in 1998 by Jean Bertran (Lucon / Paris) and Bjorgulf Claussen (Oslo University). One of the key ideas of the Working Group was to bring unemployment research closer (and perhaps back) to the research into employment. The underlying assumptions were that the increased precariousness of the work and need for restructuring lead to a situation where occupational transitions, including unemployment become more and more common; that transitions are potentially stressful for unemployed persons and those facing job insecurity and that the increased requirement to adapt should also be facilitated by the occupational health care system. This is a social concern relevant both for jobseekers, employees and for employers to maintain the workforce healthy and employable. In the course of the past ten years, the following aspects were identified and discussed in various conferences of the SC: health situation of the unemployed, the repercussions of the precariousness of work on the health of the workforce, the new demands on the organisations in regard to corporate social responsibility in the process of restructuring, and preferably closer links between labour market and social policy as well as occupational health policy. In the course of the past ten years, there has been increased focus on the quality of re-employment and job insecurity in the SC. At present it has become obvious that the simple dichotomy of employment vs. unemployment is inadequate in the globalized post-modern labour markets.

We see that the heydays of the research into unemployment and health are in the past in Europe as well as in other Western countries. In all Western countries the policy and research focus shifted in the 1990s from unemployment to more differentiated labour market concerns. At the same time the interest in links between unemployment, job insecurity and health seems to have diminished and been replaced by purely economic concerns. It is unclear how the actual recession caused by the financial and economic crisis will influence the scientific and socio-political agenda in various countries. According to some very preliminary estimates, the crisis will make world-wide approximately 20-50 million people redundant, which highlights the continuous importance of the focus of the SC Unemployment, Job Insecurity and Health. The necessity to adapt to the globalization of the market, goods and services has stimulated the restructuring of companies and organisations in all countries, sectors and branches. Economic restructuring has already in many ways transformed the nature of jobs and work and has increased the need for flexibility of the workforce. The increasing amount of transitions during the life course - into and out-of work, between jobs - may challenge the well-being, motivation and health of individuals. In compliance with the idea of lifelong learning, people now have to update their education and vocational skills throughout their work career in order to maintain their status in the labour market. The development changes also the ways in which generations of young people make their transition from school to work. Their work careers are often characterized by discontinuity, and they may find themselves overeducated and under-employed. Youth employment is recognized as one of the key risks of social exclusion and sometimes exacerbated by e.g. discrimination and ethnic segregation. Employees who try to return to work after longer absence from work or disabled persons willing to take up working, too, have difficulties in reintegrating themselves into work. Senior workers have to face challenging work changes as they try to keep up with the new developments and stay healthy and motivated before their transition to retirement. This means that there is an increasing need of work ability programmes, based on human resource management and interventions of occupational health. Despite increasing literature on work transitions, well-being and health, there are many open questions, and there is very little research on coping in work transitions and on its consequences for well-being and health, work life participation and productivity in the long run. Research needs to identify the mediation processes that produce different health outcomes.

This brings growing challenges for individuals, organisations including enterprises and for societies. How individuals respond to increasing work transitions and flexibility and how this affects individual well-being and health, will greatly depend both on individual resources for coping with the transitions and on social resources for different kinds of support, e.g., resources accessed through networks, organisational practices and public policies. The challenge of societies is to develop work life in such directions that employees are provided security in the changing work life and to promote the development and implementation of knowledge to increase individual coping resources and resources for support and design corresponding policies.

Empirical evidence shows that a considerable number of dismissed people have difficulties to adapt to the new situation without any external support, and this had led, for instance, into the development of the concept of a "social convoy" in occupational transitions. This means a scheme to accompany people in the process of transition from employment to out-of-work, training schemes or job search, and finally re-entering employment. This process involves a new interaction between human resource management and the occupational health as well as between public and private employment services and psychosocial interventions and implies an extension of the organisational responsibility for dismissals beyond the actual employment as part of corporate social responsibility (CSR) in the sense of an active labour market policy including active social plans. The new role of occupational health in this framework would be the monitoring of the health of persons in transition including out-of-work as well as those at risk of unemployment.

The overarching concept for such a new balance between individual and social responsibility can be seen in the policies compensating for the reduced security of the workplace (due to increased flexibility and precariousness) with a greater security of employability. In the life course perspective this means that job careers are constituted in a different way from a traditional model: continuity and security is provided for by new means, and there are new psychosocial risks which one must be able to cope with. This should mean a better individual adaptation in the labour market with a lifelong investment in acquiring requested qualifications and competencies on the one hand and a greater openness of the various institutions accompanying the life course which produce these skills (schools, education and training, in-job based training, rehabilitation) on the other hand. In this process human resource management and occupational health face new tasks in regard to the changing nature of employment. Equity should play a crucial role: how do we attain flexicurity and at which cost? There should be strong policies to address the differences between secure vs. insecure employment, victims-of-layoffs vs. survivors-of-layoffs, and the obvious discrepancy between occupational health and rehabilitation services available in major enterprises vs. small and medium-sized enterprises.

In order to obtain and secure a better impact of organisational interventions on an institutional and individual level there should be a better integration between interventions of the employment authorities and approaches which try to bring together health promotion and labour market reintegration. The existing research into active labour market policies and various activation measures does not seem to give here conclusive advice. Too little is known about career development and health outcomes related to the transitions and the impact of various interventions. There has also been little research on the socialization or re-socialization in the workplace for the first entrants or after a re-entry. This knowledge would also be of great importance for developing interventions to promote well-being, health and productivity in the ongoing work transitions.

Traditional unemployment research has mainly focused on health effects of the victims of organisational restructuring, and there is still a controversy concerning the societal impact of unemployment found at a macro level by means of time series related to morbidity and social disorganisation. What has been widely neglected in discussions on restructuring and health, is those who remain in the company after restructuring, the so-called "survivors-of-layoffs": they experience considerable stress levels as well due to the changed requirements, new task designs with new routines and increased job insecurity. We should also focus on the managers responsible for organizing the process, and revitalize the discourse on the fate of the families of the victims ("victims-by-proxy") and of the survivors as well as the communities in which the restructuring occur. If we want to preserve the key features of a European social model as reflected in labour market and employment relations under the new demands of a globalized competition we must not forget the individual effects of restructuring on the workforce which will show a considerable long-term impact on the competiveness of the economy as well: growth, competitiveness and employment go hand in hand. This understanding broadens the perspective from a unilateral shareholder perspective to a more balanced view on the interests of all stakeholders involved in the process of economic adaptation to the globalized economy.

There is empirical evidence that restructuring processes which neglect these issues often produce a vicious circle of restructuring leading into a loss of productivity. The health aspect of restructuring of the labour markets, labour market policies and enterprises should be considered an investment in the future at the social and enterprise levels in the same way as health is generally recognized as a key value and resource at the individual level. This understanding will bring still new stakeholders into the fore and have an impact on health insurance systems. The change of the labour markets in the industrialized countries due to globalization means increased focus on developing countries and countries in transition. In order to understand the new problems related to the new health risks, we must establish new dialogue on the labour market development and health with researchers on a global scale. The SC Unemployment, Job Insecurity and Health has made serious efforts to reach out to research from outside the Western countries, but much still remains to be done in this respect.

This volume brings finally together a selection of papers presented in the Third Conference of the SC Unemployment, Job Insecurity and Health in Bremen in 2004[2], organised by the Institute for Psychology of Work, Unemployment and Health (IPG) of

2 After the conferences in Paris in 1998 (published as Claussen, B. & Bertran, J. (eds.). (1999). International Archives of Occupational and Environmental Health (Special issue "Unemployment and Health"), 72 (Suppl.), S20 - S22) and in Adelaide in 2001 (published as Kieselbach, T., Winefield, A., Boyd, C. & Anderson, S. (eds.). (2006). Unemployment and Health. International and Interdisciplinary Perspectives. Bowen Hills: Australian Academic Press.

the University of Bremen together with Simo Mannila from STAKES Finland and the financial support of the German Research Foundation (DFG) and the Federal Association of Health Insurance funds (BKK BV) in Germany. The conference with 80 participants from 28 countries was very successful bringing together some of the key researchers into the links between unemployment and health; persons whose scientific activity covers several decades and who have had an important influence on the constitution of the research agenda in the field of unemployment and health. Under globalization and changing research environment, the new texts of these major authors give an opportunity to establish a link between the tradition of our Scientific Committee and today's new challenges.

The anthology contains also a large bulk of work by authors relatively unknown to the English-speakers. This reflects another aim of the Bremen conference: reaching out for new research, sometimes scientifically high-level but known only to native scientific circles, sometimes reflecting a rapidly developing national stream of research which deserves to be made public also internationally.

Employment and Health in the Enlarged EU – A Call for Action

Lennart Levi

1 Employment in the European Union

Policies have been adopted in some of the EU 27 countries to meet the employment target set by the European Commission: as nearly as possible to 70 per cent by 2010. This will be a challenging task for transition economies because of their sharp *decline* in employment over the decade – 7 million *fewer* people were working in 2003 than in 1993 – a decline from 58.8 per cent in 1993 to 53.5 per cent in 2003.

A particularly troubling issue is youth unemployment, which improved only slightly from 19.3 per cent in 2002 to 18.6 per cent in 2003 (ILO, 2004). In May 2009, the unemployment rate for under-25s was 19.5 per cent in the EU 27 (Eurostat, 2009). Central and Eastern Europe countries have the highest youth unemployment rates in Europe – a condition related to such aggregate reasons as lack of demand, as well as to structural reasons, such as educational mismatching, lack of adequate information and mobility constraint problems (Lubyova, 2003).

Unemployment can be a matter of health or disease, life or death (Mathers & Schofield, 1998). The unprecedented mortality upsurge experienced by many European economies in transition during the 1990s was very probably the result of an acute adjustment crisis overlaying a slow long-term deterioration of health. The crisis was fuelled by a massive increase in psychosocial stress induced first and foremost by unanticipated rises in unemployment, turnover and job insecurity, but also by the erosion of the family, mounting distress, migration and rising social stratification (Cornia, 2002).

There is strong circumstantial evidence that unemployed people have worse physical and mental health and higher mortality than people who remain employed. Marmot (2004) offers four scientifically supported interpretations but favours the fourth one. According to him, unemployed people have a worse health than those employed because:

- people who are sick find it difficult to get and hold a job;
- their upbringing and low level of skills and psychological resources lead to their being both unemployed and sick;
- unemployment leads to poverty and poverty leads to poor health; and
- unemployment represents loss of a social role and all the things that go with it.

Given the present rates of unemployment, underemployment (and over-employment!), there is an urgent need for integrative approaches to minimize unemployment and underemployment, minimize over-employment, too, promote the good job, humanize workforce restructuring, and counteract the pathogenic effects of un-, under- and over-employment.

2 EU policies

Most of these approaches were addressed in a report of the EU High Level Group on the future of social policy in an enlarged European Union. The Group's Recommendations covered a wide range of policies: (1) Employment; (2) Social protection; (3) Social inclusion; (4) Demography; (5) Combining all instruments for improved governance, and (6) External dimension (European Commission, 2004a).

In the present context, its recommendations related to employment are of particular interest. Its first priority is to extend working life; the second to implement life-long learning, and the third to address economic restructuring.

The notion of the European Employment Strategy is enshrined in the Treaty Establishing the European Community. Article 125 states that Member States shall work towards developing a coordinated strategy for employment. Under Article 128.3, every Member State must provide the Council and the Commission with an annual report on the principal measures taken to implement its employment policy in the light of the Employment Guidelines (Swedish Government, 2003). The Swedish Report quotes the Employment guidelines adopted by the Council of the European Union, which include three overarching objectives:

- full employment,
- quality and productivity at work,
- social cohesion and inclusion.

There are also ten specific guidelines, which are priorities for action:

- developing and implementing active and preventive measures for the unemployed and inactive,
- encouraging job creation and entrepreneurship,
- addressing change and promoting adaptability and mobility in the labour market,
- promoting development of human capital and lifelong learning,
- increasing the labour supply and promoting active ageing,
- promoting gender equality,
- promoting integration of, and combating discrimination against people at a disadvantage in the labour market,
- 'making work pay' through incentives to enhance job attractiveness,
- transforming undeclared work into regular employment,

- addressing regional disparities in employment.

3 More and better jobs

The European Union's basic strategy has been summarized in its Lisbon Agenda in the formula more and better jobs (European Commission 2004b). More jobs mean higher level of employment and inclusion, and less unemployment and its consequences. Better jobs means higher quality of working life, greater well-being, and less work-related health and other problems.

Important components to promote the latter objective are included in two major occupational health models – the Demand-Control Support model (Karasek & Theorell, 1990; Johnson & Hall, 1988), and the Effort-Reward Imbalance model (Siegrist, 1996). Although primarily designed for categorization and study of work-related stressors as related to health, they can probably be applied to the absence of work as well, i.e. to unemployment.

Demands (i.e., work-load) can be too high, but also too low. Control can be too low both with regard to conditions of employment, and unemployment. Similarly, support can be lacking in both types of conditions. Effort can be higher than is compatible with the preservation of health, both at work (due to over-employment or over-involvement) and in (fruitless and frustrating) job-seeking. And reward can be low under both circumstances.

The employment policies of the enlarged European Union do consider all this in both contexts, one way or another. But severe deficiencies exist with regard to the knowledge needed for evidence-based policy formulations. There are gaps between existing knowledge and its translation into policies; between policies (see above) and their actual implementation; and between implementation and evaluation, whose outcome should be fed back as new knowledge, forming a basis for improved policy formulations.

4 Future goals

With regard to *full employment*, the EU aim was an overall employment rate of 67 per cent in 2005 and 70 per cent in 2010; for women of 57 per cent in 2005 and 60 per cent in 2010; and for older workers (55 to 64 years) of 50 per cent in 2010.

The second EU objective, *improving quality and productivity at work*, encompassed:

- intrinsic quality of work;
- life-long learning and career development;
- gender equality;
- health and safety at work;
- flexibility and security;
- inclusion and access to the labour market;

- work organisation and work-life balance;
- social dialogue and worker involvement;
- diversity and non-discrimination; and
- overall work performance.

Within the third EU objective, *strengthening social cohesion and inclusion*, employment policies were to facilitate the participation in employment through promoting access to quality employment for all women and men who are capable of working.

This being said, unemployment in the enlarged EU remains rather high (21.5 million in May, 2009; ILO 2009) and also varies significantly across EU regions, ranging from 3.2 per cent in the Netherlands to 18.7 per cent in Spain.

An effective approach to the employment-unemployment and quality of working life areas necessitates a *systems* approach, very different from the prevailing fragmented piecemeal approaches.

It must be:

- interdisciplinary,
- intersectoral,
- sustainable, and
- a combination of top-down and bottom-up.

References

Cornia, G.A. (2002). The forgotten crisis – transition, psychosocial stress and mortality over the 1990s in the former Soviet block. In E. Ziglio, L. Levin, L. Levi & E. Bath (Eds.), *Investment for health. Studies on social and economic determinants of population health, 1* (pp. 32-54). Copenhagen: WHO Regional Office for Europe.

Dooley, D., Fielding, J. & Levi, L. (1996). Health and unemployment. *Annual Review of Public Health, 17*, 449-465.

European Commission (2004a). *Report of the high level group on the future of social policy in an enlarged European Union.* Retrieved from http://ec.europa.eu/employment_social-/news/2004/jun/hlg_social_elarg_en.pdf.

European Commission (2004b). *Social agenda, issue no. 10.* From http://ec.europa.eu/employment_social/social_agenda/pdf/social_agenda10_en.pdf

Eurostat (2009). Euro area unemployment up to 9.5 %. EU27 up to 8.9%. STAT/09/97, 2 July 2009

ILO. (2009) Global Employment Trends – Update, May, 2009.

ILO. (2004). *Global employment trends for youth.* Retrieved from http://www.ilo.org/public-/english/employment/strat/download/getyen.pdf.

Johnson, J.V. & Hall, E.M. (1988). Job strain, workplace social support and cardiovascular disease: A cross-sectional study of a random sample of Swedish working population. *American Journal of Public Health, 78*, 1336-1342.

Karasek, R. & Theorell, T. (1990). *Healthy work – stress, productivity and the reconstruction of working life.* New York: Basic Books.

Lubyova, M. (2003). *Youth employment and employability in the CIS.* Geneva: ILO.

Marmot, M. (2004). *Status syndrome.* London: Bloomsbury.

Mathers, C.D. & Schofield, D.J. (1998). The health consequences of unemployment: The evidence. *Medical Journal of Australia, 168*, 178-182.

Siegrist, J. (1996). Adverse health effects of high-effort-low-reward conditions. *Journal of Occupational Health Psychology, 1*, 27-41.

Swedish Government (2003). *Sweden's action plan for employment 2003*. Retrieved from http://www.sweden.gov.se/sb/d/2025/a/19850.

1. BACKGROUND AND THEORETICAL DEVELOPMENT OF UNEMPLOYMENT RESEARCH

Plea for a Renewed Research Agenda on Unemployment and Health

Ralph A. Catalano

Introduction

Some forty years have passed since Harvey Brenner (1967) brought new life to the long dormant effort to understand the health effects of economic contraction. Despite much work in the ensuing decades, our field has produced disappointingly little certainty on which, if any, of the hypothesized health effects of economic contraction we can accept or reject. I believe that this circumstance arises not from an ambiguous reality (i.e., these effects, after all, either do or do not appear in nature) but rather from a poorly organised research effort. More specifically, I argue that the empiricists among us have failed to agree on what constitutes a complete and satisfying research program for testing hypothesized effects in a systematic and timely fashion. I also argue that we spend too much effort on contributing to current policy debates, and too little on building an enduring scientific literature that can contribute over the long run to the political debate over economic policies.

I describe below the elements of what I consider an intellectually satisfying research program. I expect that this description will elicit deserved criticism, but I hope that it will also motivate others among us to offer their opinions on the matter. Such a conversation could lead us to adopt a research program that yields more timely, useful, and satisfying science than has been the case over the past three decades.

1 Elements of a satisfying research program

I argue that a research program should include at least six types of research to satisfy the combined curiosity of the audience interested in the health effects of economic contraction. These types can be arrayed in a conceptual space with columns defined by the types of method typically employed in our work and by rows defined by whether the dependent variable measures the incidence of disease or of treated disease.

The first method includes individual level analyses that measure the association between well defined psychiatric or somatic illnesses and experiences known to be more common during times of economic contraction. Many contributors to, and critics of, our field believe that such "risk factor epidemiology" yields more certainty than any other approach short of random assignment. This belief arises, in part, from the fact that risk factor epidemiology typically employs widely accepted, if not widely understood, analytic conventions. Risk ratios derived from individual level data strike many as more dependable than, for example, time-series associations.

Risk factor epidemiology also has drawbacks when used to address the question of whether contracting economies affect the incidence of illness in populations they support. One arises from the circumstance described so well by Rose (2001) in his seminal differentiation of "sick individuals" from "sick populations." We know that regional economies act as environmental stressors much like ambient temperature. The virulence of the ambient economy varies over time from low, when opportunity expands enough to meet the expectations of the population, to high when either economic contraction denies a growing fraction of the population the means to realize expectations, or when very fast economic growth poses safety and other hazards. Risk factor analysis, however, does not measure population response to varying doses of an ambient stressor. It, rather, compares groups with and without a risk factor. As Rose (2001) noted, in a population in which everyone, for example, smokes, risk factor epidemiology would attribute lung cancer to host susceptibility.

In our field, risk factor epidemiology has tended to focus on job loss even though economic contraction stresses the entire population through many other mechanisms. A reasonable person might conclude from this work that job loss increases the risk of several studied outcomes. This fact, however, does not imply that a contracting economy will induce a change in incidence consistent with the risk factor findings. Risk factor research, in other words, does not allow us to draw inferences regarding the "net effect" of economic contraction on a population. Contracting economies affect everyone they support although relatively few persons in such economies lose jobs. Other connections, continued employment with lower income, for example, may affect the dependent variable differently than does job loss.

Another shortcoming of risk factor epidemiology in our field arises from the fact that an experience presumably inflicted by the economy and subsequent diagnosis of illness can both result from the progression of pre-existing conditions. Although work in our field has gone to great lengths to control for such selection in the case of job loss (e.g., see Eliason and Storrie, 2004), the fact remains that a sceptic can always cite selection as a plausible source of error.

I argue that a well-designed research program can compensate for the drawbacks of risk factor epidemiology by including the second type of method commonly found in our collective work - time series studies of the association between the performance of an economy and the incidence of illness. This approach appropriately treats economic contraction as an ambient exogenous phenomenon and, unlike individual level epidemiology, estimates net effects of economic dynamism. These aggregate time series models can be understood as like those that estimate the yield of illness attributable to changing of doses of such environmental phenomena as air pollution or ambient temperature.

Aggregate time-series studies may also reduce scepticism arising from selection. Time-series associations between economic indicators and subsequent measures of population health unlikely arise because increased illness at time *t* caused regional economies to contract at time *t-n*. In fact, the argument that modest changes in population health typically found in most time-series studies can cause change in regional economies seems implausible even if the changes occur simultaneously.

The aggregate time-series studies in our field, however, have drawbacks that impede wide acceptance. The lack of a methodological convention, for example, implies that each article presents a unique analytic strategy (Laport, 2004). This circumstance inevitably raises the question of whether differences in findings reflect differences in method or in the phenomena under study.

A second impediment to wider acceptance of the aggregate time-series approach arises from confusion over the ecological fallacy (Robinson, 1950). It probably remains true that some readers, and perhaps authors, of aggregate time series papers make the erroneous assumption that an association discovered between two characteristics of a population (e.g.,, unemployment rate and per capita consumption of alcohol) will generalize to similar characteristics in individuals in that population (e.g.,, wanting but not finding work and the risk of alcohol abuse).

Both approaches to empirical research in our field, therefore, have weaknesses that reduce certainty. But the strengths of each approach compensate for the other's weakness. A program with both types of research can be intellectually satisfying if the work appears well done and if we have rules for what can be inferred from the various possible combinations of results.

Researchers contributing to our field can be separated for the most part into those who use risk factor epidemiology or aggregate time-series methods. These two groups often compete with each other in the attempt to explain the relationship, if any, between economic contraction and health. As someone who has tried to contribute both types of work to the field, I long ago came to the conclusion that the two types of work complement each other by compensating, as noted above, for each other's weaknesses. We, however, lack rules for drawing inferences when the findings from the two approaches diverge and converge. I suggest we develop such rules and offer the following intuitive, but not necessarily compelling, examples.

1. The two types of analyses will converge when the effect of an individual level risk factor induced by a contracting economy (e.g., job loss) predicts an outcome similar to that of another experience (e.g. fear of job loss) inflicted by economic contraction on other segments of the population. Examples of such circumstances include the association of job loss to treated depression among individuals (typically attributed to the decision to seek treatment for increased symptoms) and the time-series association between contraction of labour markets and use of mental health services (often attributed to attempts by persons with chronic illness to avoid job loss).

2. The two types of analyses will report opposite effects when an experience (e.g.,, job loss) inflicted on individuals by economic contraction affects the risk of some outcome while other experiences (e.g. fear of job loss) inflicted by a contracting economy on larger populations have a countervailing effect. An example of such an outcome could be alcohol related accidents. These may go up among job losers but down among those who, because they fear job loss, reduce any behaviour, including alcohol use, that makes them targets for dismissal.

3. Aggregate time series analysis will report an effect but risk factor epidemiology will not when the experience studied (e.g., job loss) by the latter has no effect but other experiences (e.g., fear of job loss) inflicted by a contracting economy on other segments of the population do have an effect.

4. The possibility that risk factor epidemiology reports an effect but aggregate time series analysis does not would be special case of 2 above in that the countervailing effects would "balance out" such that the net effect in the population would be 0.

A third type of method, which I will call "cross-level" analysis, in a complete research program uses risk factor epidemiology to test the possibility that job or financial stressors have effects contingent on the ambient economic environment. David Dooley and I (Catalano, Dooley, Novaco, Wilson & Hough, 1993), for example, used panel data from the Epidemiologic Catchment Area study to test the hypothesis that job loss in contracting economies has a different effect on the risk of alcohol abuse or anti-social behaviour than job loss in stable or expanding economies. We found no support for the hypothesis but these were hardly definitive tests and the theory deserves further testing for a wider array of outcomes.

The conceptual space I alluded to above has rows defined by whether the dependent variable measures the incidence of illness or of treated illness. Much of the work in our field has used the incidence of treated disorder as a surrogate for incidence of disorder. While this work often confuses the two, we should not lose sight of the fact that knowing the response of treated disorder to economic dynamism has value even if it tells us little about the association between economic dynamism and disorder. Modelling such a relationship at the very least could help care providers anticipate the need for their services. Knowing the effect, if any, of economic contraction on the demand for treatment could also further the cost/benefit analyses that presumably inform our choices among economic policies.

2 Alcohol use and abuse: An example of a satisfactorily studied health effect?

The research has produced disappointingly few examples of research programs that should, by the standard I suggested above, satisfy the curiosity of our research community. While each of us might have different candidates for programs that come close, I am most familiar with that focusing on alcohol use and abuse. I briefly summarize that program below using a few representative findings.

Risk factor epidemiology reports that, in panel data, involuntary job loss increases the risk of clinically significant alcohol dependence and abuse among those with no history of such illness (Catalano, Dooley, Wilson, & Hough 1993; Dooley, Catalano & Hough, 1992). Dooley and Prause (2004) have reported that the relationship appears to change over the lifespan with later episodes of job loss less likely to elicit clinically significant abuse. Binge drinking also reportedly increases in the multiple cross-sections of the general population and not just the unemployed, when the economy contracts (Dee, 2001).

Some of the same literature (i.e., Catalano, Dooley, Wilson & Hough, 1993; Dooley, Catalano & Hough, 1992) reports that alcohol abuse predicts subsequent job loss and that clinically significant disorder declines among individuals who remain employed in industries that shed employees (Catalano, Dooley, Wilson & Hough, 1993). This has given rise to the "inhibition" hypothesis that persons, who, during times of economic contraction, fear job loss, will reduce their use of alcohol in the hope of avoiding dismissal (Catalano, Novaco & McConnell, 1997).

Individual level research into alcohol consumption, as opposed to abuse and dependence, presents a different picture. Some reports that involuntary job loss, as opposed to unemployment, predicts increased use (e.g., Kasl & Cobb, 1982; Dooley & Prause, 2004), but other work reports no association (e.g., Iverson & Klausen, 1986), or even an inverse relationship (Ettner, 1997), leading reviewers to infer no robust effect (Temple, Fillmore, Hartka, Johnstone, Leino & Motoyoshi, 1991).

Brenner's early work (1975) reported an ecological, aggregate time series association between economic contraction and alcohol consumption. Recent aggregate level research based on more contemporary methods, however, reports that alcohol consumption declines during times of economic contraction (Ruhm, 1995).

What, if anything, do these results tell us about the possible effect of a contracting economy on the incidence of alcohol abuse and dependence? I think the work suggests that incidence probably declines due to the inhibition effect and to a simple income effect in which the fear or experience of lost wages reduces consumption of other than necessities. Job loss or income loss among those who use alcohol to cope with adversity, however, probably increases their risk of alcohol abuse and dependence above that of similar persons who remain employed.

Research in the second round effects of alcohol abuse, seem consistent with this interpretation of the findings. As implied by the inference that consumption per se decreases with lost income, alcohol related accidents appear to decline with economic contraction (Ruhm, 1995). Behaviours associated with clinically significant drinking including state removal of children from homes with abusive parents (Catalano, Lind, Rosenblatt & Attkisson, 1999; Catalano, Lind, Rosenblatt & Novaco, 2003;) and coerced treatment for grave disablement (Catalano, Novaco & McConnell, 1997; 2001) appear to increase with job loss until the losses reach high levels at which point the relationship, consistent with the inhibition theory noted above, inverts.

3 Organizing a satisfying research program

I suggest that our field could produce much more certain and intellectually satisfying research if we formed groups that included researchers with different methodological skills but shared interest in a health outcome. Members of these groups would not have to share *a priori* expectations of the associations but would have to agree *a priori* what would be inferred from various patterns of findings. Members of each group, depending on training and interest, would then commit to pursuing risk factor epidemiology, aggregate time series analyses, or cross level studies. The studies could focus on true or treated disorder. The overall effort would fill a conceptual space such as the following.

Table 1. Conceptual Space for a Research Programme

Dependent Variable	Risk Factor Epidemiology	Aggregate Time Series	Cross Level
Incidence of Disorder			
Incidence of Treated Disorder			

Filling the cells of the table would not be easy or cheap. Doing initial work in Scandinavia makes intuitive sense because the existing registries there could save researchers time and costs. As demonstrated by papers presented at our conference (see, for example, the registries described in Eliason & Storrie, 2004), these registries allow researchers to link data bases describing contacts with the health system with those describing labour market experiences.

I further suggest that we soon form a team focusing on very low birth weight. This would seem an important dependent variable for several reasons including that it causes much pain and suffering in the human community by significantly increasing the risk of infant mortality and of many developmental disabilities (Wise, Wampler & Barfield, 1995). Other reasons to study very low birth-weight include that treated and true incidences do not differ significantly in Scandinavia and in much of the developed world. We also know that the biological antecedent of very low birth weight, premature delivery, can be induced by the corticosteroids associated with the stress response (Hedegaard et al., 1996; Hobel et al., 1999; Lockwood, 1999), that natural selection may have conserved the response (Trivers & Willard, 1979), and that contracting economies may trigger it (Catalano, 2003; Catalano & Bruckner, 2005). The literature, moreover, already includes reports that could begin to populate the risk factor epidemiology (Dooley & Prause, 2004; 2005) and aggregate time series cells (Catalano, Hansen & Hartig, 1999).

4 Science, policy, and the social construction of illness

In closing, I plead two arguments that deserve much more attention than we can give them in this setting. First, I attribute our relatively weak organisation, and therefore limited success, as scientists over the last 30 years in part to our response to the moral imperative. Current policy debates, whatever they may be, too easily distract us from the enduring problem of understanding the association between economic forces and the incidence of real or attributed illness. My own experience among politicians leads me to believe that the arguments we make as scientists rarely, if ever, sway those who make public policy. Policy makers either act on beliefs held as faith, or respond to factions that inexorably pursue narrow interests until checked by countervailing groups. In these circumstances, our findings, which lack sufficient certainty to warrant broad based respect, simply become "cant" when convergent with faith or interest and are otherwise ignored or disparaged.

I plead that we repartition our efforts such that more go to contributing to the enduring literature and less to current policy debates. I understand that this may be seen as indifference to the plight of victims. I ask, however, that we consider that our science lacks sufficient certainty to affect current policy, and will affect future policy only if it yields greater certainty concerning outcomes of importance to broad constituencies. This, of course, implies that we accept the possibility that more certain findings will not necessarily serve the interests with which we, as citizens, identify.

I understand that much of the motivation and many of the resources to continue our work come not from the need to satisfy researcher curiosity but from the moral imperative. I also know that my plea to spend less time advocating and more time studying suggests a preference for the former over than latter motivation. Thirty years of observing the field, however, have led me to the conclusion that neither personal satisfaction nor social justice has been well served by the episodic, disjointed nature of our work. I argue our efforts would have served both science and humanity better had we spent more time and effort devising and completing cooperative, focused research programs and less contributing to policy debates.

In closing I also plead that we remember that our work has a tradition not only for studying the effects of economic dynamism on the physiology or behaviour of those who experience or fear adversity, but also its effects it on our tolerance for those who are behaviourally of somatically different. Harvey Brenner (1973) reminded us in his seminal book on mental hospital admissions that economic opportunity may affect the social construction of illness and aggregate time series research has supported his suspicion (Catalano & Kennedy, 1998; Catalano, Novaco & McConnell, 1997; 2002).

We should probably pay more attention to the proposition that greater collective wealth means greater capacity to "treat" the "ill." Capacity, of course, rarely goes unused, implying that we will find ill persons to treat. Greater wealth may also allow us to change our physical and social environments such that persons with other than modal physical and behavioural characteristics can function in them. Could this not reduce our need to label other than modal persons "ill?" And could not expanding economies create circumstances in which we need the labour of persons with other than modal behaviour or physiology? Would not this circumstance reduce our penchant for labelling such persons ill? Could not economic contraction have the opposite effect and thereby increase the incidence of diagnosed illness without affecting the incidence of illness?

The above considerations imply, first, that we add another row to our table. Studies of the effect of economic dynamism on our individual, organisational, and collective tolerance for illness would populate the cells in the new row. Second, these considerations imply that much of our felt need to contribute to political debates may arise not from the science we do on the physiological or psychological effects of adversity inflicted by the economy, but by the as yet less well studied possibility that the performance of the economy affects our tolerance for those afflicted by all manner of adversity regardless of source.

References

Brenner, M. (1973). Economic change, alcohol consumption and heart disease mortality in nine industrialized countries. *Social Science and Medicine, 25,* 119-132.

Brenner, M. (1973). *Mental illness and the economy*. Cambridge, MA: Harvard University Press.

Brenner, M. (1975). Trends in alcohol consumption and associated illness: Some effects of economic changes. *American Journal of Public Health, 65*, 1270-1291.

Brenner, M.H. (1967). Economic change and mental hospitalization: New York State, 1910-1960. *Social Psychiatry, 63,* 180-188.

Catalano, R. & Bruckner, T. (2005). Economic antecedents of the Swedish sex ratio. *Social Science and Medicine, 60,* 537-543.

Catalano, R. (2003). Sex ratios in the two Germanys: A test of the economic stress hypothesis. *Human Reproduction, 18,* 1972-1975

Catalano, R. & Kennedy, J. (1998). The effect of unemployment on disability caseloads in California. *Journal of Community and Social Psychology, 8,* 137-144.

Catalano, R., Dooley, D., Novaco, R., Wilson, G. & Hough, R. (1993). Using ECA survey data to examine the effect of job layoffs on violent behavior. *Hospital and Community Psychiatry, 44,* 874-878.

Catalano, R., Dooley, D., Wilson, G. & Hough, R. (1993). Job loss and alcohol abuse: A test using data from the Epidemiologic Catchment Area Project. *Journal of Health and Social Behavior, 34,* 215-226.

Catalano, R., Hansen, H. & Hartig, T. (1999). The ecological effect of unemployment on the incidence of very low birthweight in Norway and Sweden. *Journal of Health and Social Behavior, 40,* 422-428.

Catalano, R., Lind, S., Rosenblatt, A. & Attkisson, C. (1999). Unemployment and foster home placements: Estimating the net effect of provocation and inhibition. *American Journal of Public Health, 89,* 851-856.

Catalano, R., Lind, S., Rosenblatt, A. & Novaco, R. (2003). Economic antecedents of foster care. *American Journal of Community Psychology, 32,* 47-56.

Catalano, R., Novaco, R. & McConnell, W. (1997). A model of the net effect of job loss on violence. *Journal of Personality and Social Psychology, 72,* 1440-1447.

Catalano, R., Novaco, R. & McConnell, W. (2002). Layoffs and violence revisited. *Aggressive Behavior, 28,* 233-247.

Dee, T. (2001). Alcohol abuse and economic conditions: Evidence from repeated cross-sections of individual-level data. *Health Economics, 10,* 257-270.

Dooley, D. & Prause, J. (2004). *The Social Costs of Underemployment.* Cambridge: Cambridge University Press.

Dooley, D. & Prause, J. (2005). Birth weight and mothers' adverse employment change. *Journal of Health and Social Behavior, 41, 141-155.*

Dooley, D., Catalano, R. & Hough, R. (1992). Unemployment and alcoholism in 1910 and 1990: Drift versus social causation. *Journal of Occupational and Organisational Psychology, 65,* 277-290.

Eliason, M. & Storrie, D. (2004, September). *Does job loss shorten life?* Paper presented at the ICOH Conference "Precarious Work and Persistent Unemployment – Research and Policy Issues", Bremen, Germany.

Ettner, S.L. (1997). Measuring the human cost of a weak economy: Does unemployment lead to alcohol abuse? *Social Science and Medicine, 44,* 251-260.

Hartka, E., Johnstone, B., Leino, E.V., Motoyoshi, M., Temple, M.T. & Fillmore, K.M. (1991). A meta-analysis of depressive symptomatology and alcohol consumption over time. *British Journal of Addiction, 10,* 1283-1298.

Hedegaard, M., Henriksen, T., Secher, N., Hatch, M. & Sabroe, S. (1996). Do stressful life events affect duration of gestation and risk of preterm delivery? *Epidemiology, 7,* 339-345.

Hobel, C., Dunkel-Schetter, C., Roesch, S., Castro, L. & Arora, C. (1999). Maternal plasma corticotropin-releasing hormone associated with stress at 20 weeks' gestation in pregnancies ending in preterm delivery. *American Journal of Obstetrics and Gynecology, 180,* 257-264.

Laporte, A. (2004). Do economic cycles have a permanent effect on population health? Revisiting the Brenner hypothesis. *Health Economics, 13,* 767-779.

Lockwood, C. (1999). Stress-associated preterm delivery: The role of corticotropin-releasing hormone. *American Journal of Obstetrics and Gynaecology, 180,* 264-266.

Robinson, W. (1950). Ecological correlations and the behavior of individuals. *American Sociological Review, 15,* 351-357.

Rose, G. (2001). Reiteration: Sick individuals and sick populations. *International Journal of Epidemiology, 3,* 427-432.

Ruhm, C. (1995). Economic conditions and alcohol problems. *Journal of Health Economics, 14,* 583-603.

Trivers, R.L. & Willard, D.E. (1973). Natural selection of parental ability to vary the sex ratio of offspring. *Science, 179,* 90-92.

Unemployment and Health: What are the New Challenges and Opportunities for Health Systems in Europe?

Carole J. Maignan

Introduction

When speaking about the labour market and health, and in particular the health sector, the relationship between the two is often seen as unidirectional and straightforward: better health at work and more job security are important to reduce the pressure on the health system[1]. Unemployment or insecure jobs provoke stress and depression, which burdens the health system with additional costs. The common belief that failures in the job market (long-term unemployment, wrong skills matching, job insecurity, unsafe labours, etc.) have a repercussion on the health system, either directly when considering work accidents for example, or indirectly when long term unemployment or job insecurity have consequences on individual behaviour and health. But the interrelationships between the labour market and the health sector are much more complex and rich in opportunities and challenges. Not only because the labour market can express situations as different as full time employment or long term unemployment leading to social exclusion, marginalisation and poverty and therefore leading to different health consequences, but also because the health system itself, even if often under-estimated, is an intrinsic part of the economic and development system and can therefore contribute actively to the labour market and health. It is not the role of the health system to bear responsibility for poverty and marginalisation, but it is its responsibility to adapt its services and the whole system in general to changing conditions in the labour market: high levels of unemployment, longer periods of unemployment, precarious jobs, etc. require different types of services. By this we do not mean developing more mental health units in order to help depressed unemployed or stressed workers, but that the role of the public health system can make a difference by acting on the labour supply and demand, by using genuine intersectoral strategies and finally be a role model employer and health promoter.

Causes of ill health are often second on the government agenda due to the complexity of the issues and the lack of appropriate evidence on the one hand, and the ne-

1 Note that we speak here of the usual way of looking at the relationship between work and the health system. The relationship between health and labour market should of course include the reverse causality of health on labour force participation.

cessity of taking political risks and adopting long-term vision on the other hand (Marmot, 2004). Unhealthy environment or lifestyles have consequences both in rich and poor countries. In Europe, the severest problems of material deprivation have been solved (in the sense that infant mortality in the poorest groups in rich countries is low and infectious diseases are relatively reduced). But there are still major social differences, as in the case of life expectancy, for example. In addition, life expectancy is more closely related to income distribution than to other health indicators (Wilkinson, 1992). The example of Britain in 1990's is particularly telling: a 9-year gap in life expectancy between the highest and lowest social classes (almost 20 years for the US!).

Consequently, something needs to be done to tackle the causes of ill health. Campaigns for health promotion (as the "smoking kills" campaign) are not enough and do not deal with the social and economic determinants of health. There is a need for Investing for Health and Development[2] and to recognise that action across sectors is required to improve health. This in turn produces economic and social benefits and reduces the demand for health care.

1 Determinants of health

Tackling major health determinants has a great potential for reducing the burden of disease and promoting the health of the general population. Health determinants can be categorised as: personal behaviour and lifestyles; influences within communities which can sustain or damage health; living and working conditions and access to health services; and general socio-economic, cultural and environmental conditions.

1.1 Lessons from the past

The decline in death rates and infectious diseases in the last two centuries were more due to changes in environment, nutrition, supply of clean water, sewage disposal, and economic situation than to strictly "medical" measures. This is not to diminish the role of health professionals, but to insist on the special role of other determinants (not directly within the health system) in influencing and informing health.

Several studies point out the deterioration of health in Europe, especially Eastern Europe, in certain segments of the population. There exists indeed a relation between economic performance, income distribution and the health status of a nation. The higher the per capita income and the greater the income distribution equality, the greater the likelihood of living longer and healthier is. Studies in developing countries have shown that a 10% increase in income per capita corresponds to a 3.5% fall in child mortality rates (World Bank, 1993).

In today's Europe and Eastern Europe in particular, economic ups and downs in conjunction with rising unemployment, job insecurity and low paid jobs lead to widening income gaps and rising social inequality. This has created a rising proportion of people living in relative poverty. Are these changes also influencing health outcomes?

2 This mission is actually the one of the WHO Venice Office, which is aimed at clarifying, analysing and promoting better ways to invest for health and development.

Wilkinson (1996) confirms the necessity to bring social and economic issues up front in order to better understand their effect on health. The lessons can be summarized as follows:

1. Above a certain level of wealth, it is not the richest societies which have the best health but those that have the smallest income difference between rich and poor.
2. Increased inequality imposes economic, social and psychological burdens which reduce the well-being of the whole society.
3. Investment in social capital increases economic efficiency in the long run while reducing inequality, and therefore avoids the potential choice between greater equity and economic growth.

In the next section, we will provide a description of health determinants and how they are organised.

1.2 Health determinants: An overview

Health determinants are often grouped in the four following categories:

- *Lifestyle*: individual life-style is very dependent on other health determinants such as culture, employment, education, income and social and community networks. These factors must be considered in any effort to change the life-style of individuals.

Figure 1. Health determinants

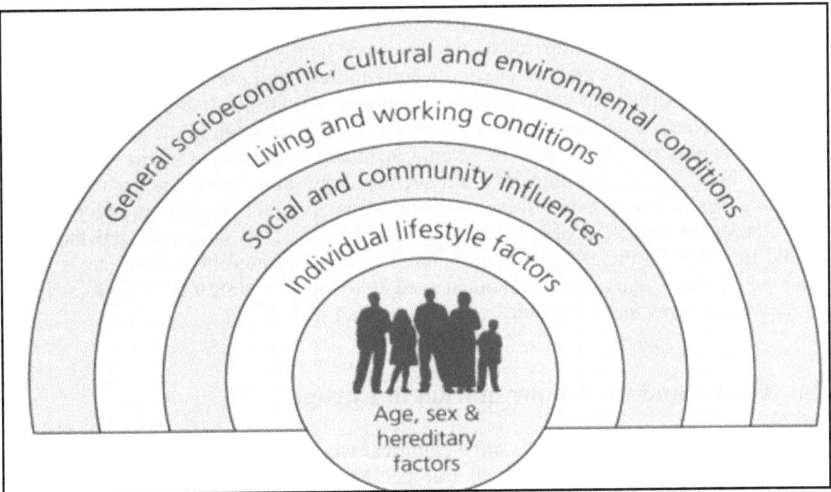

Source: Dahlgren G and Whitehead M (1991) Policies and strategies to promote social equity in health. Stockholm, Institute for Futures Studies.

- *Socio-economics*: the relationship between health and social and economic factors has been well established. Socio-economic status is an important determinant of health status, and also health status is an important determinant of socio-economic development.
- *Environment*: the physical environment is an ecosystem with which humans are in constant interaction and with which the stability of their well-being is inextricably enmeshed.
- *Genetics and screening*: genetic factors are health determinants, which extend far beyond the scope of public health interventions. The field of genetics will in future years become more and more visible.

Key determinants of health lay outside the direct influence of health and social care: education, employment, housing, and environment. The diagram above presents the determinants of health in terms of layers of influence, starting with the individual and moving to wider society. If we go deeper into the details of these determinants with a more policy-oriented perspective, we see that "non-health sector" is most important in determining health (Wamala & Lynch 2002).

Work life is a macro-level determinant of health but it is also implicitly present at the micro-level under individual socio-economic factors. Several studies look at the impact of unemployment or bad work conditions on health. For example, recent evidence from the Second report on the Health of Canadians demonstrates that employment has a significant effect on a person's physical, mental and social health. Paid work provides not only money but also a sense of identity and purpose, social contacts and opportunities for personal growth. When a person loses these benefits, the results can be devastating to both the health of the individual and his or her family. Unemployed people have a reduced life expectancy and suffer significantly more health problems than people who have a job. Conditions at work (both physical and psychosocial) can have a profound effect on people's health and emotional well-being. Participation in the wage economy, however, is only part of the picture. Many Canadians (especially women) spend almost as many hours engaged in unpaid work, such as doing housework and caring for children or older relatives. When these two workloads are combined on an ongoing basis and little or no support is offered, an individual's level of stress and job satisfaction is bound to suffer. In the 1996-97 survey, more women reported high work stress levels than men in every age category. Women aged 20 to 24 were almost three times as likely to report high work stress than the average Canadian worker[3].

2 Health and the labour market in Europe

In this section, we review briefly some empirical work on unemployment, social exclusion, temporary work and health in Europe. New trends regarding employment in

3 But the question is also whether they are really more stressed or more likely to admit to be stressed (see Bertrand and Mullainathan, 2001) or have a lower stress limit.

Europe are long-term youth unemployment (Kieselbach, 2003) and temporary employment (Storrie, 2002).

There is a large and growing literature investigating the relationship between employment and health but reviewing it is not the scope of this chapter[4]. A number of suggestions are important to bear in mind for research though[5]: the effect of economic hardship stemming from unemployment is separate from the effect of unemployment per se, unemployment may affect the health of different sub-groups of the population differently, the social matrix in which unemployment is imbedded is to be considered, a complex set of modulators of the association between health status and unemployment may be at work (financial strains, social support, psychological factors and contexts). In the graph below we summarise the possible negative effects of unemployment:

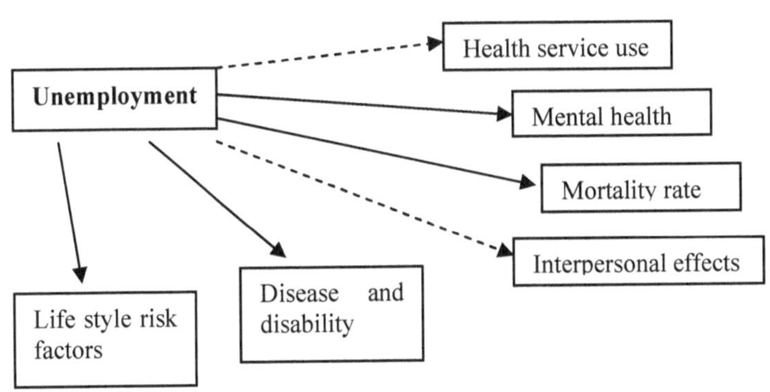

Comments on the different effects of unemployment:

Mortality: The association between unemployment and premature mortality emerges with remarkable consistency whether aggregate data or individual data is examined. Moreover, the longer the period of unemployment, the higher the mortality rate. Unemployment is specifically associated with suicide and cardiovascular disease[6].

Mental and physical health: There is ample evidence that unemployed individuals experience higher levels of psychological distress than do their employed counterparts. They also suffer from higher rates of diagnosable disorders such as depression, panic and substance abuse. These effects can vary according to the different population segments: for example, Whelan and McGinnity (2000) look at the impact of unemployment

4 For a review of evidence to suggest that employment protects and fosters health, see for example: Lavis et al., 2000; Dooley et al., 1996; Kasl and Jones, 2000.

5 Béland et al., 2001.

6 See Avison, 1998; Kuhn et al., 2004.

on well-being in Europe and report the general tendency for unemployment to have a less severe psychological impact on women, young and those with partners. However, the association between unemployment and physical illness is ambiguous. Several studies suggest that while there may be no correlation between the two during good economic times, job loss does contribute to symptoms of physical illness during economic recessions. We should mention here the existence of policies aimed at reducing unemployment levels by switching unemployed or future unemployed to the disabled group – by providing individuals incentives to report poor health. These policies have been used a lot in the Netherlands and the UK – for example the distribution of disabled benefits throughout the country is drastically higher in economically depressed areas. The high numbers on disability and incapacity were used during the Thatcher era to massage the unemployment figures and they are high in former mining areas, for instance - GPs signed people onto disability allowance because they got more money than on unemployment. However, there is increasingly high employment and low birth rates in the UK and these workers are now increasingly important for the economy. Bringing them back to work involves linking up programmes between the NHS, general practice and the employment organisations (see Section 4.1). Many small scale projects are underway to try to model how this might be done. Also technology is increasingly opening up options - telework etc.

Health service use: Efforts have been made to draw conclusions about links between unemployment, elevated rates of illness and health care utilization. A key factor appears to be the variation in access associated with different health systems. In some countries, job loss is likely to reduce access to health care even though need may increase. In others with more universal access, unemployment does seem clearly related to greater use of health services.

The health of other family members: interpersonal effects: Evidence on the effects of unemployment on spouses and children is focused in most studies on job loss among men. There is substantially less knowledge about the effects of unemployment on women and their families. The most comprehensive studies about women may have been done on poverty, not unemployment, and the adverse consequences of multiple risk factors on children's well being. Nevertheless, research does indicate that spouses of unemployed workers experience increased emotional problems. Furthermore, children, especially teens whose parents are unemployed, appear to be at higher risk of emotional and behavioural problems[7].

The impact of re-employment: The little research available in this area suggests that recovery after re-employment is neither immediate nor complete. The physical and mental health problems that are the consequences of unemployment persist[8].

In the next two sections, we look at two current phenomena observed in the European labour market in particular, i.e. long-term unemployment and temporary employment.

7 See O'Neill & Sweetman, 1998.

8 Clarck & Oswald (1994).

2.1 Long-term and temporary unemployment

Medium to long-term labour market exclusion (and/or household composition changes) can be the cause of poverty. The vicious circle of unemployment, non-permanent employment, and unemployment once again is also closely linked with job quality, unemployment and social exclusion (Taylor, 2002). In general, both permanent unemployment and social exclusion are highly persistent, whereas other states of the labour market benefit from large amounts of mobility. Ermish et al. (2001) find that children born in poverty have a reduced educational attainment and their probability of labour market participation as an adult is also lower[9]. A recent EU-funded research project focuses on the difficulties of young people entering the labour market or having lost their employment after their first job experience: it finds that unemployment is a central risk factor for young people in particular and threatens in the long-term the overall integration in society and can even lead to social exclusion (see Kieselbach, 2003). The project identifies that social exclusion processes can be considered as complex moderators of the health effects of unemployment. In all countries, the most important vulnerability factors that contribute to an increase of the risk of social exclusion for young unemployed people in the long-term are low qualification, passivity in the labour market, a precarious financial situation, low or missing social support and insufficient or nonexistent institutional support. On the other hand, the most important protective factor for unemployed youth is social support. However, the research also revealed a factor that has been rarely emphasised in earlier psychological studies: the importance of social origin: poverty and other social problems in the family can increase the risk of social exclusion for the youth. It is the specific combination of these vulnerability and protective factors that determines the risk of poor mental health. Moreover, these combinations have been found to vary in a structured way between societies as a result of different institutional frameworks and different cultural traditions, with major implications for the individual experience of unemployment. This is most evident in the striking contrasts between northern and southern European countries.

The lack of state welfare provision in the southern countries leads to a much more central role for the family in supporting unemployed people while integration into social networks is of great importance for youth from northern Europe. Especially due to the high level of family support, the number of youth at high risk of social exclusion in general is lower in southern Europe compared to northern Europe. Increasing individualization processes in southern European countries, however, may weaken this buffer effect of family support in the future. Given the very low level of institutional support in these societies, this would create a major need for social policy reform.

The conclusion is that unemployment scars: unstable early years in the labour market (as observed more and more frequently in Europe) can have far reaching repercussions for an individual's career, future financial stability and a nation's social security bill. Employment and the absence of it are central determinants for poverty, social exclusion or integration, and this independently of the country's system (family versus institutional support).

9 This refers also to the consequences of unemployment on the health of other family members as described in the previous section.

Another characteristic of the European labour market is the increase of short-term contracts or temporary employment. Temporary jobs are always related with lower wages, and they are used to face sudden change in workload or recession periods. D'Addio and Rosholm (2004) look at the European Household Community Panel data and find that workers in low quality employment clusters experience a higher risk of labour market exclusion. The results are quite striking: transition rates into unemployment are twice as high in the low quality employment clusters compared with high quality employment one. They also note that non-permanent jobs are more likely to exist in the public sector than in the private sector. This surprising result is due to the need to cut down on public expenses and therefore adapting the workforce to the workload with temporary contracts.

Storrie (2002) looks at temporary work agencies in the European Union. They currently employ over seven million workers (1.9% of the EU working population). These agencies are often viewed as a tool to promote flexibility in the labour market. They are meant to improve job matching and reduce frictional unemployment. On the other hand, workers employed in these agencies are more exposed to risk factors than permanent workers. The characteristics of temporary agency workers are that they are young, with a lower educational level than the average employee, and are used more often by large firms. Workers from temporary agencies were initially used mostly in the industrial sectors but are now moving to services. In Nordic countries in particular, they are very common in the health sector. For example, health and care workers constitute as much as 27% of the turnover in temporary agencies sector in Denmark. In Finland, almost 20% of all temporary agencies workers are employed in the health sector.

Health indicators of these workers differ significantly from permanent employees only in their reporting of being highly dissatisfied but having a much lower level of stress. There are two perspectives to explain this result: On the one hand, this characterizes a loss of control over working tasks due to the fact that the worker is considered as an outsider of the user firm, on the other hand, a temporary agency worker can choose how much he wants to work (explaining a high proportion of students using these agencies) and what kind of job they want (in theory), and are therefore less stressed or overloaded. Other studies underline the feeling of insecurity and dull work likely to cause stress, another disadvantage is the lack of access to training, and the occupational accidents among temporary agency workers is higher than among employees. So far the data show ambiguous conclusions in relation to working conditions in the temporary agency sector. Considering the results from this brief review of the labour market and health consequences, our question is then: what can the health system do?

2.2 Labour market supply: What can the health system do?

As we saw in the previous section, unemployment affects health in more than one negative way. Adding more funds to the health systems could be a way to deal with the increase of ill health conditions deriving from increasing unemployment or "bad" employment, but it is not sustainable nor provides any long-term solution. So what can the health system do about this issue? Several solutions can be proposed: prevent health deterioration, workplace health promotion, provide conditions to maximise productivity even under bad health conditions, adapt best cure and treatment, but also working hours, as in the case of HIV/AIDS contaminated workers for example, assist families with unhealthy members or self-dependent elderly, etc. These solutions are all at most palliative to the consequences of unemployment and bad health. But can the health system intervene directly or indirectly on the labour market to reduce unemployment and prevent health troubles coming from unemployment or under-employment?

The central problem for health care systems is that they are financially unsustainable given the technological progress made in medicine (more and more costly cures) and the demographic evolution. One important aspect is that health care systems do not produce health, or in other terms do not have any impact on population health determinants in order to lower the demand pressure on their services (Harrison & Ziglio, 1998). Nevertheless, population-based social policy changes, rather than individual medical interventions, brought about most of the health improvements achieved between 1841 and 1935. Improved housing, safe clean accessible water, better nutrition, better family planning, better information and more disposable income were key factors. Improvements in medical science played a very minor role. Despite this, at the inception of the UK National Health System for example, over 90% of the health care budget was strictly dedicated to the health care of individuals[10].

But the health system has more strings to its bow than one might think. Indeed, the health care system is a major purchaser of goods and services, but most often this buying power is not systematically utilised to the best effect in the region. Strengthening the alignment of demand and supply within the region, rather than through external sources, could be a powerful driving force for regenerating the local economy, reducing the risk of poverty and enhancing the environment, including the social one. However the lack of shared data and knowledge management between these sectors means that demand for goods and services in the public sector are not currently well connected to existing local supply.

Likewise, pro-active employment programmes in low-income areas could contribute to the strengthening of local economies and create better conditions for population health. The analysis could explore, for example, how to meet the demand for increasing the numbers of health and social care workers particularly in caring for the elderly, a trend that is being fuelled by the demographic changes in the age structure of European regions in general. Attention to the issue of immigrant employment, which has the potential to address both significant employment shortages in the health and social care sector, and improve the health of immigrant families and their fuller integration into the community can for example be a win-win situation. In the next section, we try to illustrate this particular aspect of the role of the health system in the labour market.

As mentioned earlier, in many European countries the health system is one of the largest employers and a major purchaser of goods and services. The potential to harness this economic power positively is enormous, therefore, there is a pressing need to enable national, regional and local governments to examine how their health system can be better utilized to create opportunities for social and economic development, reduce damages to the environment, reduce health inequalities, and in the long term free up resources to create conditions for good population health.

---

10 This is also because prevention is relatively cheap while treatment is expensive. The issue here is also the marginal return of both types of intervention. Indeed, an additional euro invested in socio-economic determinants of health might have a higher return than an additional euro invested in health care.

Recruitment and Employment Policies

Unemployment leads to an increase in ill health in both men and women. In most European countries, health systems have a substantial problem in recruiting the staff they need. In some inner city areas, the health service is the single largest employer, and yet many local people lack basic skills and qualifications for health service jobs. There are now moves to recognise this and to increase local recruitment and provide pre-employment and lifelong learning opportunities. *Pro-active employment programmes* in low-income areas can help strengthen local economies and reduce health inequalities and poverty. In England, the National Health System 'skills escalator' aims at providing access points at every level of training to ensure a constant stream of recruits through the system (Figure 1).

Figure 2. The career skills escalator

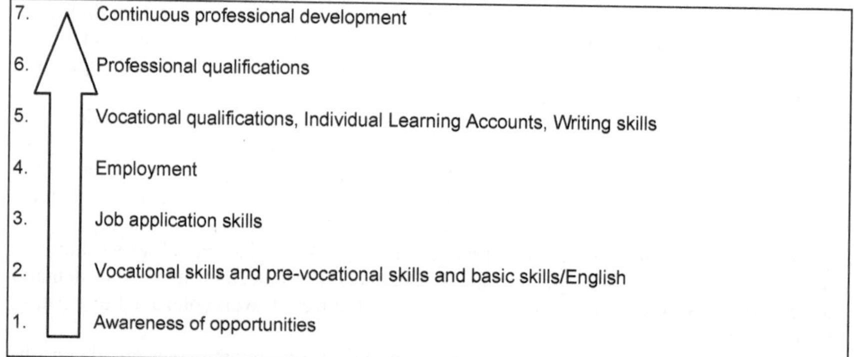

7. Continuous professional development
6. Professional qualifications
5. Vocational qualifications, Individual Learning Accounts, Writing skills
4. Employment
3. Job application skills
2. Vocational skills and pre-vocational skills and basic skills/English
1. Awareness of opportunities

Source: NHS, Employment improves your health, Regeneration and employment in health and social care, 2002.

The responsibility for leading on Frameworks for Regional Employment and Skills Actions (FRESAs) in partnership with government departments responsible for work, pensions, employment, skills, trade and industry has been given to the Regional Development Agencies in England. The London FRESA has a flagship programme to address the needs of the health sector – the NHS Pan-London Skills Escalator. This aims at enabling the health sector to access personnel with the right skills and give employees access to progression routes. This programme will provide support to target young people and those from hard to reach or excluded communities to gain the appropriate qualifications to enter the health support professions, and will provide conversion courses for immigrants and refugees with non-UK qualifications. The programme is run in partnership with all the health sector employers in London, higher education institutions and the unemployment agencies. Successful approaches like this will not only achieve longer-term solutions to health care recruitment problems, but will also increase the earning power and health status of deprived communities through increased employment. Nearly half of NHS jobs in London, for example, do not require specific entry qualifications: these posts could become entry points for which the unemployed could be trained and starting points for career development once employed. New useful skills can also be considered as a change in demographics: age and nationality of patients might require knowledge of culture, language, caring styles, etc. In theory, the career skills escalator would let unemployed people join in and provide them with the minimum skills to reach the vacancies requirements. Once employed, they should have the option to move on up to higher skills levels. By providing access points at every level of skill or training, employers can ensure a constant stream of new recruits coming in and moving through the system.

In the longer run, bringing jobs to deprived communities would also improve health in the community. The health system wins twice: it resolves its staff shortage in the long run following possibly costly investment for training, and lowers health problems of the area. The local area wins twice: a lower level of unemployment and better health.

Another health sector level example from Spain describes a healthcare initiative for medical and other healthcare personnel in coping with stress and burnout, recognised as a major cause of long-term absenteeism. It outlines a set of interventions that are very successful in promoting a return to work among a group of workers who suffer from high stress levels.

A regional example from the private sector in Spain is Programa d'atenció integral al metge malalt (PAIMM - Integrated intervention programme for ill medical staff). Doctors, along with everyone else, need diagnosis and treatment of their illnesses.

Figure 2. Integrated intervention for ill medical staff[11] – PAIMM (Spain)

A regional example from the private sector in Spain is Programa d'atenció integral al metge malalt (PAIMM - Integrated intervention programme for ill medical staff). Doctors, along with everyone else, need diagnosis and treatment of their illnesses. As an occupational group, they have the fifth highest incidence of psychological morbidity in Spain. A professional body, the Consell de Collegis de Metges de Catalunya (Catalonian Council Medical Association), recognised that occupational stress was taking its toll on the profession and on individual members. Furthermore, patients were exposed to what might be a secondary risk: doctors were continuing to practise despite health conditions, which could impair their judgement and competence.

PAIMM seeks to help doctors who suffer psychological problems or addictive behaviours that can interfere with their professional practice. Experience shows that early intervention can facilitate the doctor, even while in treatment, to stay at work. From 1999 to June 2000, PAIMM considered 179 cases and accepted 170 of these. According to preliminary results, among the 72 doctors treated, 98% are stable or abstinent during the first seven months after treatment. (US estimates for a similar US programme report rehabilitation at around 80% two years later).

As mentioned in the introduction, the health system is also a major purchaser of goods (not only one of the largest employers). This will be the subject of the next section.

Procurement policies and employment

In the EU, each year public authorities spend 16% of the EU GDP on goods, services and works, around 1500 billion euro. This includes schools, hospitals and national and

11 European Foundation for the Improvement of Living and Working Conditions, 2004

local administrations. The potential to harness this purchasing power positively to invest in health, environment and economic development, in particular in employment, is enormous. The health sector in particular is a huge consumer of goods and services. In the UK, The King's Fund explored the relationship between health improvement and sustainable development in Claiming the Health Dividend (2002), demonstrating how investment in promoting health, addressing the determinants of ill-health, employment, poverty, environmental pollution and social exclusion could contribute to reducing demand on health services. They describe how a 'virtuous circle' can be achieved of interaction between the sectors by focussing on health spend in the areas of: employment; purchasing policy and the procurement of goods and services; the management of waste, travel and energy, and the design and construction of new buildings.

Food in local procurement

Purchasing power[12] extends across the whole range of goods and services, but just to take food as an example, the Sustainable Development Commission (2004) states the argument succinctly: "How food is served, prepared, purchased and produced can have a huge impact on the health of individuals, communities and the environment. Through nutritional standards, catering and procurement, health systems have the power to speed patient recovery, build a healthy workforce, strengthen local communities, bring jobs to poor neighbourhoods, cut road traffic, and promote sustainable farming." It is therefore recognised that food can become a therapeutic process for recovering health. Buying locally and respecting cultural habits are also part of this process. The NHS in the UK spends £300 million on food each year: this can 'buy' more investment in the economy if sourced locally – every £10 spent on organic local food generates £25 for the local economy, but only £14 if spent in a supermarket[13]. Local procurement would also reduce the need for spending on transport and packaging waste. The aim is also to reduce unemployment in the farmer and agro industrial sectors[14].

Textiles

In areas other than food, similar approaches are being applied. For example, Leicester in the East Midlands is one of the centres of textile production with a lot of small businesses run by ethnic minority groups. Despite high demand for textiles for bedding and uniforms, only 5% was sourced from the local area; local firms could not understand how to do business with the NHS, so 'meet the buyer' events have been set up to establish links to match local supply with demand and skills in tendering.

Responsible buying

Another positive influence that the health sector can have is not only to buy from local enterprises but also to buy from firms, organisations, cooperatives, etc. which have adopted a policy of social responsibility: by reducing pollution, by creating healthy working conditions, flexible working time for its employees, etc. The health system can

---

12 The examples included are drawn largely from the UK, but actions undertaken within the UK framework are transferable to other health systems whether relating to strategic initiatives, or local innovative projects.

13 But it is more expensive and breaks the budget constraints of the NHS. By centralising purchases, NHS should be able to get better prices than if each hospital bargains on its own. It also creates a monitoring of costs on the quality of goods purchased and prevents corruption,

14 For further examples, consult the King's Fund report for the HDA, 2004.

actually play a role in favouring these types of enterprises with a pro-health policy as well. Indeed, inasmuch as corporate social performance (or corporate citizenship) may be desirable for a society as a whole, it is still very seldom embraced by organisations unless it yields concrete business benefits. While research is still failing to examine the impact corporate citizenship may have on important stakeholders such as investors, community leaders or job recruits, it has been shown that customers and employees benefit from desirable outcomes when the organisations they are buying from or selling to are "proactive citizens". In Maignan et al. (1999) proactive citizenship is found to be systematically associated with enhanced levels of employee commitment, customer loyalty and business performance[15]. Employee commitment is likely to engender greater job satisfaction and motivation, lower levels of absenteeism and turnover. Customer loyalty diminishes the propensity of consumers to seek information on other brands and generate positive word of mouth. These desirable outcomes of employee commitment and customer loyalty may in turn improve the overall competitive position of the business and may at least partially explain the positive association observed between corporate citizenship and business performance. The health sector, by being a responsible buyer, can motivate its suppliers (organisations, producers, distributors, etc.) to create a healthy environment for its employees but also to better the socio-economic determinants of health of the population.

Knowledge-based economic development

The health sector can also be important in developing commercial exploitation of research. Indeed, teaching hospitals with their relationships with universities are not only a place where medical students can train, but are also fostering research, breading spin-offs and incubators (especially in the pharmaceutical sector). Indeed, public health systems are continuously reshaped by new scientific developments, by the introduction of new products on the markets and by new policies at regional, local and national levels. University-industry-government relations increasingly provide the knowledge infrastructure of society. Could this "triple-helix"[16] economic model work with health systems? Could Health, University, and Government be the three helixes for regional development to take off? Or is this triple helix of state, university and industry missing an essential fourth helix[17], which is the public, or the public health system? A trilateral series of relationships is emerging in regions at different stages of development and with different inherited socio-economic systems and cultural values. As regions seek to create a self-reinforcing dynamic of knowledge-based sustainable economic development, the three institutional spheres are each undergoing internal transformations, creating hybrid organisations such as technology centres and incubators and the health sector should not stay outside. In the Northwest region of the UK for example, many industries from very diverse sectors (textile, biotechnology, chemistry, techno-environment, agro-alimentary, sport, building and energy) are sustained by regional development agencies, where the public health sector is predominant. A recent case-study as part of the OECD 'knowledge-intensive service activities" (KISA) project focuses on identifying carriers of and barriers to systemic innovation in health care[18]. It specifically explores the role of

15 Evaluated in terms of return on assets, return on investments, profits growth and sales growth.
16 See Etzkowitz, 2002 or Leydesdorff and Etzkowitz, 2001 for example.
17 Mehta (2002)
18 Kivisaari et al. (2004)

KISA in enhancing innovation. In this study, systemic innovation refers to changes in the integrated system of health care practices, services, technologies, and organisation that together form a new mode of operation. The findings relate to the carriers of and barriers to systemic innovation in Finnish specialised health care and the role of KISA in innovation. The case study suggested that some aspects concerning regional environment and management style contributed to innovation. From the point of view of regional environment, important carriers were the regional political ethos favouring collaboration for innovation, the substantial regional networks of competence, and a well-balanced relationship between the university level hospital and hospital district. As to managerial practices, the carriers incorporated the ability to create strategic conversation, corporate management's open-mindedness and strong support for experiments, as well as recruitment of tenacious and skilful champions for systemic innovation.

3 Conclusions and policy implications

In this chapter, we reviewed some challenges and opportunities facing health systems in Europe in terms of unemployment. Indeed, diverse forms of unemployment and employment have been identified as the determinants of health or ill health. The European labour market is diverse and is changing: new forms of contracts, of occupations and unemployment exist and the health system has a role to play in minimizing the negative effects on health. Its role can be much more important than one might think. In the fourth section we gave several examples of how the health systems can contribute to its local and national economic and social development. In particular, it can create jobs in deprived areas and regions and keep them in productive sectors going through an economic downturn. Pro-active employment programmes, partnership with development agencies, special training, and initiatives to promote return to work are some actions where the health system can make a difference. In a more indirect manner, by using a clever procurement policy, the health system can also promote health at work in other sectors (its suppliers) and contribute to local employment by buying locally, responsibly and foster research and knowledge-based activities.

In 2003, in "The Solid Facts", the WHO described a healthy employment policy as needing three goals: to prevent unemployment and job insecurity; to reduce the hardship suffered by the unemployed; and to restore people to secure jobs. In addition, it outlined in detail some minimum requirement levels in terms of job security, education and unemployment benefits:

- Government management of the economy to reduce the highs and lows of the business cycle can make an important contribution to job security and the reduction of unemployment.
- Limitations on working hours may also be beneficial when pursued alongside job security and satisfaction.
- To equip people for the work available, high standards of education and good retraining schemes are important.
- For those out of work, unemployment benefits set at a higher proportion of wages are likely to have a protective effect.
- Credit unions may be beneficial by reducing debts and increasing social networks.

These are useful recommendations but the precise role of the health system is not directly and clearly mentioned. The heterogeneity of settings in Europe, the dynamic socio and economic determinants of health in a diversified context, the precariousness of work contracts are all characteristics which should motivate the health system to innovate and contribute directly and actively to the labour market. The originality of this paper is that it provides examples where the health system works with different sectors to create a more "flexible-friendly" system, where the health sector aims at being a role model employer and health promoter and where health systems are actually seen to contribute to the local economy and ensure a long-term view of population health.

References

Avison, W.R. (1998). The health consequences of unemployment. In National Forum on Health (Ed.), *Determinants of health: Adults and seniors* (pp. 3-41). Ottawa: Editions MultiMonde.

Beale, N. & Nethercott, S. (1985). Job-loss and family morbidity: A study of a factory closure. *Journal of the Royal College of General Practitioners, 35*, 510-514.

Béland, F., Birch, S. & Stoddart, G. (2001). *Unemployment and health: Contextual level of influences on the production of health in populations* (SEDAP Research Paper N°54). Retrieved from http://socserv.mcmaster.ca/sedap/papers01.htm.

Bertrand, M. & Mullainathan, S. (2001). *Do people mean what they say? Implications for subjective survey data. American Economic Review, 91*(2), 67-72.

Boback, M., Blane, D. & Marmot, M. (1998). *Social determinants of health: Their relevance in the European context* (Review of Health promotion and Education Online, Verona Initiative). Retrieved from www.rhpeop.org/ijhp-articles/e-proceedings-verona/1/index.htm.

Borghi, V. & Kieselbach, T. (2003, May). *Submerged economy in southern Europe and youth unemployment.* Paper presented at the conference "Informal/undeclared work: Research on its changing nature and policy strategies in an enlarged Europe", European Commisssion, DG Research and Employment and Social Affairs, Brussels, Belgium.

Brenner, M.H. & Mooney, A. (1983). Unemployment and health in the context of economic change. *Social Science and Medicine, 17*, 1125-1138.

Brenner, M.H. (1977). Health costs and benefits of economic policy. *International Journal of Health Services, 7*, 581-623.

Brenner, M.H. (1979). Mortality and the economy: A review, and the experience of England and Wales, 1936-1976. *Lancet ii*, 568-573.

Brenner, M.H. (1987). Economic change, alcohol consumption and heart disease mortality in nine industrialized countries. *Social Science and Medicine, 25*, 119-132.

Brenner, M.H. (1987). Relation of economic change to Swedish health and social well-being, 1950-1980. *Social Science and Medicine, 25*, 183-195.

Clark, A.E., Georgellis, Y. & Sanfey, P. (2001). Scarring: The psychological impact of past unemployment. *Economica, 68*, 221-241.

Clarck, A.E. & Oswald, E.J. (1994). Unhappiness and unemployment. *Economic Journal, 104*(424), 648-659.

Coote, A. (Ed.). (2002). *Claiming the health dividend: Unlocking the benefits of NHS spending*. Retrieved from http://www.kingsfund.org.uk.

Cornia, G.A. & Paniccia, R. (2000). *The mortality crisis in transitional Economies*. Oxford: Oxford University Press.

Currie, J. & Madrian, B.C. (1999). Health, health insurance and the labour market. In O. Ashenfelter & D. Card (Eds.), *Handbook of Labour Economics* (vol. 3, pp. 3309-3416). Elsevier.

D'Addio, A.C. & Rosholm, M. (2004). Determinants and characteristics of temporary employment in Europe. *Brussels Economic Review, Editions du DULBEA, Université libre de Bruxelles, Department of Applied Economics (DULBEA), 48*(1-2), 13-41.

Dahlgren, G. & Whitehead, M. (1991). *Policies and strategies to promote social equity in health.* Stockholm: Institute of Future Studies.

Domenighetti, G. (1998). *Health effects of fear of unemployment among employees in the general population* (Review of Health promotion and Education Online, Verona Initiative). Retrieved from http://www.rhpeop .org/ijhp-articles/e-proceedings-verona/1/index.htm.

Dooley, D., Fielding, J. & Levi, L. (1996). Health and unemployment. *Annual Review of Public Health, 17*, 949-65.

Ermish, J., Francesconi, M. & Pevalin, D. (2001). *Outcomes of children poverty* (DWP Research Report No. 158). Leeds: CDS.

Etzkowitz, H. (2002). *The triple helix of University-Industry-Government implications for policy and evaluation.* Stockholm: Science Policy Institute.

European Commission (2001). *Guidelines for environment-friendly procurement.* Retrieved from http://simap.eu.int/.

European Commission (2001). *Public procurement: Guidelines for taking social issues into account.* Retrieved from http://simap.eu.int/.

European Commission (2004). *Handbook on green public procurement.* Retrieved from http://europa.eu.int/comm/internal_market/publicprocurement/key-docsen.htm.

European Foundation for the Improvment of Living and Working Conditions (2004). *Employment and disability: Back to work strategies.* Retrieved from http://www.eurofound.eu.int-/publications/files/EF04115EN.pdf.

European Science Foundation (2004). *Social variations in health expectancy in Europe* (ESF Scientific Programme, 1999-2003, Final Programme Report). Retrieved from http://www.uni-duesseldorf.de/health/FinalReport.pdf.

Ferrie, J.E., Marmot, M., Griffiths, J. & Ziglio, E. (1999). *Labour market changes and job insecurity: A challenge for social welfare and health promotion* (WHO Regional Publications, European Series, N° 81). Copenhagen: WHO Regional Office for Europe.

Ferrie, J.E., Shipley, M.J., Marmot, M.G. et al. (1995). Health effects of anticipation of job change and non-employment: Longitudinal data from the Whitehall II study. *British Medical Journal, 311*, 1264-1269.

Fontaine, F. (2004). *Do workers really benefit from their social networks?* (IZA Discussion Paper N° 1282). Retrieved from http://papers.ssrn.com/sol3/papers.cfm?abstract_id=586765.

Frese, M. & Mohr, G. (1987). Prolonged unemployment and depression in older workers: A longitudinal study of intervening variables. *Social Science and Medicine, 25*(2), 173-178.

Harrison, D. & Ziglio, E. (1998). *Social determinants of health: Implications for the health professions.* Copenhagen: WHO.

Head, J., Martikainen, P., Kumari, M., Kuper, H. & Marmot, M. (2002). *Work environment, alcohol consumption and ill-health: The Whitehall II Study* (UCL for the Health and safety Executive, Contract research Report, 422/2002). Retrieved from http://www.hse.gov.uk/research-/crr_pdf/2002/crr02422.pdf.

Health Canada (2002). *A report on mental illness in Canada.* Retrieved from http://www.phac-aspc.gc.ca/publicat/miic-mmac/index.html.

Health Canada (1999). *Statistical report on the health of Canadians.* Retrieved from http://www.statcan.ca/english/freepub/82-570-XIE/82-570-XIE1997001.pdf.

Health Canada (1999). *Toward a healthy future: Second report on the health of Canadians.* Retrieved from http://www.hc-sc.gc.ca/ahc-asc/alt_formats/cmcd-dcmc/pdf/media/releases-communiques/1999/hofc2ebk2.pdf.

IDeA/SDC (2004). *Developing sustainable procurement as a shared priority – vision to reality.* Retrieved from http://www.sd-commission.org.uk/publications.php?id=158.

Jahoda, M. (1982). *Employment and unemployment.* Cambridge: Cambridge University Press.

Jochelson, K., Delap, C. & Norwood, S. (2004). *Regional mapping of claiming the health dividend activities in the NHS* (A King's Fund report for the HAD). Retrieved from http://www.kingsfund.org.uk/current_projects/archive/corporate_citizenship/index.html.

Kasl, S. & Jones, B.A. (1998). The impact of job loss and retirement on health. In L.F. Berkman & I. Kawachi (Eds.), *Social Epidemiology*. Oxford: Oxford University Press.

Kessler, R.C., Turner, J.B. & House, J.S. (1989). Unemployment, reemployment, and emotional functioning in a community sample. *American Sociological Review, 54*, 648-657.

Kieselbach, T. (2003, May). *Long-term youth unemployment and social exclusion: The role of the submerged economy*. Paper presented at the Conference "Informal/undeclared work: Research on its changing nature and policy strategies in an enlarged Europe", European Commisssion, DG Research and DG Employment and Social Affairs, Brussels, Belguim.

Kieselbach, T. (2003, June). *Youth unemployment and health: Future perspectives of counteracting the psycho-social effects of unemployment*. Paper presented at Warsaw High School for Social and Economic Sciences (WSSE), Warsaw, Poland.

Kivisaari, S., Vayrynen, E. & Saranummi, N. (2004). *Knowledge-intensive service activities in health care innovation, Case Pirkanmaa* (VTT Tiedotteita, Research Notes 2267). Retrieved from http://www.vtt.fi/inf/pdf/tiedotteet/2004/T2267.pdf.

Kuhn, A., Lalive, R. & Zweimuller, J. (2004). *Does unemployment make you sick?* Retrieved from http://www.iza.org/conference_files/prizeconf2004/253.pdf.

Lavis, J., Mustard, C., McLeod, C., Farrant, M. & Payne, J. (2001). *Labour-market experiences and health: Systematic review of cohort studies* (Institute for Work and Health). Retrieved from http://www.cher.ubc.ca/PDFs/Labourmarketahealth.pdf.

Leydesdorff, L. & Etzkowitz, H. (2001). The transformation of university-industry-government relations. *Electronic Journal of Sociology*. Retrieved from http://www.fractal.org-/Training-Interactie-Management/University-industry-government.htm.

Maignan, I., Ferrel, O.C. & Hult, G.T. (1999). Corporate citizenship: Cultural antecendents and business benefits. *Journal of the Academy of Marketing Science, 27*(4), 455-469.

Maignan, I. & Ferrel, O.C. (2001). Corporate citizenship as a marketing instrument. Concepts, evidence and research directions. *European Journal of Marketing, 35*(3), 457-484.

Marmot, M. (2004). *Status syndrome*. Bloomsbury: London.

Mehta, M.D. (2002). *Regulating biotechnology and nanotechnology in Canada: A post-normal science approach for inclusion of the fourth helix*. Retrieved from http://arts.usask.ca-/policynut/mehta-nus-paper.pdf

Migration of health workers (Lee and Yach, Globalization and Health, 2004)

NHS, Health Development Agency. (2004, June). *The evidence about work and health*. (HAD-Briefing, n°18, Working for health opportunities in employment).

NHS, London Regional Office. (2002, January). *Employment improves your health: Regeneration and employment in health and social care*.

Office of the Deputy Prime Minster. (2004). *Environmental exclusion review for neighbourhood renewal unit* (ODPM – Summary Report).

O'Neill, D. & Sweetman, O. (1998). Intergenerational mobility in Britain: Evidence from unemployment patterns. *Oxford Bulletin of Economics and Statistics, 60*(4), 431-449.

Organisation for Economic Co-operation and Development (2000). *Greener public purchasing: Issues and practical solutions*. Retrieved from http://www.oecd.org.

Organisation for Economic Co-operation and Development (2003). *The environmental performance of public procurement: Issues of policy coherence*. Retrieved from http://www.oecd.org.

Polanyi, M. (2002). *Employment and working conditions – a response*. Retrieved from http://www.phac-aspc.gc.ca/ph-sp/phdd/overview_implications/05_working.html.

Riley, C. et al. (Eds.). (1995). *Releasing resources to achieve health gain*. London: Radcliffe Medical Press and USA Office of Technology Assessment.

Schwefel, D. (1986). Unemployment, health and health services in German-speaking countries. *Social Science and Medicine, 22*(4), 409-430.

Storrie, D. (2002). *Temporary agency work in the European Union*. Retrieved from http://www.eurofound.europa.eu/ewco/reports/TN0408TR01/TN0408TR01.pdf.

Sustainable Development Commission (2004). *Healthy futures: Food and sustainable development*. Retrieved from http://www.sd-commission.gov.uk/healthyfutures.

Sustainable Development Commission (2004). *Healthy futures: Progress in practice*. Retrieved from http://www.sd-commission.gov.uk/healthyfutures.

Taylor, M. (2002). *"Labour market transitions in the context of social exclusion: A study of the EU.* University of Essex, UK: Institute for Social and Economic Research.

Towle, A. (1998). Changes in health care and continuing medical education for the 21st century. *British Medical Journal, 316*, 301-304.

Wamala, S.P. & Lynch, J. (Eds.). (2002). *Gender and social inequities in health – a public health issue.* Lund: Studentlitteratur.

Warr, P. & Jackson, P. (1985). Factors influencing the psychological impact of prolonged unemployment and re-employment. *Psychological Medicine, 15*, 795-807.

Warr, P. (1987). *Work, unemployment and mental health.* New York: Oxford University Press.

Whelan, C. & McGinnity, F. (2000). Unemployment and satisfaction: A European analysis. In D. Gallie & S. Paugam (Eds.), *Welfare regimes and the experience of unemployment.* Oxford: Oxford University Press.

WHO, Alliance for Health Policy and Systems Research. (2004). *Strengthening health systems: The role and promise of policy and systems research* (Global Forum for Health Research).

WHO (2003). *Social determinants of health. The solid facts.* Retrieved from http://www.who.dk/document/e81384.pdf

Wilkinson, R.G. (1992). Income distribution and life expectancy. *British Medical Journal, 304*, 165-168.

Wilkinson, R.G. (1996). *Unhealthy societies.* London: Routledge.

Winkelmann, L. & Winkelmann, R. (1998). Why are the unemployed so unhappy? *Economica, 65*, 1-15.

World Bank (1993). *Investment in health. The World Bank in action.* Washington DC: World Bank.

Introducing Four Psychologies of Unemployment and their Implications for Intervention

Kesi Mahendran

Introduction

Much of the 'activity' involved in the active labour market policies (ALMP), which characterize the UK's New Deal intervention programme, is focused on increasing the activation level, through regular attendance at training and guidance centres, in order to increase the 'employability' of the unemployed person. The New Deal was one strand of the flagship Welfare to Work policy of the New Labour government when it came to power in 1997. There was to be no option to live a life 'inactive' and in receipt of benefits for those who were able to be employed.

The New Deal intervention was built on an international evidence base suggesting that increased 'employability' would lead to employment outcomes, reduce the 'dependency culture', create social inclusion and offset health problems. By the late 1990's converging evidence had led to a consensus that there is a relationship between unemployment and health. Murphy and Athanasou, in reviewing the relationship between unemployment and mental health, carried out a meta-analysis of 16 longitudinal studies. They posed the question "does job loss on average, affect the mental health of the unemployed?" (Murphy & Athanasou, 1999, p. 88). Mental health was chiefly measured using the General Health Questionnaire or the Hopkins Symptom Checklist. They found "good support for the claim that job loss on average has a negative impact on the psychological well-being of the unemployed" (ibid., p. 88).

This chapter draws attention to four psychologies of unemployment and explores their implications for intervention. When one intervenes in unemployment, whether as a policy-maker, clinician, applied scientist or researcher, as knowledge-creator, it is worth being aware that these psychologies of unemployment are implicitly at play. Outlined are three instances, termed 'actions', where a proposed intervention was challenged using a dialogical psychology of unemployment. The chapter introduces two organisations, the Young Person's Centre (YPC) and Strategic Delivery (SD)[1], who collaborated

1 The names of the organisations, locality and participants have been changed in order to ensure confidentiality. The author would like to thank Stirling University's Psychology department for funding the study and the Scottish Executive for funding the presentation of this paper at the ICOH conference in Bremen. Thanks also to Luke Cavanagh and Ken Bryan for reading and discussing drafts of this manu-

with me on participatory action research for sixteen months between 1999 and 2001. Both organisations delivered the New Deal by providing training and guidance to unemployed people in Central Scotland.

Interventionists today, such as the state or delivery organisations, frequently take as their starting point the clearly recognised relationship between unemployment, in terms of job loss and gain, and health, in terms of GHQ and similar measures. All four psychologies of unemployment are to found within the evidence base, used by interventionists, where the evidence base consists of all the social knowledge in the public sphere, such as research, evaluations, good practice examples and social theorising that surrounds the social problem.

A difficulty here is that the relationship reviewed by Murphy and Athanasou (1999) has become narrowly defined and decontextualised. This chapter aims to broaden the discussion by providing a review of the field and setting the relationship between unemployment and health into the contemporary context. A context where one's health and sense of wellbeing rests on dynamic self-other relations and one's employment chances relate to one's ability in navigating training providers as well as the flexible labour market.

All four psychologies explain the relationship between unemployment and health in social psychological terms. They are the *agency-deprivation* psychology of unemployment, *social perception* psychology of unemployment, *self-perception* psychology of unemployment and finally a *dialectical* psychology of unemployment. Indebted to Bakhtin, the dialogical approach developed in this study, is a particular form of dialectical psychology of unemployment.

The research evidence presented here is built upon and reacts to these psychologies; in particular, in taking a dialogical approach it places the locus of unemployment in the interaction of *all* the people who are involved in the unemployed experience, including researchers and workers in the field of unemployment, as well as unemployed people themselves.

1 Four psychologies of unemployment

It is chiefly a matter of how unemployment is conceptualised that determines the point at which psychologists have intervened in the phenomenon. In particular the questions that the interventionists have chosen to ask, to some the question becomes not just "what does unemployment do to people?" but also "what do people do about unemployment?" (Svensson & Starrin, 1989, p. 14).

The dominance of a supply-side orientation has meant that studies generally have concerned themselves with the question, "what does unemployment do to people?" Figure 1 sets out each psychology of unemployment and the associated intervention strategy. The first tradition, *agency-deprivation* accounts, locates unemployment within the macro-economic situation and explains the socio-psychological states of individuals and families in terms of actor-environment relations. Here, the locus of unemployment

---

script. The views expressed are the authors alone and do not reflect the views of the Scottish Executive.

Figure 1. Four psychologies of unemployment and associated intervention strategies

Psychology of Unemployment	Locus of Unemployment	Sub-types	Intervention strategy
Agency-deprivation	External - Interaction of actor and environment	1.Latent functions 2.Agency-restriction	Intervening in the family context or community
Social perception of the event	External – critical life event	1.Narratives on the event 2. Coping	Reinterpretation of the event of becoming unemployed
Self-perception	Internal – selfhood	1. Behaviour 2. Personality 3. Self 4. Job-search skills	Counselling and psychological training on self-esteem, self-efficacy and employment motivation
Dialectical	Multi-level interaction, self-other relations and social knowledge	1. Self-other 2. Discourses 3.Social representations 4. Dialogical	Multi-level innovations into the social relationship of unemployment

exists outside the individual. A classic example is Jahoda, Lazarsfeld and Zeisel's ethnographic study of the Austrian village Marienthal in the mass depression of the 1930s. Such accounts exist along a continuum which, at one extreme emphasizes the power of social institutions and, at the other, emphasizes the individual's agency and the social forces which restrict it. The agency-restriction account argues that the agency of unemployed people is frequently restricted by future-insecurity, by information and material poverty, and "by the very social relationship which is unemployment" (Fryer, 2000, p. 19).

Without delving too deeply into these two positions what is critical here is that studies may focus on communities in the same way as the work of Jahoda or they may focus on social support networks that surround a family and agency-restriction (McGhee & Fryer, 1989; Fryer & Fagan, 1993) but all locate unemployment as external to the individual's psychology. The focus of both the agency-restriction and latent-deprivation accounts is the dualisms of the actor and their environment or agency-structure. In this sense, they operate as two sides of the same ontological coin and in this conceptualisation they have been housed together.

Today it is common for researchers in unemployment to set out the latent-deprivation account and then challenge it with the agency-restriction account. Creed and Macintyre (2001) describe the "latent deprivation model" and the "agency restriction model" as having "dominated the research and applied efforts in the area of unemployment and mental health" (Creed & Macintyre, 2001, p. 324). Using a longitudinal data set of 3,500 people within Sweden, Nordenmark and Strandh (1999) assessed the latent-deprivation and agency "models", they reconciled the tensions between them using Doug Ezzy's status-passage model. Ezzy (2000) states, in his account, the challenge is to explore the interplay between actively interpreting individuals and social institutions. A strength of this analysis is that it locates unemployment in the wider economic context. Nordenmark and Strandh (1999) conclude that the differences in mental well-being between unemployed people are due to variances in economic and psychosocial needs.

Both the second and third psychologies of unemployment are best understood as existing within the same frames of reference as agency-deprivation accounts. The second psychology of unemployment, *social perception of the event* accounts concentrate on the unemployed person's ability to 'reinterpret the stressor' as a feature of emotionally-focused coping to maintain positive psychological well-being. Returning to the status-passage model, Ezzy (2000) explores the interaction between agency and deprivation in examining the extent to which interviewees systematically under or overestimate the significance of social forces in the story they tell about how they came to be unemployed. In his model much of the psycho-social health of unemployed people is understood in terms of the narratives they create, such as "heroic agency narratives" or "stories of tragic fate" (Ezzy, 2000, p. 121). Ezzy is aware that "narrative identities are not creations of isolated individuals. They require intersubjective support in order to be plausible and sustainable" (ibid., p. 131). Here such narratives are seen to perform rhetorical devices, which help unemployed people to understand the past and look to the future. Though clearly a psychological intervention, the locus of unemployment again becomes an *external* phenomenon, something the subjects can interpret and construct in different ways through expressive writing. Ezzy (2000) does not focus on the phenomenon of unemployment itself and how different groups are constructing it rather his focus is on the individual subject, in particular the "self as a rhetorical project" (ibid., p. 130; see also Spera, Buhrfiend and Pennebaker, 1994).

The third approach, which has clear implications for intervention, locates unemployment within the individual and explains the psychology of unemployment in terms of an individual's self-perception. This third type ranges from: a focus on an individual's behaviour, e.g. understanding people's job search behaviour, time-management or meaningful activity; to exploring the intrapsychic world of the unitary subject using psychometric tests on confidence, self-esteem and self-efficacy (Eden & Alviram, 1993); to cognitive-behaviour therapies which activate and explore the motivations of "the subject"[2] (Proudfoot, Guest, Carson, Dunn & Gray, 1997; Proudfoot, Gray, Carson, Guest & Dunn, 1999).

It is this notion of psychological well-being which has led *self-perception* accounts to intervene in the individual psychologies of unemployed people and to focus on selfhood and intra-psychic states. Here the locus of unemployment shifts to within the unitary subject and subject-phenomenon relations are removed. Interventions are more radical in setting up psychological training and behaviour modification, where participants do not develop cognitions and perceptions of the critical life event of their unemployment, but are instead requested to delve deep into themselves to find the cause and potential solution to their unemployment.

---

2 Frequently conceptualised as the experimental subject.

A psychometric intervention which is rapidly becoming a citation classic is Eden and Aviram's (1993) two-and-a-half week workshop into general self-efficacy (GSE) to boost the chances of reemployment.

McPolin (1999) cites Eden and Aviram (1993) in his intervention within a Belfast project. Pre-participants on a youth training programme and ex-participants on the same programme were tested on a Multifactorial Achievement-Motivation Scale (MAM) and the Career-Decision Making Self-Efficacy Scale (CDMSE). Ex-participants were predicted to score higher than pre-participants on the CDMSE and MAM and, whilst this occurred with the CDMSE and parts of the MAM, such studies, which use self-report questionnaires, warrant a closer look.

A question on the CDMSE scale asks young people to indicate on a scale of 1 to 9 "how confident are you that you can define the type of lifestyle you want to live?". The MAM asks whether the respondent agrees with the statements, "basically I am a lazy person" and "I like to be busy all the time" (McPolin, 1999).

Self-perception account's also house clinical psychological interventions, such as Proudfoot et al.'s (1997) study in the United Kingdom. Proudfoot and colleagues also citing Eden and Aviram (1993), use an "occupational training programme based on the principles of CBT[3] (…) to help people identify and modify their attributional style" (Proudfoot et al., 1997, p. 97). Proudfoot et al. (1999), though stating clearly that the long-term unemployed people they work with are a "non-psychiatric group", comment that "nearly 3 times as many CBT as control participants found full-time employment. Not only does this represent substantial psychological and financial benefits for the individual's concerned, there are potential societal benefits as well reduced health service usage (…) decreased welfare costs and increased tax revenues" (Proudfoot et al., 1999, p. 42).

Proudfoot et al.'s (1997) study does not provide evidence of decreased welfare costs and simply assumes this along with the other assumptions they make around attrition and the type of employment the long-term unemployed people actually took. The difficulties with this use of psychological training are firstly that the locus of unemployment becomes the self-concept and secondly that the methodology, whilst appealing to many interventionists because of its 'pre' and 'post' metrics, relies so heavily on self-report measures rooted in self-other relations. The knowledge claims of this approach often go further than the psychological in suggesting a social and economic solution.

The final psychology of unemployment locates unemployment neither within the individual's behaviour, personality type or degree of employability, nor in a structural account of the power of social institutions, but rather in a dialectical account. Such *dialectical* accounts attribute a degree of power to social relations e.g. self-other relations and emphasize the role of the social knowledge which circulates around the unemployed. Often phenomenological, but always situated in the socio-historical context, these accounts are indebted to sociology. They are often informed by the *communicative turn* within psychology, though they should not be understood as confined to discursive post-modernist or social constructionist accounts.

In this final psychology of unemployment, there is room to critically examine not just the unemployed person's experiences in terms of self-other relations but also the

3 Cognitive-Behaviour Therapy

common rationalities of the practitioners and organisations that work within the unemployment industry i.e. how they make sense of the world they work in. Social knowledge on 'unemployment', the 'culture of poverty' and 'the dependency of the welfare recipient' is understood as impacting on unemployed people's everyday lived realities, in particular their health, dialectically. This critical appreciation is evident in the work on shame by Bengt Starrin (i.e., Starrin, Rantakeisu, Forsberg & Kalander-Blomqvist, 2000) and examinations of discourses around unemployment (Drewery, 1998; Straehle, Weiss, Wodak, Muntigl & Sedlak, 1999).

The *dialectical* contribution is not confined to critique; psychologists are also engaged in direct interventions. These tend to intervene at the level of the organisation, its processes and the social knowledge implicit and sometimes explicit in policy. In these approaches unemployed people, who today are frequently required to regularly attend some sort of intervention, are less likely to risk viewing their own attributional style or level of self-esteem as the root of their unemployment.

There are two interventions that can be usefully placed here: firstly the evaluation of the SPRITE[4] project, within the United Kingdom, which provided non-employed people with access to Information Technology (IT) in three local community centres. This action research study involved a multi-level evaluation of the processes, people, organisations and products of the SPRITE project and the wider implications of the introduction of IT to further community developments. The study provided accessible research products for the project (Cassell, Fitter, Fryer & Smith, 1988, p. 94).

Such an intervention is dialectical in as much as the locus of unemployment becomes the interaction of the employed and unemployed people, organisations, communities and the council. In Canton Ticino, Switzerland, a multi-level intervention was also attempted where the strategies are to "train and raise awareness on the psycho-social issues for those who "deal" with the unemployed, awake public opinion to the reality of the labour crisis (diminish stigmatisation and raise solidarity towards the unemployed)" and "Empowerment in the form of life skills, information on services and promote networking" (Villaret & Gianinazzi, 1999, p. 26).

The dialogical approach is a form of dialectical approach which systematically analyses the role that dialogue is playing at four levels. Firstly all the actors in the unemployment industry are in a face-to-face situated dialogue with other actors. Decision-making proceeds on the basis of this subject-subject dialogue, whether that is between the unemployed person and the personal advisor at the training service, the manager and the personal advisor, or the senior manager and the policy-maker.

As Habermas (1989) outlines when we communicate we move within the life-world, we cannot step out of it, and there are three worlds at play, the objective world, that is the world that is consensually recognised, the normative world, the world inter-subjectively recognised and finally the subjective world that is the private world which has only privileged access. Communicative action occurs within "horizons of meaning which shift with the theme that is thrown into relief by themes articulated through goals and plans of action" (Habermas, 1989, p. 128). When we debate there are shifting horizons and moments of intersubjective understanding and shared frames of reference. This will be evident in the three interventions outlined below.

4 Sheffield People's Resource for Information Technology.

At a second level there is the dialogue of the dialogical and relational self. Hermans, extending a Bakhtinian analysis, explains that "the dialogical self is based on the assumption that there are many *I*-positions that can be occupied by the same person. The *I*-position, moreover, can agree, disagree, understand, misunderstand, oppose, contradict, question, challenge or even ridicule the *I* in another position" (Hermans, 2001, p. 249).

The dialogical self, can take the voice of another person, or the position of an imaginary audience e.g. the public. The approach analyses who the self is in a dialogue with and the ways these different voices populate people's speech.

Dialogue is at play at a third level, when the individual, as a social agent, is in a dialogue with the public sphere (in the sense that Habermas used it). In this study the YPC and SD are understood as "thinking" organisations, where members are actively engaged in developing their own lay theories or social representations (Moscovici, 2000, p. 22) to inform their practice at work, using information from the public sphere. Social representations refer to what is 'between our heads' the socially shared knowledge, which, because it is shared, can be objectified as Moscovici explains: "these representations become established in behaviour and in relationships, from which they derive an enduring and, so to speak, external existence" (Moscovici, 1990, p. 77). The social representations or common thinking on unemployed people take on a life of their own. People do not passively internalise the socially shared knowledge they encounter, they construct autonomously in order to answer their own questions. They are however vulnerable to persuasive ideological discourse or what Moscovici (1990) calls hegemonic social representations.

Finally there is dialogue, worthy of analysis, within the ways people employ words – what Bakhtin called dialogism of words in use (Bahktin, 1981, 1986). Words-in-use in this field, such as 'person-centred' or 'employability', involves extra-linguistic features i.e. connotations. In the rhetorical exchanges, which characterise the communication in the unemployment industry, the creative use around particular 'buzz' words may even come to overshadow the normative term i.e. a dictionary definition of the term.

Though rarely articulated in academic texts or scientific journals, each of the four psychologies of unemployment outlined is born out of the norms and conventions of different traditions with its own epistemological considerations and stipulations about how to create knowledge. As Israel explains and contemporary researchers are increasingly aware, such accounts *construct* in two senses (Israel, 1972, p. 192-3). Firstly they do not reflect a reality they construct an interpretation of it, they 'confer meaning' upon the data collected in scientific inquiry and the concepts used to frame the data collection. For example, research in the 1980s tended towards concepts such as the 'work ethic' and 'unemployability' and in the 1990s and today it is around concepts such 'employability' and 'social inclusion'. The second sense in which they can construct is that they 'bring about' a reality. Psychologies of unemployment, once disseminated into various arenas partly construct the social knowledge on unemployment, the experience of being unemployed and inform policy-makers in their development of state interventions.

Whilst naturally the interventionist is impatient for action, a careful consideration of these theoretical sensitivities will reveal insights into the implications that psychologies of unemployment have for intervention into people's transitions towards the labour market, and equally the relationship between unemployment and health (Mahendran, 2000).

2 Converging aims and developing an approach with the YPC and SD

It is now time to balance out some of the scientific lines of inquiry and conceptual development on unemployment with the socio-political realities of being unemployed in Britain today. The first step of a dialogical approach is to locate unemployment within the context of the modernisation of poverty. This is characterised by a number of features, the new *'active' welfare state*, which has emerged over the last ten years, the 'Welfare to Work' legislation, the rise of the concept 'employability' in relation to full employment and the shared social knowledge on welfare recipients that exists in the public sphere.

The increasing dominance of a flexible labour market means, for all but a small minority, working life is characterised by transition. A facet of this flexible labour market is that training and guidance organisations, have built up around successive government's policies on welfare to work.

The Young Person's Centre (YPC) was a public-sector organisation working with young people aged between 16 and 24 delivering the New Deal and Skillseekers programmes. These young people were referred to the centre from other training and guidance services, the careers service or directly from school. They were also referred from a private-sector organisation, Strategic Delivery (SD) which worked principally with the long-term unemployed aged 18 and above. SD took referrals from the Careers Service but also from the Employment Service (now known as Job Centre Plus) a one-stop shop assessing employment opportunities and eligibility for welfare benefits. Both organisations cover Moultrie and Sommerville where 14% of the population is aged 16-24 and less than 1% is of African or Asian descent. In 1999 the Tarbert region, which encompasses Moultrie and Sommerville, has slightly lower claimant count unemployment at 4% than the national average in Scotland at 6%.

In order to remain viable in an atmosphere of competitive tendering, these service-providers are constantly in transition and must deliver measurable outcomes – either a job or quantifiable progress towards one. Equally Welfare to Work policies are no different to other public policies in so far as they are developed, in part, by policy-makers and other change agents from the social knowledge or evidence base about 'what works' in relation to the agreed desirable outcomes. In this field the question is now narrowly focused on what works to increase 'employability'.

In 1998 Brussels developed a European Employment Strategy with four key pillars: Improving employability, developing entrepreneurship, encouraging adaptability and strengthening equal opportunities. It is from this point that the notion of 'employability' has emerged as a key concept that underpins the current supply-side orthodoxy of training programmes. According to Hillage and Pollard (1998) 'employability' can be understood as: "the capacity to move self-sufficiently within the labour market to realise potential through sustainable employment. For the individual, employability depends on the knowledge, skills and attitude they possess, the way they use those assets and present them to employers and the context (e.g., personal circumstances and labour market environment) within which they work" (Hillage & Pollard, 1998, p. 1).

Employability has become less about the right skills and attributes for specific employers and more about a flexible attitude to the labour market and one's ability to be independent within it; to maintain one's transferable skills and competencies; to manage one's own portfolio and self-presentation; and to fully exploit one's access to social networks in an on-going lifetime process. Work is viewed as consisting increasingly of short–life contracts and projects and British policy discussions talk less about full employment and more in terms of 'full employability' (Finn, 1999, p. 3).

The interventions outlined begin a much needed dialogical analysis of the contemporary use of the concept 'employability' emphasizing the interaction between individuals, whether potential employees, employers or practitioners, working within unemployment; such actors, as discussed, are engaged in self-other relations which are in a dialectical relation with the social knowledge that exists around them. Social knowledge on 'young people' or 'the long-term unemployed' or 'the sort of employee a customer would like to be served by'.

Unlike the pervasive self-perception approach to intervention a dialogical intervention does not start with the unemployed individual's psychology but rather the communicative interactions of all the actors in the system it places unemployed people within. Here 'employability' becomes a multi-level concept. The nature of this pragmatic science can be illustrated using three problems that occurred during the study where an initial self-perception solution was challenged through debate and replaced by a dialogical solution.

It is a requirement of the current New Deal legislation that in order to receive welfare benefits a person must undertake one of four options: (i) subsidised work, (ii) self-employment, (iii) full-time education/training or (iv) working on the 'environmental task force'/in the voluntary sector. Recently this has been extended beyond the 'claimant' unemployed to people who receive 'inactive' or incapacity/ disability benefits.

Each organisation in the first *access* phase of the fieldwork, articulated their aims. The YPC had the overarching aim to move from a programme-centred approach to a person-centred approach in order to meet the needs of the young person. As JM, one of the senior managers put it in our first meeting: "The whole approach changed with the new government and the New Deal from being project-centred to being person-centred though the 'on your bike' element from the Tories is still there" (July 1999).

At SD a senior manager explained to me that with the long-term unemployed the likelihood of a sustainable job (defined by New Deal as over 13 weeks) as a result of a New Deal intervention was 25%. SD stated that their aim was to measure 'distance travelled' towards employability through soft indicators for the 75% that were unlikely to get a job for 13 weeks or more.

In both organisations the research consisted of three phases; firstly an *access* phase between March 1999 and December 1999, which I have in part begun discussing. In this phase I gave presentations on a social psychological conceptualisation around *why* there was a relationship between unemployment and health involving the idea of self-other relations, identity and the concept of *psycho-social space* which I had developed in my initial analysis of the evidence base.

These early meetings with both management teams built up trust and rapport and put the research relationship on a conceptual footing. It is during these months of discussion that we isolated a key area of transition that I could collaborate on; the transition to a person-centred approach at the YPC and the measurement of soft indicators at SD.

In the second *action* phase (December 1999 - July 2000) a cohort of clients were followed through their 13-week Individually-Focussed Gateway assessment at both organisations and interviews were carried out with front-line staff and management. In the third *co-analysis* phase (January 2001 – April 2001) I returned to the YPC only with my findings and engaged in further co-analysis with management and a small number of young people. In total 39 separate observations, 16 recorded interviews, 13 action meetings, 2 presentations and one co-analysis discussion group was carried out. The research was iterative with two phases of reflection and analysis.

An objective of my collaborative action research was that data was returned quickly, in real-time, to both organisations as useable research products for their transition. It was in the second phase that all three 'actions' presented occurred. In each case they relate to tools and processes being developed by the YPC and SD to use in the individually focused gateway process. The Gateway is the first 13-week programme that prepares the unemployed person for one of the four New Deal options or, in the case of Skillseekers[5], for a vocational programme which leads to a Vocational Qualification (VQ).

3 Using a dialogical approach to develop alternative solutions

In each of the three 'actions' the organisation had in mind a proposed solution built on a self-perception psychology of unemployment (see Figure 2). Through one-to-one dialogue and debate with a key decision-making individual, the proposed solution was challenged using a dialogical approach to arrive at an alternative solution.

The actions summarised here depart from a conventional action research approach in one important respect; we did not agree a new process carry it out for a stated period then evaluate it. Nevertheless the dialogical solutions presented here I would argue are preventative insofar as they attempt to reduce the psychosocial impact of being an unemployed user of a training and guidance service.

3.1 The introduction of anger-management sessions

In January 2000 I began following a cohort of 14 young people at the YPC through the pre-vocational programme and regularly having meetings with manager PE. He explained that the programme was currently being re-developed so that each young person would have an individually-tailored programme and invited me to attend all development meetings. The programme consisted of individual sessions that related to the goals and aspirations of the young person and a set of core-sessions. The core sessions included job-orientation, Information Technology, CV writing, support & guidance, anger-management, confidence building and 'knowing your skills'.

5 Skillseekers was the welfare programme for school-leavers aged between 16 and 18. It has now been re-developed into the programme Get Ready for Work.

Figure 2. Three problems, their original solutions and the processes envisaged

Action	Problem	Self-perception solution	Process
Anger-Management Sessions	Disruptive & inappropriate behaviour	Focus on emotions and behaviour	Outsourced anger-management sessi-ons
Rationalisation of Information Flow through the YPC	Too much information gathered from client with overlaps and inconsistencies	Referral agencies and the YPC create a new tool	Self-assessment questionnaire
Measuring Soft Indicators	Low outcomes in sustained employment	Measure other outcomes	Gateway self-assessment questionnaire

I asked PE why anger-management sessions were being introduced. He explained that young people had a lot of time to mill around and disruptions had occurred including serious fighting and threatening behaviour. The solution it was decided would be to introduce anger-management sessions by using an outsourced specialist organisation.

He said he would welcome the opportunity to discuss their introduction. This took the form a recorded action interview. As is illustrated in the extract below PE was frequently talking about 'working relationships' and how young people needed to develop the appropriate way of behaving in order to be able to sustain a job. This, he stated, was the aim of the anger-management sessions:

> *PE: No, it was really about trying to identify exactly what I do want to do with these sessions [the anger-management sessions] you know and what are they about.*
>
> *KM: I would be interested in knowing where it started. In knowing where that anger-management came from.*
>
> *PE: I mean a lot of them have got issues about ...*[6] *I mean in some cases you know how they respond and responding inappropriately and becoming aggressive (...). Originally it was about trying to get through to them that that kind of reaction isn't going to achieve the goal. It's like you could argue in some cases it's like how I deal with my six year old son you know, 'you doing this is not going to change it' you know 'change the way you come to me and we can work round it' or whatever' but 'If I said no and you scream and shout and kick and thump isn't going to make me suddenly go OK let's say yes' and I think that a lot of the young people that we deal with have, still have that problem you know, that because we have said no to something or because it doesn't fit with ... They are*

6 ... Is used to denote a pause in the speech and (...) to denote a section that has been removed.

not prepared to listen to an argument they think that if they shout loud enough and react badly enough they will get us to change our mind.

KM: I think they think they might be able to wear you down (laughs)

PE: Yerr they might be (laughs) but I think there were issues like that that I felt that we needed to explore and but I suppose I have moved more towards about you know expectations in the workplace, how to behave in the workplace how would they respond if things aren't going well and if they are not happy with things. (...) And I accept that often things will be wrong in the workplace that they need to be able to speak up about it.

KM: Yerr but it is how

PE: But it is how to do it, an appropriate way to do it. (...) But it is about maybe giving them the skills to try and negotiate (July 2000)

During such dialogue we debated the concept person-centred and the notion of employability and a working consensus emerged that it was less a question of anger-management and more a question of developing a module on 'working-relationships'. This module was integrated into the job-orientation part of the core programme.

3.2 Rationalising information flow

DDG: At the moment there are a number of systems (...) would like for there to be one system that is person centred. It is about identifying the needs of trainees, we talk about the need to record progress (April 2000).

In the second action at the YPC there were inconsistencies in the different systems key-workers (trainers) used to gather information on the young person (trainee). Further, information was being gathered which had already been gathered by the referral body e.g. the Careers Service or school. DDG explained, in the first of three recorded sessions, that a new person-centred system was needed to measure progress and set out agreed action points between the young person and their key-worker. In her first solution, a self-assessment questionnaire was developed to be used with the young person in the introductory session.

Using a flow diagram I fed back to DDG my evidence from following the January 2000 cohort of 14 young people through their Gateway programme. I explained that the first session consisted of an interview of around twenty minutes followed by an application form which could take around forty minutes. The form had many of the sections of a conventional job application form: 'qualifications' 'previous work experience', 'referees' etc. In nearly all cases the young person could not fill in the form because so many sections were irrelevant to them. This was followed three days later by the young people filling in an 'Individual Training Plan' (ITP) where again most sections could not be filled in by the young people. The ITP seemed to be designed to meet the needs of the main funder of the programme the Enterprise Agency. I had pointed out to senior management that these tools and processes were not particularly person-centred and upon their advice this was one reason DDG had approached me.

We began by co-developing the information flow diagram and DDG and I debated all stages of the young person's movement through the organisation from initial 'Referral' to the 'Exit & After-Care' stage. After this first session she developed the aforementioned self-assessment questionnaire. This had three sections (i) person skills which may affect you (ii) core skills and (iii) skills you might need in the workplace. In each section the young person was asked a question where they had to rate themselves. Questions such as "are you good at handling your money?" "Are you a couch potato? "Do you easily lose your temper?" and "do you feel confident about writing a letter?".

I suggested that, though there were some questions which related to the young person's *external* barriers to employment e.g. "do you have difficulties organising transport?" most of the questions were related to perceived *internal* barriers to employment.

We discussed how DDG saw this tool being used and the extent to which employability itself could be related to a person's internal attributes or personality traits. I suggested that using the assessment when the young person had just arrived in the organisation would be rather daunting. Through recorded debate and discussion DDG and I came to the working consensus that the initial meetings with the young unemployed person served three distinct functions which needed to be recorded in an assessment tool: (i) an opportunity for the young person to express their concerns, goals, issues etc; (ii) an opportunity for the YPC to measure and baseline the core skills that the Scottish Qualifications Authority and the Enterprise Agency required of them and; (iii) and an on-going document which recorded key workers concerns and issues in relation to the young person and thus evidenced progress. It was agreed that these should form three sections of a booklet and that this along with an open unstructured interview between the young person and key-worker to promote actual dialogue, should replace the current inductions tools. This formed the recommendation placed in DDG's report to the senior management team.

3.3 Measuring soft indicators

In the final action Strategic Delivery (SD) had also developed a self-assessment questionnaire and again I had three action interviews with a senior manager (NE) in charge of the New Deal for long term unemployed. NE wanted to develop assessments, which measured softer indicators or distance travelled towards employability.

NE's original self-assessment questionnaire had eight sections such as (i) *Getting and using skills,* which had statements such as "I can organise myself to get things done on time", (ii) *Understanding and Valuing yourself,* with statements such as "I feel relaxed when meeting new people", "I would like to be my own friend – I think I am a good one". There were ten statements in each section with the response alternatives 'I am confident', 'I am generally OK' and 'I would like to improve'. The unemployed client was required to go through the statements choosing from the three alternatives in each case.

A further section had sub-sections entitled "your voice, your appearance, your temperament". "Your appearance", for example, had five closed response alternatives 'very fashionable', 'scruffy', 'untidy', 'neat' and 'grubby'. In discussing this assessment, I drew attention to the personal probing questions and NE and I debated the nature of employability and the idea of both internal and external barriers to employment.

Figure 3. Three problems, dialogical solutions and processes used

Action	Terms of shared frame of reference	Alternative dialogical solution	New Process
Anger-Management Sessions	Person-centred approach & Employability	Focus on working relationships	Introduction of "Working Relationships" module
Rationalisation of Information Flow	Person-centred approach & Employability	Assessment and interview through dialogue in context	Interview & Goal-setting questionnaire
Measuring Soft Indicators	Employability	Measure distance travelled in context	"What about me?" assessment questionnaire

I shared some research papers I had and our discussions in the second meeting centred on the 75% of participants of the programme that the New Deal projected were unlikely to achieve a sustainable job. I argued that self-assessment questionnaires such as NE's original solution were likely to increase the psycho-social impact of being unemployed, not only because they were intrusive but also because they could be read as encouraging the unemployed person to see themselves as the cause of their unemployment – the risk inherent in a self-perception approach to intervention.

In this extract NE explained that society expected people who were unemployed to do something in exchange for receiving state benefits and the organisation needed to assess clients in order to help them choose the appropriate of the four New Deal options:

> *NE: We feel a great deal of sympathy for somebody who has been forced in the system if you like, they are claiming benefits society deems that they should then go to get a job, (...). The Employment Service, I am not blaming them, but a large organisation, then, who might well pay their insurance stamp, or pay their benefits turns round and says to them "right sorry you are now x amount of time, long-term unemployed, you have to go into a provision which is designed to get people off their fat backsides and into a job" and that is the perception that they may well have. It causes them stress and we understand that.*
>
> *KM: Do you think you could build that into your delivery?*
>
> *NE: We do unofficially build it into our delivery not officially. (...) We are not policy makers here, we are not saying to somebody "we think you should work we think you shouldn't work". That's the design of the pilot (...) There is another argument which goes how do you make that assessment because we don't have an ability to screen, that's why we do it unofficially.*
>
> *KM: To be realistic I am fully aware that there are funding issues there and we started this interview by saying that the measured outcome was employment, and I am not so naïve as to expect these sorts of strategic partnerships to then start taking on, you know health issues and holistic issues, but if you can build in an appreciation of it, that I would think would be a positive step for-*

ward. In the way that it is delivered, not really saying to people "right we want to exempt you from the New Deal" because I think that becomes very very hard and it becomes a political nightmare, it is more about=

NE: =I know what you mean about building in an appreciation of it, we have run workshops on mental health issues, for our providers, and our adult guidance staff, to be able to identify that and also to identify where a client is likely to have difficulties so we can recognise that client needs an additional amount of support. (...) If there was scope to do something like this again, another pilot, I would have a lot more flexibility in the delivery model, in terms of identifying where clients are not actually going to benefit from this kind of support (March 2000)

After this meeting NE then developed a new measure, entitled "What about me?". In the introduction page of the seven-page booklet it states "The 'what about me' exercise is in two parts. The first part 'Where am I now?' looks at external factors that may or may not affect you when you look for work. The second part 'This is me' is based on all the things we know employers tend to look for in new recruits in terms of personal attitudes, enthusiasm, getting on with people etc" (SD "What about me?" Gateway Assessment Tool). The booklet's covers, "my income, transport issues, care requirements, my health, my housing situation, my criminal record, my work history and my qualifications and skills". The "This is me" section includes "how I work with others", "my communication skills", and "where are all the jobs – do I know?". There are no questions on personal appearance, voice or temperament. In each section the client can choose from four statements.

The three actions outlined demonstrate how interventions into unemployment are increasingly a matter of self-other relations. The self-perception psychology of unemployment is currently dominating the training and guidance sector in Scotland. Organisations are particularly interested in measures of self-perception such as self-esteem and self-efficacy. In each case a dialogical psychology is used to moderate the individual-focus of the solution by setting it in a wider context and encouraging the organisation to emphasize self-other relations and external barriers (see Figure 3). This sets the unemployed person's behaviour, when on such compulsory programmes, not only in the context of working relationships but also in the limited actual services the organisation can provide to support them.

4 Features of a dialogical approach to intervention

The starting point of a dialogical intervention is multi-level it occurs in the processes used, the organisational change and in collectively changing the social knowledge which surrounds the phenomenon. The study explores the contradictions inherent in two street level organisations which are part of a system that simultaneously compels a person to attend a service then describes that service as offering a person-centred approach. (A fuller account of the YPC's transition to a person-centred approach is given in Mahendran, 2003). I did not arrive with a solution based on the conventions of my psychological tradition but, together with people in the organisations, co-created one through actual dialogue and an analysis of dialogue.

The three actions demonstrate the four levels of dialogue (i) the dialogue between the self and the other; (ii) the dialogue between the individual as social agent and the public sphere; (iii) the features of the dialogical self and (iv) the dialogism of words-in-use. There is the dialogue between myself and the manager as decision-maker but perhaps more critically we are often discussing the first interpersonal encounter between the unemployed person as client and their personal advisor (at SD) or key-worker (at the YPC) and the possible terms of reference for that discussion.

A dialogical understanding of well-being is dynamic and relational. As Starrin and colleagues are arguing the stigma and health problems during unemployment can be analysed in terms of the interpersonal encounters unemployed people face during their unemployment. In this respect the encounter the unemployed 'client' has with the personal advisor or key-worker is important and worthy of analysis.

Further there is evidence of the dialogue between the subject as a social agent and the public sphere. The decision-maker is as an argumentative-debater who debates with me from the knowledge available in the public sphere and with closer analysis of these communicative interactions it is possible to see the social representations of the 'welfare recipient', 'young people' and so on.

The dialogical self is also apparent where PE engages in dialogue by introducing the *I*-position of the manager, but also the *I*-position of the parent in dialogue with his son. NE introduces the *I*-position of the SD manager in a dialogue with the Employment Service. There are insights to be gained by a transparent and systematic analysis of such *I*-positions in the communicative interactions of organisations. Organisations such as the YPC and SD can be viewed as being *multivoiced*, not just the voices of the client, staff, and managers, but also the voices that populate their communicative interactions. With a dialogical approach it becomes possible to demonstrate the extent to which organisations feel themselves to be in a dialogue with the government, policy-makers and the socio-political discourse.

Finally it is possible to precisely track words such as 'person-centred', and 'employability', thematically analyse the dialogism in their use over time and reveal their implications. In the presentation of initial findings report to the YPC in February 2001, I simply highlighted all the terms that had been used to describe the young people and the front-line staff.

The three actions hopefully demonstrate the extent to which welfare legislation is currently highly individualised, set within a new economic orthodoxy and organisations are constantly in transition. Organisations are required to remain in a close dialogue with the terms of government legislation and central funding bodies such that they fail to hear the voice of the actual clients they aim to serve. An authentic person-centred approach rests on the development of tools and processes to create the space for actual dialogue between service-providers and unemployed people as individuals. Rather than working with the dominant frames of reference, which, as discussed throughout this chapter, frequently derive from policy discourse and its relationship to academic evidence, they would need to consider the frames of reference actually used by such individuals.

5 Conclusion – adding to the conceptual currency

Psychologists and other interventionists are knowledge creators, creating new realities according to the research carried out within the norms of their traditions. Policy discourse is formulated, in part, by such social knowledge which exists in the evidence base and in the public sphere. This social knowledge creates the frames of reference, the terms of the debate, which become the measured outcomes used by delivery organisations in the relentless requirement to attract funding. If interventionists and knowledge creators focus on self-esteem then the solution to unemployment will become a matter of self-esteem. Policies will arise around intervening into a person's sense of self in the periods when they are unemployed. Targets regarding employability will become about changes in unemployed clients' self-esteem while 'distance travelled' toward employability will be narrowly conceived as improvements in reporting of self-esteem or self-efficacy which at best can perhaps lead to employment for a small minority.

It is possible for knowledge-creators to shift their understanding of self-other relations as well as the terms used in framing their studies. This can, in turn, shift the policy discourse and new outcomes can become legitimised. Both the organisations discussed here operate at the interface between unemployed people and their local labour markets. The actions described above reflect small-scale changes. Further shifts in the frames of reference, in particular the conceptualisation of 'employability', could allow organisations to spend their time working with employers and local economic development agencies. They could create opportunities for their clients, job outcomes and impact on the economies of their regions.

Within the Scottish Executive, the devolved government of Scotland where I am now based, I have worked with the Effective Intervention's Unit as a part of their reference group to update their 'Moving On' booklet. This booklet is for organisations and practitioners working with unemployed drug and alcohol users during their transition towards employment. One of the first changes we made was to redefine employability for such organisations to highlight both internal and external barriers. It now states: "Employability entails achieving a match between the abilities, attitudes and capabilities of an individual, the needs, expectations and attitudes of employers and the demands of the current local labour market conditions" (Moving On: Update EIU, 2003, p. 7).

Such a broadening of definition, which has been widely adopted, permits new interventions and practices with the unemployed person, employers, sector skills academies etc. It permits the development of micro-demand strategies within local labour markets and means that all of these can legitimately attract funding. A dialogical approach can work in a bottom-up way working with unemployed people, and the services that exist and also in a top-down way working directly in dialogue with government policy makers.

It is hoped that this charcoal sketch of a dialogical psychology being used to successfully challenge a self-perception psychology of unemployment, has drawn attention to the different psychologies of unemployment within the evidence base which are consequently at work in the debates of policy-makers and the practices of front-line organisations. Interventions if they become dialectical, allow unemployment itself to be understood within a labour market which is characterised by transition. By re-orienting the locus of intervention to the dialogical interaction of all the components of the system such practical interventions stand the possibility of reducing the psycho-social impact of being unemployed.

References

Bakhtin, M.M. (1981) *The dialogic imagination: Four essays*. Austin and London: University of Texas Press.

Bakhtin, M.M. (1986) *Speech genres and other late essays*. Austin, Texas: University of Texas Press.

Cassell, C., Fitter, M., Fryer, D. & Smith, L., (1988). The development of computer applications by non-employed people in community settings. *Journal of Occupational Psychology, 61*, 89-102.

Creed, P.A. & Macintyre, S.R. (2001) The relative effects of deprivation of the latent and manifest benefits of employment on the well-being of unemployed people. *Journal of Occupational Health Psychology, 6*, 324-331.

Drewery, W. (1998). Unemployment: What kind of problem is it? *Journal of Community & Applied Social Psychology. 8*, 101-118.

Eden, D. & Aviram, A. (1993). Self-efficacy training to speed reemployment: Helping people to help themselves. *Journal of Applied Psychology, 78*, 352-360.

Ezzy, D. (2000). Fate and agency in job loss narratives. *Qualitative Sociology, 23*, 121-134.

Finn, D. (1999). *From full employment to employability: New labour and the unemployed*. Paper presented at the Social Policy Association 32nd Annual Conference, 'Social Policy and Social Insecurity', Roehampton College, 22-23rd July.

Fryer, D. (2000). Unemployment and mental health: Hazards and challenges of psychology in the community. In K. Isaksson., C. Hogstedt., C. Eriksson & T. Theorell (Eds.), *Health effects of the new labour market* (pp.11-25). New York: Kluwer Academic/Plenum Publishers.

Fryer, D. & Fagan, R. (1993). Coping with unemployment. *International Journal of Political Economy. 23*, 95-120.

Habermas, J. (1989). *The theory of communicative action*, (vol. 2). Cambridge: Polity Press.

Hermans, H.J.M. (2001). The dialogical self: Towards a theory of personal and cultural positioning. *Culture & Psychology, 7*, 243-281.

Hillage, J. & Pollard, E. (1998). *Employability: Developing a framework for policy analysis*. Institute for Employment Studies: DfEE noRR85.

Israel, J. (1972). Stipulations and construction in the social sciences. In J. Israel & H. Tajfel (Eds.), *The context of social psychology*. London: Academic Press

Jahoda, M., Lazarsfeld, P. & Zeisel, H. (1933; 1972). *Marienthal: The sociography of an unemployed community*. London: Tavistock Publications.

Mahendran, K. (2000). Book review of the ICOH working group 'Unemployment and Health'. *Journal of Community & Applied Social Psychology, 10*, 167-8.

Mahendran, K. (2003). The transition of a Scottish young person's centre – a dialogical analysis. In Grant C.B. (Ed.), *Rethinking communicative interaction* (pp.235-256). Amsterdam/Philadelphia John Benjamins.

McGhee, J. & Fryer, D. (1989). Unemployment, income and the family: An action research approach. *Social Behaviour, 4*, 237-252.

McPolin, N. (1999). *Summary paper activities programme: Self-efficacy and achievement motivation among unemployed and marginalized youth*. Available from Upper Springfield Development Trust, Norglen Gardens, Belfast BT11 8EL

Moscovici, S. (1990). The generalised self in mass society. In H.T. Himmelweit & G. Gaskell (Eds.), *Societal Psychology*. London: Sage.

Moscovici, S. (2000). The phenomenon of social representations. In G. Duveen, (Ed.), *Social representations, explorations in social psychology*. Cambridge: Polity Press.

Murphy, G.C. & Athanasou, J.A. (1999) The effect of unemployment on mental health. *Journal of Occupational and Organisational Health, 72*, 83-99

Nordenmark, M. & Strandh, M. (1999). Towards a sociological understanding of mental well-being among the unemployed: The role of economic and psychosocial factors. *Sociology, 33,* 577-597.

Proudfoot, J., Guest, D., Carson, J., Dunn, G. & Gray, J. (1997). Effect of cognitive-behaviour training on job-finding among long-term unemployed people. *The Lancet, 350,* 96-100.

Proudfoot, J., Gray, J., Carson, J., Guest, D. & Dunn, G. (1999). Psychological training improves mental health and job-finding among unemployed people. *International Archives of Occupational & Environmental Health, 72* (Suppl.), 40-42.

Spera, S.P., Buhrfiend., E.D. & Pennebaker, J. W. (1994). Expressive writing and coping with job loss. *Academy of Management Journal, 37,* 722-733.

Starrin, B., Rantakeisu, U., Forsberg, E. & Kalander-Blomqvist, M. (2000). Understanding the health consequences of unemployment – the finance/shame model. In T. Kieselbach (Ed), *Youth unemployment and health.* Opladen: Leske & Budrich.

Straehle, C., Weiss, G., Wodak, R., Muntigl, P & Sedlak, M. (1999). Struggle as metaphor in European Union discourses on unemployment. *Discourse & Society, 10,* 67-99.

Svensson, P.-G. & Starrin, B.(1989). A WHO perspective on unemployment, poverty and quality of working life. In B. Starrin, P-G. Svensson & H. Winterberger (Eds.), *Unemployment, poverty and quality of working life.* Berlin: Sigma.

Villaret, M. & Gianinazzi, A. (1999). Health consequences of unemployment: “Disoccupazione & salute” is a prevention program at the local level. *International Archives of Occupational & Environmental Health. 72* (Suppl.), 40-42.

Mortality and Unemployment in the Context of Macroeconomic Change: The International Evidence

M. Harvey Brenner

Introduction

A series of time-series studies on mortality in relation to macroeconomic change and unemployment rates in the 1970s and early 1980s, dealing with the United States and the United Kingdom, set the stage for a considerable European and American research literature on unemployment and health. This derived literature is classically "epidemiological" in nature in that the research designs and samples are at the individual level of analysis. Several excellent and up to date summaries of this literature are now available (Tausky and Piedmont, 1967/68; Catalano, 1991; Dooley, Catalano and Wilson 1994; Hallsten, Grossi & Westerlund, 1999; Kasl and Jones, 2002). All of these summaries come to virtually the same conclusion, namely that unemployment is an important risk factor to mortality and morbidity – especially if the unemployment is of long duration. Several of the outstanding epidemiological studies on unemployment and health include Martikainen and Valkonen, 1996; Merva and Fowles, 1999; and Moser, Fox, Jones and Goldblatt, 1986.

In this chapter we review data from a series of studies performed for the European Commission during 1998-2004 examining the time-series relations between mortality, on the one hand, and both economic growth and unemployment, on the other. The principal methodological innovations in these studies involved (1) the use of distributed lag analysis for all principal predictors and (2) the use of the error correction method (ECM) developed by Granger and Engle (2009) which also involves analysis of annual changes. Additionally reviewed are cross-sectional studies of mortality, also done for the European Commission, involving industrialized countries, with the emphasis on economic growth and unemployment. In these studies economic growth appears essential to increasing life expectancy, whereas unemployment rates are routinely related to elevated mortality rates. Explanations of these phenomena are discussed in relation to current debates on the importance of income equality to health.

1 Unemployment and mortality

The unemployment rate is the second most important mortality predictor, especially in its long-term impact on mortality over at least a decade. It needs to be made very clear that estimation of the unemployment-mortality relationship without a distributed lag may not provide the researcher with a significant positive relationship. Indeed, especially in countries like the United States, if only the zero-lag relation is estimated (without appropriate controls), it is possible to observe an inverse relationship between unemployment and mortality. The knowledgeable researcher will immediately recognize that such a relationship is spurious – and is due to serious confounding – since nearly all literature at the individual level of analysis shows unemployment to be an important risk to poor health and higher mortality. In addition, we have recently discovered that if *any* significant and meaningful relationship between unemployment and overall mortality exists at the zero lag, it appears to be in the United States. This relationship involves only extremely short-term unemployment, specifically less than 5 weeks that is inversely related to mortality. Such extremely short-term unemployment implies the situation of a "decoupled" economy, where relatively mild recession coexists with continued growth in major industries. In fact, this extremely short-term unemployment may actually indicate a relative improvement in the work situation and career possibilities of employees who move to the industries that are characterized by higher growth.

These time-series models, while extending their range over another 10-15 years (1984-2000), and covering another dozen countries, are essentially extensions of original work done in the 1970s and early 1980s. An entirely new question is whether these relations, involving economic growth and unemployment as mortality predictors, can be seen in cross-sectional analysis when large numbers of countries are compared. In this paper we examine two databases, the first covering the OECD countries and the second involving all of Europe and including countries that were formerly part of the Soviet Union.

Among the OECD countries we observe that GDP per capita is the central inverse predictor of mortality in virtually all age-groups. Additionally, the unemployment rate for persons with a secondary school education (i.e., medium skilled employees) is positively related to mortality for populations in the age range 50-75. For the European and former Soviet countries GDP per capita is also the central predictor of mortality reduction. Furthermore, among these latter countries, the unemployment rates are significant predictors of higher mortality in nearly all age-groups. Equally important, both the unemployment rate for people with secondary education, and for people with tertiary education, is positively related to mortality. Most especially, in the European and former Soviet sample of countries, it is the unemployment rate among those with tertiary education that is the most outstanding mortality predictor. Among some of the eastern European countries and former Soviet states we are of course dealing with relatively impoverished societies with low concentrations of high technology industries. This means that there is little room for the employment of highly educated people, who then may hesitate to take lower level employment, and thus remain unemployed for relatively long periods. Altogether, however, an economy at high levels of unemployment is one in which a significant part of the workforce is left out – also indicating a situation of considerable economic inequality.

Figure 1 shows the relationship between age-specific mortality rates and the independent variables of GDP per capita, the unemployment rate for the population with secondary education, based on the OECD database for 1995. It shows the multivariate relationship for the age-group 45-49. Similar graphs can be shown for mortality rates in the age groups 50-54, 55-59 and 60-64. It can be seen that both GDP per capita (with a negative sign), and the unemployment rate (with a positive sign), are significantly related to mortality.

Figure 1: Multivariate relation of economic variables to age-specific mortality rates for 1995, 26 industrial OECD countries

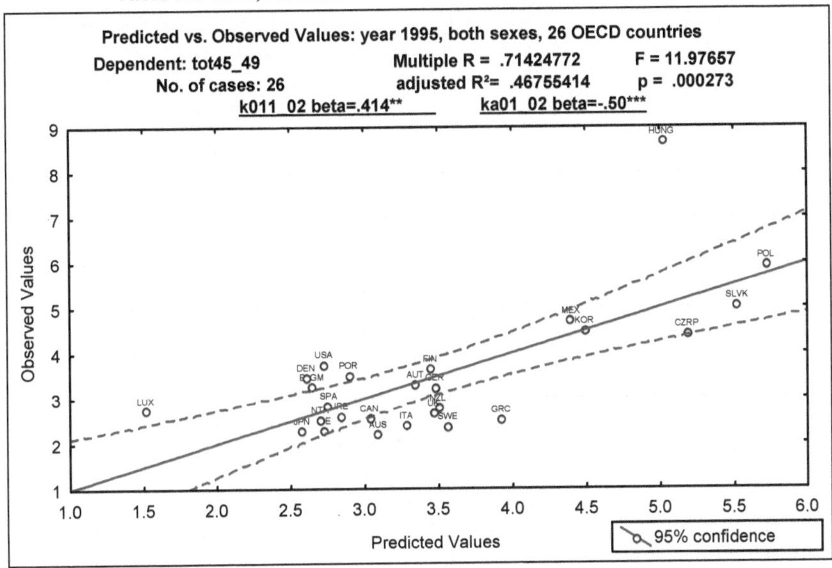

Legend

Independent variables:

- Ka01 _02: GDP per capita in international dollars, in purchasing power parity (ppp)
- K011 _02: unemployment rate for population with secondary education as percent of total unemployment rate

Dependent variables:

- tot45-49: mortality rates for age-group 45-49

** p< .01, *** p < .001

2 'Socioeconomic status' of populations

An aim of these projects was to examine the implications of research findings on the "social determinants of health" for explaining mortality differentials among industrialized countries.

The epidemiological literature has emphasized the predominant significance of socio-economic status – income, occupational status and educational level – as an essential source of understanding the differential health status and longevity of individuals (Mackenbach, Kunst, Cavelaars, Groenho & Geurts, 1997; Smith, Hart, Blane, Gillis & Hawthorne, 1997). The prime question is whether "socioeconomic status" can be said to characterize the populations of different nations and, if the answer is affirmative, then one would like to determine whether indicators of "national socio-economic status" predict mortality differentials among countries.

Further, if measures of socioeconomic status can indeed describe characteristics of entire nations, and these are related to their populations' longevity potential, then it becomes a matter of considerable policy significance as to what the levels of these indicators actually are. Policies concerning the enhancement of these economic and social predictors then become the crucial foundation of "health policy" – of which health services financing is only a single (but highly significant) component. In that case, principal health policies must concern economic growth (and recession), educational attainment, the industrial structure of nations, and overall 'social protection' of economically and socially vulnerable segments of national populations.

3 Economic growth and employment

In general, the principal prior findings of the epidemiological literature, at the individual level of analysis, on social determinants of health can be reproduced at the national level and applied to the EU countries – as well as to virtually all industrialized societies.

The economic level of living (real GDP per capita) and economic productivity, in conjunction with the related sectoral distribution of industrial development, are the major sources of improvement in a society's health and longevity – even in the advanced industrial nations. Both sexes are seen to benefit (in somewhat different ways) from these developments in national wealth development and occupational structure.

In addition to measuring the economic (material) level of living and employment productivity the unemployment rate serves as a principal indicator of below capacity-utilization, relatively low economic growth and, typically, recession. These data inform us that GDP per capita alone is an insufficient indicator of the material well-being of society. The extent of employment versus unemployment partly indicates the extent of distribution of the material wealth produced by the overall economy. The strong findings of the relationship between high unemployment rates and higher mortality, in cross-sectional analyses comparing countries, are consistent with the very large body of individual level epidemiological findings demonstrating higher morbidity and mortality among the unemployed as compared with employed people.

4 Significance of productivity and technology

What are the factors that make material wealth and productivity in a society so essential to health? The first group of material factors includes, of course, the most basic and obvious: namely nutrition, sanitary engineering and housing.

Second, national wealth is the basis for investments in improved working conditions including reduction of exhausting work; the creation of comfortable and climate controlled work environments; the development of ergonomics; the reduction of accidents; the control of chemical carcinogens and environmental toxins, and air particulates.

Third, national wealth serves as the financial foundation of social protection programs that counteract the effects of catastrophes and poverty. These include pensions in older age, health care and rehabilitation, unemployment insurance, accident insurance and social welfare.

Fourth, national wealth is the basis for investment in future productivity including the provision of general and specialized education (i.e., "human capital"), investments in science and technology, both of which are, independently, sources of improved health.

Fifth, with respect to older and frail or disabled populations, national wealth is a source of investment in specialized means of adaptation, including elevators, climate control systems and specialized access to buildings, roads and transportation.

Sixth, national wealth is the basis for investment in communications, media development and transportation. These three items, paradigms of the 20th century, have been of essential importance in the promotion of interpersonal linkage and social integration. Such social integration, in the epidemiological literature, is clearly an important element in health enhancement in so far as it promotes the reinforcement and frequency of human relationships.

Now, the above items all refer to 'real' things – i.e., physical items that are objectively observable. Does inequality enter into this picture? It clearly does, as does the presence or absence of these items in the lives of individuals. Thus, from the standpoint of distribution of these items that are important for health and life span, the *absolute* availability of these items and the *inequality* of availability of these items are virtually identical. In the vast majority of cases, differences in absolute wealth and inequality of wealth are semantically equivalent. From the standpoint of emotional differences based on 'relative deprivation', the same distinction occurs. Thus, if I am objectively deprived of the money required for, say, transportation and telephone use – and I subjectively require these things – then I am both objectively deprived and perhaps further deprived (emotionally) *relative to* other persons who possess these things.

There are also a great many goods and services which may not be directly required for health and life, but provide considerable convenience or comfort, e.g., the services of attorneys, craftsmen, financial analysts and child care providers. On a subjective level, the deprivation of these services among people who greatly desire them would represent emotional loss or relative deprivation. Such emotional loss might well be aggravated by comparison with other persons who are able to utilize them. However, in all cases of even these subjectively desirable things, the things themselves are real and purchasable; indeed they are frequently rather expensive.

Now, there are significant cases in which the issue of relative deprivation is somewhat more idiosyncratic to the individual. For example, these might occur when one does not have some "things" that a family member possesses, for example a high income and high prestige form of employment. But even in these broader life situations of comparison that seems to be deeply subjective and emotional, we are nevertheless in a situation that the subjective deprivation relates to an objective phenomenon – in this case the presence of a well remunerated and highly reputed occupation.

To conclude the present issue of absolute deprivation versus "inequality" (or relative deprivation), it appears that the theoretical distinction is very hard to make for people living in modern industrial societies; indeed, there may be little or no practical distinction. And in terms of our finding of the central importance of real wealth, productivity per capita and employment to a nation's health, we infer that: (1) absolute differences in wealth and other resources are *equivalent* to inequality, and (2) subjective feelings of relative deprivation based on the sense of inequality refer ultimately to the deprivation of real, measurable things that are often important in the maintenance of health.

A final question on inequality deals with the impact of increased national wealth (GDP) per capita as it would affect different income groups in the population. The basic assumption in both economics and epidemiology is that the same amount of additional money obtained by higher versus lower income groups would benefit the lower income groups more in terms of improved health. On a policy level, for example, this would argue on behalf of progressive income taxation (whereby higher income groups pay proportionally higher tax rates). With respect to the "normal" progression of increased GDP per capita (due to increasing productivity gains) the serious question is whether the added annual GDP per capita is more beneficial to higher or lower income groups. This is an empirical question. It is possible, first of all, that increased GDP per capita simply entails that all groups earn more and do so in a manner that is roughly proportionate to their initial level of wealth. In that case there is no change in the relative positions of different economic groups – all groups gain and health is improved in the whole population; that will be the usual situation. There are periods however, when, despite increases in GDP per capita, the lower or lowest socioeconomic groups will not gain, and may even lose employment and income due to obsolescence of skill and technology and/or the movement of employment to lower-wage countries. This is not so uncommon in Western industrial societies which are exemplified by relatively higher rates of technological change and export of capital. In those conditions, it would appear vital that social protection mechanisms vigorously enter the policy situation, namely to provide significant financing in the areas of unemployment insurance, health care expenditures and social welfare.

The key point is that additional GDP per capita and productivity normally result in the reduction of poverty. When poverty occurs, it will be experienced both absolutely and in relative – i.e., inequality – terms. In those situations where GDP per capita is increasing, but specific population groups are actually moving into poverty, industrial societies have a special responsibility for social protection to prevent damage to health and loss of life in the valuable groups.

5 Material well-being or inequality as the principal predictor of national health

There exist a substantial number of studies dealing with populations in the United States and Europe on the relation between income and mortality differentials. While income has been the last of the three principal socioeconomic indicators to be examined in epidemiological studies, it has moved to the forefront of social epidemiology and, as of the present time, commands the key point of debate in much of modern epidemiological research. First of all, the finding is consistent among nearly all studies of this subject that income is strongly inversely related to mortality at the individual level of analysis.

This means that in population samples, we are able to discriminate with considerable accuracy the differential mortality probabilities of individual persons in accordance with their real income level. There is a qualification of these findings in that it is sometimes observed, especially in studies with finely categorized income distinctions, among persons in the highest levels of income there appears to be a lower proportionate decrease in mortality than is true at moderate or low income levels.

In the early 1990s an extraordinary new interest arose, not in further defining or explaining the basis for the income-mortality relationship, but rather in taking a new perspective altogether. The new perspective dispensed with the concept of income per se and focused on the idea of income inequality or, more broadly, the relative inequality of the material level of living among individuals. It was strongly suggested that in modern industrial societies it is not *absolute* income differentials that truly influence health or mortality, but rather the subjective sense of relative deprivation based on the inequalities of income among individuals or population groups.

The initial empirical studies were introduced by Wilkinson (1992) in which he presented evidence to show that, among nine western industrial societies, income inequalities (within those societies) were strong predictors of higher mortality rates. Interestingly, the original data displayed a relationship in which the proportion of total societal income earned by the "lower" 70 percent of the population was inversely related to national mortality rates. Despite the fact that the displayed relation introduced considerable theoretical and interpretive problems, the original data display and its argumentation was the basis of an extensive literature into the relation between income inequality and health at national, regional and metropolitan levels. While the initial Wilkinson data presentation was convincing to some, the inferences led to incoherent conclusions. First of all, the usual nature of the "economic inequality" argument is that it refers particularly to the lowest income groups as compared to middle and higher income groups, although such findings were not obtained in the Wilkinson study. Indeed, it could be argued that if the lower 70 percent of population income groups had a relatively egalitarian distribution of income among themselves, but a sharp income gap as compared with the upper 30 percent, then the relative income experience of the lower 70 percent would be of minimal deprivation since their comparison would most likely be among income groups somewhat higher and lower than themselves (i.e., within that 70 percent).

Despite the anomalous character of some of the original findings, many specialists examining the epidemiological SES-health relationship quickly became convinced that the correct understanding of the inverse relation between income and mortality should probably be based on the subjective experience of inequality itself rather than the increment of material gain in "real" things (goods and services). While a large number of studies followed from the original Wilkinson paper, we will identify some of the most frequently cited. In the United States, Kawachi (1997) presented data for US states, purporting to demonstrate a positive relationship between income inequality (concentration) and mortality, but not between absolute income levels and mortality. He argued that the correct theoretical interpretation was not so much along the subjective, relative deprivation lines of Wilkinson, but rather that the lack of relative income equality pointed to the absence of integral social relations among income groups and a lower level of "social capital" (i.e., trust) and social integration in general. Thus, along the classical sociological theories of Durkheim and Coleman, Kawachi's position has been that it is social integration in general that promotes population health.

Similar findings for US states using measures of income inequality and mortality were obtained by Lynch, Smith, Kaplan and House (2000). These findings served to stimulate the scientific field considerably, but subsequent studies in which Lynch et al. try to bring together Canadian and US populations at the urban level failed to replicate the initial US-state observations. As a result, the authors concluded that the relationships were spurious and not to be taken seriously for policy purposes. They went on to examine additional populations, including those in Western Europe, and again were unable to find any relations between measures of income inequality and mortality rates. Other important studies in the United States (Fiscella & Franks, 2000) and a major study across a large body of UN member countries (1997) failed to support the income inequality/mortality hypothesis regardless of the combination of countries investigated. At that point, certainly the weight of the evidence clearly pointed to the absence of a relationship between income inequality and health. Further, when the racial distribution of US states and the unemployment rate are controlled, no relationship between income inequality and mortality is found, but a very strong inverse relationship between median income per capita and US state-specific age-adjusted mortality rates can be measured.

In the present study, across 29 OECD countries and across 22 Western and Eastern European countries, real per capita GDP is a major inverse predictor of the age-adjusted mortality rate. We can come to the conclusion that, on the basis of the overall empirical literature, no robust relation has been found between income inequality and mortality among European or OECD countries, while the present study indicates very substantial inverse relations between real GDP per capita and age-adjusted mortality rates.

How shall we interpret these data? Is it correct to say that "income inequality" actually is not related to mortality among European and OECD countries? Our interpretation of this material is *not* that the relation between income inequality and mortality is truly absent, but that it is actually *present* in the basic inverse relation between *absolute* real per capita income and mortality. Since this appears to be contrary to some of the current epidemiological thinking, it seems worthwhile to provide a more detailed explanation. We provide this explanation because the issue is, in terms of policy, politics and moral philosophy, perhaps the single most outstanding conundrum in the health sciences.

The current arguments are most recently related to the Nobel Prize winning work of the economist Amartya Sen (1999, 2001) for whom, in principle, people should enjoy equal basic capabilities – i.e., resources should be distributed in such a way that each person is able to exercise the same set of capacities (for instance, is physically mobile, can feed and clothe himself or herself). This proposal has the advantage that it is sensitive to differences in individuals' needs; but it fails to offer a comprehensive conception of equality in the sense that it sets a minimum standard that everyone should achieve rather than identifying an overall distribution. Nevertheless, Sen's formulations through a substantial number of writings have had an immense influence on questions of distribution in macroeconomics, social welfare theory and development economics. And, while they have been applied largely to policies regarding developing countries, they have had considerable influence on the sociological and social psychological study of industrial societies.

Nevertheless, it appears that there is no agreed answer to the question 'In what respect should people be judged more or less equal?" The same applies to the question of measurement. Suppose we have to decide which of two income distributions is somewhat more egalitarian: what criterion should we use? Should we consider the range, the dispersion from the mean, etc.? (Dasgupta, Sen & Starrett, 1973). In particular, how far, if at all, should our measure reflect our concern about the welfare implications of inequality: should we give greater weight to inequalities at the bottom end of the distribution than to those at the top, on the grounds that the former matter more than the latter? This shows that the measurement issue is not merely a technical one, but reflects disagreement about the precise idea of equality that any proposed measure ought to capture.

A second major influence is a purely empirical one. It is that both in the United States and the United Kingdom (and to some extent in Scandinavia) where income inequality has been measurable over historical time, there has been and continues to be a substantial expansion of income inequality since the 1970s. This movement toward greater income inequality has been extensively analyzed and has been linked to the increasingly greater value of higher education for the achievement of medium to high income levels in societies that emphasize science and high technology in their production and services industries. Thus, education appears to be considerably more valuable in raising productivity rates after the 1980s than before. This would suggest a substantial policy commitment toward investment in education which, in principle, should both improve economic growth (and thus the material welfare per capita of the whole society) and the income distribution within the society.

The third influence seems to arise out of the well established sociological concept of "relative deprivation". Like the narrower term of poverty, deprivation can be viewed in absolute or relative terms. *In this formulation*, absolute deprivation refers to the loss or absence of the means to satisfy the basic needs for survival – usually identified as food, clothing, and shelter. The term relative deprivation refers to deprivations experienced when individuals compare themselves with others: that is, individuals who lack something compare themselves with those who have it and in so doing feel a sense of deprivation. Consequently, relative deprivation not only involves comparison, it is also usually defined in subjective terms. The concept is intimately linked with that of a comparative 'reference group' – the group with whom the individual or set of individuals compare themselves – the selection of reference group being crucial to the degree of relative deprivation.

Sociological debates have tended to focus on subjectively experienced relative deprivation. In the field of social policy, however, externally assessed material and cultural deprivations have been the focus. One important issue has been the extent to which deprivation is transmitted from one generation the next. In this context, the idea of a cycle of deprivation has been employed to refer to the intergenerational transmission of deprivation, primarily through family behaviors, values and practices.

Clearly, then, the sociological sense of relative deprivation refers to new subjective comparisons of the possessions and attributes of the individual as against his or her social context as defined in terms of reference groups. This conception is highly individual-based and tends to deemphasize the possible role of social 'classes' or income groupings *across* which (rather than within which) individual group members compare one another.

In social-psychological theory and research, yet another concept of relative deprivation, and thus subjective inequality, has prevailed. It appears to have originated with the psychologist L. Berkowitz who formulated the famous "frustration-aggression" theory. The theory argues that it is not absolute deprivation per se that motivates aggressive behavior (or stress-related behavior altogether), but rather the occurrence of an obstacle or impediment to the achievement of a goal which, otherwise, would have been very likely to be achieved. The 'relative deprivation', then, refers to thwarted expectations. This type of deprivation describes a very broad set of circumstances in which a 'rational' person is behaving in such a way (the means) so as to achieve a particular end. This person expects the end, or goal, to be achieved through the means being acted upon. However there arises some event, either by accident or human design, that makes the goal either somewhat less or entirely unachievable, despite considerable investment or effort on the part of the subject. The meaning of relative deprivation, here, is that the person is deprived of achieving a goal that he or she expected to accomplish.

As in the previous case of the sociological reference group, the sense of deprivation is thought to rely upon things that are not 'real'; namely, in the reference group case it is a matter of subjective interpretation and in the case of expectations, it is what the individual imagines to be the probable future. But the standard of comparison in both cases is again something not quite 'material'. Thus, in the expectations case, economic or social inequality as judged between people need not refer to the contemporaneous situation, but rather to a future state that does not yet exist. A typical example would be the comparison of a student of one of the professions (law, medicine, engineering, architecture, science etc.) who has a similar income and style of life to a semi-skilled industrial worker. While the income of the professional student is absolutely comparable to that of a semi-skilled worker at the relevant moment, the lives of these individuals will develop in very different ways, such that the graduate student will *normally* become a professional earning many times what the semi-skilled worker will make. In that case we have a concurrent identity of absolute incomes, but a subjective assessment of a future of income deprivation and very large income inequality in the situation of the semi-skilled worker.

What is the conclusion of this complex theoretical discussion? First of all, it may not be possible to develop society-wide measures of relative deprivation or economic inequality that are based on their classic sociological or social-psychological meanings. Secondly, from the standpoint of economics relative deprivation, or the sense of inequality, does rely ultimately on the use of both material things and 'services' – which are equally real – and not simply subjective interpretations. To be direct, if I possess knowledge derived from higher education, or an expensive automobile that are not possessed by my neighbor, then my neighbor will be absolutely deprived of these things and may well *have the sense of* being additionally deprived *in comparison with me*.

Now, the additional subjective sense of relative deprivation may also occur because my neighbor realizes that the higher education and the more expensive status symbol (the car) may enhance other derived and valued items such as social respect – which he may feel he does not possess to the same extent. Nevertheless, in the economic conception the ultimate comparison is of real things that are attributes or possessions of some people but not others. Indeed, it is possible for any real possessions or attributes to be, at the same time, a source of both absolute and relative deprivation, which means that the relative deprivation is simply the emotional aspect of the perception of not possessing the relevant material good or 'service'.

We can go as far as to say that where important differences lie in income and wealth or occupational prestige, it will probably be true that such absolute differences also engender an emotional sense of inequality and relative deprivation. Indeed, such deprivations often provide the basic motivation for the achievement of higher income, more highly skilled occupations and higher levels of education. Virtually by definition, absolute income differences imply inequalities. That means that people are unequal to each other in their basic incomes. Now, since our data demonstrate that there is a significant inverse relationship between income level and mortality rates, this clearly implies that where people are substantially unequal to one another in income, it is probable that the lower income individuals will also *feel* a sense of deprivation, which may lead to a further reduction in life expectancy. In sum, absolute deprivations imply real inequalities (such as inequalities of income, education and occupation) and often have major implications for health in themselves. Additionally, those same real inequalities are often preconditions to a psychological sense of deprivation in which emotions such as anger, sadness, injustice and stress may give rise to increased levels of illness based on neuro-endocrine pathways.

Unemployment and economic inequality

The present paper has provided extensive evidence, for both OECD and European countries, that unemployment rates are a prime predictor of differential country-specific mortality rates in industrialized societies. There is also considerable evidence from the macroeconomic literature (especially in the field of finance) that unemployment is a valid measure of some aspects of economic inequality over historical time (Brenner 1995). In addition, Amartya Sen has argued that *economic* inequality overall – and especially that pertaining to employment and unemployment – is more important for the understanding of economic differences among individuals and nations that *income* inequality alone, e.g., as represented by the Gini Index.

From a commonsense point of view, clearly, the unemployed and those who have left the labour force as discouraged workers, are certainly in a sufficiently small minority to be regarded as suffering inequality. This type of inequality can either be thought of as involving the lack of meaningful human work, or the separation of employees from their colleagues (the absence of social integration), or, as a result of long-term unemployment, the incidence of poverty. In the poverty situation, of course, we return to the precise issue of extreme income inequality. As to whether such economic inequality based on unemployment more accurately represents absolute or relative deprivation will depend on individual subjective assessment. From the standpoint of the general population, however, the unemployed probably suffer both absolute and relative deprivation.

6 Current debates and intellectual trends

The argument presented above states that economic growth is the bedrock of improvements in health and life expectancy in industrialized as well as developing countries. The argument is that there is no substantial diminution in the contribution of economic growth and development to societal health as a result of the economic growth process even among comparatively wealthy societies. Indeed, there is evidence that the beneficial impact of economic growth on health is considerably greater in advanced industrial societies as compared to developing countries.

It is true that, at the individual epidemiological level higher income and other aspects of socioeconomic status (occupational skill, education level, power/authority in the work organisation and prestige) are related to higher levels of health and life expectancy. And it is true that these individual level epidemiological data lend support – and are a part of the positive relationship – between real per capita GDP and national health levels. However, it is a large, and relatively common mistake, to infer that the only reason GDP per capita is positively related to national health is due to the economic position of individuals. There are much more powerful reasons, at the macro-level, that underlie the basic national-level relationship between GDP per capita and health.

The broadest reason that national income and wealth is the foundation for national health levels relates to the capacity of GDP to provide investment. This investment will be in the production of scientific and technological knowledge, the diffusion of that knowledge into technologies which enhance occupational safety and health, food safety, environmental safety, transportation safety and generalized public health safety in the areas of infectious and waterborne diseases. At the most fundamental level economic growth provides for investment in nutrition, sanitation, housing, health care and climate control. At more advanced levels, GDP growth provides for investment in secondary and tertiary care, and rehabilitation, for the degenerative diseases – cardiovascular diseases, diabetes and malignancies. Economic growth provides the basis for investment in education at primary, secondary, tertiary and professional levels, including the sciences, humanities, law, accounting and the managerial professions. GDP growth also provides the basis for investment in the minimization of poverty and economic inequality – through the social welfare state in general, and specifically through insurance systems for unemployment, impoverishment, catastrophic illness, disability and frailty in older age.

Indeed, if we recognize that the fundamental sources of expansion in life expectancy have come from scientific developments, as well as knowledge development and education, and the material level of living, there is no escape from the inference that economic growth and development are the sources – both in direct consumption and investment.

In the last decade and a half, there has nevertheless developed, in epidemiological and public health circles, a position that either explicitly denies – or at a minimum, neglects – the importance of economic growth and development to population health. This new frame of reference asserts that it is economic inequality within nations that dominates the levels of health and life expectancy that are observed in country comparisons. This view appears to have arisen from a new interpretation of the standard inverse relation, at the individual level, between income level and mortality patterns for nearly all ages and for the vast majority of illnesses. The new interpretation is that in wealthy country populations the significance of income levels lies not in the function of value of what can be purchased with money, but rather the symbolic/status value of what these purchases can bring. For example, it is not important that the purchase of a higher priced automobile can bring greater safety, speed, mileage, design attractiveness, or comfort, but rather that it has greater 'status' or prestige value. The idea then is, in an intensely competitive world for social status that the higher status value of the item brings the purchaser greater prestige – and thus less psychological stress than if a cheaper less prestigious model were purchased at a lower price. This presumably occurs because, in modern industrialized societies, the true "value" of the person is unknown, and what impresses the onlooker are "status symbols" recognized for their relatively high price or exclusivity.

This view, commonly referred to, as involving "relative deprivation" as distinct from absolute deprivation now has taken on as important a position as the 'functional' or 'material' importance of the value of purchased goods and services. One of the conventional means of measuring the extent of income inequalities – rather than simply the range of income differentials in a society – is the Gini Index which measures the concentration of incomes at high and low levels of an income structure. Using such measures, the argument has been made that rather than the national – or average – income per capita of a society, it is the inequality of distribution of incomes that is the principal source of differential in health and life expectancy among societies. The literature on this subject, over the past 15 years has involved intense debate with empirical evidence being used in both sides, but with no clear resolution.

6.1 The single-variable mentality

In the meantime, as this debate has raged, the primary influence of economic growth and development for industrialized countries' health has all but disappeared from discussion. The reason for this neglect is not only that it appears to contradict a view which emphasizes economic inequality, but that it embodies a view that perhaps the dominant effects of economic growth actually lead to health damage rather than enhancement.

Thus, it is often assumed that high rates of consumption of alcohol, tobacco, carbohydrates and animal fats are features of affluent societies, especially in periods of long-term growth in GDP per capita. But, these assumptions are empirically inaccurate as shown in both cross-sectional and time-series analyses. Further, it may well be true that toxic emissions and environmental pollution increase in relation to economic growth, but this will be less true in higher income countries, and wealthier firms which possess the financial means of modernizing industrial technology for the sake of improving occupational and environmental safety and health.

Despite the intensity of debate over whether income inequality within countries influences comparative mortality among industrialized countries, recent evidence, and the work of my own research team, does show that the Gini Index is a significant explanatory factor in specific mortality rates. However, GDP, unemployment and several 'lifestyle' and environmental factors are also statistically significant predictors of mortality. The Gini Index appears to show its influence predominantly in the age groups under 40. For this reason, the total age-standardized mortality rate, and overall life expectancy do not show the influence of the Gini Index, since mortality is obviously more frequent over age 50. It is of interest that frequently the Gini Index and other epidemiologically important predictors do not show significant relations to mortality without controls for GDP per capita and factors related to it (e.g, per capita government expenditure, tertiary education).

Indeed, economic inequalities are a feature of the process of economic growth itself. At any time, the technological structure of society is the foundation of its industrial and occupational structure – and thus, in general, the range of employment available.

The question then becomes, "who will fill these different positions?" – a process of "ranking and sorting"of potential job competitors along the lines of Blau and Duncan (1977), Davis and Moore (1945), and the structuralists who emphasized dual and segmented labour markets. The issue of who obtains which job then largely rests on the nature of competition. In modern industrial societies, matters of fairness are thus subsumed under principles of achievement associated with competition, with an emphasis on fairness as meritocracy and equality of opportunity. But partly because of the assumption that economic growth and development are independent of unemployment, income inequalities, cultural patterns and environmental pollution, the "single-variable mind" begins to emerge in the research community, i.e. referring to those whose research careers focus on cultural 'life-style' risks to health who omit economic growth, shocks and the environment.

Those who concentrate on the "economy" believe they can forget about all else, even when their own key hypotheses involve (say) alcohol or tobacco, their research designs do not include such data (Ruhm, 2000; Granados, 2005). These examples of mis-specify research models using a single-variable mentality altogether forget the basic rule of multivariable factorial causality – that lack of appropriate controls can lead to gross over-space or under-estimation of the impact of the independent variable which is bearing the main hypothesis.

6.2 Policy model 2005

In the Bremen ICOH conference of 2004, several simple cross-sectional models were presented by the author showing the relation between GDP per capita and unemployment, on the one hand and life expectation, on the other, using data for the year 1995. We now have data from the World Health Organisation, World Bank, ILO, FAO and other international organisations up to 2005. In order to update the 1995 relationships a decade later, a more inclusive "policy model" pertaining to the effects of the international economy on age-standardized mortality rates is presented. In this policy model GDP and the unemployment rate are retained, but now the more specific focus is on GDP per employee (a measure of productivity) and the unemployment rate specifically for adult males, since this is a somewhat more precise predictor than the undifferentiated unemployment rate. This model also controls for the potential impact of high inflation rates in some countries by the use of the Consumer Price Index.

In comparison to 1995, there are now two major additions in this policy model of 2005, namely self-employment as a proportion of total employment, and total national public expenditures on health as a percentage of the GDP. The self-employment rate is intended as a measure of entrepreneurship in small and medium-size enterprises (SMEs). In traditional economic thinking entrepreneurship is the "fourth" factor of production and is essential to innovation and risk-taking in market economies. It is theoretically the source of continuous renewal of the technology and structure of the economy and thus at least part of the basis for continuous long-term growth. The self-employment rate in SMEs is also the source of a relatively high rate of job creation. And, it is a traditional source of absorption into employment of persons unemployed as a result of recession or economic restructuring (including downsizing, offshoring, delocalization and outsourcing). Thus self-employment is the traditional safety net for professional, managerial, skilled and unskilled workers who have lost employment.

There is also considerable theory that contrasts the SME with the large bureaucratic work organisation, with its intensely hierarchical and impersonal employment structure. The hierarchical component itself is a source of enlarged socioeconomic inequalities within the firm's internal labour market. Scholars also point to the importance of the exceptional autonomy of the self-employed, obviously involving control over the extent of work demands as well as the pace, scheduling and intensity of work (Karasek & Theorell, 1990). All of these factors lead to the inference that the rate of self-employment is important to economic growth and renewal, economic survival of the unemployed, and social capital – and for these reasons would provide important sources of societal coping with economic shocks.

It is assumed by some health policy experts that health services expenditures, as a proportion of GDP, should bear some relation to the intensity, and the technological sophistication, of the health care utilized by a population. The further assumption would be that the intensity and sophistication of health care should, in principle, be associated with decreased mortality rates (other things equal).

There is an equally vocal group of health policy specialists who contend that health care expenditures need not bear any relationship to improved health. Especially intense are critiques from United States health policy specialists who contend that, in the case of privately funded health care typically involving profit-making insurance companies, the cost of care bears little relation to its quality or intensity, since much of private health care expenditure is either medically unnecessary or occurs too late in life to significantly influence life expectancy. On the assumption that health care expenditures, at least in the public sector, have some basic relationship to the volume of health care resources utilized by the society, in relation to total national product, the hypothesis is offered that publicly financed health care expenditures are related to reduced age-standardized mortality rates (other things equal).

Figure 2 shows the explanation of variance in age-standardized mortality rates among 43 countries on the basis of five variables: GDP per person employed, male self-employment as percentage of male unemployment, public health expenditures as percentage of total GDP, consumer price index, and male adult unemployment rate as percentage of male adult labour force. All variables are statistically significant, and the adjusted explanation of variance is .828. High income countries include those of Western Europe, North America and the Pacific. These countries have the lowest mortality rates ranging from approximately 350 to 515 deaths per 100,000. There is a middle income cluster essentially involving Eastern Europe and Mexico, with mortality ranging from 650 to 850 deaths per 100,000. And there is a relatively low income cluster, somewhat overlapping Eastern Europe, with mortality ranging from approximately 1,100 to 1,200 deaths per 100,000.

Figure 2: Relation of economic variables to age adjusted total mortality, 43 European, CIS and highly developed non-European countries, year 2005

R = .92116159 R^2 = .84853868 Adjusted R^2 = .82807093

F(5,37) = 41.457 p < .00000 Std.Error of estimate: 81.688

N=43	Beta	Std.Err.	B	Std.Err.	t(37)	p-level
Intercept			981.7666	149.4709	6.56828	0.000000
GDP per person employed in 1990 Geary-Khamis (Groningen Growth and Development Centre)	-0.504813	0.109621	-0.0069	0.0015	-4.60506	0.000047
Male self-employment as percentage of total male employment (ILO, Key Indicators of the Labour Market, 6th ed.)	-0.413554	0.075217	-10.3958	1.8908	-5.49817	0.000003
Public health expenditure as percentage of total GDP (WHO, OECD)	-0.351582	0.094072	-40.5185	10.8414	-3.73738	0.000626
Consumer price index (base year=2000) (ILO, Laboursta)	0.234575	0.079955	1.9671	0.6705	2.93385	0.005719
Male adult unemployment rate as percentage of male adult labour force (ILO, Key Indicators of the Labour Market, 6th ed.)	0.174517	0.075564	6.7215	2.9103	2.30951	0.026599

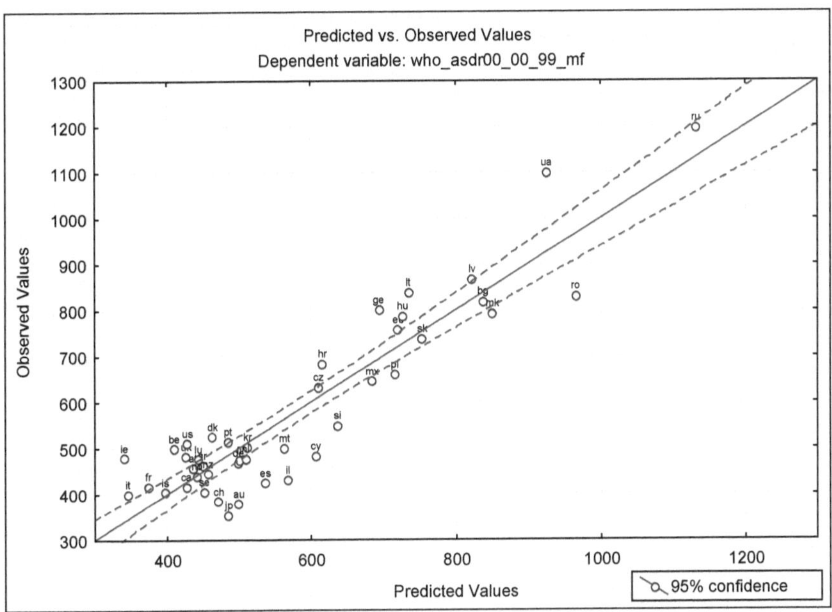

Country Legend: Australia (au), Austria (at), Belgium (be), Bulgaria (bg), Canada (ca), Croatia (hr), Cyprus (cy), Czech Republic (cz), Denmark (dk), Estonia (ee), Finland (fi), France (fr), Georgia (ge), Germany (de), Greece (gr), Hungary (hu), Iceland (is), Ireland (ie), Israel (il), Italy (it), Japan (jp), Korea (kr), Latvia (lv), Lithuania (lt), Macedonia (mk), Malta (mt), Mexico (mx), Netherlands (nl), New Zealand (nz), Norway (no), Poland (pl), Portugal (pt), Romania (ro), Slovak Republic (sk), Russia (ru), Slovenia (si), Spain (es), Sweden (se), Switzerland (ch), Turkey (tr), Ukrane (ua), United Kingdom (uk), United States (us).

6.3 Historical time lags

Despite work over several decades showing lagged relations between national economic changes (in GDP and unemployment rates) and mortality, a tendency has emerged to estimate these relations without any lag – whether cross-sectional or time-series analyses are used.

The two-three-year lag between unemployment rates and coronary mortality at the national level, was initially found in 1971 (Brenner, 1971). The time-series methodology continued to advance over the years with the increasingly sophisticated use of distributed lag analyses. Lag estimates running six years were demonstrated in 1975 between unemployment and both cardiovascular and total mortality (Brenner, 1975), and 10-year lags were published in 1979 (Brenner, 1979a,1979b). These findings were replicated at the national level (Gravelle, Hutchinson & Stern, 1981). In the 1980s a series of epidemiology studies at the individual level were published using U.K. data and also observing a 10-year lag between unemployment and elevated mortality rates (Moser, Fox, Jones & Goldblatt, 1986).

Continuing in the macro-economic tradition, Merva and Fowles (1999) replicated the original 2-year lagged relation between recession and heart disease mortality using pooled cross-sectional analysis with a database of U.S. metropolitan areas. However, Ruhm (2000) failed to use lagged analysis in a pooled cross-sectional study of U.S. states and was thus unable to reproduce the Merva and Fowles results.

In the short-term of one year it is possible to find relations between employment loss and increased mental hospitalization and suicide (Brenner, 1973; 1975; 1984). The same findings are reproduced in many modern day epidemiological studies. However, for the cardiovascular diseases, while one can find a short-term relation based on unemployment (with a correctly-specified model), the more robust relations are found with distributed lags (2-10 years) in time-series analyses.

With respect to GDP per capita in relation to mortality, the consistent findings at the national level with time-series analysis are based on distributed lags over 0-10 years. The distributed lag approach maintains the fluctuating structure of GDP, avoids treating GDP as merely a time trend, and shows the relation in annual changes (first differences) (Brenner, 2005). Despite the consistency of such findings over the last three decades, there are examples of times series research on the GDP-mortality relation which omit any lag analysis and are often unable to replicate the standard inverse findings (Granados, 2005).

6.4 Future directions

We can expect to see increasing convergence between epidemiological and national level findings, as our macro-research methodologies improve. That is, with greater scientific understanding of the analytic tools and the mechanisms of these relationships, the individual or micro level findings will have greater correspondence to the macro level findings. The scientific community is rejecting the view that the macro and micro levels are two different, parallel worlds operating independently of each other, and is welcoming a more interconnected multi-level perspective.

Over time we can expect more complex modelling in which variation in mortality or ill health is explained by a number of factors including those in the spheres of national income, industrial development, labour force participation, life-styles, environment and health care. But such a development assumes researchers really have the knowledge of how to incorporate into their models at least most of the key factors that influence health status with the appropriate time lags and control variables. This means that our training must be fine-tuned to give young professionals the tools to specify models, control appropriate variables, and think beyond one point in time analyses to comprehend and use time lags in order to better understand current factors that will influence future conditions.

Such modelling will, I believe, demonstrate that the healthiest societies are those that are able to produce high levels of goods and services (GDP) per capita. The equally crucial question is the extent to which this societal wealth is invested in social protection, minimization of socioeconomic differentials and unemployment rates, and the development of science and education. Wealth is needed to support society's scientific and social programs; those programs, in turn, make it possible to produce further sustainable growth.

The societies that can combine high wealth, social protection and human capital will tend to be those with the highest levels of life expectation. At present these include Japan, Switzerland, Australia, the Scandinavian countries and the Netherlands. Quantitative analyses, among countries and over historical time, should continue to bear this out.

References

Blau, P.M. & Duncan, O.D. (1967). *The American Occupational Structure*. New York, N.Y.: John Wiley & Sons, Inc.

Brenner, M.H. (1971). Economic changes and heart disease mortality. *American Journal of Public Health, 61*, 606–11.

Brenner, M.H. (1973). *Mental Illness and the Economy*. Cambridge: Harvard University Press.

Brenner, M.H. (1975). Trends in alcohol consumption and associated illnesses: Some effects of economic changes. *American Journal of Public Health, 65*(12), 1279-1292.

Brenner, M.H. (1979a). Mortality and the national economy: A review and the experience of England and Wales, 1936-1976. *The Lancet*, 568-573.

Brenner, M.H. (1979b). *Unemployment, economic growth, and mortality. The Lancet 8117*, 672.

Brenner, M.H. (1984). *Estimating the Effects of Economic Change on National Health and Social Well-Being*. Joint Economic Committee of the U.S. Congress. Washington, D.C.: Government Printing Office.

Brenner, M.H. (1995). Political economy and health. In B.C. Amick, S. Levine, A.R. Tarlov, & D. Chapman Walsh (Eds.), *Society and health* (pp. 211-246). Oxford University Press: New York.

Brenner, M. H (2005). *Commentary*: Economic growth is the basis of mortality rate decline in the 20th Century – Experience of the United States 1901-200. *International Journal of Epidemiology, 34*, 1214-1221.

Catalano, R. (1991). The health effects of economic insecurity. *American Journal of Public Health, 81*(9), 1148-52.

Dasgupta, P., Sen, A. & Starrett, D. (1973). Notes on the measurement of inequality. *Journal of Economic Theory, 6(2)*, 180-187.

Davis, K. & Moore, W.E. (1945). Some principles of stratification. *American Sociological Review, 10*, 242-249.

Dooley, D., Catalano, R., & Wilson, G. (1994). Depression and unemployment: Panel findings from the epidemiologic catchment area study. *American Journal of Community Psychology, 22*(6), 745-65.

Fiscella, K. & Franks, P. (2000). Quality, outcomes, and satisfaction. Individual income, income inequality, health, and mortality: What are the relationships? *Health Services Research, 35(1)*, 307-318.

Granados, J.A.T. (2005). Increasing mortality during the expansions of the US economy, 1900-1996. *International Journal of Epidemiology, 34*(6), 1194-1202.

Gravelle, H.S., Hutchinson, G., & Stern, J., (1981). Mortality and unemployment: A critique of Brenner's time-series analysis. *The Lancet, 2*(8248), 675-9.

Hall, S.G. (2009). An application of the Granger & Engle two-step estimation procedure to United Kingdom aggregate wage data. *Oxford Bulletin of Economics and Statistics, 48* (3), 229-239.

Hallsten, L., Grossi, G., & Westerlund, H. (1999). Unemployment, labour market policy and health in Sweden during years of crisis in the 1990's. *International Archives of Occupational and Environmental Health, 72* (Suppl), S28-30.

Karasek, R. & Theorell, T. (1990). *Healthy Work: Stress, Productivity, and the Reconstruction of Working Life* (pp. 89–103). New York, NY: Basic Books.

Kasl, S.V. & Jones, B.A. (2002). The impact of job loss and retirement on health. In L.F. Berkman & I. Kawachi (Eds.), *Social Epidemiology* (pp. 118-136). New York: Oxford UP.

Kawachi, I., Kennedy, B.P., Lochner, K., & Prothrow-Stith, D. (1997). Social capital, income inequality, and mortality. *American Journal of Public Health, 87*, 1491-1498.

Lynch, J.W., Smith, G.D., Kaplan, G.A. & House, J.S. (2000). Income inequality and mortality: Importance to health of individual income, psychosocial environment, or material conditions. *British Medical Journal, 320*, 1200-1204.

Mackenbach, J.P., Kunst, A.E., Cavelaars, A.E.J.M., Groenho, F., & Geurts, J.J.M. (1997). Socioeconomic inequalities in morbidity and mortality in Western Europe. The EU working group in socioeconomic inequalities in health. *Lancet, 349*, 1655-1659.

Martikainen, P. & Valkonen, T. (1996). Excess mortality of unemployed men and women during a period of rapidly increasing unemployment. *Lancet, 348*, 909-912.

Merva, M. & Fowles, R. (1999). Economic outcomes and mental health. In R. Marshall (Ed.), *Back to Shared Prosperity: The Growing Inequality of Wealth and Income in America* (pp. 69-75). New York, London: M.E. Sharpe.

Moser, K.A., Fox, A.J., Jones, B.R., & Goldblatt, B.O. (1986). Unemployment and mortality: Further evidence from the OPCS longitudinal study 1971-81. *The Lancet, 1*, 365-7.

Ruhm, C.J. (2000). Are recessions good for your health? *Quarterly Journal of Economics, 115*(2), 617-650.

Sen, A. (1999). *Development as Freedom*. Oxford: Oxford University Press.

Sen, A. (2001). Economic progress and health. In D. Leon, G. Walt, (Eds.), *Poverty, Inequality and Health - An International Perspective* (pp. 333-345). Oxford: Oxford University Press.

Smith, G.D., Hart, C., Blane, D., Gillis, C., & Hawthorne, V. (1997). Lifetime socioeconomic position and mortality: Prospective observational study. *British Medical Journal, 314*, 547.

Tausky, C. & Piedmont, E.B. (1967-68). The meaning of work and unemployment: Implications for mental health. *International Journal of Social Psychiatry, 14*(1), 44-9.

Wilkinson, R.G. (1992). Income distribution and life expectancy. *British Medical Journal, 304*, 165-168.

Economic Antecedents of the Swedish Sex Ratio

Ralph A. Catalano[1] *& Tim Bruckner*

Introduction

The ratio of male to female live births (i.e., the secondary sex ratio) declines in human populations coping with war (Graffelman & Hoekstra, 2000), natural (Fukuda, Fukuda, Shimizu, Yomura, & Shimizu, 1996) and manmade (Lyster, 1974) disasters, and collapsing economies (Catalano, 2003). Perhaps the most elaborate, and controversial, theory invoked to explain this phenomenon argues that natural selection has imparted a mechanism to females that allows them to increase their chances of grandchildren by aborting male foetuses when stressed (Trivers & Willard, 1973). The work posits that relatively stressful environments weaken a gravid female and her offspring. The reproductive success of a congenitally weak son supposedly drops more precipitously than that of a weak daughter because weaker males must physically compete with stronger males from other birth cohorts for females. Female offspring during periods of population stress, therefore, should increase a parent's yield of grandchildren. Natural selection supposedly favors and conserves any mutation that allows stressed females to abort weak male foetuses and become available to conceive a daughter or, assuming improved environmental circumstances, a stronger son (Trivers & Willard, 1973).

Much controversy remains over the theory that change over time in the secondary sex ratio results in part from "strategic" offspring sex selection (Krackow, 2002; Post, Forchhammer, Stenseth & Langvatn, 1999; Brown & Silk, 2002). Research, however, supports the associated arguments that stress increases the risk of foetal death (Hansen, Moller, & Olsen, 1999) and that male foetuses are more likely to die than female foetuses (Mizuno, 2000; Hassold, Quillen & Yamane, 1983; Byrne & Warburton, 1987). Other research also suggests that stressed males may father fewer males than otherwise expected due to changes in sperm characteristics (Fukuda, Fukuda, Shimizu, Yomura & Shimizu, 1996; Fukuda, Fukuda, Shimizu & Moller, 1998) and to changes in frequency and timing of coitus (James, 1999; Krackow, 1995; Lazarus, 2002).

The seminal work in this field (Trivers & Willard, 1973) speculated that to have fewer goods and services than needed or desired might sufficiently stress human popu-

---

1 University of California at Berkeley, School of Public Health, 322 Warren Hall, #7360 Berkeley, California 94720-7360, USA, Corresponding Author: Professor Ralph Catalano, Ph.D. Email address: rayc@uclink4.berkeley.edu

lations to lower the secondary sex ratio. This speculation implies that the observed secondary sex ratio would decline when a population experiences an economy in which households consume fewer goods and services than expected from history. There have been no direct tests of this hypothesis. This may reflect the fact that the early literature converted the longitudinal hypothesis into a cross-sectional test of the association between socio-economic status and the sex ratio (Shapiro, Schlesinger & Nesbit, 1968; Erickson, 1976; Rostron & James, 1977). The results of this work do not converge. The approach, moreover, does not offer a compelling test of the reduced consumption hypothesis. Being relatively poor may be more stressful than being relatively well off, but the experience cannot be assumed equivalent in either the nature or degree of stress induced by a declining economy that reduces consumption by the population below the expected level. The cross sectional research may directly address the hypothesis that relatively poor families have fewer male offspring than their wealthier counterparts, but it provides little evidence for determining whether or not reduced consumption of needed or desired goods and services affect the sex ratio.

One longitudinal test of the reduced consumption hypothesis appears in the literature. Data describing post war secondary sex ratios in the former East and West Germany support the inference that economic decline coincides with lower sex ratios. Catalano (2003) reports that sex ratios in the former East Germany declined to a post war low in 1991 when industries formerly subsidized and sheltered by the old regime had to face market competition for the first time in nearly a half century. The statistical significance of the decline survived controlling not only for trends, cycles, and other forms of autocorrelation in the East German sex ratio, but also for variance shared with the West German ratio.

The East German test may not be compelling for several reasons. These include that the political and institutional chaos of the period may have triggered the biological antecedents of low sex ratios whether or not the economy sunk to record lows. The test, moreover, uses an "interrupted time-series" design that specifies economic stress as a binary variable implying that the population experiences no economic stress for most of the observations and a constant level for the remainder. The East German findings would be more compelling if replicated as a dose response relationship over an extended period in an otherwise stable society. We offer a test based on 129 years of economic and sex ratio data from Sweden, a country that has enjoyed a relatively high level of political and social, if not economic, stability.

1 Data

Measurements of the capacity of Swedish households to consume goods and services determined the years subjected to analyses. The longest, internally consistent, time series measuring this construct begins in 1862 and ends in 1991 (Institute for International Economic Studies, 2003; Statistics Sweden, 1997). The data gauge the annual percentage change in the value, expressed in 1985 prices, of goods and services consumed by private households. The annual changes ranged from a minimum of -11.98 percent to a maximum of 14.6 percent and averaged 2.5 percent (standard deviation = 4.05). Figure 1 shows the variable plotted over the test period.

Figure 1. Percentage change in total private consumption (in 1985 Kroner) of goods and services for the 129 years beginning 1862.

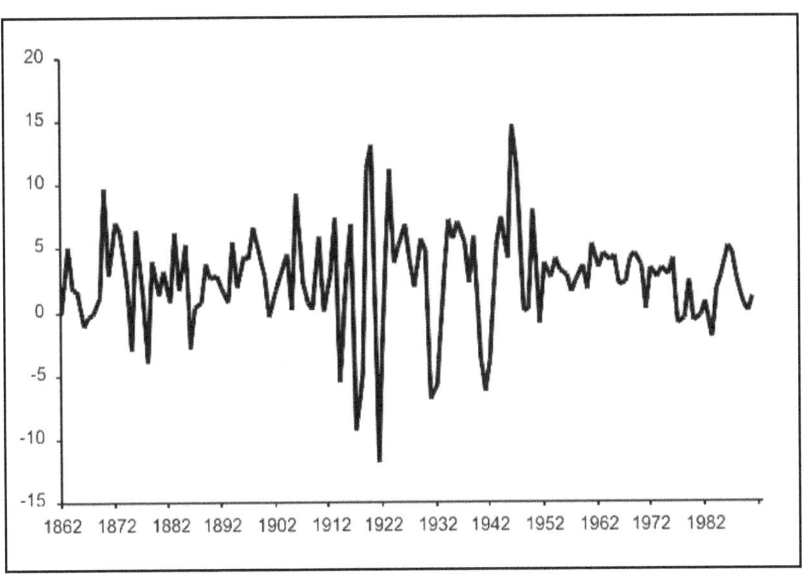

We acquired counts of male and female live births in Sweden for the same years from the Human Mortality Database (University of California, Berkeley & Max Plank Institute for Demographic Research, 2003). Male births ranged from a minimum of 43,794 to a maximum of 71,924 and averaged 60,496 (standard deviation = 8,501). The fewest female births was 41,201 and the most was 68,256 while the average was 57,219 (standard deviation = 8,098). Figure 2 shows the secondary sex ratio plotted over the test period.

Figure 2. Swedish sex ratio for the 129 years beginning 1862

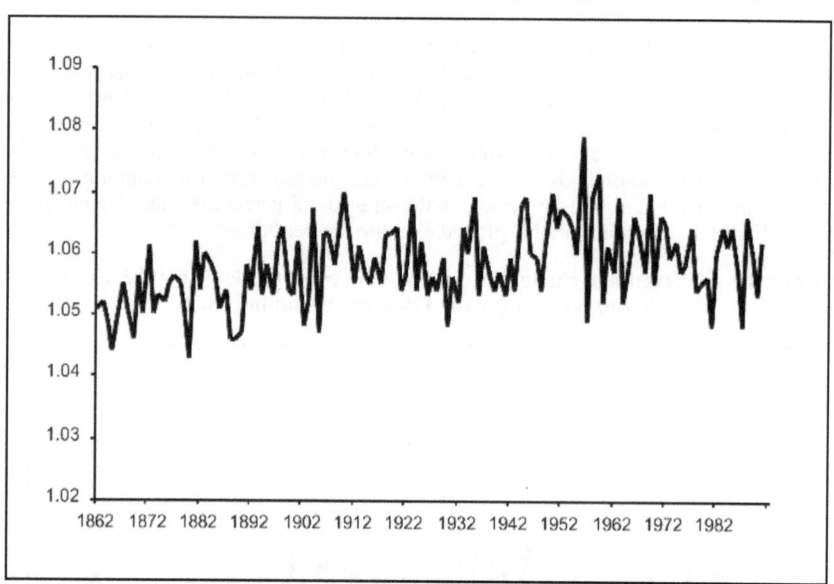

2 Methods

The economic stress theory implies that the observed Swedish sex ratio should fall below that expected under the null hypothesis (*i.e.*, under the assumption that consumption has no effect) when the observed values of consumption rise above their expected values. The controlled experiment derives expected values under the null hypothesis from an unbiased control group. No such groups allow us to estimate the expected value of the Swedish secondary sex ratio under the null hypothesis.

The lack of an unbiased control group typically leads to the assumption that the mean of the dependent variable's observations is the expected value under the null hypothesis. Sex ratios, however, often exhibit autocorrelation including trends, cycles, and the tendency to remain elevated or depressed after high or low values (Gini, 1955). Autocorrelation complicates any correlational test because the expected value of autocorrelated series is not their mean.

Epidemiologists have devised methods for deriving expected values under the null hypothesis for autocorrelated time-series (Catalano & Serxner, 1987). These methods "decompose" a series into expected and unexpected components by applying pattern recognition routines to its observed values. The test then hinges on whether the unexpected component in the dependent variable varies, as implied by theory, with changes in the independent variable.

Our test proceeded through the following steps.

1. Using software from Scientific Computing Associates (Liu & Hudak, 1994), we decomposed the percentage change in private consumption variable into statistically expected and unexpected components. More specifically, we applied the methods devised by Dickey-Fuller (Dickey & Fuller, 1979) and Box-Jenkins (Box, Jenkins & Reinsel, 1994) to identify and express autocorrelation in time series. The estimated values of the best-fitting Box-Jenkins model can be thought of as the expected component of the modeled series while the differences between the observed and estimated values are the unexpected component. The general Box-Jenkins model applied to the percentage change in total private consumption is as follows:

$$\left(1-\phi B^{q}\right)\left(\frac{Z_{t-1}-Z_{t}}{Z_{t}}\right)=c+\left(1-\theta B^{p}\right)a_{t} \quad [1]$$

Z_t is value, in millions of kronor, of total private consumption in Sweden in year t.
C is a constant.
B^p and B^q are the values of a_t at month t-p or t-q.
qis the moving average Box-Jenkins parameter.
fis the autoregressive Box-Jenkins parameter.
a_t is the unexpected value at month t.

2. We used the same methods (Dickey & Fuller, 1979; Box, Jenkins & Reinsel, 1994) to decompose the Swedish sex ratio into statistically expected and unexpected components.

3. We added the unexpected component of the private consumption variable (i.e., a_t in equation 1) to the equation resulting from step 2. We also added the first lag of the variable to insure that we detected any delayed effect.

4. We inspected the residuals from the equation formed in step 3 for autocorrelation to ensure that the estimated parameters were efficient. We added Box-Jenkins (Box, Jenkins & Reinsel, 1994) parameters to the equation if we discovered additional autocorrelation. Our test equation, therefore, was as follows:

$$\left(\frac{M_t}{F_t}\right)=C+\left(\omega_0+\omega_1 B^{n}\right)x_t+\frac{\left(1-\theta B^{p}\right)}{\left(1-\phi B^{q}\right)}a_t \quad [2]$$

M_t is the number of males born in Sweden during year t.
Ft is the number of females born in Sweden during year t.
C is a constant that, for most developed countries, will be approximately 1.05.
X_t is the unexpected component of change in private consumption (i.e., a_t in equation 1) in year t.
B^n is the value of X_t at month t-n.
ω_0 and ω_1 are the estimated parameters for the X_t.
θ is the moving average Box-Jenkins parameter.

ϕ is the autoregressive Box-Jenkins parameter.
a_t is the error term at month t.

5. We estimated the test equation again.
6. We rejected the null hypothesis if the coefficient for the private consumption variable exceeded twice its standard error.

3 Results

Step 1, decomposing the percentage change in annual private consumption, yielded the following best fitting Box-Jenkins univariate model in which Z_t is the percentage change in private consumption, measured in constant 1985 kronor, from year t-1 to year t and a_t is as described above (Box, Jenkins & Reinsel, 1994).

$$\left(1+.2275B^2+.3383B^5\right)\left(\frac{Z_{t-1}-Z_t}{Z_t}\right)=4.032+a_t$$

The parameters were all at least twice their standard errors. The residuals, or unexpected component of the series, exhibited no autocorrelation through 36 lags. The coefficients suggest that percentage change in private consumption exhibited "echoes" such that unusually high or low values were followed 2 and 5 years later by similarly elevated or depressed values although of declining size.

Table 1. Coefficients (standard errors in parentheses) of models for the sex ratio in Sweden for the years 1862 through 1991.

Parameter	Initial Model Coefficient	Final Model Coefficients
Constant	1.0577** (.0015)	1.0584** (.0010)
Autoregression at Lag 2	.2804** (.0803)	.2095** (.0859)
Autoregression at Lag 3	.3405** (.0787)	.2515** (.0850)
Percentage Change in Private Consumption with no lag		.0002* (.0001)
Percentage Change in Private Consumption at lag 1		.0003** (.0001)

*$p<.05$, two-tailed test; **$p<.01$, two-tailed test

Step 2 above yielded the results shown in the first column of table 1. Consistent with observations over long periods in all societies with such data, the constant indicates that the typical annual sex ratio (i.e., 1.057) in Sweden favors males. The two autoregressive parameters at lag 2 and 3 suggests oscillation such that movements away from the constant tend to be followed 2 and 3 years later, although at attenuated levels, with opposite movements.

Steps 3 through 5 above yielded results shown in the second column of table 1. The estimated parameters imply that, as hypothesized, the sex ratio and private consumption moved similarly around their expected values over 129 years. The association peaked in the year after changes in private consumption.

We conducted several additional analyses to gauge the stability of the association. First we added a second lag to the equation to determine if the association continued. Our theory would not predict an association beyond the first lag. None was found.

Second, we applied the methods of Chang, Tiao, and Chen (1988) to determine if outliers in the two series could have induced the associations we discovered. Only one outlier was detected and controlling its influence on the estimations did not change the results.

Third, we transformed the sex ratio into its natural logarithm to determine if systematic variability in the series' variation could have induced the association. The results of the test remained the same with the obvious exception of the metric of the effect coefficient.

While using the longest continuous time series of sex ratio for which matching economic data could be had allowed us to optimize external validity, it raises the question of whether the association appeared because of strong effects in one era but not others. To test this possibility we separated the series into its first and last 65 years and conducted the tests described above separately on each series. While both tests supported the hypothesis, a potentially important difference did emerge. Only the one year lagged effect was significant in the first 65 years whereas only the synchronous effect (i.e., no lag) was significant in the second 65 years.

To gauge how robust our findings might be, we converted our test equation for the last 65 years into one in which the dependent variable was the number of males born and the independent variables included the number of females born. Following the steps described above, we estimated that each 1% increase in private consumption above expected levels was associated with the birth of approximately 25 males more than expected from history and from female births. The obverse, of course, could also be inferred. For every 1 percent drop below expected levels of consumption, Swedish women gave birth to 25 fewer males than expected from history and the number of girls born.

Based on the above results, we estimated how many men in the Swedish population over the last 65 years "owed" their existence to good economic times. We estimated this number by summing the positively signed percentage changes in private consumption over the last 65 years of our data and multiplying the result by 25. The product was 2,217. Given the relatively low death rates at all ages in Sweden, a large fraction of these men are likely to be living at the time of this writing.

We, of course, could also estimate how many men were not born over the same period due to worse than expected economic times. Using the logic described above, we estimated that number at 1,999.

An intuitive reaction to these findings could be to ask if years in which well understood phenomena lowered consumption coincided with a lower sex ratio. Most persons would accept that the global economy suffered its worst contraction of the 20th century in the early 1930's. Our private consumption variable clearly indicates the depth of this "Great Depression" in that the years 1930, 1931, and 1932 exhibit the steepest (i.e., -7.5%, -7.6%, and 5% respectively) sustained decline in our data. We created a "Great

Depression" variable scored 1 for the years 1930, 1931, and 1932 and 0 otherwise. We substituted this for the private consumption variable in our test equation and estimated the coefficients over the last 65 years of the data. Results suggest that approximately 470 ($p > .05$, two tailed test) of the 1,999 "missing" males alluded to above can be attributed to these three years.

4 Discussion

We find the hypothesized association between the consumption of goods and services and the Swedish sex ratio over the 129 years beginning in 1862. This finding converges with the report that the East German sex ratio reached its lowest observed level during the economic crisis of 1991 (Catalano, 2003). The combined findings support the theory that human responses to ambient stressors includes mechanisms that reduce the likelihood of conceiving a male or increase the chances of spontaneously aborting a male foetus.

Our data cannot discriminate between factors that influence sex at conception and mechanisms that act on the foetus *in utero*. Doing so requires further research on sperm characteristics, timing of fertilization, and foetal loss under stressful and other circumstances (Lazarus, 2002). In addition, time-series analysis of preterm and very low-birth weight infants during periods of economic stress may elucidate the contributions of each mechanism to the reduction in sex ratio. While several factors at conception (such as sperm characteristics and frequency and timing of coitus) may influence the human sex ratio, they are not believed to affect the incidence of preterm birth and very low birth weight (Fukuda, Fukuda, Shimizu & Moller, 1998). Research, however, reports an increase in premature delivery among stressed mothers (Hedegaard, Henriksen, Secher, Hatch & Sabroe, 1996; Hobel, Dunkel-Schetter, Roesch, Castro & Arora, 1999; Lockwood, 1999). An excess of preterm or very low-birth weight infants that coincides with a reduced sex ratio would provide additional evidence that, in periods of economic stress, factors acting *in utero* influence the secondary sex ratio.

The association we report is net of autocorrelation and therefore cannot spuriously arise from trends, cycles, or other forms of autocorrelation shared by the sex ratio and private consumption. Nor can the association arise from any unspecified variable that affects female foetuses equivalently to males. The possibility, of course, remains that our findings arise not from the mechanisms we offer, but from a circumstance in which movements in private consumption away from its expected values coincide by chance over 129 years in Sweden with movements in an inaccessible third variable that causes change in the sex ratio. The validity of such rival hypotheses can be judged only by the law of parsimony. We know of none as parsimonious as the mechanisms we offer.

Our results have implications beyond supporting the theory that the human sex ratio can react to ambient social stressors including economic change. Much ecological research reports that populations adapting to worsening economies manifest increased incidence of stress-related illness (Catalano, 1991; Dooley, Fielding, & Levi, 1996; Jin, Shah, & Svoboda, 1997). Authors of this work, including seminal work concerned with birth outcomes (i.e., Brenner, 1973), typically argue that the illness should be accounted in the cost benefit assessments that presumably guide policy choices (Catalano, 2000; Brenner, 1979; Smith, 1992).

Considerable controversy remains over whether this research should influence economic policy. The biological mechanisms connecting contracting economies to individual health apparently remain too vague to overcome scepticism induced by the "ecological fallacy" (Lew, 1979). While much individual level research reports that undesirable job and financial events frequently precede psychological and somatic illness, the threat of reverse causation through selection makes the work less than compelling (Prause & Dooley, 2001).

We believe our results contribute to this debate for at least four reasons. First, much theory (Trivers & Willard, 1973; Brown & Silk, 2002; Lazarus, 2002) as well as basic (van Schaik & Hrdy, 1991) and epidemiologic (Fukuda, Fukuda, Shimizu, Yomura, & Shimizu, 1996; Fukuda, Fukuda, Shimizu, & Moller, 1998) research describes mechanisms that connect the sex ratio to ambient population stressors. Second, an association between the sex ratio and the consumption of goods in services cannot arise from selection. Third, the sex ratio may "trace" the incidence other outcomes of gestation, such as very low birthweight and spontaneous abortion (Forchhammer, 2000; Hobel, Dunkel-Schetter, Roesch, Castro, & Arora, 1999), that can induce psychological distress and their sequelae in parents as well as behavioral and somatic morbidity in the infant. Indeed, research in Norway and Sweden has discovered associations between economic contraction and the incidence of very low weight (i.e., <1500 grams) births (Catalano, Hansen & Hartig, 1999). Fourth, the external validity of sex-ratio based tests can be more readily assessed than tests based on unique, short-term, assessments of pathology in research samples from highly developed countries (e.g., Catalano, Dooley, Wilson, & Hough, 1993; Catalano, Dooley, Novaco, Wilson, & Hough, 1993). Whether the findings from this type of exceptional and expensive work describe other times and societies cannot be assessed without equally exceptional and expensive research at other times and places. The growing availability of data describing birth cohorts over long spans of time in many societies (University of California, Berkeley & Max Planck Institute, 2003) makes replication of tests like ours comparatively easy.

The behavioral and biological antecedents of the sex ratio have been more fully described than those of many other health outcomes thought to vary with the economy. Filling the relatively few, albeit large, gaps in what we already know about the serial connections among economic policy, labor market dynamics, individual behavioral adaptations to the labor market, the biological sequelae of such adaptations, and the sex ratio could provide a general model for understanding the implications of economic policy for population health.

References

Box, G., Jenkins, G. & Reinsel, G. (1994). *Time series analysis: Forecasting and control* (3rd ed.). London: Prentice Hall.

Brenner, M.H. (1973). Fetal, infant and maternal mortality during periods of economic stress. *International Journal of Health Services, 3*, 145-59.

Brenner, M.H. (1979). Mortality and the national economy. A review, and the experience of England and Wales, 1936-76. *Lancet, 2* (8142), 568-573.

Brown, G.R. & Silk, J.B. (2002). Reconsidering the null hypothesis: Is maternal rank associated with birth sex ratios in primate groups? *Proceedings of the National Academy of Sciences of the United States of America, 99*(17), 11252-11255.

Byrne, J. & Warburton, D. (1987). Male excess among anatomically normal fetuses in spontaneous abortions. *American Journal of Medical Genetics, 26*(3), 605-611.

Catalano, R. & Serxner, S. (1987). Time series designs of potential interest to epidemiologists. *American Journal of Epidemiology, 126*(4), 724-731.

Catalano, R. (1991). The health effects of economic insecurity. *American Journal of Public Health, 81*(9), 1148-1152.

Catalano, R., Dooley, D., Novaco, R.W., Wilson, G. & Hough, R. (1993). Using ECA survey data to examine the effect of job layoffs on violent behavior. *Hospital & Community Psychiatry, 44*(9), 874-879.

Catalano, R., Dooley, D., Wilson, G. & Hough, R. (1993). Job loss and alcohol abuse: A test using data from the Epidemiologic Catchment Area project. *Journal of Health and Social Behavior, 34*(3), 215-225.

Catalano, R., Hansen, H.T. & Hartig, T. (1999). The ecological effect of unemployment on the incidence of very low birthweight in Norway and Sweden. *Journal of Health and Social Behavior, 40*(4), 422-428.

Catalano, R. (2000). Economic factors and stress. In G. Fink (Ed.), *Encyclopedia of stress, vol. 2.* (pp. 9-15). New York: Academic Press.

Catalano, R.A. (2003). Sex ratios in the two Germanies: A test of the economic stress hypothesis. *Human Reproduction, 18*(9), 1972-1975.

Chang, I., Tiao, G. & Chen, C. (1988). Estimation of time series parameters in the presence of outliers, *Technometrics, 30,* 193-204.

Dickey, D. & Fuller, W. (1979). Distribution of the estimators for autoregressive time series with a unit root. *Journal of the American Statistical Association, 74,* 427-431.

Dooley, D., Fielding, J. & Levi, L. (1996). Health and unemployment. *Annual Review of Public Health, 17,* 449-465.

Erickson, J.D. (1976). The secondary sex ratio in the United States 1969-71: Association with race, parental ages, birth order, paternal education and legitimacy. *Annals of Human Genetics, 40*(2), 205-212.

Forchhammer, M.C. (2000). Timing of foetal growth spurts can explain sex ratio variation in polygynous mammals. *Ecology Letters, 3,* 1-4.

Fukuda, M., Fukuda, K., Shimizu, T., Yomura, W. & Shimizu, S. (1996). Kobe earthquake and reduced sperm motility. *Human Reproduction, 11*(6), 1244-1246.

Fukuda, M., Fukuda, K., Shimizu, T. & Moller, H. (1998). Decline in sex ratio at birth after Kobe earthquake. *Human Reproduction, 13*(8), 2321-2322.

Gini, C. (1955) Sulla probabilita che x termini di una serie erratica sieno tutti crescenti (o non decrescenti) ovvero tutti decrescenti (o non crescenti) con applicazioni ai rapporti dei sessi nelle nascite umane in intervalli successivi e alle disposizioni sessi nelle fratellanze umane. *Metron, 17*(3–4), 1-41.

Graffelman, J. & Hoekstra, R. (2000). A statistical analysis of the effect of warfare on the human secondary sex ratio. *Human Biology, 72,* 433-445.

Hansen, D., Moller, H. & Olsen, J. (1999). Severe periconceptional life events and the sex ratio in offspring: Follow up study based on five national registers. *British Medical Journal, 319*(7209), 548-549.

Hassold, T., Quillen, S.D. & Yamane, J.A. (1983). Sex ratio in spontaneous abortions. *Annals of Human Genetics, 47*(1), 39-47.

Hedegaard, M., Henriksen, T.B., Secher, N.J., Hatch, M.C. & Sabroe, S. (1996). Do stressful life events affect duration of gestation and risk of preterm delivery? *Epidemiology, 7*(4), 339-345.

Hobel, C.J., Dunkel-Schetter, C., Roesch, S.C., Castro, L.C. & Arora, C.P. (1999). Maternal plasma corticotropin-releasing hormone associated with stress at 20 weeks' gestation in pregnancies ending in preterm delivery. *American Journal of Obstetrics and Gynecology, 180*(1 Pt 3), S257-263.

Institute for International Economic Studies. *Swedish Macro Data Set.* http://www.iies.su.se/~perssont/ (data downloaded on January 14, 2003).

James, W.H. (1999). The status of the hypothesis that the human sex ratio at birth is associated with the cycle day of conception. *Human Reproduction, 14*(8), 2177-2178.

Jin, R.L., Shah, C.P. & Svoboda, T.J. (1995). The impact of unemployment on health: A review of the evidence. *Canadian Medical Association Journal, 153*(5), 529-540.

Krackow, S. (1995). The developmental asynchrony hypothesis for sex ratio manipulation. *Journal of Theoretical Biology, 176*(2), 273-280.

Krackow, S. (2002). Why parental sex ratio manipulation is rare in higher vertebrates. *Ethology, 108,* 1041-1056.

Lazarus, J. (2002). Human sex ratios: Adaptations and mechanisms, problems and prospects. In I. Hardy (Ed.), *Sex ratios: Concepts and research methods* (pp.287-311). Cambridge, UK: Cambridge University Press.

Lew, E. (1979). Mortality and the business cycle: How far can we push an association? *American Journal of Public Health, 69,* 782-783.

Liu, L. & Hudak, G. (1994). *Forecasting and time series analysis using the SCA Statistical System.* Oak Brook, Illinois: Scientific Computing Associates.

Lockwood, C.J. (1999). Stress-associated preterm delivery: The role of corticotropin-releasing hormone. *American Journal of Obstetrics and Gynecology, 180*(1 Pt 3), 264-266.

Lyster, W. (1974). Altered sex ratio after the London smog of 1952 and the Brisbane flood of 1965. *The Journal of Obstetrics and Gynecology of the British Commonwealth, 81,* 626–631.

Mizuno, R. (2000). The male/female ratio of fetal deaths and births in Japan. *Lancet, 356*(9231), 738-739.

Post, E., Forchhammer, M.C., Stenseth, N.C. & Langvatn, R. (1999). Extrinsic modification of vertebrate sex ratios by climatic change. *American Naturalist, 154,* 194 204.

Prause, J. & Dooley, D. (2001). Favourable employment status change and psychological depression: A two-year follow-up analysis of the National Longitudinal Survey of Youth. *Applied Psychology, 50*(2) 282-304.

Rostron, J. & James, W.H. (1977). Maternal age, parity, social class and sex ratio. *Annals of Human Genetics, 41*(2), 205-217.

Shapiro, S., Schlesinger, E. & Nesbit, R. (1968). *Infant, Perinatal, Maternal, and Childhood Mortality in the United States.* Cambridge, MA: Harvard University Press.

Smith, R. (1992). "Without work all life goes rotten". *British Medical Journal, 305*(6860), 972.

Statistics Sweden (1997). Statistika meddelanden, 1980-1995: Appendix 1. *Nationalräkenskaper* (N 10 SM 9601), Table 1:2.

Trivers, R.L. & Willard, D.E. (1973). Natural selection of parental ability to vary the sex ratio of offspring. *Science, 179*(68), 90-92.

University of California, Berkeley & Max Planck Institute for Demographic Research. *Human Mortality Database.* www.mortality.org or www.humanmortality.de (data downloaded on January 14, 2003).

van Schaik, C.P. & Hardy, S.B. (1991). Intensity of local resource competition shapes the relationship between maternal rank and sex ratios at birth in cercopithecine primates. *American Naturalist, 138,* 1555-1562.

Unemployment and Health: A Gender and Life Course Perspective

Anne Hammarström[1]

Introduction

The aim of this article is to present a gender and life course perspective in the context of empirical research on unemployment. The article addresses the following five questions: What is the importance of early unemployment for future ill health? Is unemployment worse for adults than for young people? What about the impact of the length of unemployment? The link between unemployment and ill-health: selection or exposure? What is the impact of economic cycles?

Each of these questions will be analysed separately below. First, the research methods are described, and after that, each question is analysed with the point of departure in different publications within a project carried out at Umeå University, Sweden.

1 Method

In a prospective cohort study, the Luleå cohort consisting of all school leavers from compulsory school in an industrialised town in the north of Sweden (n = 1.083) was followed from 1981 until 1995, i.e. for 14 years. The cohort of school-leavers was 16 years old at the beginning of the study and was followed independently of their post-school activities (e.g. further studies, employment, unemployment) until the age of 30 (see Figure 1). Three follow-ups were made at the ages of 18, 21 and 30. The attrition rate was extremely low. Of the original 1,083 pupils in the sample, almost all were reached in the first survey and 1,068 (98.6% of the original sample) participated in at least two of the surveys. For 1,044 individuals (96%) complete data from all four surveys were available for the whole 14 years period (547 men and 497 women).

1 I would like to acknowledge my collaborators from Umeå University; first of all Professor MD Urban Janlert and also MD PhD Mehmed Novo and PhD student Ieva Reine. I would also like to acknowledge our main international collaborator Professor AH Winefield, University of South Australia, Adelaide.

Figure 1. **The design of the 14 year follow-up of the Luleå cohort as well as of "the boom group" (the 21 years old of the Luleå cohort) and the "recession group" – a cross-sectional study of 21 years old in 1995 (sampled from all school leavers in 1989) (from Novo 2000).**

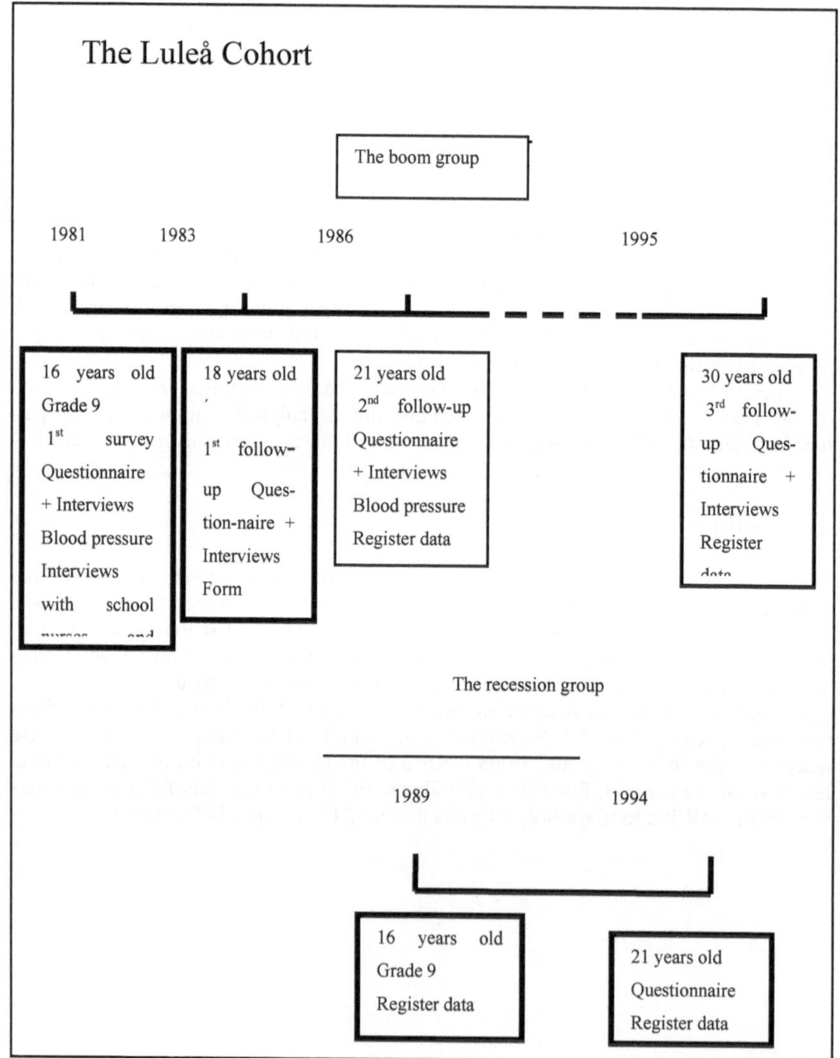

The unemployment figures in the country as a whole were quite low during the 1980s, which can be seen in Figure 2. Especially during the period from 1981 until the end of the decade there was a boom. In the beginning of the 1990s there was a dramatic increase in the rate of unemployment, which still persisted at the 14 year follow-up of the cohort in 1995.

Figure 2. The unemployment rate in Sweden during 1981 until 2003 for men and women.

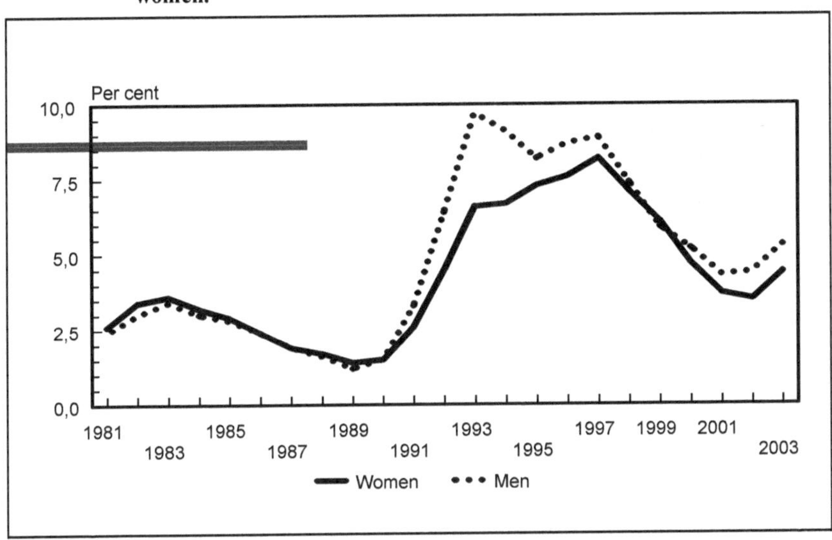

As we wanted to compare young people´s life courses during boom vs. recession, a new cohort of school leavers was investigated. All persons who attended or should have attended the final year of compulsory school in the municipality of Luleå in 1989 were investigated at the age of 21 years in the recession year of 1994. This "recession group" included 898 persons and the response rate in this study was 90% (see Figure 1). The "boom group" consisted of the Luleå cohort at age 21.

Data collection

Data in the Luleå Cohort were collected by group questionnaires – at the ages of 16 and 18 during school hours and at the ages of 21 and 30 the participants were invited to meet with their former classmates, see Figure 1. Those who could not come (and those at the age of 18 who had finished school) received a mailed questionnaire. Additionally, personal or telephone interviews were conducted with participants who did not fill in the questionnaire (because of reading and writing difficulties, etc.), in which the interviewer read the questions and response categories exactly as written in the mailed questionnaire. The "recession group" received a mailed questionnaire (see Figure 1).

The questionnaire consisted of around 90 questions regarding e.g. social background, work situation, health, family situation, leisure time. The questionnaire was almost identical at all investigations and it has been validated in previous research (Hammarström 1986, Hibell & Jonsson 1981, Thorslund & Wärneryd 1985).

Aditionally, in the Luleå cohort personal interviews were conducted with all form teachers (n=65) of each individual pupil by the project leader (AH). The interviewed teacher had been responsible for the class during the last three last years of the compulsory school. The interviews were performed in a quiet room during school time. The teacher was interviewed with a previously validated questionnaire (Sundelin & Vuille 1975), consisting of 35 questions regarding the pupils' situation at school, e.g., absence, popularity, reading and writing difficulties.

The final marks in compulsory school at age 16 were taken from the school register for both cohorts. Data were registered from each participant's school health record regarding height and weight at age 16 for the Luleå cohort. The school health care was free of charge and built up around regular medical check-ups performed by a nurse, who also was available most of the school days for emergency cases and a doctor.

The study was approved by the Research Ethics Committees at Uppsala and Umeå Universities.

2 What is the significance of early unemployment on later ill health?

This chapter is based on an earlier publication by the author (Hammarström & Janlert 2002). The short-term consequences of unemployment among young people (e.g. Winefield et al 1993, Hammarström et al 1988) as well as among adults (e.g. Janlert 1991) have been well documented in the unemployment literature. However, only a few longitudinal studies have addressed the problem of long-term effects of unemployment (Winefield et al 1993, Morrell et al 1994. Moser et al 1987) and many of them have methodological shortcomings. Therefore, one of the crucial questions in unemployment research remains, whether unemployment has only a short-term negative effect on health or is there also an effect many years after the unemployment has come to an end.

Aim

The aim of this long-term follow-up study of young men and women was to analyse potential effects of early unemployment on adult health symptoms and health behaviour.

Unemployment measures

The data on the length of unemployment were taken from a set of questions where the participants were asked to report how many months and weeks they were unemployed, employed, studying or participating in labour market programmes since the last follow-up.

Three types of employment patterns, defined in the following way, were compared to each other:

- *Early unemployment*: > 0.5 years in total of unemployment between the ages of 16 and 21, irrespective of later unemployment.
- *Late unemployment*: >1.5 years in total of unemployment between the ages of 22 and 30 with unemployment < 0.5 years between the ages of 16 and 21
- *Reference group*: Not included in any of the unemployment groups above, i.e. < 0.5 years of unemployment between the ages of 16 and 21 and < 1.5 years of unemployment between the ages of 22 and 30.

A composite index of psychological symptoms was constructed, consisting of nervous and depressive symptoms on a four-grade scale – from 0 (never) to 3 (constantly). The nervousness component of the psychological index consisted of five items about restlessness, lack of concentration, worries, palpitation and anxiety, and their frequency during the last twelve months. The depressive component of the psychological index consisted of two questions regarding depression and sleeping problems. This index (range 0 to 21) was dichotomised at the 75th percentile and the proportion over the cutpoint is presented below as the segment suffering from psychological symptoms.

Results

One way of illustrating the findings is to show the development of symptoms over time in the three different employment groups (Figure 3).

Figure 3. Frequencies of psychological symptoms (<75th percentile) among men at different ages in three employment groups.

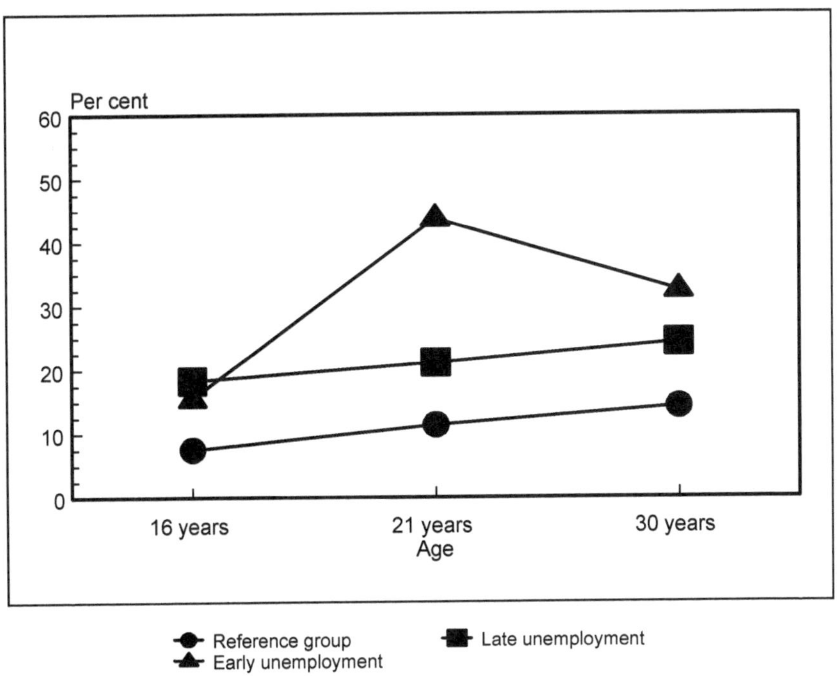

The figure shows that the group with early unemployment had more symptoms than the two other groups, and also at the outset of the study this early unemployed group was worse off. The chi square test for trend was not significant, except for the increase in the reference group.

The model incorporated early and late unemployment and selection, i.e. the value regarding symptoms and habits for each of the tested outcomes at the beginning of the study (age 16) on individual level. Besides, working-class affiliation among parents was used as a background social indicator.

In order to analyse the change in patterns a multivariate logistic regression analysis was performed for psychological symptoms at age 30 (Table 1). The model incorporated early and late unemployment after controlling for working class background as well as for the potential impact of health-related selection, i.e. psychological symptoms at age 16 before entering the labour market.

Table 1. Multivariate logistic regression analyses. Odds ratios (OR) and 95% confidence interval [CI (95%)] for psychological symptoms at age 30 for early and late unemployment, working-class background as well as health-related selection (psychological symptoms at age 16).

		Psychological symptoms			
		Men		Women	
		OR	CI (95%)	OR	CI (95%)
Early unemployment	No	1		1	
	Yes	2.6	1.4-4.7	1.9	1.0-3.5
Late unemployment	No	1		1	
	Yes	1.5	0.9-2.6	1.5	0.8-2.6
Selection	No	1		1	
	Yes	3.7	2.0-6.6	3.2	2.0-5.1
Working class background	No	1		1	
	Yes	1.2	0.8-1.9	0.8	0.5-1.3

The table shows that health-related selection as measured by symptoms at the start of the study was of crucial importance for both men and women. However, even after controlling for this an effect of early unemployment remained. Working-class background was of minor importance in the analysis.

Conclusion

Even after controlling for initial symptoms as well as working-class background and late unemployment, the experience of early unemployment showed a significant explanatory effect on psychological symptoms for both young men and women after a follow-up of 14 years. Thus, youth unemployment constitutes a significant public health problem, which to a certain extent remains until adult age.

3 Is unemployment worse for adults than for young people?[2]

Growing unemployment research has proved the association between unemployment and ill health not only for middle-aged men but also for middle-aged women and young people (e.g. Janlert, 1991; Winefield, 1995). A comparison between different ages indicates that the

2 This part of the article is based on a publication by Reine, Novo & Hammarström, 2004.

health consequences of unemployment seem to differ, and, for example, health behaviour is more affected by unemployment in younger age (Hammarström & Janlert 1994). However, in spite of these obvious differences there is a lack of actual studies that analyse the impact of age (cf. Vuori & Veselainen, 1999; Stronks et al., 1997).

The objective of the study described in this chapter was to analyse whether the association between ill health and unemployment differs between young and adult men and women. The results of the analysis are given in Table 2.

The table shows that long-term unemployment was associated with psychological symptoms in all groups, except for adult women. The OR was highest among young men.

Poor financial position was related to poor psychological health among adult men and women. Being working class was related to poor psychological health among young men alone. Low control was related to poor psychological health among young women and adult men. Unemployed relatives were also related to poor psychological health among adult men and young women.

The lack of association between long-term unemployment and poor psychological health among adult women might be explained by their potentially unsatisfactory employment that affects health to the same extent as unemployment (Gonäs 1994). It is reasonable to believe that women in work might also be at higher risk of having psychological symptoms. Low-status occupations, wage gap, little opportunities to control the work situation, high levels of stress together with a threat of downsizing among women

Table 2. Logistic regression for having poor psychological health (90th percent of the index) among men and women in different age groups.

MEN	Young adults (age 21)			Adults (age 30)		
		Multivariate			Multivariate	
	OR	95.0%	CI	OR	95.0%	CI
Long-term unemployment[3]	4.79	2.24	10.23	2.29	1.08	4.85
Have children	2.02	0.67	6.09	0.89	0.51	1.55
Poor financial position	1.08	0.54	2.16	1.82	1.02	3.22
Working class (own)	3.32	1.22	9.01	0.96	0.54	1.71
Working class (parents)	0.84	0.42	1.69	1.03	0.58	1.82
Low control	1.98	0.98	4.01	4.24	2.43	7.41
Unemployed relatives	1.16	0.35	3.88	2.11	1.21	3.66
WIS[4]	0.98	0.94	1.03	0.98	0.95	1.01

3 Continuously unemployed for six months or more during the previous five-year period
4 Work Involvement Scale

WOMEN	Young adults (age 21)			Adults (age 30)		
		Multivariate			Multivariate	
	OR	95.0%	CI	OR	95.0%	CI
Long-term unemployment	2.71	1.24	5.88	1.38	0.55	3.44
Have children	0.68	0.24	1.97	0.74	0.43	1.27
Poor financial position	1.74	0.84	3.63	2.78	1.54	5.02
Working class (own)	1.45	0.72	2.92	1.19	0.70	2.04
Working class (parents)	0.93	0.47	1.82	0.78	0.46	1.30
Low control	1.99	1.01	3.93	1.48	0.88	2.51
Unemployed relatives	2.51	1.02	6.31	1.27	0.74	2.17
WIS	0.99	0.94	1.04	1.01	0.97	1.04

of all ages characterises the current Swedish labour market. Besides, women still have the main responsibility for domestic work which may contribute to increased stress among employed compared to unemployed women.

Conclusions

The study showed that the association between long-term unemployment and psychological symptoms was stronger for young persons as compared to adults even after controlling for potential confounders. The results highlight the need in health promotion to prevent unemployment, especially among young people.

4 What about the length of unemployment?[5]

Few studies have been able to analyse the importance of the length of unemployment. Will a plateau effect be reached or is there a cumulative association, i.e. the longer the exposure to unemployment the worse is the deterioration of health?

The aim of the study was to analyse the relationship between the length of unemployment (from age 16 until age 21) and the changes in nervous and depressive symptoms during this period. The results are shown in Figure 4.

When comparing the odds ratios for an increase in nervous complaints or depressive symptoms we see that the longer the unemployment period, the higher the odds ratio for increases in symptoms during the five year follow-up period. Regarding nervous complaints there was a fairly uniform increase with increasing unemployment length. For depressive symptoms higher odds ratios appeared first after an unemployment period of one year or more.

There were no more pronounced gender differences, except for depressive symptoms in relation to long-term unemployment, where young women showed the highest increase.

5 This part of the article is based on a publication by Hammarström & Janlert, 1997.

Figure 4. Odds ratios for the increase in (a) nervous complaints and (b) depressive symptoms compared to the length of unemployment

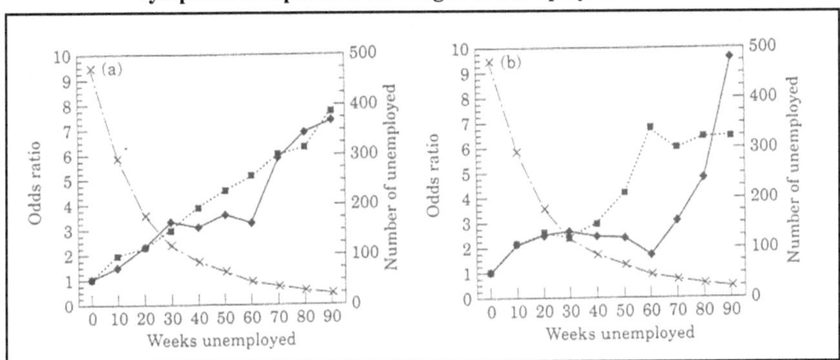

Note: Men = broken line, women = unbroken line; the right y-axis represents the number of unemployed persons (x).

Conclusion

A cumulative association was found between the increase in psychological ill health and the length of unemployment for both young men and young women. The same trend was found at age of 30. Thus, we can generally conclude that the longer the unemployment, the worse the health development.

5 Associations between unemployment and ill-health: A question of selection or exposure?

The question about whether the relation between unemployment and ill health could be related to selection or exposure has been much debated in unemployment research (Winefield, 1995; Novo, 2000).

Figure 4 indicates a strong exposure effect between unemployment and mental ill-health. In contrast, figure 3 and table 1 pointed out that the selection effect was stronger than the exposure (Hammarström & Janlert, 2002). In a multiple linear regression analysis of the possible impact of unemployment on alcohol consumption from age 16 until age 21 a stronger correlation for exposure than for selection was found (Janlert & Hammarström, 1992). Even so, early alcohol consumption (i.e., selection) was significantly correlated with future alcohol consumption.

Thus, our study gives support to the hypothesis that both selection and exposure are important in explaining the association between unemployment and ill-health. A possible interpretation that needs more research is that the exposure is stronger in young ages while the selection is stronger among adults.

6 Has the economic cycle any impact?

Mehmed Novo's thesis (Novo 2000) had the aim to analyse whether economic cycles influenced the health of young men and women. The thesis was based on a comparison of the cohort of young people in the boom and in the recession group (see Figure 1).

The main conclusion from his thesis was that the economic cycle had little or no impact on the health of the unemployed. Early unemployment research argued for the health impact of high unemployment levels in society (Brenner 1980). However, Brenner's research was based on longitudinal, aggregate level research in where unemployment over time was used instead of individual unemployment. These kinds of studies cannot be used to interpret causal relationships. Instead the results are mostly used as a hypothesis generator for later individual research. Another problem with these kinds of studies is the lack of robustness which means that when the studies are repeated with the same variables on the same population but during a different time period, the results will not be the same. This is a serious problem, but not all replications show this instability.

However, Novo also analysed the importance of economic cycle for the health of employed. A comparison of somatic symptoms during boom and recession among unemployed, on the one hand, and not unemployed (consisting mainly of those being employed, in studies or in labour market measures), on the other hand, showed interesting results, see Figure 5 (Novo et al., 2001) (Novo et al., 2000).

The figure shows that the worst health situation was found among the unemployed young women during both boom and recession. The unemployed did not have less somatic symptoms during recession. However, during recession the non-unemployed young women had significantly more somatic symptoms than during boom ($p<0.001$). No similar results were found among young men. The difference in symptoms between the unemployed and the not unemployed young women was significant during boom ($p<0.001$) but not during recession ($p=0.166$). Thus, economic cycle seemed to influence the health situation of those young women who are not unemployed – a finding which was further supported by a multiple regression analysis. The analyses showed that compared to the boom, the recession was associated with more ill health among young women, even after adjustment was made for possible confounders. Among the group of non-unemployed the deterioration of health among women was most pronounced for those employed and in labour market programmes. Among students, both men and women reported more symptoms during recession. The regression analyses also showed that the effects of unemployment in society on young people's health may be manifested by pessimism about the future, high demands and financial problems. Women's lack of control over the work situation may also be an important contributing factor to ill health during recession.

Figure 5. Somatic symptom index among unemployed and not unemployed young women in boom and recession.

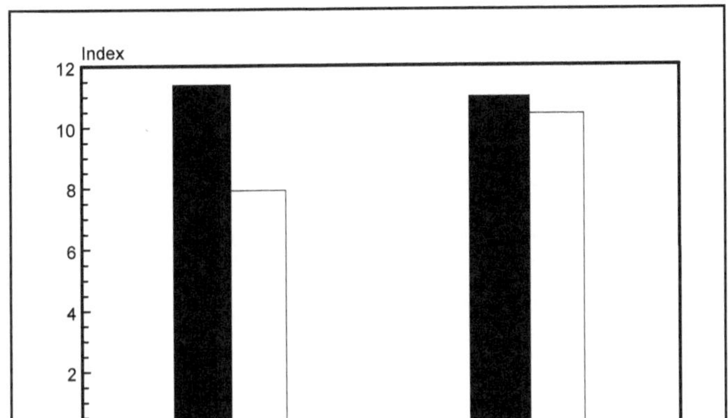

An explanation offered for the poorer health among women during a recession could be the gendered face of rationalisation. The risk of strained work situations and deteriorated health is greater when cuts are made in female-dominated branches compared to male-dominated branches. Thus, for historical and macroeconomic reasons, the female labour market seems to be harder hit by a recession.

7 Final conclusions

Final conclusions are presented as answers to the questions that rose in the introduction.

- The health consequences of youth unemployment remain to a certain extent in adult age.
- Thus, youth unemployment causes a significant public health problem, which to a certain extent remains in adult age.
- The health of young people seems to be harder hit than the one of adults.
- The longer the exposure to unemployment, the worse are the health consequences.
- Both the mechanisms of selection and exposure work in parallel. However, it would seem that exposure dominates among young people, while selection dominates in adult age.
- The business cycle had a low impact on the unemployed, but recession was linked to deteriorated health among women in work and in labour market measures.

This article has intended to answer some of the questions in unemployment public health research. There still remain a lot of questions to be analysed. Unemployment still constitutes a major public health problem. Even if we know much more today than we did a century ago about the relationship between unemployment and ill-health, a lot of research remains to be done.

References

Brenner, M.H. (1980). *Estimating the social costs of youth unemployment problems 1947-1976. The Vice President's task force on youth unemployment.* Center for Public Service, Brandeis University, Waltham.

Gonäs, L. (1994). *Transformation of the welfare state and its labour markets.* Stockholm: The Swedish Institute of Working Life Research.

Hammarström, A. (1986). *Ungdomsarbetslöshet och ohälsa. Resultat från en två- årsuppföljning. [Youth unemployment and ill-health. Results from a two-year follow up]* Stockholm: Karolinska Institute; 1986. [In Swedish, summary in English].

Hammarström, A., Janlert, U. & Theorell, T. (1988). Youth unemployment and ill health: Results from a 2-year follow-up study. *Social Science and Medicine, 26*, 1025-33.

Hammarström, A. & Janlert, U. (1994). Unemployment and change of tobacco habits: A study of young people from 16 to 21 years of age. *Addiction, 89*, 1691-1696.

Hammarström, A. & Janlert, U. (1997). Nervous and depressive symptoms in a longitudinal study of youth unemployment – selection or exposure? *Journal of Adolescence 1997, 20* (3), 293-305.

Hammarström, A. & Janlert, U. (2002). Early unemployment can contribute to adult health problems – results from a longitudinal study of school-leavers. *International Journal of Epidemiology and Community Health, 56* (8), 624-630.

Hibell, B. & Jonsson, E. (1982). *En undersökning av elevers alkohol-, narkotika-, tobaks- och sniffningsvanor våren 1981. [An investigation into pupils' alcohol, narcotics, tobacco and sniffing habits in Spring 1981].* Stockholm: Skolöverstyrelsen. [In Swedish]

Janlert, U. (1991). *Work deprivation and health. Consequences of job loss and unemployment.* Sundbyberg: Karolinska institutet, Dept of Social Medicine (Thesis).

Janlert, U. & Hammarström, A. (1992). Alcohol consumption among unemployed youths. *British Journal of Addiction, 87*, 703-714.

Morrell, S., Taylor, R., Quine, S., et al. (1994). A cohort study of unemployment as a cause of psychological disturbance in Australian youth. *Social Science and Medicine, 38*, 1553-64.

Moser, K.A., Goldblatt, P.O., Fox, A.J. et al. (1987). Unemployment and mortality: Comparison of the 1971 and 1981 longitudinal study census samples. *British Medical Journal, 294*, 86-90.

Novo, M. (2000). *Young and unemployed – does the trade cycle matter for health? A study of young men and women during times of prosperity and times of recession.* Umeå: Department of Public Health and Clinical Medicine, Umeå University.

Novo, M., Hammarström, A. & Janlert, U. (2000). Health hazards of unemployment – only a boom phenomenon? A study of young men and women during times of prosperity and times of recession. *Public Health, 114*, 25-29.

Novo, M., Hammarström, A. & Janlert, U. (2001). Does high level of unemployment influence the health of those who are not unemployed? A gendered comparison of young men and women during boom and recession. *Social Science and Medicine, 53* (3), 293-303.

Thorslund, M. & Wärneryd, B. (1985). Methodological research in the Swedish surveys of living conditions. Problems of measurement and data collection. *Social Indicator Research, 116*, 77-95.

Stronks, K., van de Mheen, H., van den Bos, J. & Mackenbach, J. (1997). The interrelationship between income, health and employment status. *International Journal of Epidemiology, 26*, 592-600.

Sundelin, C. & Vuille, J.C. (1975). Health screening of four-year-olds in a Swedish county. 1. Organisation, methods and participation. *Acta Paediatrica Scandinavica, 64*, 795-780.

Vuori, J. & Veselainen, J. (1999). Labour market interventions as predictors of re-employment, job seeking activity and psychological distress among the unemployed. *Journal of Occupational and Organisational Psychology, 72*, 523-538.

Winefield, A. (1995). Unemployment: Its psychological costs. In C.L Cooper & I.T. Robertson (Eds.), *International Review of Industrial and Organisational Psychology* (pp.169-212). London: Wiley.

Winefield, A.H., Tiggeman, M., Winefield, H.R. et al. (1993). *Growing up with unemployment. A longitudinal study of its psychological impact.* London, New York: Routledge.

Gender Aspects of Unemployment and Health in East and West Germany

Cornelia Bormann

Introduction and aims

Since the reunification of Germany in 1991, former East Germans have faced profound changes in their living and especially in their working situation. Employment rates have changed drastically. Unemployment did not officially exist in the former GDR; the government was responsible for offering jobs and women participated on a much higher level in paid work outside of the own home than in West Germany. In 2004 more than 4.3 million persons were officially registered as unemployed in Germany. The unemployment rate is approximately 11% for the whole country, with unemployment being 8.7% in West Germany and 19.5% in East Germany. In both parts of Germany, men and women face unemployment on a different level. In West Germany 9.4% of men and 8.0% of women were unemployed in 2002, in East Germany the rates were 19.4% for both sexes. Although the differences in most of the living conditions between East and West Germany have seemed to narrow, the field of employment and unemployment continues to see differences which even after 15 years of reunification are still very large and on the increase.

Research has shown that health deteriorates during unemployment (Catalano, 1991; Hammarström, 1994; Paul & Moser, 2001), but that deterioration is quite different for men and women. Most studies show a stronger relationship between psychic as well as somatic illness and unemployment for men than for women (Bormann, 2004). This result is explained by the different socially accepted family roles women have as mother and housewife as compared to men, and that this role protects their health in the case of unemployment (Ensminger & Celentano, 1990). Other studies show that women express the same health consequences of unemployment as men when their work orientation is similar (Morrison et al., 2001; Leeflang et al., 1992). These studies suggest that unemployment is a serious problem for men and women.

The question as to whether unemployment among men and women in East and West Germany has the same health effects will be analyzed here. Because of the very high unemployment rates for men and women in East Germany it is expected that East Germans suffer from unemployment more than West Germans.

1 Methods

The data of the cross-sectional German Health Survey from 1998/9 with 7,124 participants have been used (public use file). It is a representative sample of the residential population aged between 18 and 79 years and was carried out in 120 sample points in West and East Germany. The response rate for men and women was almost similar (61.5% vs. 61.4%); in East Germany it was with 63.9% slightly higher than in West with 60.2%. In order to study possible selection biases, an analysis of non-respondentshas been carried out. According to this analysis, since non- and respondents differ in particular items such as education or age over 60 years, differences in relation to the health status are negligible (Das Gesundheitswesen, 1999). Further information on differences of non-respondents and respondents in East and West Germany, which might help in interpreting the results are not available.

The survey includes a physical examination with focus on cardiovascular risk factors and a self-administered questionnaire on various aspects of health, health behaviour, living conditions and quality of life. For the present analyses, persons aged 25 to less than 65 years have been selected (n = 5,393). The analysis is focused on subjects who say that they are officially registered as unemployed at the labour office, i.e. persons who have no paid work at the moment, who are looking for a new job with at least 15 hours time of work per week and who can accept it without any reservation and who get money from the insurance fund against job loss. Persons who are registered at the labour office and participate in a labour market programme are not considered to be unemployed. That means that a substantial number of persons who are willing to work but do not satisfy the official criteria of being registered as unemployed are missing from the statistics. This is in particular a problem for women (e.g., house-wives who want to enter the labour market but who have not been engaged in remunerative work before) and leads to an underestimation of unemployment rates among women. Here a wider concept of work and unemployment would be necessary but the relevant indicators for such a wider concept (e.g., domestic or care work) are not included in the German Health Survey as this was not the main objective of the data collection.

The selected indicators for ill health are:

- self-perceived health status,
- deterioration of the health status within the last 12 months,
- prevalence of intense psychic and somatic complaints and disorders (list from Zerssen (1976) – a standardized instrument with a norm sample of the German population for comparison),
- severe pains and disorders within the last 4 weeks before the survey (one item of the SF-36),
- great limitations in daily work because of health problems,
- reduction of social contacts due to health problems and
- frequency of utilization of physicians.

Descriptive analyses were carried out in the first step. Significance was tested for men in East and West Germany and women in East and West Germany. For estimating the significance of the results, the chi-square test was used. A p-value < 0.01 has been chosen to be statistically significant for men in East Germany vs. men in West Germany and women in East Germany vs. women in West Germany. In the second step, a logistic regression analysis was executed, which besides employment status also includes socio-demographic factors such as age, school education, occupational status, income and caring for children under 15 years of age as possible confounding or independent factors. The independent variables were chosen because of their significance in explaining the probability of ill health according to other studies (Catalano, 1991; Hammarström/Janlert, 1997; Leana/Feldmann, 1991; Morrison et al., 1991). It might be the case that some of the independent variables, e.g., education or income, interact with unemployment. But if unemployment remains a significant factor in the regression analysis and other socio-demographic factors do not, the independent influence of unemployment can be assumed. The results are represented by odds ratios. An odds ratio demonstrates the ratio between the probability that an event will occur – here less good or bad health - and the probability that it will not occur. A 95% confidence interval is used as a measure of significance in the logistic regression analysis.

2 Results

Table 1 shows the results of different subjective health indicators for unemployed women and men in both parts of Germany.

For almost all health dimensions, unemployed men and women from West Germany demonstrate higher prevalence rates than subjects from East Germany. The highest rates in general can be found in West German unemployed men. Within this group the rates of all health aspects are much higher in West Germany than in East Germany and all of the values are significant. When comparing women in both parts of Germany, higher rates can also be observed in West Germany than in East Germany except for the items 'severe pains and disorders' and 'the usage of a physician within the 4 weeks before the survey', but the value is significant only for the 'high utilization of doctors within 4 weeks before the survey'. The differences between the two groups of women are not as marked as they are between men and the values are very often not significant. Comparing these 4 groups with fulltime employed men and women in East and West Germany, the unemployed groups always demonstrate much higher prevalence rates than the employed (Bormann, 2004). The adverse health effects of unemployment can be confirmed by these results for both sexes and both parts of Germany.

In order to estimate the probability of the employment status besides other socio-demographic factors as an explaining factor for the health situation, a step-by-step backward-oriented logistic regression analysis was carried out separately for men and women. In the first step, all independent variables were included in the procedure. In the further steps, the non-significant independent variables were eliminated in order to determine the variable, which had the highest significant impact on the health status.

Table 1. Prevalence rates of subjective health dimensions of unemployed men and women in East and West Germany (in percent)

Health dimensions	Unemployed Men		Unemployed Women	
	East Germany N = 100	West Germany N = 88	East Germany N = 158	West Germany N = 74
Perceived health status as less good or bad	15.0	34.1	18.4	26.4
Deterioration of the health status compared to the year before the survey	17.0	19.3	14.0	24.3
Severe pains and disorders	3.0	19.8	14.0	10.8
Great limitations because of complaints and health problems	4.1	18.2	9.6	9.6
Remarkable frequency and intensity of complaints and disorders	7.0	25.0	18.6	26.0
Intense reduction of social contacts because of health problems	3.0	5.7	5.7	10.8
High utilization of doctors within 4 weeks before the survey	38.0	45.5	51.6	41.9

First, the prevalence of the independent variables in both parts of Germany for men and women is shown in Table 2.

There exist more differences between East and West than between men and women. This includes in particular school education and income and has to be considered when interpreting the results.

The results concerning self-perceived health – a very valid health indicator (Idler & Benyamini, 1997) – for which the values of the descriptive analysis have been significant in all 4 groups, confirm the high vulnerability of West German men to unemployment.

In West Germany, the probability of having a less good or bad health status is almost two times higher for unemployed men than for men employed full-time. But West German men, who experienced a longer duration of unemployment in the past, also demonstrate a probability of ill health, which is more than two times higher than men employed full-time (see Table 3). In East Germany, the employment status is not associated with the probability of less good or bad health for men. Only age remains a dominant and significant predictor for ill health there. Men older than 40 years show a much higher probability of suffering from bad health than younger men. This is true for both parts of Germany, but to a greater extent for East Germany. In West Germany, little or no school education and low income are independent predictors for a less good or bad health status. The possibility of confounding with the status of unemployment has to be considered.

Very similar results can be observed for women (see Table 4 later).

Table 2. Prevalence of the independent variables for men and women in both parts of Germany (in percent)

	Women		Men	
Independent variables	**East** N = 728	**West** N = 1.462	**East** N = 696	**West** N = 1.146
Employment status:				
Unemployed	14.1	6.2	22.7	6.5
Employed with an experience of unemployment in the past which lasted longer than 6 months	9.6	4.9	15.8	5.5
Employed part-time	----	----	16.4	45.4
Employed full-time	76.3	88.9	45.1	42.7
Age:				
25 – 29 years	10.2	10.1	8.9	12.3
30 – 39 years	33.8	32.8	32.2	31.2
40 – 49 years	27.5	26.9	29.2	29.5
50 – 59 years	26.2	26.4	28.3	24.0
60 – 64 years	2.3	3.8	1.4	3.0
School education:				
No or low education	20.1	47.3	15.4	45.4
Moderate education	57.6	24.4	61.7	30.5
High educational level	22.3	28.3	22.9	24.1
Occupational status:				
Low status	47.6	34.3	29.2	22.1
Medium status	26.5	30.4	54.0	61.4
High status	25.9	35.2	16.9	16.5
Income:				
Less than 2,000 DM/month	13.4	6.3	22.6	9.7
2,000 - < 3,500 DM/month	42.0	27.9	36.1	28.5
3,500 - < 4,500 DM/month	23.8	22.5	21.5	22.6
4,500 DM and more/month	20.8	43.2	19.8	39.3
Children under 15 years:				
No	61.8	62.4	61.6	66.5
One or more	38.2	37.6	38.4	33.5

Table 3. **Odds ratios and 95% confidence interval of the health status of men aged 25-65 years perceived as less than good or bad in association with social factors in East and West Germany (last step of the logistic regression analysis - backward elimination procedure)**

Independent variables	East Germany		West Germany	
	OR	CI (95%)	OR	CI (95%)
Employment status:				
Unemployed			**1.806**[1]	**1.006-3.242**
Employed with an experience of unemployment in the past which lasted longer than 6 months			**2.199**[1]	**1.169 – 4.135**
Fulltime Employed			1.000	
Age:				
25 – 29 years	0.573	0.091-3.596	0.591	0.280-1.249
30 – 39 years	1.000		1.000	
40 – 49 years	**2.816**[1]	**1.023-7.757**	1.535	0.977-2.411
50 – 59 years	**5.126**[3]	**1.919-13.758**	**3.663**[3]	**2.384-5.626**
60 – 64 years	3.711	0.421-32.703	**2.438**[1]	**1.124-5.290**
School education:				
No or low education			**2.334**[3]	**1.513-3.601**
Moderate education			0.874	0.508-1.505
High educational level			1.000	
Occupational status:				
Low status				
Medium status				
High status				
Income:				
Less than 2,000 DM/month			1.254	0.668-2.353
2,000 - < 3,500 DM/month			1.000	
3,500 - < 4,500 DM/month			0.709	0.455-1.104
4,500 DM and more/month			**0.602**[1]	**0.401-0.905**
Children under 15 years:				
No				
One or more				

Significance level: [1] ≤ 0.05
[2] ≤ 0.01
[3] ≤ 0.001

Table 4. Odds ratios and 95% confidence interval of a suboptimal or bad perceived health status of women aged 25-65 years in association with social factors in East and West Germany (last step of the logistic regression analysis - backward elimination procedure).

Independent variables	East Germany		West Germany	
	OR	CI (95%)	OR	CI (95%)
Employment status:				
Unemployed			**2.277**[2]	**1.240-4.181**
Employed with an experience of unemployment in the past which lasted longer than 6 months			0.905	0.425-1.925
Employed part-time			0.709	0.478-1.052
Employed full-time			1.000	
Age:				
25 – 29 years	1.332	0.179-9.897	0.427[1]	**0.199-0.918**
30 – 39 years	1.000		1.000	
40 – 49 years	**5.289**[1]	**1.510-18.524**	1.437	0.917-2.253
50 – 59 years	**11.162**[3]	**3.261-38.213**	1.320	0.821-2-124
60 – 64 years	10.259	0.973-108.17	1.386	0.494-3.888
School education:				
No or low education			**2.010**[2]	**1.214-3.328**
Moderate education			1.292	0.750-2.224
High educational level			1.000	
Occupational status:				
Low status				
Medium status				
High status				
Income:				
Less than 2,000 DM/month			0.739	0.410-1.335
2,000 - < 3,500 DM/month			1.000	
3,500 - < 4,500 DM/month			**0.554**[1]	**0.340-0.902**
4,500 DM and more/month			**0.472**[3]	**0.303-0.905**
Children under 15 years:				
No				
One or more				

Significance level: [1] ≤ 0.05
[2] ≤ 0.01
[3] ≤ 0.001

For West German women, the status of unemployment is significantly associated with the probability of a suboptimal or bad health status. The odds ratio is more than twice as high than for fulltime employed women and it is even higher than for men. The experience of a longer period of unemployment in the past has no effect. Age is again a very dominant predictor for the health status in East Germany. Here the odds ratios increase dramatically when age is over 40 years. In West Germany, age is relevant for women aged 25-29 years only: they show half of the risk of elderly women. In West Germany, the probability of showing a less good or bad health status is significantly associated with little school education. Women with a high income have a much lower risk. Caring for young children is not a predictor for health status in either part of Germany.

3 Discussion

The study shows for West German men and women a positive relationship between unemployment and ill health. For men the experience of a longer period of unemployment in the past is also related to a bad health status. We expected a higher vulnerability for East Germans. This hypothesis could not be confirmed.

Further analyses have been conducted concerning other health dimensions (see Table 1) and with a second data set, the Healthcare Access Panel, with a much larger number of participants (n = 31,395) (Bormann, 2004). The results demonstrate that for West German men and women, unemployment is related to the probability of ill-health; this effect is significant in West, but not in East Germany (see also other health indicators, Bormann, 2004). It is remarkable that in addition to employment status, income and level of school education are independent predictors for perceived bad health in West German men and women, but not for East Germans. There only age is a significant factor. This leads to the question as to whether the theory of social class is adequate for analyzing social and health inequalities in East Germany.

When discussing the above mentioned result of a higher vulnerability of West Germans by unemployment, methodological considerations must first be taken into account. There may be various biases, which could influence the results.

First, a selection bias has to be considered: Compared with the official unemployment statistics, unemployed persons are underrepresented in the German Health Survey (Bormann, 2004). But according to the analysis of non-respondents, there is no suggestion that non-respondents differ in their health status from respondents (Das Gesundheitswesen, 1999), and that this could explain differences between men and women in East and West Germany. Furthermore, different rates of long-term unemployment in East and West could explain the observed differences in health. More long-term unemployed people in West, possibly the most marginalized ones, might have participated in the survey than those in East Germany, although the official rates of long-term unemployment in both parts of Germany were quite similar at the time of the survey. This could have influenced the result. This hypothesis could not be analyzed because the number of long-term unemployed people within the survey was too small.

Another bias to be considered is the recall bias. For the employment status one's present situation was surveyed. It can be expected that this occupational information can be recalled with useful accuracy, as other studies have shown (Novo, 2000). As the main health indicator for describing the health situation in general the self-perceived health status was used in this analysis. This indicator, which is a valid predictor for morbidity, mortality and survival (Idler & Benyamini, 1997), reflects the feelings, ideas and beliefs held by the individuals concerning their health. It provides an indication of the general feelings of persons about their own state of health in relation to their expectations. As the German Health Survey is a cross-sectional study, a recall-bias cannot be expected. This is rather a problem in longitudinal surveys.

A confounding bias, which occurs when two factors are closely associated and the effects of one confuses the effects of the other, can be expected for the independent variables, especially for employment status, school education, occupational status and income. The correlation between the independent variables may be stronger in East than in West Germany. But as Novo stated, it is difficult to decide whether a factor is a confounder or a determinant (Novo, 2000). And there is no indication from other studies that such differences exist between East and West Germany. Here more research is needed.

The results of the presented analyses demonstrate a strong relationship of unemployment and ill health for West German men and women, which seems to be an independent factor compared with other social variables. We hypothesized that East Germans would suffer most from unemployment and ill health. This could not be verified by these data. A possible explanation of the fact that East German unemployed persons do not suffer that much of a bad health situation as West Germans, although they have a higher rate of unemployment, could be that the very high involvement in unemployment in all social groups in East Germany does not lead to such strong stigmatization and discrimination as in West Germany, where it is very often said that unemployed persons are responsible for their own situation and that they could work if they really wanted to. But with the data of the health survey, this hypothesis could not be analyzed.

Another explanation could be that East Germans have a different understanding of (bad) health than West Germans. In other studies it could be shown that these differences do not exist for subjective health indicators (Angermeyer & Matschinger, 1999 and Potthoff, 2003).

According to these results, the differences between both parts of Germany seem to have a stronger effect on the relationship of health and unemployment than gender differences. But this should be further analyzed by using more and different health indicators.

4 Conclusion

With this analysis more questions than answers concerning the gender perspective of unemployment and health could be presented.

Much more research is necessary in order to understand the different links between unemployment and ill-health in men and women in East and West Germany.

Such research should focus on the following aspects:

- Gender specific differences in labour orientation, division of labour, job demands, role configurations, social norms concerning masculinity and feminity, living concepts at different life courses and the role of moderating factors should be included in such research activities;
- Research should not focus on the status of unemployment only, but also include a complete work history. At least the time of job insecurity – as the step before becoming unemployed - should be studied. The health situation and different coping

strategies between men and women should also be analyzed for this period in order to elucidate the underlying mechanisms and to develop better preventive strategies;

- Research is needed on the interpretation of and attitudes towards work and unemployment by society, including the processes of stigmatization and discrimination. This could help to understand the differences between East and West Germany.

Until now most unemployment research (in Germany) has been based on quantitative, cross-sectional surveys. For a deeper understanding, more qualitative and longitudinal research is needed. It is time to go beyond descriptions. We need to proceed to explanations and to a theory of the correlation between unemployment and ill-health.

References

Angermeyer, M.C. & Matschinger, H. (1999). Lay beliefs about mental disorders: A comparison between the western und the eastern parts of Germany. *Social Psychiatry and Psychiatric Epidemiology, 34,* 275-281.

Bartley, M. (1994). Unemployment and ill health: Understanding the relationship. *Journal of Epidemiology and Community Health, 48,* 333-337.

Bormann, C. (2004). *Geschlechtsspezifische Aspekte zum Zusammenhang zwischen Erkrankungen und Erwerbstätigkeit mit besonderer Fokussierung auf die Arbeitslosigkeit in den alten und neuen Ländern Deutschlands.* [Gender aspects in the context of diseases and employment with focus on unemployment in the old and new states of Germany]. Regensburg: Roderer Verlag.

Bormann, C. & Kneip, H. (2002). Arbeitslosigkeit und Gesundheit bei Frauen im Vergleich der alten und neuen Länder der Bundesrepublik Deutschland. Problemaufriss und Präventionsmöglichkeiten. [Unemployment and health of women in comparison oft he old and new states of Germany] In A. Trojan & H. Döhner (Eds.), *Gesellschaft, Gesundheit, Medizin. Erkundungen, Analysen und Ergebnisse* (pp. 115-124). Frankfurt a.M.: Mabuse.

Catalano, R. (1991). The health effects of economic insecurity. *American Journal of Public Health, 81*(9), 1148-1152.

Catalano, R. (1993). Gesundheitseffekte wirtschaftlicher Unsicherheit: Ein analytischer Überblick. In T. Kieselbach & P. Voigt (Hrsg.), *Systemumbruch, Arbeitslosigkeit und individuelle Bewältigung in der Ex-DDR* (pp. 84-94). Weinheim: Deutscher Studien Verlag.

Das Gesundheitswesen (1999). *Schwerpunktheft: Bundesgesundheitssurvey 1998 – Erfahrungen, Ergebnisse, Perspektiven* [Special issue: German Health Survey 1998 – experiences, results and prospects]. *61. Jahrgang*. Stuttgart: Thieme.

Ensminger, M.E. & Celentano, D.D. (1990). Gender differences in the effect of unemployment on psychological distress. *Social Science and Medicine, 30*(4), 469-477.

Hammarström, A. (1994). Health consequences of youth unemployment-review from a gender perspective. *Social Science and Medicine, 38,* 699-709.

Hammarström, A. (1999). Why feminism in public health? *Scandinavian Journal of Public Health, 27,* 241-244.

Hammarström, A. (2002). What could a gender perspective mean in medical and public health research? In S.P. Wamala & J. Lynch (Eds.), *Gender and social inequities in health. A public health issue* (pp. 21-41). Lund: Studentliteratur.

Hammarström, A. & Janlert, U. (1997). Nervous and depressive symptoms in a longitudinal study of youth unemployment-selection or exposure? *Journal of Adolescence, 20*(3), 293-305.

Idler, E.L. & Benyamini, Y. (1997). Self-rated health and mortality: A review of twenty-seven community studies. *Journal of Health and Social Behaviour, 38,* 21-37.

Laubach, W., Mundt, A. & Brähler, E. (1999). Selbstkonzept, Körperbeschwerden und Gesundheitseinstellungen nach Verlust der Arbeit - ein Vergleich zwischen Arbeitslosen und Beschäftigten anhand einer repräsentativen Untersuchung der deutschen Bevölkerung. [Self-concept, disorders and attitudes towards health after loss of employment – a comparison of unemployed and employed persons on the basis of a representative surves of the German population] In A. Hessel,

M. Geyer & E. Brähler (Hrsg.), *Gewinne und Verluste sozialen Wandels. Globalisierung und deutsche Wiedervereinigung aus psychosozialer Sicht* (pp. 75-92). Opladen: Westdeutscher Verlag.

Leana, C.R. & Feldman, D.C. (1991). Gender differences in responses to unemployment. *Journal of Vocational Behaviour, 38*(1), 65-77.

Leeflang, R.L., Klein-Hesselink, D.J. & Spruit, I.P. (1992). Health effects of unemployment – II. Men and women. *Social Science and Medicine, 34*(4), 351-363.

Mohr, G. (Ed.). (1993). *Ausgezählt - Theoretische und empirische Beiträge zur Psychologie der Frauenerwerbslosigkeit.* [Knocked out: Theoretical and empirical results of the psychology of unemployment in women]. Weinheim: Deutscher Studien Verlag.

Morisson, T., O'Connor, W.E., Morrison, M. & Hill, S. (2001). Determinants of psychological well-being among unemployed women and men. *Psychology and Education: An International Journal, 38*(1), 34-41.

Novo, M. (2000). *Young and unemployed – does the trade cycle matter for health?. A study of young men and women during times of prosperity and times of recession.* Unpublished dissertation, Umea University.

Paul, K. & Moser, K. (2001). Negatives psychisches Befinden als Wirkung und Ursache von Arbeitslosigkeit: Ergebnisse einer Metaanalyse. [Negative psychic health feeling as effect and cause of unemployment: results of a meta-analysis] In J. Zempel, J. Bacher & K. Moser (Eds.), *Erwerbslosigkeit. Ursachen, Auswirkungen und Interventionen* (pp. 83-110). Opladen: Leske & Budrich.

Potthoff, P. (2003). Subjektive Erlebniskomponenten von Gesundheit im Ost-West-Vergleich. [Subjective components of health experiences in comparison of East and West Germany] *Forum Public Health, 11*(38), 11-12.

Schwarzer, R., Hahn, A. & Fuchs, R. (1994). Unemployment, social resources, and mental and physical health: A three-wave study on men and women in a stressful life transition. In G. Puryear Keita & J. Hurrell (Eds.), *Job stress in a changing workforce. Investigating gender, diversity, and family issues* (pp. 75-87). Washington: American Psychological Association.

Zerssen, D. van (1976). *Die Beschwerden-Liste. Manual.* [List of complaints and disorders. Manual]. Weinheim: Beltz.

The Explanation of Unemployment Effects in the Latent Function Model of Marie Jahoda - A Critical Assessment

Alois Wacker

Introduction

Jahoda developed her latent function model of employment and unemployment at the beginning of the 1980s, mainly as a result of the analysis and interpretation of the findings of unemployment research in the 1930s (Jahoda & Rush, 1980; Jahoda, 1982). As one of the researchers and the main author of the classic study "Marienthal: the sociography of an unemployed community" (1933) her aim was to determine whether "the enormous social changes" that had resulted in large sections of the community becoming unemployed had altered the current social meaning and experience of unemployment (Jahoda, 1982, p. 6).

According to Jahoda, unemployment has two main consequences for the individual: (1) reduction of income and the risk of poverty, and (2) exclusion from the material and social context provided by a workplace. Jahoda believed that the "physical deprivation" represented by unemployment in the 1930s had turned to "relative deprivation" in the 1980s, but that the *psychological* impact of unemployment was much the same at both time periods (Jahoda, 1982, p. 38). In her view, exclusion from employment produces emotional distress and poor mental health, because basic human needs are no longer satisfied. These needs are the need for "*structured* rather than empty time, for *purposeful activity* rather than feeling useless, for a *defined place in society* rather than being an outcast, for *having an entry into the web of society* rather than being excluded from participation" (Jahoda, 1986, p. 10; emphasis added). As positive features these aspects are mirrored in her so-called "categories of experience" or "latent" functions of employment.

These categories of experience are bound to the social institution of employment and, according to Jahoda (1982), have not changed since the beginning of the industrial revolution. In a powerful, intergenerational process of socialization "people have learned to satisfy [their needs] within an industrial environment" (Jahoda & Fryer, 1986, 111). To illustrate the overwhelming, and sometimes negative, influence of society on the need-articulation of the individual she commented: "It may well be an unintended consequence of industrialism that the all-powerful social institution of employment which pervades the entire way of life in our societies has undermined the ability of millions to find satisfaction in work outside this institution" (Jahoda, 1986, p. 10). Elsewhere she has predicted that it may take several generations ("about 500 years") to change this dominant socialization and need pattern of work society (Jahoda & Ernst, 1981, p. 72).

Sociologists and psychologists have welcomed Jahoda's approach as offering a sound theoretical explanation for the negative psychological consequences of job loss and unemployment that have been documented in many empirical studies. Hartley and Fryer (1984, p. 15) honoured Jahoda‘s deprivation approach as offering the “most influential framework for understanding unemployment”, and Fryer spoke of “the most substantial theoretical progress in the unemployment field” brought about by Jahoda‘s deprivation theory (Fryer, 1986, p. 4). Ten years later Gallie and Marsh praised her model as the “perhaps most influential account of the processes by which unemployment undercuts people‘s psychological stability” (Gallie & Marsh, 1994, p. 16; see also Feather, 1997, p. 38).

For Jahoda the study of the psychosocial effects of *un*employment was not an end in itself, but an opportunity to “reveal the psychological *meaning of work*” in general (Jahoda, 1986, p. 10; author's italics.). Therefore, it is not surprising that her approach has also entered the chapters on work motivation in textbooks on organisational psychology (e.g., Furnham, 1997, p. 255ff.; Foster, 2000, p. 307ff.).

A reconstruction of Jahoda's central assumptions and statements

In Jahoda's (1982) influential book, in which she summarized the results of her research into experiences of unemployment during the 1930s and 1980s, she argued very strictly that “as an *unintended* though inevitable consequence of its own purposes and organisation” any kind of employment enforces these five “categories of experience” - the core dimensions of her latent function model - for the benefit of the employed (Jahoda, 1982, p. 39). In her controversy with Fryer (see references) she clarified: “The exclusion from employment is not a psychological but a *social* event; automatically and inevitably former employees lose access to the structural properties of employment whether or not they were aware of the fit between them and their psychological needs. Objectively they are deprived” (Jahoda, 1986, p. 27).

In these two statements we find the key elements of her thinking about employment and unemployment. Firstly, Jahoda demonstrated a strong preference for sharp contrasts and dualistic concepts: employment-unemployment, manifest-latent, and society-individual. Her explanation of unemployment effects is the result of this kind of dualism. “Employment” and “unemployment” represent polar opposites, with no transitional states in between. But although the unemployed may be seen as a single social group in relation to the labour market (and, according to Burchell, (1992), even this is debatable), the unemployed are not a homogeneous group, as their life situations vary considerably. Thus, to know that somebody is unemployed may mean comparatively little from a psychological and sociological point of view.

In Jahoda's view, employment and unemployment are social institutions that either enable or prevent the satisfaction of important psychological needs. As a model of person-situation-interaction the latent function model corresponds with other person-environment-fit-models (e.g. French, 1973). At least on the background of the discussion in the social sciences in Germany her interpretation of the general positive function of institutions seems somewhat puzzling. For example, in the tradition of the Frankfurt School of critical theory (Adorno, Horkheimer), institutions are perceived as repressive; they force individuals to conform to the norms of the mainstream society and perpetuate irrational societal structures. In Jahoda's thinking however, notwithstanding unemployment as a notable exception, social institutions are seen far more positively. In her writing she stresses their supportive and protective function for the average individual although she admits that in the modern world of work "alienation conditions of unemployment" still exist (Jahoda, 1986, p. 10).

Jahoda examines institutions as objective elements of the social structure. In a long historical process societies established and shaped institutions to fulfil objective purposes which were beyond the capacity of individuals to create on their own (e.g. schools for the education of the next generation). Social institutions answer to societal demands and give structure to individual lives. So resourcefulness is their key feature in Jahoda's thinking.

As Jahoda conveys her main arguments at a general and abstract level, it is not easy to discover their precise meaning. In her argumentation she alternates between a sociological and a psychological perspective, between the psychosocial functions of work in general and of employment in particular (cf. Hartley & Fryer, 1984, p. 17). She strictly separates the categories of experience from their empirical content, that is, from the subjective quality of the experiences within these categories. In a philosophical sense she speaks about the categories of experience as the enabling conditions for experience, not about the experience itself (Kant, 1798/1970).

In the first of her interviews with Fryer on unemployment (see Jahoda & Fryer, 1986), Jahoda stresses the importance of social institutions for human behaviour: "I do think that participation in institutions such as employment has a compelling influence on the nature of experience [that is] enforced by the environment." Then she continues: "But enormous individual differences in the *interpretation* of these experiences exist" (Jahoda & Fryer, 1986, p. 110). In this respect her model resembles the well-known stress-strain-model in work psychology: psychological states are determined by special features of the environment, in the case of the unemployed by the consequences of their exclusion from employment. In Jahoda's thinking, access to the categories of experience is strictly bound to an objective situation: employment status. With some consequence she therefore can conclude: "*Objectively* they [the unemployed] are deprived" (Jahoda, 1986, p. 27).

From an empirical as much as from a psychological point of view the argument "unemployment as objective deprivation" seems to be a problematic statement. If we take it seriously, there would be no need for empirical unemployment research at all: All unemployed are deprived by definition – through the objective situation of being out of work; no empirical evidence is required. Sometimes Jahoda argues this way: "... *automatically* and *inevitably* the former employees lose access to the *structural properties of employment* whether or not they were aware of the fit between them and their psychological needs" (Jahoda, 1986, p. 27). But in order to find a link to the negative psychological effects of being unemployed she introduces her concept of universal needs: "As human beings, we all have some very basic needs which we do not formulate to ourselves ... They are taken for granted that they do not engage the explicit conscious thought of a person as long as the world is reasonably normal" (Jahoda & Fryer, 1986, p. 110). Only their universality guarantees, on the conceptual level, that we all suffer from the withdrawal of the latent functions of employment.

But, as latent functions do not normally engage our explicit thought, how can we be sure that their withdrawal produces a decline in well-being? The implication is that what unemployed people say about their personal experience does not allow a valid conclusion to be drawn about their needs. As Fryer puts it, "The quality of experience within each category is irrelevant" with regard to the effects of the employment status on mental health (Fryer, 1986, p. 10).

Much of the discussion and confusion about Jahoda's position may be due to this uneasy mixture of sociological and psychological thinking and argumentation. On the one hand, as a sociologist, she speaks of the five categories of a "*compulsory* experience" which produces "*inevitable* consequences"; on the other hand, as a psychologist, she speaks of the unemployed as "*feeling* deprived" (Jahoda, 1982, pp. 39, 48).

With reference to the latent functions of employment Jahoda speaks of the "*unintended* consequences" of employment. This concept was borrowed from Merton's (1968) classic distinction between manifest and latent functions. Merton's distinction concerned the problem of the "neither intended nor recognized" and therefore "unanticipated consequences of purposive action" (Merton, 1936, p. 894). In sociology a critical discussion of this distinction and its usefulness took place in the 1980s and 1990s (e.g., Campbell, 1982; Boudon, 1990; Elster, 1990). In his re-analysis Campbell suggested that Merton's concept of latent consequences was ambiguous and loosely defined, because of the multiple meanings given by him to his manifest-latent distinction (Campbell, 1982, p. 33).[1] From an empirical point of view Elster concluded that "It [the manifest-latent-distinction] does not seem to have generated much good empirical research" (Elster, 1990, p. 129). Campbell criticises its theoretical inconsistency: "The concepts [Merton] created have served less to clarify than to obscure and confuse, ... the contrast between the manifest and the latent ... is set more in an action theory than a functionalist context" (Campbell, 1982, pp. 41, 43).

The original version of the "manifest-latent"-distinction implies two criteria that Merton appeared to use interchangeably. Thus, "manifest" referred to both the intended *and* recognized consequences of action", while "latent" referred to the unintended *and* unrecognized consequences of action. (Campbell, 1982). Several sociological critics subsequently queried whether the dimensions of intentionality and recognisability should not be separated, thus creating four, instead of two, categories with which to classify the consequences of action: intended and recognized, intended and unrecognized, and so on (Campbell, 1982, p. 34).

Jahoda partly avoided this issue by referring to only one of the two criteria, intentionality (e.g., Jahoda, 1982, p. 39): The categories of experience "are latent ... only in the sense that they follow from actions not intended to produce them" (Jahoda & Rush, 1980, p. 48). But the main methodological problem remains: two *collective* phenomena (i.e., employment and unemployment as social institutions) are linked with the problem of the intentionality of *individual* actions.

In any case the model requires an *actor* or *agent* who intends to attain a particular outcome through his/or her purposive action (cf. Merton, 1936, p. 895; Merton, 1957). In her conversation with Fryer (Jahoda & Fryer, 1986) Jahoda identified "the employer" as this actor: "What I mean by talking of latent consequences of employment is that the deliberate, overt, admitted purpose of those who organise employment is ... the creation of goods and services ... But, in order to meet that purpose, the *employer* has to impose certain time experiences, goals, and so on, upon everybody in the enterprise" (Jahoda & Fryer, 1986, pp. 110f.). In contrast, Creed and Machin (2002, p. 1208) in their interpretation of Jahoda's theory see the *employees* themselves as the actors in Jahoda's formulation: "People pursue employment to attain manifest benefits but while employed profit from the latent benefits". Although the identity of the actor changes in these contradictory interpretations, in both cases the distinction is between conscious intention and actual consequences or functions (meaning 1 of the manifest-latent distinction; Campbell, 1982).

---

1 Merton's four best known distinctions are between: (1) conscious intention and actual consequence; (2) commonsense knowledge and sociological understanding; (3) official and unofficial aims in organisation; and (4) surface meaning and a deeper reality (Campbell, 1982, p. 33)

It is difficult enough to identify individual motives, but it is nearly impossible to identify all the intentions that may lead to the foundation of a social institution as a result of the efforts of many different actors. In a joint publication, Jahoda and Rush (1980), created an even more complex model by distinguishing between the levels of society and the individual, and between constructive and destructive functions. Thus, if we were to build a three-dimensional model we would get a cube with at least 20 cells, each cell filled with a different combination of latent functions! In a less systematic manner, Jahoda and Rush illustrate only a few of the possible combinations (Jahoda & Rush, 1980).

How can we identify the latent functions? In a short text on "manifest and latent functions" Jahoda (1995, p. 317) admits that "the identification of latent consequences" is a "difficult and ambiguous" task (see also Jahoda & Rush, 1980, p. 48). No accepted scientific method of identifying them does exist; it is a mode of interpretation. Furthermore, what does "latent" mean in this context? It seems to be a mixture of the second and fourth meaning in Campbell's terminology; a 'deeper' understanding of the sociological expert in contrast to commonsense knowledge of ordinary people.

Jahoda illustrated her way of interpretation with an example: "When an unemployed man says 'I miss the people at work....' he speaks [in the eyes of a sociological observer, A.W.] about the absence of a category" (Jahoda, 1982, p. 39). Apart from the problem that she uses a single remark as a piece of evidence, a systematic analysis would require examination, not only of the losses associated with unemployment, but also of the potential resources available to compensate for their loss (e.g., friends, family, membership in organisations etc.).

Jahoda's search for latent functions also seems to imply the introduction of a normative element. Her social background, and her socialization in the Austro-Marxist movement in Vienna, may have influenced what some of her critics call her glorification of work and employment (e.g., Hartley & Fryer, 1984; Brief, Konovsky, Goodwin & Link, 1995; Wacker, 2001). As Jahoda abstracts from the qualities and conditions of working life, employment per se becomes a "good" institution (O'Brien, 1986, p. 246).

Normative questions, of course, are difficult to discuss. Lewin's (1920) comment seems to provide a more adequate account of the experience of work (rather than employment) than Jahoda's account. Lewin distinguished between two "faces of work" in humankind's experience, namely work as a burden, and work as an opportunity for personal agency and development. Like Freud (1930) Lewin was convinced that our attitude to work is inevitably ambivalent. As a more general basis of mental health and psychological well-being Lewin (1920, p. 134) suggested not employment but engagement in meaningful activity: "Every human being, who is not ill or old, seeks a field of activity [Wirkungsfeld]", which may be either inside or outside the labour market and employment system.

Beside the general inconsistencies and problems of the latent function model there are other features to be discussed separately.

1 The theoretical status of Jahoda's latent function model

Jahoda herself labelled her model as an "approach". "Deprivation theory" (Fryer, 1992; Feather, 1997), "Jahoda's hypothesis" or "Jahoda's framework" (Gershuny, 1994), "latent needs theory" (Furnham, 1997), "latent deprivation model" (Creed & Evans, 2002) or "latent function model" (Fryer, 1986), and "functional approach" (Nordenmark

& Strandh, 1999) are other terms found in the literature to describe Jahoda's model. Jahoda herself was uncomfortable with the term "theory". "I never liked what I wrote about unemployment being called a theory ... I am certainly not offering a theory, I am uttering a thought", Jahoda answered in an interview with Fryer (Jahoda & Fryer, 1998, p. 93). The variety of labels given to her model signals that the theoretical status and aims of her thought are ambiguous.

In her book she paraphrases the Kluckhohn statement on personality: "In some respects every unemployed person is like every other unemployed person [level 1]; in some respects every unemployed person is like some other unemployed person [level 2]; and in some respects every unemployed person is like no other unemployed person [level 3]" (Jahoda, 1982, p. 48; Kluckhohn, Murray & Schneider, 1953). Jahoda's latent function model is at level 1. And it is due to its abstract and general character that it is difficult to derive empirically testable hypotheses from it. To illustrate this point we may go back to the example already cited: "When an unemployed man says 'I miss the people at work ...' he speaks about the absence of a category" states Jahoda (1982, p. 39). But if he does not say that he misses his former colleagues this simply means, according to Jahoda, that he is not "aware of the fit between [the categories of experience] and [his] psychological needs" (Jahoda, 1986, p. 27). In this way Jahoda immunizes her model against empirical falsification.

Some critics reproach Jahoda for being unable to explain the differential impact of unemployment on different groups of people. Previous job experiences, duration of unemployment, age, position in job history, gender, financial situation, and basic personality, are all factors that may influence and colour the individual's actual experience of unemployment. As already shown Jahoda does not deny these differences; but for her they are only variations of the central message: Unemployment reduces access to categories of experience necessary for the maintenance of mental health.

In some respects Jahoda's model confronts us with the long-standing tension between group tendency and individual variation. In her (1982) book she clearly admits that her approach is at the "most general level that inevitably ignores ... the individual level" (Jahoda, 1982, p. 48). And in a conversation with Fryer she admits: "Of course, there are individual differences" among the unemployed (Jahoda & Fryer, 1986, p. 110). The latent function model is a model of the effects of unemployment in general, not a look at its differential impact.

2 Structural categories of experience in employment and unemployment

Everyone who tries to identify the structural dimensions of work experience has to ask at least two questions: Do other relevant categories of experience exist which are not in her model?[2] Are all the categories of equal importance, or are some of them more important or central than others? Jahoda did not systematically reflect on these questions, however, judging from most of her writing the five latent functions seem to be of equal importance. Nevertheless, it seems puzzling that – in the same context - she named two other universal human needs in her book, namely "to make sense of events" and "for

2 In one of Jahoda's conversations with Fryer she named "control" as a sixth function, so that one may conclude that her catalogue is open for revision and supplements (Jahoda & Fryer, 1986, p. 110).

some degree of personal control over one's immediate environment" (Jahoda, 1982, pp. 69ff). Neither of these is a core component of her model.

Is Jahoda's five-component model complete? Definitely not. Warr's vitamin model, for instance, proposes nine environmental features, only some of which correspond to Jahoda's categories (Warr, 1987).

Are all components of equal importance to well-being? Although opinions vary, the consensus appears to be that this is not so. Nordenmark and Strandth (1999, pp. 582-583) argue that satisfaction of the "need for a social identity" is a key component of psychological well-being in societies in which "employment is the norm" and that the impact of unemployment on well-being depends on how important the individual regards employment as being to his or her personal identity. Fryer (1986, 1992) has argued that agency restriction, the loss of control over one's life circumstances, is the central component of the experience of unemployment. Several studies (e.g. Feather, 1997) have also found income reduction and financial hardship to be responsible for reduced well-being among unemployed people. Other authors stress the influence of activity level on well-being in everyday life (e.g. Haworth, 1997).

The main difficulty that Jahoda's latent function model poses for empirical researchers is, as Gershuny (1994) put it, that her "various categories of experience are only impressionistically defined and there is nothing like an agreed or tested set of indicators" (p. 215). In the course of her sociological analysis Jahoda might have argued that there is no need for operationalisation as the categories of experience are analytical tools on a conceptual level, not necessarily conscious components of the individual's mind. But Jahoda welcomed Miles attempt to measure the degree of access to the five components (Miles, 1983).

3 Distinction between latent and manifest functions of employment

As already suggested, Jahoda's distinction between latent and manifest functions creates problems that are not easy to resolve. But, to be fair, one has to take note of the context of her argument. Comparing the experience of unemployment in the 1930s and 1980s she was puzzled by the similarity of the psychological reactions of the unemployed. How could this constancy in reactions be explained? Conditions of employment had improved; the work ethic had altered; people were better educated than they had been in the past; and the standards of living of the employed and the unemployed had improved, and so on. In her perspective the question was: Which factor stayed constant so that it could explain this similarity in the experience of unemployment within differing social and economic contexts? Jahoda's answer pointed to the social institution of employment. Her relative devaluation of the role of financial and material circumstances stems from her comparative historical point of view: She herself saw the starvation of the unemployed people in the thirties and the improvement of social security measures in the eighties. Therefore she stressed the psychological impact of unemployment as a historical constant.

Most of her critics saw her thesis that the effects of unemployment are mainly a consequence of the latent function deprivation; i.e. psychological in nature. They therefore argued with some justice that the role of material and financial deprivation in contemporary unemployment is underestimated in her approach (e.g., Fryer, 1992).

Social security measures have improved in most industrialized countries since the 1930s: To become and to stay unemployed today does not mean absolute poverty and starvation as it did then. Nevertheless, empirical unemployment research from a variety of countries has demonstrated that measures of financial and material insecurity and loss, the so-called individual manifest functions in Jahoda's latent-function model, are the best single predictors of psychological ill health associated with unemployment (e.g., Kessler, Turner & House, 1987; Whelan, 1992). In a late interview with Fryer she admitted that, contrary to her model: "Nobody can ever deny the importance of manifest functions" (Jahoda & Fryer, 1998, p. 90). However, she herself did not examine this side of contemporary unemployment experience.

A confusing consequence of Jahoda's manifest-latent distinction is that, by definition, what is latent cannot also be manifest. But many people today who are asked why they have taken up employment refer to so-called latent functions (e.g. social recognition). Even the unemployed women in the Marienthal study told their interviewers about the wonderful social life they had enjoyed in the factory (Jahoda, Lazarsfeld & Zeisel, 1933/2002, p. 76). If we take Merton's (1968) actor model seriously we would have had to ask these women before they started their work in the factory of Marienthal whether they *intended* to enlarge "the scope of (their) social experience into areas less emotionally charged than the family life" (Jahoda, 1982, p. 59). The only thing we can say for sure is that, looking back, these unemployed women recognized very clearly what they had lost. Thus, the usefulness of categorizing the intentions of actors in this way does not seem to be high.

4 Latent functions of employment and mental health

While it is rather easy to agree upon the negative effects of long-term unemployment it is not so easy to agree which factors are responsible for the worsening of the mental health of the unemployed: withdrawal of the so-called latent functions of employment, or financial deprivation. Jahoda herself saw the problem but did not suggest a solution. "...since unemployment is as a rule coupled with a significant drop in the standard of living, relative poverty, not the lack of a job may fully account for the state of mind" (Jahoda, 1986, p. 168).

Referring to the latent consequences of paid work Jahoda argues that "some of the latent consequences of employment ... are the unintended daily provision of categories of experience …", thus referring to employment as a social institution. However, she incorporates a psychological dimension to her argument when she adds "… which most people value and seem to need for their well-being" (Jahoda, 1995, p. 318). Jahoda and Rush (1980) claimed that: "There is abundant evidence that beyond financial problems, unemployment of more than a very short period is psychologically destructive *because* of the absence of the latent consequences of employment ..." (Jahoda & Rush, 1980, p. 13). This, of course, is a strong thesis, not easily confirmed or rejected. Meta-analytic research has only confirmed that the decline in well-being of the unemployed is largely due to unemployment experience in general comparing employed and unemployed people (e.g. Murphy & Athanasou, 1999; Paul & Moser, 2001).

Jahoda and Rush argue that the "*desirable* latent consequences of employment are psychologically enriching" (Jahoda & Rush, 1980, pp. 13, 48). Quoting Freud, Jahoda maintains that "institutional arrangements ... protect man ... from being swamped by extreme emotionality" (Jahoda, 1982, p. 60). The integration of individuals into the institution of employment therefore becomes a necessity for psychological health. But work psychology and work medicine have produced abundant evidence that whether this is true or not depends on the interplay between the concrete circumstances of the workplace and personality (e.g. Siegrist & Dragano, 2006). So Fryer is right to stay sceptical: "It would indeed be remarkable if the ... conditions which have evolved in the twentieth century industrialized world ... happened to guarantee or maximise psychological health" (Fryer, 1986, p. 17).

Several authors draw attention to the fact that the latent function model is situation-centred: "The individual's mental well-being is largely seen as the result of structural factors outside the individual" (Nordenmark & Strandh, 1999, p. 578; see also Fryer, 1986 and Haworth & Ducker, 1991). As these structural features of employment are described as a bundle of psychologically necessary and useful resources their absence produces the negative effects on health in the case of unemployment.

Such a general thesis, of course, is not well suited to explain the differential impact of unemployment on psychological health which has become of a focus of recent research. As Shams and Jackson (1993, p. 343) state: "... the most productive line of research has been to incorporate factors which act as both mediators and moderators of the link between life-events and well-being ... Research attention has switched therefore ... *to identifying those factors which differentiate between people who cope more or less well with such experiences*". As empirical research has already demonstrated, not all people experience employment as beneficial and some of the unemployed cope quite well with their situation (e.g., Feather, 1997).

In his vitamin model Warr (1987) made three important changes to Jahoda's approach. Firstly, he dropped the distinction between manifest and latent functions. Secondly, his nine environmental features refer both to employment *and* unemployment. And thirdly, Warr argues that unemployment can be "bad" or "good", just as employment can be "bad" or "good" (Warr, 1987). So it is up to empirical research to identify the conditions which make unemployment "good" or "bad".

5 The exclusive status of employment as a support of mental health

Although Jahoda admits "that some types of jobs under some modern conditions are psychologically destructive" (Jahoda, 1982, p. 43), in most of her work she portrays employment as the basic source of psychological health in adult life. She goes even further: *All* people outside the labour market - housewives, the retired, the mentally and physically handicapped – "experience similar psychological deprivation" to that experienced by the unemployed (Jahoda & Rush, 1980, p. 48). But unemployment research does not support this claim. As the Australian researcher Ezzy (1993, p. 41) puts it: "Simplistic identifications of work with 'good' and unemployment with 'bad' are manifestly inadequate as explanations of observed variations in the effects of unemployment on mental health".

If one considers the evidence on adaptation to retirement one finds that dissatisfaction appears to be a result of poor health or inadequate finances, rather than the absence of employment itself (e.g., Haworth, 1997, p. 10; Lang-von Wins, Mohr & von Rosenstiel, 2004). Most of the retired find alternative ways to satisfy the needs that Jahoda attributes rather exclusively to employment, through participation in committee work or further education. In countries where early retirement regulations exist many employees decide to leave their job as soon as they can afford it economically. So Jahoda's general thesis is not in correspondence with the findings of empirical research on retirement, at least in Western industrialised countries (e.g. Haworth, 1997, pp. 10 ff.; Kiefer, 1997).

Another critical point in Jahoda's model relates to the controversy about whether out-of-employment activities can compensate for job loss. The results of several studies suggest that meaningful activities in general have an anti-depressive effect and that this effect is not bound to job activities (e.g., Haworth, 1986, 1997; Pelzmann, 1988). Swedish sociologists mention parenthood as an example (Nordenmark & Strandh, 1999, p. 578).

Research on retired people, good copers among the unemployed (e.g., Fryer & Payne, 1984) and looking back at some periods of Jahoda's own life (Jahoda, 1997), suggests that activities outside of employment do offer access to the categories of experience relevant in Jahoda's latent function model.

These findings are in accordance with Lewin's position cited earlier. Involvement in personally meaningful tasks - activities in Lewin's sense - can provide an alternative means for obtaining the benefits associated with the latent functions of employment.

6 Employment and unemployment - a sharp contrast?

As previously mentioned, Jahoda distinguishes sharply between the life situation of the employed and that of the unemployed. This seems to be inadequate in the light of changing labour market conditions. Burchell (1994), for instance, argued that the distinction between "employed" and "unemployed" is too simple to describe all the different labour market positions in modern economy. Job insecurity has spread into many trades. "Labour market insecurity covers a much broader range of situation than that of unemployment, although unemployment is likely to constitute an extreme form" (Burchell, 1994, p. 18).

7 Conclusion

As documented in literature and shown in this article Jahoda's latent function model on the one hand was a important step in formulating an integrative frame of reference for unemployment research; on the other hand it is open to many objections in general and in detail. Nevertheless it stimulated research in the field of psychological and sociological unemployment research and suggested new research questions.

In some respects Jahoda's latent function model seems to be a variation of the old adage "Man does not live by bread alone". Her approach has its merits in highlighting the role of employment as a way in which modern industrial societies manage the difficult task to integrate the majority of the population in their adult life. Employment is still the main institution of social integration in the modern world.

But Jahoda did not notice that the labour market and employment system itself has changed and is still changing in a dramatic way so that the difference between the employed and the unemployed population is not so clear-cut as it was in the past.

Only a minority of social scientists and psychologists will rigidly adhere to her thesis that employment is the only source of experiences that can satisfy the so-called latent needs. Voluntary engagement in work-like activities, serious leisure activities, family life, committee work or education - all activities that can give meaning and structure to the individual life - can serve as an alternative to employment. Therefore, the sharp contrast between employment and other areas of life is blurred.

Jahoda's approach cannot explain the differential impact of age, gender, the position in the lifespan, duration of unemployment, previous job experience and so on. However, one has to admit, that this was never her intention. She was not looking for those factors "which differentiate between people who cope more or less well with" the experience of unemployment (Shams & Jackson, 1993, p. 142). So her mode of explanation is in a way too abstract and too simple, but at the same time open for more sophisticated extensions.

Empirical attempts to measure the access to Jahoda's latent categories, although questionable in details, give some support for her general thesis that unemployment - on the average - does reduce the degree of access to the types of experiences described in the latent function model (e.g., Haworth, 1987, 37ff.; Creed & Machin, 2002, p. 1210), although the correlation between access measures and measures of psychological well-being generally is rather weak. According to Gallie and Marsh (1994, p. 18) Jahoda's model can "provide only a limited part of an overall explanation of the psychological effects of unemployment". Recent research has demonstrated that the separation and de-evaluation of the economic side of unemployment was a deficiency of Jahoda's model.

The idea of some sociologists that individual behaviour is mainly a function of current social pressures is unrealistic. Job history, previous experiences, future time perspective, personality traits will all determine how people react to the situation of unemployment. Jahoda once called it one of the main weaknesses of the Marienthal study that the researchers did not succeed in separating the personality influences from the situational influences (Jahoda & Kreutzer, 1983). But in her later latent functional model this separation is missing, too.

There exist further merits of her approach: In her endeavour to make the suffering of unemployment understandable to the public she was an advocate of the unemployed in society. Her latent function model helps to a better understanding of the costs of societal change.

References

Boudon, R. (1990). The two facets of the unintended consequences paradigm. In J. Clark, C. Modgil & S. Modgil (Eds.), *Robert K. Merton - consensus and controversy* (pp. 119-128). London: Falmer Press.

Brief, A.P., Konovsky, M.A., Goodwin, R. & Link, K. (1995). Inferring the meaning of work from the effects of unemployment. *Journal of Applied Psychology, 25*(8), 693-711.

Burchell, B. (1992). Towards a social psychology of the labour market: Or why we need to understand the labour market before we can understand unemployment. *Journal of Occupational and Organisational Psychology, 65*(4), 345-354.

Burchell, B. (1994). The effects of labour market position, job insecurity, and unemployment on psychological health. In D. Gallie, C. Marsh & C. Vogler (Eds.), *Social change and the experience of unemployment* (pp. 213-230). Oxford: Oxford University Press.

Campbell, C. (1982). A dubious distinction? An inquiry into the value and use of Merton's concepts of manifest and latent function. *American Sociological Review, 47*, 29-40.

Creed, P.A. & Evans, B.M. (2002). Personality, well-being and deprivation theory. *Personality and Individual Differences, 33*, 1045-1054.

Creed, P.A. & Machin, M.A. (2002). The relationship between mental health and access to the latent functions of unemployment for unemployed and underemployed individuals. *Psychological Reports, 90*, 1208-1210.

Elster, J. (1990). Merton's functionalism and the unintended consequences of action. In J. Clark, C. Modgil & S. Modgil (Eds.), *Robert K. Merton - consensus and controversy* (pp. 129-135). London: Falmer Press.

Evans, S.T. & Banks, M.H. (1992). Latent functions of employment: Variations according to employment status and labour market. In C.H.A. Verhaar & L.G. Jansma (Eds.), *On the mysteries of unemployment: Causes, consequences and policies* (pp. 281-295). Dordrecht: Kluwer.

Ezzy, D. (1993). Unemployment and mental health: A critical review. *Social Science and Medicine, 37*(1), 41-52.

Feather, N.T. (1997). Economic deprivation and the psychological impact of unemployment. *Australian Psychologist, 32*(1), 37-45.

Foster, J.J. (2000). Motivation in the work place. In N. Chmiel (Ed.), *Introduction to work and organisational psychology* (pp. 302-326). Oxford: Blackwell.

French, J.R.P. (1973). Person role fit. *Occupational Mental Health, 3*, 15-20.

Freud, S. (1963). *Civilisation and its discontents* (Collected Works, Vol. 21). London: Hogarth.

Fryer, D. (1986). Employment deprivation and personal agency during unemployment: A critical discussion of Jahoda`s explanation of the psychological effects of unemployment. *Social Behaviour, 1*, 3-23.

Fryer, D. (1992). Psychological or material deprivation: Why does unemployment have mental health consequences? In E. McLaughlin (Ed.), *Understanding unemployment: New perspectives on active labour market policies* (pp. 103-125). London: Routledge.

Fryer, D. & Payne, R. (1984). Proactive behaviour in unemployment: Findings and implications. *Leisure Studies, 3*, 273-295.

Furnham, A. (1997). *The psychology of behaviour at work. The individual in the organisation.* Hove: Psychology Press.

Gallie, D. & Marsh, C. (1994). The experience of unemployment. In D. Gallie, C. Marsh & C. Vogler (Eds.), *Social change and the experience of unemployment* (pp. 1-30). Oxford: Oxford University Press.

Gershuny, J. (1994). The psychological consequences of unemployment: An assessment of the Jahoda thesis. In D. Gallie, C. Marsh & C. Vogler (Eds.), *Social change and the experience of unemployment* (pp. 213-230). Oxford: Oxford University Press.

Hartley, J. & Fryer, D. (1984). The psychology of unemployment: A critical appraisal. In G.M. Stephenson & J.H. Davis (Eds.), *Progress in applied social psychology* (Vol. 2, pp. 3-30). Chichester: Wiley.

Haworth, J.T. (1986). Meaningful activity and psychological models of non-employment. *Leisure Studies, 5*, 281-297.

Haworth, J.T. (1997). *Work, leisure and well-being.* London: Routledge.

Haworth, J.T. & Ducker, J. (1991). Psychological well-being and access to categories of experience in unemployed young adults. *Leisure Studies, 10*, 265-274.

Haworth, J.T. & Evans, S.T. (1987). Meaningful activity and unemployment. In D. Fryer & P. Ullah (Eds.), *Unemployed people - social and psychological perspective* (pp. 241-267). Milton Keynes: Open University Press.

Jahoda, M. (1982). *Employment and unemployment - A social-psychological analysis.* Cambridge: Cambridge University Press.

Jahoda, M. (1995). Manifest and latent functions. In N. Nicholson (Ed.), *Encyclopedic Dictionary of Organisational Behaviour* (pp. 317-318). Oxford: Blackwell.

Jahoda, M. (Engler, S. & Hasenjürgen, B.). (1997). Biographisches Interview mit Marie Jahoda. In S. Engler & B. Hasenjürgen (Eds.), *"Ich habe die Welt nicht verändert". Lebenserinnerungen einer Pionierin der Sozialforschung* (pp. 101-185). Frankfurt a. M.: Campus.

Jahoda, M. & Ernst, H. (1981). "Arbeitslose haben alles Recht der Welt, über ihre Lage unglücklich zu sein" - Das Psychologie-heute-Gespräch mit Marie Jahoda. *Psychologie Heute, 8*(12), 71-76.

Jahoda, M. & Fryer, D. (1986). The social psychology of the invisible: An interview with Marie Jahoda. *New Ideas in Psychology, 4*, 107-118.

Jahoda, M. & Fryer, D. (1998). The simultaneity of the unsimultaneous: A conversation between Marie Jahoda and David Fryer (1996). *Journal of Community & Applied Social Psychology, 8*(2), 89-100.

Jahoda, M. & Kreuzer, F. (1983). Franz Kreuzer im Gespräch mit Marie Jahoda. In F. Kreuzer (Ed.), *Des Menschen hohe Braut - Arbeit, Freizeit, Arbeitslosigkeit. Franz Kreuzer im Gespräch mit*

Marie Jahoda, fünfzig Jahre nach der Untersuchung "Die Arbeitslosen von Marienthal" (pp. 7-33). Wien: Deuticke.

Jahoda, M., Lazarsfeld, P.F. & Zeisel, H. (2002). *Marienthal: The sociography of an unemployed community (1933). With a new introduction by Christian Fleck.* New Brunswick/London: Transaction.

Jahoda, M. & Rush, H. (1980). Work, employment and unemployment - An overview of ideas and research results in the social science literature. *SPRU Occasional Paper Series No. 12 University of Sussex.*

Kant, I. (1798/1970). Kritik der praktischen Vernunft. In W. Weischeldel (Eds.). Werke (vol. 10, pp. 399-690). Darmstadt: Wissenschaftliche Buchgesellschaft.

Kessler, R.C., Turner, J.B. & House, J.S. (1987). Intervening processes in the relationship between unemployment and health. *Psychological Medicine, 17*, 949-961.

Kiefer, T. (1997). *Von der Erwerbsarbeit in den Ruhestand. Theoretische und empirische Ansätze zur Bedeutung von Aktivitäten.* Göttingen: Hogrefe.

Kieselbach, T. (1988). Arbeitslosigkeit. In R. Asanger & G. Wenninger (Eds.), *Handwörterbuch der Psychologie* (4th ed., pp. 42-51). München: pvu.

Kluckhohn, C., Murray, H.A. & Schneider, D.M. (Eds.). (1953). *Personality in nature, society, and culture.* New York: Knopf.

Lang-von Wins, T., Mohr, G. & von Rosenstiel, L. (2004). Kritische Laufbahnübergänge: Erwerbslosigkeit, Wiedereingliederung und Übergang in den Ruhestand. In H. Schuler (Eds.), *Organisationspsychologie - Grundlagen und Personalpsychologie.* (pp. 1113 - 1189). Göttingen: Hogrefe.

Lewin, K. (1981). Die Sozialisierung des Taylorsystems (1920). *Gestalt Theory, 3*, 129-151.

Merton, R.K. (1936). The unanticipated consequences of purposeful social action. *American Sociological Review, 1*, 894-904.

Merton, R.K. (1968). *Social theory and social structure (1957)* (2nd ed.). New York: The Free Press.

Miles, I. (1983). Adaption to unemployment? In Science Policy Research Unit (Ed.), *SPRU Occational Paper Series* (Vol. 20). Falmer, Brighton: University of Sussex.

Murphy, G.C. & Athanasou, J.A. (1999). The effect of unemployment on mental health. *Journal of Occupational and Organisational Psychology, 72*, 83-99.

Nordenmark, M. & Strandh, M. (1999). Towards a sociological understanding of mental well-being among the unemployed: The role of economic and psychosocial factors. *Sociology, 33*(3), 577-597.

O'Brien, G. E. (1986). *Psychology of work and unemployment.* Chichester: Wiley.

Paul, K. & Moser, K. (2001). Negatives psychisches Befinden als Wirkung und als Ursache von Arbeitslosigkeit: Ergebnisse einer Metaanalyse. In J. Zempel, J. Bacher & K. Moser (Eds.), *Erwerbslosigkeit. Ursachen, Auswirkungen und Interventionen* (Vol. 12: Psychologie sozialer Ungleichheit, pp. 83-110). Opladen: Leske + Budrich.

Pelzmann, L. (1988). *Wirtschaftspsychologie - Arbeitslosenforschung, Schattenwirtschaft, Steuerpsychologie* (2nd ed.). Wien: Springer.

Shams, M., & Jackson, P.R. (1993). Religiosity as a predictor of well-being and moderator of the psychological impact of unemployment. *British Journal of Medical Psychology, 66*, 341-352.

Siegrist, J. & Dragano, N. (2006). Berufliche Belastung und Gesundheit. In C. Wendt & C. Wolf (Eds.), *Soziologie der Gesundheit (KZfSS-Sonderheft Bd. 46).* (S. 109 - 124). Wiesbaden: VS Verlag für Sozialwissenschaften.

Wacker, A. (2001). Was fehlt, wenn die Arbeit fehlt? Arbeitslosigkeit aus sozialpsychologischer Perspektive. In U. Becker, F. Segbers & M. Wiedemeyer (Eds.), *Logik der Ökonomie - Krise der Arbeit. Impulse für eine solidarische Gestaltung der Arbeitswelt* (pp. 51-64). Mainz: Matthias-Grünewald-Verlag.

Warr, P. (1987). *Work, unemployment and mental health.* Oxford: Clarendon Press.

Whelan, C.T. (1992). The role of income, life-style deprivation and financial strain in mediating the impact of unemployment on psychological distress: Evidence from the Republic of Ireland. *Journal of Occupational and Organisational Psychology, 65*(4), 331-344.

Measuring the Access to the Manifest and Latent Functions of Employment Among Middle-Aged Portuguese Unemployed

Marta Sousa-Ribeiro & Joaquim Luís Coimbra

Introduction

In Portugal and in several industrialised countries, unemployment is one of the most serious problems people and governments face at the present. Poorer psychological and physical wellbeing among the unemployed when compared to employed have been demonstrated both in several cross-sectional and longitudinal studies (e.g., McKee-Ryan, Song, Wanberg & Kinicki, 2005) and have also been shown to be mainly consequential to unemployment, but not the result of a downward trajectory of those with poorer health drifting into unemployment (Creed & Macintyre, 2001).

The research into the psychological effects of unemployment is extensive and numerous reviews of this research have been published (e.g., Hanisch, 1999; McKee-Ryan et al., 2005; Murphy & Athanasou, 1999; Warr, 1987). However, in Portugal, the research in this area is still scarce and insufficient in taking into account the dimension of this problem: According to EUROSTAT, the Portuguese unemployment rate in May 2006 was 7.5%, which means that around 442 000 people were out of work, 43% of them long-term unemployed (i.e. unemployed for more than one year). The present study aims to provide a better comprehension of the relation between unemployment and psychological well-being among older unemployed Portuguese in the light of the Latent Deprivation Model theorised by Marie Jahoda (1982).

1 Theoretical framework

1.1 Latent deprivation model (Jahoda, 1982)

In this approach, employment is considered to provide both manifest (associated with income) and latent (time structure, social contacts outside of the immediate family, collective purpose and effort, regular activity and social status and identity) benefits to people. These latent benefits are viewed by Jahoda (1982) as reflecting human needs. The author states that when individuals become unemployed, they not only lose the manifest benefits of employment but are also deprived of the latent benefits, and that it is the loss of the latent benefits that primarily accounts for a decline in psychological well-being after job loss. Jahoda argues that although there are other institutions that provide the access to one or more of the manifest and latent functions, only the social institution of paid employment combines them all. This model and Jahoda´s work in general (Jahoda et al., 1981) have greatly stimulated research on the psychological effects of unemployment.

Several studies have demonstrated that compared with employed people, the unemployed show higher levels of deprivation in manifest and latent benefits and psychological distress (e.g., Creed & Reynolds, 2001; Miles, 1983). Creed & Muller's (2006) findings also go in this direction, although significant differences were not found regarding social support. Reduced access to the latent and manifest benefits have been shown to be associated with poorer psychological wellbeing (Creed & Evans, 2002; Creed & Macintyre, 2001; Evans & Haworth, 1991), and positive correlations between the access to the latent functions of employment and satisfaction with life have also been demonstrated (Evans & Banks, 1992). Some functions appeared to be more important than others in predicting psychological distress. Creed & Macintyre (2001) found financial deprivation to be the most important predictor, followed by the deprivation of status, time structure and lastly, collective purpose. Creed and Muller (2006) found that financial deprivation followed by deprivation of status were the only predictors of psychological distress, and in the study by Wanberg, Griffiths and Gavin (1997), psychological distress was only predicted by financial deprivation and the deprivation of time structure.

1.2 Agency restriction model (Fryer, 1986)

Another model, the Agency Restriction Model, was developed by Fryer as a counterpoint to the latent deprivation model of Jahoda. Although recognizing the significance of the five latent benefits identified by Jahoda, he refutes the centrality of the social institution of employment in providing these benefits, and also criticizes the primary importance that Jahoda attributes to the latent in contrast to the manifest benefits. According to Fryer, unemployment generally results in financial distress, which reduces the capacity to plan a meaningful future, restricts personal agency and, consequently, conduces to a decline in psychological well-being (Fryer, 1986). Some research has supported Fryer's approach. For example, Creed and Evans (2002) found stronger associations between the manifest functions and wellbeing than between latent functions and well-being. In the study by Hoare and Machin (2004), latent benefits accounted for a significant 14% of the variance in psychological distress, but only time structure emerged as a significant single predictor. After controlling for other key predictors (including financial deprivation), the latent benefits were unable to significantly predict psychological distress. These findings are consistent with the results by Creed and Klisch (2005). In their study, only financial strain was able to predict psychological distress: time structure, social support, collective purpose, activity and status were not significant individual predictors. Brief, Konovsky, Goodwin and Link (1995) also found that only financial deprivation - and not the deprivation of the latent benefits - was negatively related with subjective wellbeing (as measured with the Satisfaction with Life Scale).

1.3 Employment commitment

Employment commitment is a psychological variable that refers to the importance an individual sets on work (Kanfer, Wanberg & Kantrowitz, 2001) and it has been identified as an important moderator of the negative psychological effects of unemployment. For example, Wiener, Oei and Creed (1999) found that psychological distress of unemployed participants was predicted by employment commitment, while Vansteenkiste,

Lens, De Witte and Feather (2005) found that employment commitment negatively predicted satisfaction with life.

1.4 Perceived age discrimination

A growing number of older workers are being prevented from participating in work because of prejudice and discrimination, despite of the antidiscrimination legislation adopted unanimously by the European Union member states in 2000 (European Directives 2000/43/CE and 2000/78/CE), which prohibits discrimination in employment and training on the grounds of age and other individual characteristics. Age discrimination in employment consists of those decisions made by an employer at all stages of the employment process about an individual that are solely based on his/her chronological age (Sargeant, 2001). Several authors have argued that older workers often face age discrimination in the labour market, even when their competence and qualifications are comparable to those of younger workers (e.g., Warr, 1998). Therefore, many individuals at this age consider their reemployment chances to be scarce, feeling hopeless about finding another job. Thus, the distress experienced by middle-aged individuals following job loss may be related with the intensity with which they perceive to be discriminated by the labour market on the grounds of age and, consequently, the distress may also be related to how they perceive their chances of obtaining another comparable job, or even any other job (Broomhall & Winefield, 1990).

2 Research goals and hypotheses

The first goal of the study was to examine the relative contributions of the latent and manifest benefits of employment in predicting psychological distress and satisfaction with life in a sample of middle-aged Portuguese adults. The second goal was to examine whether the latent and manifest benefits of employment can be obtained also by the unemployed through the participation in a full-time training course. The third goal was to examine the contributions of the employment commitment and perceived age discrimination in predicting psychological distress and satisfaction with life in a sample of middle-aged unemployed Portuguese adults.

Hypotheses

1. Deprivation of the latent benefits of employment is significantly correlated with psychological distress and satisfaction with life.
2. Deprivation of the manifest benefits of employment is significantly correlated with psychological distress and satisfaction with life.
3. Unemployed participants will show greater deprivation of both the latent and manifest benefits of employment, higher levels of psychological distress and lower levels of satisfaction with life than the employed participants.
4. Unemployed participants attending a vocational training course will show less deprivation of the latent benefits of employment, lower levels of psychological distress, and higher levels of satisfaction with life than the other unemployed group.
5. In the unemployed participants, after controlling for demographic variables (course attendance, gender, family characteristics, socio-economic level and

length of unemployment), the latent and manifest benefits of employment, employment commitment and perceived age discrimination will be significant predictors of both the psychological distress and satisfaction with life, with the manifest benefit of employment accounting for the largest amount of the variance in the well-being variables.

3 Research methods

3.1 Sample

The participants for the study were a convenience sample of 312 individuals, comprising three groups, according to their occupational status: (1) consisted of 132 individuals (42% of the total sample) who were employed full-time; (2) consisted of 94 individuals (30% of the total sample) who were unemployed and not enrolled at the time of the study in any kind of vocational training course; (3) consisted of 86 individuals (28% of the total sample) who were unemployed and enrolled in a full-time vocational training course leading to a professional qualification. Socio-economic level of the participants was determined by the academic attainment and the employment status on the current (employed sample) or latest (unemployed samples) job. The latter indicator was considered to be more accurate than the educational attainment, as in many cases there is not a direct correspondence between education and the employment status: for example, a considerable number of individuals with very low academic qualifications, work(ed) in managerial positions. Eleven individuals belonging to the above group 1, three individuals belonging to group 2 and two individuals belonging to group 3 did not report their academic attainment or current/past job. Table 1 describes the socio-demographic characteristics for the three subgroups.

3.2 Procedure

The data from unemployed persons was gathered by anonymous questionnaires from jobseekers in five job centres (Group 1) and two training centres (Group 2). These participants were all registered for work with the Portuguese National Employment and Vocational Training Institute in Porto Metropolitan area and received public income support. Employed participants (Group 3) worked in the same geographical area as these centres, and they were approached via their employers at their workplace (8 small/medium size companies participated in the study) or directly by the researchers. Information about the non-responses to the questionnaire is not available.

3.3 Measures

The questionnaire developed included socio-demographic questions and the following measures: the Latent and Manifest Benefits (LAMB Scales: Muller et al., 2005), the 12-item version of the General Health Questionnaire (GHQ: Goldberg, 1972), the Satisfaction With Life Scale (Diener et al., 1985; Neto, 1997), the Employment Commitment Scale (Jackson et al., 1983) and the discrimination subscale of the Age Related Risks Scale (Bailey & Hansson, 1995). The latest two scales were administered only to the two unemployed groups.

Latent and Manifest Benefits (LAMB) Scale (Muller et al., 2003). For the purpose of the research, this scale was first translated from English to Portuguese. The Portuguese version was piloted for language use and clarity, and minor changes were taken as needed to ensure the items were appropriate in the Portuguese context, while the original meaning was retained. The preliminary results of the validation of the Portuguese version are presented in the results section. The original scale consists of six subscales each with six bipolar items measured on a seven-point Likert scale and it was used to assess the latent and manifest benefits theorized by Jahoda: Collective purpose, time structure, enforced activity, status, social contact and financial strain. In the original measure, scores for each subscale ranged from 7 to 42, with higher scores indicating more deprivation in the access to the latent benefits.

Twelve-item version of the General Health Questionnaire (GHQ: Goldberg, 1972). This measure has been widely utilized and recommended for use as an estimate for psychological distress in the study of employment and occupational-related issues (e.g., Banks et al., 1980; Warr, 1987). Participants were asked questions about how they felt recently on a variety of variables. For instance, one sample item is "Have you recently felt able to enjoy your normal day-to-day activities?" measured on a four-point scale from 0 to 3, and using anchors such as "More than usual", "Same as usual", "Less than the usual" "Much less than the usual". The possible range for this scale is 0-36, with higher scores showing more psychological distress. This measure was translated and adapted to Portuguese using the same procedures described above. The internal reliability coefficient for the total sample in this study was 0.87.

Satisfaction with Life Scale (SWLS: Diener et al., 1985; Neto, 1997). The Portuguese version of the SWLS (Neto, 1997) was used to assess the global life satisfaction. This is a five-item measure, and each item is scored on a seven-point Likert-type scale, so the possible range is from 5 to 35 with higher scores indicating more satisfaction with life. For instance, a sample item was "In most ways my life is close to my ideals". The internal reliability coefficient for the total sample in this study was 0.79.

Table 1. Subgroup characteristics

Variable	Group 1 Employed	Group 2 Unemployed not attending a training course	Group 3 Unemployed attending a training course
Number of Participants	132	94	86
Gender (N female, %)	57 (43%)	59 (63%)	61 (71%)
Age (years) – Mean; Standard Deviation	47.4 (5.10)	47.9 (5.60)	47.6 (6.00)
Family Characteristics			
Single; Divorced; Widowed without dependents (N, %)	10 (7,6%)	14 (15%)	10 (12%)
Single; Divorced; Widowed with dependents (N, %)	5 (3,8%)	17 (18%)	22 (26%)
Married / Living with someone without dependents (N, %)	11 (8,3%)	15 (16%)	10 (12%)
Married / Living together with someone with dependents (N, %)	106 (80,3%)	48 (51%)	44 (51%)
Socio-Economic Level			
Low (N, %)	42 (32%)	32 (34%)	31 (36%)
Medium (N, %)	25 (19%)	24 (26%)	30 (35%)
High (N, %)	54 (41%)	35 (37%)	23 (27%)
Length of unemployment (months) – Mean; Standard deviation	- -	14.9 (12.4)	31.30 (21.8)

Employment Commitment Scale (Jackson et al., 1983). This six-item measure of employment commitment was translated and adapted to Portuguese using the same procedures mentioned before. One item was added: "Even if I won a great amount of money in the lottery I would carry on working" to evaluate non-financial employment commitment. Another sample item was "Work will make me feel I'm doing something with my life", measured on a five-point Likert-type scale. Scores ranged from 7 to 35 with higher scores suggesting increasing levels of employment commitment. The internal reliability coefficient for the unemployed sample (Group 1 and Group 2) in this study was 0.60.

Discrimination subscale of the Age Related Risks Scale (Bailey & Hansson, 1995). This measure was used to assess the participants' perception of being discriminated by the employers in the grounds of age. It was, too, first translated and adapted to Portuguese by the same procedures mentioned before. As the Age Related Risks Scale was originally developed to assess the perceived obstacles to adaptive job or career changes involving middle-aged and older workers, some small changes were made to adapt it to the unemployment context. A sample item was, for instance, "My age places me at a competitive disadvantage", measured on a five-point Likert-type scale. Scores ranged from 5 to 25 with higher scores suggesting increasing levels of age discrimination perception. The internal reliability coefficient for the unemployed sample (Group 1 and Group 2) in this study was 0.84.

4. Results

4.1 Exploratory factor analysis of the LAMB scales

A principal axis factor analysis was used to examine the structure of the Portuguese version of the LAMB Scales. Three analyses were computed. The first solution identified eight factors with eigenvalues greater than 1 that were rotated using oblique (direct oblim) rotation. This solution presented a factor with a single item. Using the Gorsuch (1997) recommendation to keep only the factors having at least three salient items, this trivial factor was dropped and a second analysis was undertaken. For this second solution, the Kaiser-Meyer-Olkin measure of sampling adequacy (KMO) was .87 and Bartlett's Test of Sphericity was highly significant ($p<.0001$). From this analysis seven factors with eigenvalues greater than 1 were retained. As the original LAMB Scale consists of six subscales assessing the five latent and one manifest benefits theorized by Jahoda, a third analysis was conducted, this time specifying six factors to be extracted. Whilst presenting one more factor than the original scale, the seven-factor solution emerged as the most meaningful one and is the only one to be reported.

For item selection two criteria needed to be simultaneously met, as suggested by Lent, Hill & Hoffman (2003): The item should yield a factor loading with an absolute value greater than .50 and show a difference greater than .10 between its two highest factor loadings' values. According to these authors "(...) *these criteria were designed to seek a reasonable balance between cross-loadings and detection of coherent factors*" (p. 100). One item was dropped for not meeting these criteria and therefore the final scale consists of 34 items. Eigenvalues, percentage of variance explained, number of items, and Cronbach's internal reliabilities for both the original Latent and Manifest Benefits (LAMB) scales and their Portuguese adaptation are presented in Table 2.

Some reflections should be made upon the seventh factor, which accounts for 3.17% of the total variance. It consists of two items belonging in the original scale to the Status subscale and one item originally belonging to the Enforced Activity subscale. As this subscale showed a poor reliability ($\alpha = .58$) and does not reflect any of the five latent and one manifest benefits theorized by Jahoda, it will not be used in any of the further analyses.

4.2 Correlation analysis

According to correlation analysis, deprivation of collective purpose, time structure, enforced activity and financial deprivation were significantly positively associated to psychological distress (GHQ). Stronger correlations were found for financial deprivation followed by deprivation of time structure. No correlations were found for status and social contact. Positive significant correlations were also found between perceived age discrimination and psychological distress. Contrary to what was expected, employment commitment was not correlated to this well-being measure. Deprivation of all the manifest and latent benefits was significantly negatively associated to satisfaction with life. Stronger correlations were found for financial deprivation followed by the deprivation of collective purpose. Perceived age discrimination was also negatively related to satisfaction with life and, as in the case of psychological distress, no significant correlations were found between employment commitment and satisfaction with life.

Table 2. Eigenvalues, percentage of variance explained, number of items and Cronbach's alpha for the original and Portuguese adaptation of the Latent and Manifest Benefits (LAMB) Scales.

Factor	Example of Item	Eigenvalue		% Variance Explained		Number of Items		α	
		P.A.	O.S.	P.A.	O.S.	P.A.	O.S.	P.A.	O.S.
1. Collective Purpose	I (often/rarely) feel that I make a meaningful contribution to society	8.16	7.50	23.31	20.84	6	6	.86	.88
2. Financial Strain	My income (usually/rarely) allows me to do the things I want	3.89	3.89	11.10	10.81	6	6	.92	.92
3. Time Structure	I (often/rarely) wish I had more things to do to fill up the time in my days	3.79	2.34	10.83	6.50	6	6	.84	.78
4. Enforced Activity	My days are usually (well/not well) organised	2.09	1.60	5.97	4.43	4	6	.75	.76
5. Status	I am (often/rarely) valued by the people around me	1.49	2.60	4.26	7.22	3	6	.77	.84
6. Social Contact	I (usually/rarely) have a lot of opportunities to mix with people	1.43	3.34	4.09	9.27	6	6	.82	.89
7. ____	____	1.12	____	3.17	____	3	____	.58	____

Note. P.A. = portuguese adaptation; O.A. = original scale

4.3 Differences on key variables among the three groups

Means and standard deviations for collective purpose, time structure, enforced activity, status, social contact, financial strain, psychological distress and satisfaction with life for the three groups are presented in Table 3. A multivariate analysis of variance was conducted to determine whether there were differences among the three groups on key variables levels. As there were not equal cell sizes for the MANOVA, Pillai's criterion was used as the multivariate test of significance. Results for this analysis are presented in Table 3.

A significant multivariate effect was found, F (16,606) = 11,279, p<0.001. At the univariate level, significant effects were identified for collective purpose, F (2, 309) = 5,54, p<0.01, time structure, F (2, 309) = 38,82, p<0.001, financial strain, F (2, 309) = 35,26, p<0.001, psychological distress, F (2, 309) = 12,33, p<0.001 and satisfaction with life, F (2, 309) = 18,76, p<0.001. Post hoc multiple comparisons test of Scheffé indicated that for *collective purpose*, those who were unemployed not attending a course (Group 2) reported significantly more deprivation than those who were employed (Group 1, p<0.01), and those who were unemployed but attending a course (Group 3, p<0.05). For time structure, Group 2 reported significantly (p<0.001) more deprivation than Group 1 and Group 3. For *financial strain*, both unemployed Group 2 and Group 3 showed significantly higher levels than Group 1 employed participants (p<0.001). In relation to *psychological distress*, Group 2 participants reported higher levels than Group 1 participants (p<0.001) and Group 3 participants (p<0.01). Group 2 showed significantly lower levels of *satisfaction with life* than Group 3 (p<0.05) and Group 1 (p<0.001), and Group 3 showed significantly lower levels than Group 1 (p<0.05).

Table 3. Means, Standard Deviations and MANOVA for Collective Purpose, Time Structure, Enforced Activity, Status, Social Contact, Financial Strain, Psychological Distress and Satisfaction with Life for the three groups and the total sample.

Variable	Range	Total N = 312 M	SD	Group 1 N = 132 M	SD	Group 2 N= 94 M	SD	Group 3 N = 86 M	SD	F (2, 309)	G 1 vs G 2	G 1 vs G 3	G 2 vs G 3
1. CP	6 - 42	19.36	7.96	18.55	7.42	21.60	8.22	18.16	8.05	5.54**	1<2**	n.s.	2>3*
2. TS	6 - 42	18.68	8.66	16.18	7.75	24.57	8.28	16.06	7.18	38.82***	1<2***	n.s.	2>3**
3. EA	4 - 28	11.20	4.75	11.64	4.76	10.71	4.63	11.06	4.85	1.11	n.s.	n.s.	n.s.
4. ST	3 - 21	7.95	3.67	7.91	3.26	7.97	3.93	7.99	3.98	0.01	n.s.	n.s.	n.s.
5. SC	6 - 42	23.43	8.74	23.77	8.9	23.52	9.10	22.82	8.64	0.31	n.s.	n.s.	n.s.
6. FS	6 - 42	30.42	9.71	25.54	10.35	34.05	7.55	33.94	7.29	35.26***	1<2***	1<3***	n.s.
7. PD	0 - 36	12.48	6.21	11.25	5.57	15.04	5.98	11.56	6.61	12.33***	1<2***	n.s.	2>3**
8. SWL	5 - 35	19.29	6.93	21.61	6.52	16.18	6.03	19.13	7.17	18.76***	1>2***	1>3*	2<3*

Note. Group 1 = Employed participants; Group 2 = unemployed participants not attending a course; Group 3= unemployed participants attending a course. CP = Collective Purpose; TS= Time Structure; EA= Enforced Activity; ST= Status; SC= Social Contact; FS= Financial Strain; PD= Psychological Distress; SWL= Satisfaction with Life; n.s. = non-significant; * $p<0.05$; ** $p<0.01$, *** $p<0.001$

4.4 Predicting psychological distress and satisfaction with life

Two hierarchical multiple regression analyses were used for predicting psychological distress and satisfaction with life utilizing only the unemployed sample (group 2 and group 3 which are the target groups of the present study). In order to control for the effects of socio-economic level, family characteristics, gender, course attendance and length of unemployment in both analyses, these variables were entered into the regression as the first step. Dummy variables were created for socio-economic level (baseline = high) and family characteristics (baseline = married or married de facto with dependents). Financial strain was entered as the second step, and deprivation on the five latent employment benefits, perceived age discrimination and employment commitment were included as the third step.

The results of the first analysis show that together the variables accounted for 29% of the variance in psychological distress. Socio-economic level, family characteristics, gender, course attendance and length of unemployment (Model 1) made a significant contribution in predicting psychological distress, F (8, 156) = 2,08, $p<0.05$, and accounted for 9% of the total variance, although of these variables, only course attendance was significantly related to psychological distress. Adding financial strain was able to significantly predict psychological distress, F (1, 155) =16,68, $p < 0.001$ (Model 2) and to account for a further 9% of the total variance. Adding the five latent benefits of employment, perceived age discrimination and employment commitment (Model 3) was also able to predict psychological distress, F (7,148) =3,22, p<0.01 and to account for a further 11% of the total variance, although the significant individual predictors were only Enforced Activity and Time Structure.

The results of the second analysis show that together the variables accounted for 30% of the variance in satisfaction with life. Socio-economic level, family characteristics, gender, course attendance and length of unemployment (Model 1) did not make a significant contribution in predicting satisfaction with life, F (8,156) = 1,70, p>0.05, although accounting for 8% of the total variance. The addition of financial strain was again able to significantly predict psychological distress, F (1, 155) = 25,59, p < 0.001 (Model 2) and to account for a further 13% of the total variance. The addition of the five latent benefits of employment, perceived age discrimination and employment commitment (Model 3) was also able to predict satisfaction with life, F (7,148) = 2,55, p<0.05 and accounted for a further 9% of the total variance, although the only significant individual predictor was perceived age discrimination.

5 Discussion

The scale resulting from the adaptation of the Latent and Manifest Benefits (LAMB) Scales to the Portuguese Middle-Aged Population showed good psychometric properties. Its structure, although presenting one more factor (with three items and low reliability coefficient) is similar to the original. Nevertheless, as this is, to our knowledge, the first use of the LAMB Scales in Portugal, further research is needed to evaluate the factor structure identified in this study, using other samples - covering younger unemployed, for example - and confirmatory factor analysis procedures.

Results showed that hypothesis 1 was only partially supported, as the deprivation of all latent benefits was significantly negatively associated to satisfaction with life, but only deprivation of time structure, enforced activity and collective purpose were significantly positively correlated with psychological distress. No correlations were found between status and social contact and psychological distress. Hypothesis 2 was supported, as the deprivation of the manifest benefit (financial deprivation) was significantly negatively associated to Satisfaction with Life, and positively to psychological distress. In relation to the 3rd hypothesis, no significant differences were found between the three groups in the access to enforced activity, status and social contact showing that, independently of having or not having any formal occupation, unemployed can have access to, at least, some of the latent benefits of employment. No significant differences were found between full-time employed and unemployed attending a vocational training course in any of the latent benefits of employment. Both unemployed groups have shown greater financial deprivation than did the employed group. It can thus be assumed that both unemployment benefit and training benefit provided no protection from financial deprivation. These two kinds of material support are in some cases, much lower than the salary individuals received in their former jobs and households can be in serious financial trouble. To this we can probably add the perception of an insecure future, especially when people are not hopeful about finding another job, which is what happens with a great number of middle-aged unemployed.

Results partially support both hypothesis 3 and hypothesis 4, i.e. the relation between occupational status and psychological distress and satisfaction with life. Unemployed participants who were not attending a vocational training course have shown significantly more psychological distress than did the other two groups. However, no significant differences were found between the full-time employed participants and unemployed participants who were attending a course, showing that training courses may have some protective role against psychological distress. Expected differences were found between the three groups with relation to satisfaction with life: Unemployed participants who were not attending a vocational training course reported the lowest satisfaction with life, followed by the other unemployed group. Employed participants reported the highest satisfaction with life.

Hypothesis 4 was also partially supported as there were no differences between the employed persons and those unemployed attending a training course in the access to the latent benefits of employment. On the other hand, significant differences were found between the two unemployed groups in the access to time structure and collective purpose. It can be assumed that vocational training courses can act as providers of a greater access to time structure and collective purpose.

In relation to the relative contributions of the latent and manifest benefits of employment, employment commitment and perceived age discrimination in predicting psychological distress (hypothesis 5), financial deprivation accounted for 9% of the total variance. From the latent benefits of employment, the significant predictors were enforced activity and time structure, accounting for a larger amount of variance than financial deprivation (11%). Contrary to what was expected, neither employment commitment nor perceived age discrimination predicted psychological distress. Financial deprivation accounted for 13% of the total variance in satisfaction with life, and none of the five Latent Benefits of Employment emerged as significant predictors of this variable. The only significant individual predictor after financial deprivation was perceived age discrimination, and, once more, employment commitment was not a significant predictor. The fact of employment commitment has not emerged as a significant predictor of psychological distress and satisfaction with life may be related to the restriction of range that could have reduced the observed correlations with the well-being variables and also contributed to the scale's low alpha coefficient (.60).

6 Conclusion

The results of this study and other empirical evidence highlight the need to reconsider Jahoda's (1982) emphasis on paid employment as a provider of access to the latent benefits, and the primary role of latent benefits deprivation for the decline in wellbeing. Results showed that latent benefits of employment can also be obtained by the unemployed people, though not all to the same extent as by employed people of the same age group. Training courses, for example, can be considered as another social institution that provides daily situational experiences where people may have access to the latent benefits.

Enforced activity and time structure were the only two of the latent benefits that significantly predicted psychological distress, and accounted for a larger amount of variance than financial deprivation, giving some partial support to Jahoda's (1982) perspective. On the other hand, none of the latent benefits of employment was able to significantly predict satisfaction with life, while financial deprivation did, giving support to Fryer's (1986) assumption that it is the financial distress that primarily accounts for the decline in psychological well-being among the unemployed. Taking these findings into account, it is preferable to consider models that incorporate both latent and manifest benefit variables, providing a more complete explanation of psychological well-being in unemployed individuals than to consider one-dimensional models.

For future research, there is a need for more qualitative research in this area, as an alternative methodology to supplement quantitative findings. Longitudinal research would also allow for conclusions regarding causal relations between variables.

This study has some implications for intervention. For example, the importance of helping unemployed people to identify and participate in meaningful activities that may constitute constructive alternatives to employment and that may have a protective role against psychological distress. The role of financial deprivation also needs to be considered: it is important to help individuals to build up both short-term budgeting and long-term financial planning skills, and governments should keep unemployment benefits or other kinds of support - in goods, for example - available to the unemployed which face financial strain.

Acknowledgements

This study is part of a larger research project that is supported by the Fundação para a Ciência e Tecnologia/MCES under SFRH/BD/17516/2004. The authors would like to thank the participants at the "Unemployment and Health" Special Session, ICOH Conference, Milan, 11-16 June 2006, for helpful comments.

References

Bailey, L. III & Hansson, R. (1995). Psychological obstacles to job or career change in late life. *The Journals of Gerontology, Psychological Sciences and Social Issues, Series B, Vol.50* (6), 280-288.

Banks, M., Clegg, C., Jackson, P., Kemp, N., Stafford, E. & Wall, T. (1980). The use of the general health questionnaire as an indicator of mental health in occupational studies. *Journal of Occupational Psychology, 53,* 187-194.

Brief, A., Konovsky, M., Goodwin, R. & Link, K. (1995). Inferring the meaning of work from the effects of unemployment. *Journal of Applied Social Psychology, 25,* 693-711.

Broomhall, H. & Winefield, A. (1990). A comparison of the affective well-being of young and middle-aged unemployed men matched for length of unemployment. *British Journal of Medical Psychology, 63,* 43-52.

Creed, P. & Evans, B. (2002). Personality, well-being and deprivation theory. *Personality and Individual Differences, 33,* 1045-1054.

Creed, P. & Klisch, J. (2005). Future outlook and financial strain: Testing the personal agency and latent deprivation models of unemployment and well-being. *Journal of Occupational Health Psychology, 10* (3), 251-260.

Creed, P. & Macintyre, S. (2001). The relative effects of deprivation of the latent and manifest benefits of employment on the well-being of unemployed people. *Journal of Occupational Health Psychology, 6* (4), 324-331.

Creed, P. & Muller, J. (2006). Psychological distress in the labour market: Shame or deprivation? *Australian Journal of Psychology, 58* (1), 31-39.

Creed, P. & Reynolds, J. (2001). Economic deprivation, experiental deprivation and social loneliness in unemployed and employed youth. *Journal of Community and Applied Social Psychology, 11* (3), 167-179.

Diener, E., Emmons, R., Larsen, R. & Griffin, S. (1985). The satisfaction with life scale. *Journal of Personality Assessment, 49,* 71-75.

Erikson, E. (1980). *Identity and the life cycle.* New York : W. W. Norton & Company.

Evans, S. & Haworth, J. (1991). Variations in personal activity, access to categories of experience and psychological well-being in unemployed young adults. *Leisure Studies, 10,* 249-264.

Evans, S. & Banks, M. (1992). Latent functions of employment: Variations according to employment status and labour market. In C. Varhaar & L. Jansma (Eds.), *On the mysteries of unemployment* (pp. 281-295). Dordrecht: Kluwer Academic.

Fryer, D. (1986). Employment deprivation and personal agency during unemployment: A critical discussion of Jahoda's explanation of the psychological effects of unemployment. *Social Behaviour, 1,* 3-23.

Goldberg, D. (1972). *The detection of psychiatric illness by questionnaire.* London: Oxford University Press.

Gorsuch, R. (1997). Exploratory factor analysis: Its role in item analysis. *Journal of Personality Assessment, 68* (3), 532-560.

Hanisch, K. (1999). Job loss and unemployment research from 1994 to 1998: A review and recommendations for research and intervention. *Journal of Vocational Behaviour, 55,* 188-220.

Hoare, P. & Machin, M. (2004). *Self-Esteem, affectivity and deprivation:Predictors of well-being in the unemployed.* Paper presented at the 19th Annual Conference of the Society for Industrial and Organisational Psychology, 2-4 April 2004, Chicago.

Jackson, P., Stafford, E., Banks, M. & Warr, P. (1983). Unemployment and psychological distress in young people: The moderating role of employment commitment. *Journal of Applied Psychology, 68* (3), 525-535.

Jahoda, M. (1982). *Employment and unemployment. A socio-psychological analysis.* Cambridge University Press.

Jahoda, M., Lazarsfeld, P. & Zeisel, H. (1981). *Marienthal. The sociography of an unemployed community.* Chicago: Aldine Atherton.

Kanfer, R., Wanberg, C. & Kantrowitz, T. (2001). Job search and employment: A personality-motivational analysis and meta-analytic review. *Journal of Applied Psychology, 86* (5), 837-855.

Lent, R., Hill, C. & Hoffman, M. (2003). Development and validation of the counselor activity Self-Efficacy Scales. *Journal of Counseling Psychology, 50* (1), 97-108.

McKee-Ryan. F., Song, Z., Wanberg, C. & Kinicki, A. (2005). Psychological well-being during unemployment: A meta-analytic study. *Journal of Applied Psychology, 90* (1), 53-76.

Miles, I. (1983). *Adaptation to unemployment.* Science Policy Research Unit. Brighton: University of Sussex.

Muller, J., Creed, P., Waters, L. & Machin, M. (2005). The development and preliminary testing of a scale to measure the latent and manifest benefits of employment. *European Journal of Psychological Assessment, 21* (3), 191-198.

Murphy, G. & Athanasou, J. (1999). The effect of unemployment on mental health. *Journal of Occupational and Organisational Psychology, 72* (1), 83-99.

Neto, F. (1997). Escala de satisfação com a vida: Propriedades psicométricas numa amostra de adolescentes (Satisfaction with life scale: Psychometric properties in a sample of adolescents). In F.F. Neto (Ed.), *Estudos de Psicologia Intercultural: Nós e os Outros.* (pp. 347-359). Lisboa: Fundação Calouste Gulbenkian e Junta Nacional de Investigação Científica e Tecnológica.

Sargeant, M. (2001). Lifelong learning and age discrimination in employment. *Education and the Law, 13* (2), 141-153.

Vansteenkiste, M., Lens, W., De Witte, H. & Feather, N. (2005). Understanding unemployed people's job search behaviour, unemployment experience and well-being: A comparison of expectancy-value theory and self-determination theory. *British Journal of Social Psychology, 44,* 269-287.

Wanberg, C., Griffiths, R. & Gavin, M. (1997). Time structure and unemployment: A longitudinal investigation. *Journal of Occupational and Organisational Psychology, 70,* 75-95.

Warr, P. (1987). *Work, unemployment and mental health.* Oxford: Clarendon Press.

Warr, P. (1998). Age, work and mental health. In K. Schaie & C. Schooler (Eds.), *Impact of work on older workers* (pp. 252-296). New York: Springer.

Wiener, K., Oei, T.P. & Creed, P.A. (1999). Predicting job seeking frequency and psychological well-being in the unemployed. *Journal of Employment Counseling, 36* (2), 67-81.

Time Structure or Meaningfulness? Critically Reviewing Research on Mental Health and Everyday Life in Unemployment

Benedikt G. Rogge

Introduction

Currently, the relationship of everyday life and health receives increasing attention not only in the field of unemployment research (Waters & Moore, 2002). Whilst sociological and psychological health researchers pay more and more attention to the conducive and detrimental effects of everyday life to physical and mental health (e.g. Thoits, 2006; Singh-Manoux, Richards & Marmot, 2003), scholars from the fields of leisure and everyday life research have augmented their interest in negative life events and such variables as life satisfaction, stress, and coping processes (e.g., Iwasaki & Schneider, 2003; Kleiber, Hutchinson & Williams, 2002). The significance of everyday life to health and vice versa today clearly enjoys wide and still growing recognition.

This chapter addresses the question what particular aspects of everyday life influence mental health. One concept that has attracted academic concern is daily time structure. This especially applies to the field of unemployment research where the work of Marie Jahoda has made the concept popular. Her notion of time structure refers to the temporal regularity of things that happen in everyday life (Jahoda, 1982). In empirical research, time structure has mostly been related to mental rather than physical health. The prevalent hypothesis and oft-alleged result of this research is that time structure fosters mental health (McKee-Ryan et al., 2005).

In this chapter, by contrast, I will argue that this assumption is at odds with empirical reality. Not only do the respective studies show highly inconsistent results but those studies that seem to confirm the association of time structure and mental health actually contest it. Rather than for an 'objective' temporal regularity, as postulated by Jahoda, these investigations only deliver evidence for the meaningfulness of everyday life to be associated with mental health. In fact, time structure does not correlate at all with mental health. The reason why this construct yet continues to be regarded as a core predictor of mental health in psychological unemployment research lies in shortcomings in the respective studies. As I will show, an undifferentiated theoretical conception of time structure and, as a corollary, inadequate operationalisations and data analyses have led to severe impairments of construct validity in the investigations in question. Instead of researching time structure, these studies deal with the meaningfulness of everyday life. As qualitative studies however confirm, time structure may serve as a vehicle to facilitate the pursuit of meaningful activities with some persons. But it is the notion of meaningful everyday life that is crucial to understanding mental health in unemployment and other situations of potential deprivation.

After outlining previous theorising on time structure and mental health (section 2), I will deliver a thorough analysis of the empirical state of the art and its theoretical and methodical shortcomings (section 3). Then, I will depict research on the meaningfulness of everyday life and argue for making it a core category in the study of unemployment and mental health (section 4). Finally, I will summarise the chapter (section 5).

1 Theorising the link of time structure and mental health

1.1 The conception proposed by Marie Jahoda

Historically speaking, the link of time structure and mental health in the social sciences goes back to the 1930s. While Durkheim's *Elementary Forms* and *Suicide*, as is well-known, were fructified for the dawn of the sociology of time (Sorokin & Merton, 1937) and research on mental health (Nordenmark & Strandh, 1999, p. 581) respectively, it was yet in the field of unemployment research that time structure was explicitly connected to mental health. In the pioneering Marienthal study Jahoda, Lazarsfeld & Zeisel (1933/1975) described the erosion of time structure as one of the main deleterial effects of unemployment. The time that was freed from employment was stated to turn into a "tragic gift" since the unemployed allegedly lost the capacity to use their time (Jahoda, Lazarsfeld & Zeisel, 1933/1975, p. 83). The absence of time structure was thought to cause feelings of vacuity and apathy.

Later, Jahoda (1981, 1982) proposed her well-known 'latent deprivation model'. Beyond the manifest benefit to earn a living, in this model she postulated five latent psychosocial functions of employment to which corresponded "deep-seated needs" in all human beings (Jahoda, 1982, p. 83). The five needs she claimed were the needs for time structure, social experience, collective purpose, status and identity, and regular activity (Jahoda, 1982, p. 59).[1] Among these needs or benefits, Jahoda deemed time structure to be the most important one (Jahoda, 1982, p.85; see also Creed & Bartrum, 2006, p. 3), for she argued all human beings needed "a supportive frame within which [to] shape their individual lives" (Jahoda, 1982, p. 23).

Yet, Jahoda's comments on the character of the five "enduring needs" remained contradictory. Although she spoke of "psychological needs […] which are deeply anchored in the social norms of this society" (Jahoda, 1982, p. 91-92), she sometimes attributed them an anthropological status or a genetic foundation (Jahoda, 1982, p. 84; see also Haworth, 1997, p. 26-27).[2] This notwithstanding, the decisive point about her perspective is that time structure as much as the satisfaction of the remaining needs can only adequately be attained through the institution of employment (Jahoda, 1982, p. 58).[3] In other words, Jahoda's very concept of time structure goes back to her embracement of the temporal regularity that full time employment imposes on everyday life.

1 Later, Jahoda (Jahoda & Fryer, 1986, p. 110) also included 'control' over one's life circumstances as another latent function of employment (see Kieselbach & Beelmann, 2006, p. 16).

2 It is however noteworthy that Jahoda herself did actually not view her model as a proper theory (Wacker, in this volume). Yet, her model was clearly awarded a theoretical status by researchers (Creed & Bartrum, 2006).

3 This reminds of Arnold Gehlen's anthropological reflections on this point. Gehlen thinks of ennui as a constitutive part of human existence which should be countered by the stabilising effects of institutionalised action and work (see Lepenies, 1969, p. 232-237). Jahoda's comments resemble this thought.

Her belief that temporal regularity is required for maintaining good mental health is thus a direct corollary of her employment-centred approach (Haworth, 1986, p. 282). This is why the lack of temporal regularity in situations of non-employment was stated to entail detrimental effects on mental health.

1.2 Consequences and criticisms of Jahoda's conception

Since more than 20 years, the latent deprivation model has given rise to a truly impressive amount of empirical studies and continues to do so. Till this very day, it ranks as the most influential approach to understanding the negative mental health consequences of job loss (Creed & Bartrum, 2006). With regard to time structure, in fact, there is barely any study that does not make direct reference to the work of Jahoda (see below). Complementing Jahoda's model by the explicit claim to pay attention to the differential processes in everyday life (Wanberg & Marchese, 1994; Wacker, 1983)[4], psychological scholars currently conceive of time structure as one moderator variable that buffers the impact of unemployment on mental health. In line with Jahoda, the persistent argument is that "mental health is higher for those who are able to impose daily routines on their lives […] and to use their time in a structured way" (McKee-Ryan et al., 2005, p. 68). Today, this Jahodian idea can be found in most text books and research overviews on unemployment and mental health (see Kieselbach & Beelmann, 2006; McKee-Ryan et al., 2005; Winefield, 2002; Wanberg, Kammeyer-Mueller & Shi, 2001, p. 255-256; Hanisch, 1999).

But the association of time structure and mental health is far from being a truism. Jahoda's assumptions have attracted a variety of criticisms. The critique was essentially directed at its overemphasis on deprivation rather than agency, its romanticist view of the institution of employment, and its neglect of other life spheres such as leisure and the family (e.g. Wacker, in this volume; Ezzy, 1993; Fryer, 1986; Haworth, 1986; Artazcoz et al., 2004). In opposition to Jahoda, David Fryer's agency-restriction model focused on people's self-determination, proactivity and goal-directed behaviour (Fryer & Payne, 1984, p. 290; Fryer, 1986) instead of their psychosocial deprivation. The subsequent controversy between Jahoda and Fryer exemplified two polar positions in unemployment research (Jahoda & Fryer, 1986). Ezzy (1993) has described the two positions as a sociological reductionism and a psychological reductionism respectively. While Jahoda clearly overemphasised the status of institutions, namely of employment, Fryer fell prey to overstating the importance of self-determination and agentic processes in the individual. Today, psychological researchers attempt to integrate the two positions (e.g. Muller, Creed, Waters & Machin, 2005). In this context, however, the relevant question is: How have the criticisms of Jahoda impacted the view of time structure in unemployment research?

Ultimately, they have not at all. Whereas Fryer's suggestions, for example, stood in contrast with Jahoda's focus on employment, they did not question the importance of time structure for mental health (Fryer & Payne, 1984, p. 289). Then, Haworth (1997, 1986) while seeking to demonstrate the value of leisure, as opposed to employment, as a potential source of mental health, even adopted Jahoda's five needs as 'categories of experience'. Moreover, the recent life-facet model of unemployment by McKee-Ryan

4 Note that, contrary to some accounts of her work, the assumption of differentiality can also be found in Jahoda herself (Jahoda, 1982, p. 85), as well as with illustrations on a group of good copers in the Marienthal study (Jahoda, Lazarsfeld & Zeisel, 1975, p. 108-110; see also Fryer's comments on the subtext of unemployment research: Fryer, 1986).

and Kinicki (2002; see also Latack, Kinicki & Prussia, 1995) also takes over the notion of time structure as critical to mental health. The same goes for the scale of manifest and latent benefits that has recently been developed by Muller et al. (2005). Disagreeing with Jahoda on her exclusive accent on employment and the latent benefits, these researchers apparently agree with her view of time structure.

Only some researchers have casted doubt on the status of time structure in unemployment research (Hanisch, 1999, p. 94-95; Lang-von Wins et al., 2004, p. 1137; Mohr & Müller, 2000, p. 62), and there is only one theoretical approach that has criticised awarding time structure the status of a core predictor of mental health, namely the vitamin model by Peter Warr (1987). Although Warr's model does incorporate time structure into one of his nine features of environment, namely the feature "externally generated goals", he does not assume that time structure has a separate effect on mental health. More importantly, Warr suggests a non-linear relationship between the characteristics of the environment and mental health (Warr, 1987, pp. 211, 214, 281-283). In other words, Warr did explicitly not regard time structure as a simple moderator of mental health. Nevertheless, Warr's critique did not penetrate to the academic discourse. The claim that time structure is one – in Jahoda's model in fact: the most – crucial psychosocial explanans of mental health remained largely uncontested in the research community. Whereas Jahoda's model is extensively debated, her time structure postulate, albeit a direct derivation of her overemphasis on the institution of employment, was seldom questioned.

This is all the more astonishing as the relationship between time structure and mental health is far from insinuating a simple linear association (see Feather & Bond 1994, p. 135-136). For instance, Weber's comments on the iron cage (1986/1920), Foucault's description of the discliplinisation of time (1977) or Goffman's work on total institutions (1961) all could have enhanced the awareness that time structure may well turn into a prison rather than a 'supportive frame'. Yet, Warr's idea of a non-linear relationship between time structure and mental health (Warr, 1987, p. 112) did not come through.[5] Empirical research still, if often implicitly, follows the assumption of a linear association. Yet, it has so far failed to deliver evidence for any association at all.

2 Empirical evidence on time structure and mental health

2.1 Quantitative unemployment studies

Some unemployment researchers have operationalised time structure with standardised measures and correlated the results with scores of mental health scales. The three most popular psychometric measures shall be dealt with here. These are: the *Time Structure Questionnaire* (TSQ) by Feather and Bond (1983; Bond & Feather, 1988); the *Access to Categories of Experience Scale* (ACE) by Evans (1986; Evans & Haworth, 1991); and the recent *Latent and Manifest Benefits Scale* (LAMB) by Muller et al. (2005). Whilst the TSQ has been constructed to measure time structure only, the remaining two scales contain operationalisations of all five needs postulated by Jahoda.

The first observation to be made is that an important portion of quantitative studies has failed to find any association of time structure and mental health. After control-

5 One may speculate how this has come about. I presume it is due to a modernistic bias inherent to the respective research (see, e.g., the work of Weber). Yet, this assumption would of course require further elaboration.

ling for financial strain and future outlook (optimism/pessimism), in a LAMB-study Creed and Klisch (2005) find that the five Jahodian constructs do not exert any significant influence on mental health (see also Hoare & Machin, 2006). Likewise, in an ACE-study with young unemployed, Evans & Haworth (1991) did not find any correlation between time structure and any measure of psychological well-being. Moreover, neither Haworth & Ducker (1991) found any correlation in their ACE-study, nor did Creed and Watson (2003) in another recent LAMB-study with 386 unemployed from various age groups. Note that all these studies reported zero correlations.[6] Finally, a number of German studies could equally not find any interpretable association of the two variables (Wacker, personal communication; Kuhnert, 1999; Bleich & Witte, 1992).

Still, there is an important number of studies that suggest a positive association of time structure and mental health. Yet, these findings are distorted by a lack of conceptual clarity and a deficit of construct validity in the respective measures of data collection. This deficit is due to a theoretical misconception of time structure which has then translated into shortcomings at the level of operationalisation and data aggregation. This misconception goes back to the confusion of time structure with other time-related constructs. It is yet essential to distinguish between a variety of disparate time-related constructs, among which at least the following ones[7]: a) Time use (relates to content of activities); b) pace of life (relates to speed); c) time perception (relates to passage of time); d) temporal orientation (relates to time perspective); e) goal formation and attainment (e.g., planning, decision-making); f) time-related feelings and emotions (e.g., boredom, impatience, apathy); g) strategies of time management; and h) time structure (relates to temporal regularity, Jahoda's concept).

It is beyond the scope of this chapter to go into further detail with these constructs. The crucial point however is that, albeit some of them might be empirically interrelated, these constructs are conceptually distinct. With regard to our subject, this must particularly be held for temporal regularity and time-related emotions such as boredom, apathy, vacuity or the feeling that "time drags". Indeed, a variety of unemployment studies confirms that job loss may entail such feelings (e.g. Muller et al., 2005; Kuhnert, 2004; Brinkmann & Wiedemann, 1994; Hess, Hartenstein & Smid, 1991). And evidently, their occurrence is negatively correlated with mental health. But these feelings are *different* from time structure and must not be equated with it. As I will show in the following, this is what has yet happened in all three mentioned measures of time structure.

2.1.1 The Time Structure Questionnaire (TSQ)

The Time Structure Questionnaire (TSQ) by Feather & Bond (1983; Bond & Feather, 1988), a standardised 26 item self-report questionnaire, is the only scale that was specifically developed to measure time structure. TSQ-studies have consistently yielded positive correlations of time structure and mental health (see Feather & Bond, 1994, 1983; Bond & Feather, 1988; Waters & Muller, 2003; Waters & Moore, 2002; Jackson, 1999; Wanberg, Griffiths & Gavin, 1997).[8] The meta-analysis by McKee-Ryan et al. (2005) also predominantly draws on TSQ-studies.

The authors of the TSQ define time structure as "the degree to which individuals perceive their use of time to be structured and purposive" (Bond & Feather, 1988, p. 321). Here, at the definitional level, two distinct constructs are mixed up that are 'time

6 In the study by Creed & Watson (2003), for instance, the correlation amounted to r = -.04.
7 I draw on and extend the work of McGrath & Tschan (2004: 28; 47-48).
8 Schatz (2000, 1999) has developed and validated a German version of the TSQ, that has however not been applied in mental health related research.

structure' and 'time-related feelings' or rather 'general emotions'. This fundamental confusion is perpetuated at the level of operationalisation and, later, data aggregation. For the five TSQ-subscales, gained through factor analysis, measure entirely heterogeneous constructs. These have been labelled 'structured routine', 'sense of purpose', 'present orientation', 'effective organisation' and 'persistence'. Note that exclusively the subscale 'structured routine' – operationalised through items such as "Do you have a daily routine which you follow?" and "Do your main activities during the day fit together in a structured way?" – measures Jahoda's notion of time structure. By contrast, the subscale 'sense of purpose' – measured through items such as "Do you often feel that your life is aimless, with no definite purpose?" and "Do you ever feel that the way you fill your time has little use or value?" – clearly relates to some life satisfaction or sense of coherence construct (see e.g. Antonovsky, 1987). The others refer to 'temporal orientation' and 'strategies of time management'.

As Mudrack (1997, p. 231) has shown, the TSQ-subscales are not only theoretically but also statistically independent: They do not display any intercorrelations. Hence, an aggregation of the subscale scores to an overall TSQ-score severely reduces the construct validity and in fact renders the scores "relatively meaningless" (Mudrack, 1997, p. 231). Nonetheless, almost all TSQ-studies have drawn on such aggregate scores when correlating them with mental health scores. In other words, the correlation of time structure and mental health that has been found in most TSQ-studies *must not* be traced back to time structure but is accounted for by a host of heterogeneous constructs. Mudrack (1997) therefore explicitly criticises proceeding so and claims researchers should deal with TSQ scores at subscale level only.

But two studies have analysed the TSQ results at subscale level and their results are highly surprising: The factor 'structured routine' – standing for Jahoda's concept of time structure – did not correlate with mental health and depression scores at all. Once more, zero correlations were found (between $r = -.05$ and $r = -.10$ for psychological distress and depression; see Bond & Feather, 1988: 326; Feather & Bond, 1994, p. 131). By contrast, the subscale 'sense of purpose' did strongly correlate with the dependent variables (between $r = -.30$ and $r = -.42$ with $p < .001$ for psychological distress and depression). Thus, the correlations of the overall TSQ scores with mental health go essentially back to the 'sense of purpose' scale. So the subtext of the TSQ studies says that 'sense of purpose' is related to mental health but time structure is not. In other words: temporal regularity in everyday life is not associated with mental health, whereas the feeling that one fills everyday time in a meaningful way is. This subtext of the TSQ-studies has not been taken notice of so far. Yet, it is at odds with Jahoda's original assumption that time structure is of vital importance to mental health.[9]

2.1.2 The Access to Categories of Experience Scale (ACE)

The Access to Categories of Experience Scale (ACE) by Evans (1986) and Evans and Haworth (1991) respectively contains operationalisations of all five latent benefits as postulated by Jahoda, each subscale encompassing three items. Among the existing ACE-studies (e.g. Creed & Evans, 2002; Haworth & Paterson, 1995; Haworth & Hill,

9 Note that, in a scarcely received publication, the authors of the TSQ, Feather and Bond (1994) themselves, admit the problematic character of their subscale "structured routine". They comment they would need to "conduct a conceptual analysis of the factor that we called Structured Routine" (Feather & Bond, 1994: 136). Till this day, such an analysis has not been conducted. Nor has the conclusion been drawn that the TSQ does not deliver any evidence for the hypothesis that Jahoda's concept of time structure is relevant to mental health.

1992; Haworth & Ducker, 1991; Haworth & Evans, 1987), only few went down to the subscale level in order to analyse, inter alia, the correlation of time structure and mental health. As to the findings, ACE-studies are less consistent than the TSQ-studies described before. Neither Evans and Haworth (1991), nor Haworth and Ducker (1991) found a correlation between time structure and any of their measures of psychological well-being. Anyway, as Creed and Machin (2003) found out, a factor analysis of the ACE yields no separate factor for time structure. Thus, the ACE does actually not contain any dimension in its own right that reflects time structure. Besides, other studies could equally not replicate the factor structure suggested by Evans and Haworth (Waters & Moore, 2002; Creed & Macintyre, 2001). I will therefore not deal with ACE-studies in further detail here.[10]

2.1.3 The Latent and Manifest Benefits Scale (LAMB)

The depicted deficit in the ACE was one motive for the recent construction of The Latent and Manifest Benefits Scale (LAMB) by Muller et al. (2005). The authors also complemented Jahoda's five latent psychosocial categories by the category of 'financial strain', drawing on Fryer's agency-restriction model. Until now, only few LAMB-studies have been published (e.g. Hoare & Machin, 2006; Muller et al., 2005; Creed & Watson, 2003). Again, their findings are contradictory. In the testing study with 250 participants, Muller et al. (2005, p. 197) found the time structure subscale to display a significant correlation of $r = -.18^{**}$ with General Health Questionnaire Scores (GHQ). Likewise, Hoare and Machin (2006) report an association albeit without indicating its size.

As to the LAMB, however, a criticism similar to the one of the TSQ is in order. Resembling the TSQ-subscale 'sense of purpose', the LAMB-subscale for 'time structure' equally taps into affective measures. Albeit perhaps not correlated with those scales of negative affectivity employed by the authors, three of the six items operationalising time structure clearly collect affective appraisals from the participants. This has to do with the very formulations of the items: Two of them assess the participant's perception of discrepancies between her/his ideal and her/his real everyday life: "I often wish I had *more* things to do to fill up the time in my days" and "There is usually *too much* spare time in my day" (emphases mine). These formulations clearly grasp emotional or affective constructs. Another item belongs to the category of 'time-related feelings and emotions' (item: "time drags"), while the remaining three measure the level of general activity (items: "I usually keep busy"; "have nothing to do"; "have a lot of time on my hands"). Recalling the differentiations made earlier 'time-related feelings and emotions', 'time use', 'level of activity' and 'time structure' are definitely distinct constructs that may not be confused.

The fact that these constructs are mixed up is further illustrated by the verbal overlap in the formulations of LAMB-items with items from the Boredom Proneness Scale (Farmer & Sundberg, 1986) and the Leisure Boredom Scale (Iso-Ahola & Weissinger, 1990). Both of these scales declaredly measure an affective construct that is boredom. In the former you find the items: "Much of the time I just sit around doing nothing." and "I often find myself with nothing to do – time on my hands." (Farmer & Sundberg, 1986, p. 6), whereas the latter contains the items: "For me, leisure time just drags on and on." and "I am very active during my leisure time." (Iso-Ahola & Weissinger, 1990, p. 9). We may therefore conclude, the LAMB does not deal with temporal regularities

10 Research with a German adaptation of the ACE reported the same results (Wacker, personal communication).

but instead measures affective states. These are naturally correlated with measures of mental health. However, this statistical association does not say anything about time structure.

Summing up the critique of all previous quantitative studies on mental health and time structure in unemployment, we may state that the only psychometric measure that operationalises time structure in the sense of 'temporal regularity', as suggested by Jahoda, is the subscale 'structured routine' in the TSQ. Due to a severe lack of construct validity, the other scales do not measure time structure but time-related emotions and affective states. The TSQ-subscale 'structured routine' however is not associated with mental health (zero correlations) while the TSQ-subscale 'sense of purpose' strongly is. Drawing on aggregate TSQ-scores, previous studies have seemingly delivered evidence that time structure is associated with mental health while in fact indicating that it is not but sense of purpose is.

2.2 Quantitative studies with other populations

There is a limited number of quantitative studies with other populations than the unemployed. For instance, researchers used the TSQ to investigate the relevance of time structure to the willingness to retire, to worry and to mental health in working adults (Kelly, 2003), or to compare the unemployed's life satisfaction with the one of students and employed individuals (Martella & Maas, 2000; see also George, 1991; Schatz, 1999, 2000). Creed and Evans (2002) used the ACE to transfer Jahoda's assumptions to the population of students, relating aggregated latent deprivation scores to mental health.[11] Of course, to these studies the same critique applies as uttered above. It is however interesting that some of the authors have explicitly problematised the construct of time structure and its relation to mental health. For instance, Martella and Maass (2000, p. 1107) point out that "the general concept of time structure consists of multiple factors" and beyond "the 'perceived sense of purpose' factor may be more predictive of life satisfaction than the 'structured routine' factor". Still, as highlighted before, it was Mudrack (1997, 1999) who conducted TSQ-studies with student samples and hinted at the uselessness of aggregated TSQ-scores. Therefore, the argument that time structure is not associated with mental health, is also backed up by studies with other populations.

2.3 Qualitative studies

The notion that time structure plays a vital role for mental health had originally been taken from qualitative unemployment studies. And at first sight, there seems to be some face validity in this contention. Swinburne (1981, p. 53), for example, reported that "the role of setting a routine in maintaining self-directed activity provided a key to understanding the ease with which this problem was handled". He relates the absence of a routine to feelings of boredom, apathy, helplessness, and the loss of a sense of identity (Swinburne, 1981, p. 47). Similarly, in their study on unemployed people in a rural community in Scotland, Bostyn and Wight (1987, p. 149) suggested that depression and apathy in the participants resulted from "an absence of both money and of structure, the latter arising from their inability to differentiate one period of time clearly from anoth-

11 Another interesting result is delivered by the German psychologist Hinz (2000). With a student population, he found out that time planning is positively correlated to academic success but also to the level of stress (Hinz, 2000, p. 172-173). For several (methodical) reasons, this finding must however be regarded with caution.

er". Moreover, Heinemeier (1991) analysed the relationship between unemployment and the erosion of time structure with interview data from the 1970s, while in a small explorative study Mohr and Müller (2000) found confirmation for both a lack of time structure and a decrease of mental health in their interviewees, as Heinemeier had postulated.

However, most qualitative researchers have drawn too hasty a conclusion. They have suggested an association, if not a causal link, between time structure and mental health although they did not systematically deal with the interviewees' emotions but instead restricted themselves to providing sketchy accounts of dissatisfaction, frustration and despair. Notwithstanding the problem of causal explanation in qualitative research, they have secondly rarely looked for counter-examples that are cases in which time structure and mental health might display no association (no time structure and good mental health or conversely). But meantime, there is a number of qualitative studies that do indicate the unrelatedness of the two variables in question. Instead of time structure, they deliver evidence that the meaningfulness of everyday life exerts a crucial influence on mental health in unemployment.

3 Unemployment and meaningful everyday life

In an interview study with 100 unemployed, Kronauer, Vogel & Gerlach (1993, p. 202-203), for example, provide evidence that eroded time structures are not automatically perceived as damaging but may enable people to dispose of their time in a more self-determined manner. In a recent own study (Rogge, Kuhnert & Kastner, 2007) with German long-term unemployed men in late adulthood, we systematically sought to assess both the temporal regularities in the men's everyday lives and their subjective well-being. As was to be expected from the in-depth analysis of quantitative research given above, no association was found between time structure and well-being. Men belonging to all three different types of daily time structure reported negative *and* positive well-being. We had divided the sample into three types of temporal regularity that are 'rigidity', 'temporal frame', and 'unstructuredness'. Rather than the importance of temporal regularities, it was the meaning the men attributed to their current everyday life that proved to be critical for their well-being. This meaning appeared, of course, to arise out of a complex interaction of several factors, among which also time-related constructs, such as the time horizon (see Heinemann, 1982). Beyond, the disposability of alternative social roles, e.g. spouse, family or civic roles, the social embeddedness and the repertoire of potential activities (see also Hepworth, 1980) turned out to influence whether the men deemed their everyday meaningful or not.

Interestingly, without being aware of it, our typology of time structure strongly resembles the one developed by Burzan (2002) in an interview study with retired women and men. Burzan (2002, p. 144), as much as we did, points out that there is no clear-cut let alone linear relationship between the degree of temporal regularity in her participants' everyday lives and their emotional satisfaction. Boredom and frustration as well as a sense of well-being and content can be found in all three time structure groups with this sample of retired persons (likewise Köller, 2006, p. 208). The qualitative evidence thus also backs up the argument that time structure is not directly relevant to mental health. Time structure may function as a vehicle of meaningful undertakings enabling people to pursue activities which provide them with a feeling of purpose. Yet, a number of other contextual and individual factors are more crucial to the 'making' of meaning in everyday life than time structure. As Ezzy (2001, p. 100) has stressed, time structure as much as the other latent benefits is only important to an individual in the light of his/her subjective attribution of meaning to his/her life.

Earlier unemployment studies such as the ones by Kilpatrick and Trew (1985) and Winefield, Tiggemann and Winefield (1992) have investigated mental health and 'purposive activities' in the unemployed that is to say the 'objective' features of everyday activities. They suggested some activities (e.g. attending the church) were naturally more purposive than others (e.g. watching TV). Haworth (1986, p. 294) yet distinguished between 'purposive activities' that are activities which are normatively judged by their content and (subjectively) '*meaningful activities*' that are activities which are perceived as relevant by individuals themselves. I agree with him on pointing out the value of the latter for an individual's mental health. Indeed, an increasing tendency towards researching subjectively meaningful activities can currently be observed in unemployment research. A point in case is the development of a Meaningful Leisure Activity Questionnaire by Waters and Moore (1999), albeit still unpublished, as well as the respective quantitative studies that have been conducted by the authors (Waters & Moore, 2002; Waters & Muller, 2003). These studies underline that the meaning people attribute to their everyday life is of vital importance to mental health.

I think this development opens up promising channels for future research. For in line with Ezzy (1993, p. 49), I argue, mental health emerges as a "product of the meanings given by an individual to their objective social relationships". In other words, we may neither ignore the subjective interpretation and agency of the individual, nor the social and contextual factors that enable and limit both of them when researching mental health. Psychological unemployment researchers have often tended to fall prey to one or the other reductionism in the past. The notion of meaningful everyday life, I hold, is yet a fruitful concept for researching "the interplay between social institutions and individual agency" (Ezzy, 1993, p. 47). By this, it helps to explain the heterogeneity that can be found in the subjective experience of unemployment (Wacker, 1983). I base this suggestion on four arguments.

First, the meaningfulness of everyday life is a concept that draws attention to the individual's interpretation of and action in a situation. What people perceive as meaningful, depends on their individual preferences, values, or identity saliences. As a recent meta-analysis by Paul and Moser (2006) has shown, unemployed people differ in their employment commitment, which, in turn, impacts mental health in unemployment. Thus, their perception of the situation depends on subjective attributions and identity saliences. The notion of meaningful everyday life is however of course not restricted to the sphere of work, as is the notion of employment commitment, but enables us to look at the importance of other life spheres as well such as family, leisure or civic roles. Then, how meaningful an individual thinks his or her everyday life is, is not a stable personality characteristic (as in Paul & Moser, 2006) but subject to change and development. This is not only influenced by the individual's (cognitive) interpretation of the situation but also by his/her agentic capacities and active coping with a situation. It is this aspect that David Fryer has made prominent in unemployment research.

Second, how meaningful an individual finds his/her everyday life also depends on social and contextual parameters. The meaningfulness of activities, as has been shown in a recent qualitative study by Ball and Orford (2002), cannot be thought of as detached from the social environment for social approval and social recognition are of paramount importance for attributing meaning to whatever activity. In a quantitative study, Sheeran and Abraham (1994) have shown that this includes particularly the reflected appraisals of significant others that is how others think of me and of what I am doing. Moreover, in his insightful qualitative study, Ezzy (2001) has shown how people relate to social reference groups when trying to make sense out of their lives. Finally, the social exclusion perspective (Kieselbach, 2006) and the finances-shame model (Starrin & Jönsson, 2006) have underlined the indispensability to pay tribute to the social exclusion of unemployed persons within work society. Today, the establishment of a meaningful everyday in unemployment is supposed to take place given the sociocultural norms of Western achieving societies. These sociocultural conditions in the immediate and wider context and their impact on the unemployed individual need to be taken into account in future research.

Third, the dimension of financial strain can be linked to the concept of meaningful everyday life. As is well-known, financial insecurity and deprivation may constitute a severe menace to the actual or future meaningfulness of life in unemployment (e.g. Creed & Klisch, 2005). Some scholars have argued that financial strain constitutes one core explanation for mental health in unemployment (Fryer, 1986; Starrin & Jönsson, 2006). Notwithstanding the validity of this assumption, it may not be made absolute. For at the conceptual level, a lack of money cannot directly explain a decrease in mental health or health (Behrens, 2006). Rather, financial deprivation unfolds its evidently deleterious effects in relation to the individual's life world. This goes in particular for the successful enactment of social roles, for instance, the role of family father or mother. As Artazcoz et al. (2004) have shown in a social epidemiological study, the fulfilment of family responsibilities in married unemployed represents a major psychological burden, particularly for men. The notion of meaningful everyday life thus permits to take into account how financial deprivation damages an individual's subjective condition.

Fourth, the notion of meaningful everyday life is compatible with congruence models and can therefore easily be related to the occurrence of emotions and ultimately mental health. In unemployment research, albeit from a different perspective than I do[12], Paul and Moser (2006) have suggested to conceive of mental health on a congruence-incongruence continuum. The congruence of an individual's standards and her/his situational perceptions can be linked to positive emotion while the incongruence can be assumed to entail negative emotion (Burke, 2006; Stets, 2003). This can be depicted with the help of cybernetic models (Grawe, 2004). The potential usefulness of cybernetic models to explain mental health in unemployment has recently been underlined by Creed and Bartrum (2006). The notion of meaningful everyday life allows for looking at an individual's holistic state rather than single factors when researching mental health in unemployment. It especially fits the bill for analysing the emergence of positive and negative emotions.

For reasons of scope, these comments must remain sketchy. I have developed them in further detail elsewhere (Rogge, 2007). In a current research project, I attempt to decipher the social/contextual factors and the agentic processes taking place in unemployment to gain a better understanding of the conditions under which individuals suffer from poor mental health and under which good mental health can be maintained. I think research into these conditions is mostly required in the future, as it will help enhance our understanding of how personal suffering from unemployment comes about. Placing such emphasis on meaningful everyday life, however, does not prepare the ground for a voluntaristic justification of blaming the victims (see Kieselbach, 2003).

12 The differences to the suggestions made by Paul & Moser (2006) lie essentially in a focus on a) an individual's entire identity rather than single aspects, b) its potential malleability instead of its postulated stability, and c) the social and cultural context.

Quite the reverse, my argumentation directly links up with the inequal distribution of the capacities and resources to establish a meaningful everyday. Social epidemiology has drawn attention to the fact that as much as material resources are inequally distributed among the social strata, so are social and behavioural resources (Mirowsky & Ross, 2003). Individual agency, so the possibility to "create" a meaningful everyday life, is restricted and impaired by a great number of situational, social, structural and cultural parameters. Ezzy (1993) has postulated to look at the interplay of structure and agency in unemployment research (see also Thoits, 2006). Indeed, it is research into this interplay that will take us closer to understanding mental health problems in unemployment.

4 Summary

In this chapter, I have dealt with the problem of mental health in the everyday life of the unemployed. Following the work of Marie Jahoda, numerous psychological unemployment researchers hold that time structure that is the temporal regularity of everyday life is associated with mental health. In spite of the criticisms Jahoda's employment-centred approach has attracted, this assumption is still wide-spread in the psychological research community. In line with the theoretical argument of a small number of psychologists, among which Peter Warr, I have shown that relevant empirical research in fact does not support the notion that time structure is important for mental health. Due to severe shortcomings at the levels of conceptualisation, operationalisation and data aggregation, this notion has however survived in the psychological debates. What quantitative studies with unemployed and other samples yet evidence is the relevance of the meaningfulness of everyday life to mental health. I have advocated using this concept as a core category for the analysis of mental health in unemployment. While time structure may serve as a vehicle for pursuing meaningful activities with some persons, qualitative research has demonstrated that the meaningfulness of everyday life and thus mental health is independent from time structure. I therefore argue that the concept of meaningful everyday life is appropriate for understanding the interplay of agentic and social/contextual processes and the emergence of good and poor mental health in unemployment.

References

Antonovsky, A. (1987). *Unraveling the mystery of health. How people manage stress and stay well.* San Francisco: Jossey-Bass.

Artazcoz, L., Benach, J., Borrell, C. & Cortès, I. (2004). Unemployment and mental health: Understanding the interactions among gender, family roles, and social class. *American Journal of Public Health, 94,* 82-88.

Ball, M. & Orford, J. (2002). Meaningful patterns of activity amongst the long-term inner city unemployed: A qualitative study. *Journal of Community & Applied Social Psychology, 12,* 377-396.

Behrens, J. (2006). Meso-soziologische Ansätze und die Bedeutung gesundheitlicher Unterschiede für die allgemeine Soziologie sozialer Ungleichheit. In M. Richter. & K. Hurrelmann (Eds.), *Gesundheitliche Ungleichheit. Grundlagen, Probleme, Perspektiven* (pp. 53-72). [Health inequalities. Basics, Problems, Perspectives]. Wiesbaden: VS.

Bleich, C. & Witte, E.H. (1992). Zu Veränderungen in der Paarbeziehung bei Erwerbslosigkeit des Mannes [Changes in relationship status in unemployed men]. *Kölner Zeitschrift für Soziologie und Sozialpsychologie, 44,* 731-746.

Bond, M. & Feather, N.T. (1988). Some correlates of structure and purpose in use of time. *Journal of Personality and Social Psychology, 55,* 321-329.

Bostyn, A.-M. & Wight, D. (1987). Inside a community: Values associated with money and time. In S. Fineman (Ed.), *Unemployment – personal and social consequences* (pp. 138-154). London: Tavistock Publications.

Brinkmann, C. & Wiedemann, E. (1994). Individuelle und gesellschaftliche Folgen von Erwerbslosigkeit in Ost und West. In L. Montada (Ed.), *Arbeitslosigkeit und soziale Gerechtigkeit* (pp. 175-192) *[Unemployment and social injustice].* Frankfurt: Campus.

Burke, P.J. (2006). Identity change. *Social Psychology Quarterly, 69*(1), 81-96.

Burzan, N. (2002). Zeitgestaltung im Alltag älterer Menschen. *Eine Untersuchung im Zusammenhang mit Biographie und sozialer Ungleichheit [An examination of the relationship between biography and social inequality].* Opladen: Leske + Budrich.

Creed, P.A. & Bartrum, D. (2006). Explanations for deteriorating wellbeing in unemployed people: specific unemployment theories and beyond. In T. Kieselbach, A.H. Winefield, C. Boyd & S. Anderson (Eds.), *Unemployment and health. International and interdisciplinary perspectives.* (pp. 1-20). Bowen Hills: Australian Academic Press..

Creed, P.A. & Evans, B.M. (2002) Personality, well-being and deprivation theory. *Personality and Individual Differences, 33,* 1045-1054.

Creed, P.A. & Klisch, J. (2005) Future outlook and financial strain: Testing the personal agency and latent deprivation models of unemployment and well-being. *Journal of Occupational Health Psychology, 10,* 251-260.

Creed, P.A. & Machin, M.A. (2003). Multidimensional properties of the Access to Categories of Experience scale. *European Journal of Psychological Assessment, 19 (2),* 85-91.

Creed. P.A. & Macintyre, S.R. (2001). The relative effects of deprivation of the latent and manifest benefits of employment on the wellbeing of unemployed people. *Journal of Occupational Health Psychology, 6,* 324-331.

Creed, P.A. & Watson, T. (2003). Age, gender, psychological wellbeing and the impact of losing latent and manifest benefits of employment in unemployed people. *Australian Journal of Psychology, 55(2),* 95-103.

Evans, S.T. (1986). *Variations in activity and psychological well-being in employed young adults.* Unpublished doctoral dissertation, University of Manchester, UK.

Evans, S.T. & Haworth, J.T. (1991). Variations in personal activity, access to 'categories of experience', and psychological well-being in young adults. *Leisure Studies, 10(3),* 249 – 264.

Ezzy, D. (1993). Unemployment and mental health – a critical review. *Social Science and Medicine, 37 (1),* 41-52.

Ezzy, D. (2001). *Narrating unemployment.* Burlington: Ashgate.

Farmer, R. & Sundberg, N. (1986). Boredom proneness – the development and correlates of a new scale. *Journal of Personality Assessment, 50(1),* 4-17.

Feather, N.T. & Bond, M. (1983). Time structure and purposeful activity among employed and unemployed university graduates. *Journal of Occupational Psychology, 56,* 241-254.

Feather, N.T. & Bond, M. (1994). Structure and purpose in the use of time. In Z. Zaleski (Ed.), *Psychology of future orientation.* (pp. 121-140) Lublin: Towarzystwo Naukowe KUL.

Foucault, M. (1977). *Discipline and punish: The birth of the prison.* New York: Vintage.

Fryer, D. (1986). Employment deprivation and personal agency during unemployment. A critical discussion of Jahoda's explanation of the psychological effects of unemployment. *Social Behaviour, 1,* 3-23.

Fryer, D. & Payne, R. (1984). Proactive behaviour in unemployment: Findings and implications. *Leisure Studies, 3,* 273-295.

George, J. (1991). Time structure and purpose as a mediator of work-life linkages. *Journal of Applied Social Psychology, 21(4),* 296-314.

Goffman, E. (1961). *Asylums. Essays on the social situation of mental patients and other inmates.* Hammondsworth: Penguin.

Grawe, K. (2004). *Psychological therapy.* Seattle: Hogrefe & Huber.

Hanisch, K.A. (1999). Job loss and unemployment research from 1994 to 1998: A review and recommendations for research and intervention. *Journal of Vocational Behavior, 55,* 188-220.

Haworth, J.T. (1986). Meaningful activity and psychological models of non-employment. *Leisure Studies,* 5 (3), 281-297.

Haworth, J.T. (1997) (Ed.). *Work, leisure and well-being.* London: Routledge.

Haworth, J.T. & Ducker, J. (1991). Psychological well-being and access to 'categories of experience' in unemployed young adults. *Leisure Studies, 10,* 249-264.

Haworth, J.T. & Evans, S.T. (1987). Meaningful activity and unemployment. In D. Fryer & P. Ullah (eds.), *Unemployed people: Social and psychological perspectives* (pp. 241-267). Milton Keynes: Open University Press.

Haworth, J. & Hill, S. (1992). Work, leisure and psychological well-being in a sample of young adults. *Journal of Community and Applied Social Psychology, 2,* 147-160.

Haworth, J. & Paterson, F. (1995). Access to categories of experience and mental health in a sample of managers. *Journal of Applied Social Psychology, 25(8),* 712-724.

Heinemann, K. (1982). Arbeitslosigkeit und Zeitbewusstsein [Unemployment and time awareness]. *Soziale Welt, 33(1),* 87-101.

Heinemeier, S. (1991). *Zeitstrukturkrisen. Biographische Interviews mit Arbeitslosen [Crises of time structures. Biographical interviews with the unemployed].* Opladen: Leske + Budrich.

Hepworth, S.J. (1980). Moderating factors of the psychological impact of unemployment. *Journal of Occupational Psychology, 53,* 139-145.

Hess, D., Hartenstein, W. & Smid, M. (1991). Auswirkungen von Arbeitslosigkeit auf die Familie [Effects of unemployment on the family]. *Mitteilungen aus der Arbeitsmarkt- und Berufsforschung, 24,* 178-192.

Hoare, N. & Machin, M.A. (2006) Maintaining well-being during unemployment: The role of the latent benefits of employment. *Australian Journal of Career Development, 15,* 19-27.

Iso-Ahola, S., & Weissinger, E. (1990). Perceptions of boredom in leisure: Conceptualization, reliability, and validity of the Leisure Boredom Scale. *Journal of Leisure Research, 22,* 1-17.

Iwasaki, Y. & Schneider, I.E. (2003). Leisure, stress, and coping: An evolving area of inquiry. *Leisure Sciences, 25,* 107-113.

Jackson, T. (1999). Differences in psychosocial experiences of employed, unemployed, and student samples of young adults. *The Journal of Psychology, 133(1),* 49-60.

Jahoda, M. (1981). Work, employment, and unemployment: Values, theories, and approaches in social research. *American Psychologist, 36,* 184-191.

Jahoda, M. (1982). *Employment and unemployment: A social-psychological analysis.* Cambridge, England: Cambridge University Press.

Jahoda, M. & Fryer, D. (1986). The social psychology of the invisible: An interview with Marie Jahoda. *New Ideas in Psychology, 4,* 107-118.

Jahoda, M., Lazarsfeld, P.F. & Zeisel, H. (1975/1933). *Die Arbeitslosen von Marienthal.* Frankfurt a.M.: Suhrkamp.

Kelly, W.E. (2003). No time to worry: The relationship between worry, time structure, and time management. *Personality and Individual Differences, 35(5),* 1119-1126.

Kieselbach, T. (2003). Psychologie der Arbeitslosigkeit: Von der Wirkungsforschung zur Begleitung beruflicher Transitionen (pp. 123-141). [Psychology of unemployment: From effect research to te accompaniment of occupational transitions]. In A. Bolder & A. Witzel (Eds.), *Berufsbiographien.* Leverkusen: Leske + Budrich.

Kieselbach, T. (2006). Youth unemployment and the risk of social exclusion in six European Countries. In T. Kieselbach, A.H. Winefield, C. Boyd & S. Anderson. (Eds.), *Unemployment and health. International and interdisciplinary perspectives* (pp.233-258). Bowen Hills: Australian Academic Press.

Kieselbach, T. & Beelmann, G. (2006). Arbeitslosigkeit und Gesundheit: Stand der Forschung. [Unemployment and Health: Stake of Research] In A. Hollederer & H. Brand (Eds.), *Arbeitslosigkeit, Gesundheit und Krankheit.* (pp. 13-31) Bern: Huber.

Kilpatrick, R. & Trew, K. (1985). Life-styles and psychological well-being among unemployed men in Northern Ireland. *Journal of Occupational Psychology, 58,* 207-216.

Kleiber, D.A., Hutchinson, S. L. & Williams, R. (2002) Leisure as a resource in transcending negative life events: Self-protection, self-restoration, and personal transformation. *Leisure Sciences, 24,* 219-235.

Köller, R. (2006). *Ruhestand – mehr Zeit für Lebensqualität? Die Bedeutung von Erwerbstätigkeit und Zeiterfahrungen im Lebenslauf für die individuelle Gestaltung des Ruhestandes [Retirement – more time for quality of life? The meaning of employment and time experience in life for the individual design of retirement].* Dissertation, University of Bremen.

Kronauer, M., Vogel, B. & Gerlach, F. (1993). *Im Schatten der Arbeitsgesellschaft. Arbeitslose und die Dynamik sozialer Ausgrenzung [In the shadow of working society. Unemployed and the dynamics of social exclusion].* Frankfurt a.M.: Campus Verlag.

Kuhnert, P. (1999). *Bewältigungskompetenzen und Beratung von Langzeitarbeitslosen [Coping competencies and counselling of unemployed].* Dissertation, University of Dortmund.

Kuhnert, P. (2004). Work Life Balance trotz Arbeitslosigkeit und instabiler Beschäftigung? Paradoxie oder neue Chance? In M. Kastner (Ed.), *Die Zukunft der Work Life Balance. Wie lassen sich Beruf und Familie, Arbeit und Freizeit miteinander vereinbaren?* (pp. 141-194). *[The future of work life balance. How can job, family, work and leisure be combined with one another?]* Kröning: Asanger.

Lang-von Wins, T., Mohr, G. & von Rosenstiel, L. (2004). Kritische Laufbahnübergänge: Erwerbslosigkeit, Wiedereingliederung und Übergang in den Ruhestand. In H. Schuler (Ed.), *Organisationspsychologie – Grundlagen und Personalpsychologie* (pp. 1113-1191). *[Organisational psychology - Foundations and personnel psychology].* Enzyklopädie der Psychologie, Themenbereich D, Praxisgebiete, Serie III, Wirtschafts-, Organisations- und Arbeitspsychologie, Bd. 3. Hogrefe: Göttingen.

Latack, J.C., Kinicki, A.J. & Prussia, G.E. (1995). An integrative process model of coping with job loss. *Academy of Management Review, 20 (2),* 311-342.

Martella, D. & Maass, A. (2000). Unemployment and life satisfaction: The moderating role of time structure and collectivism. *Journal of Applied Social Psychology, 30(5),* 1095-1108.

McKee-Ryan, F.M. & Kinicki, A.J. (2002). Coping with job loss: A life-facet model. *International Review of Industrial and Organisational Psychology, 17,* 1-29.

McKee-Ryan, F.M., Song, Z.L., Wanberg, C.R. & Kinicki, A.J. (2005). Psychological and physical well-being during unemployment: A meta-analytic study. *Journal of Applied Psychology, 90(1),* 53-76.

Mirowsky, J. & Ross, C.E. (2003). *Social causes of psychological distress.* New York: Aldine de Gruyter.

Mohr, G. & Müller, K.E. (2000). Zeitstrukturierung und Zeiterleben in der Erwerbslosigkeit. Eine qualitative Studie mit langzeiterwerbslosen Industriearbeiterinnen. *Verhaltenstherapie und psychosoziale Praxis,* 2, 53-63.

Mudrack, P.E. (1997). The structure of perceptions of time. *Educational and Psychological Measurement,* 57 (2), 222-240.

Mudrack, P.E. (1999). Time structure and purpose, type A behavior, and the protestant work ethic. *Journal of Organisational Behavior,* 20, 145-158.

Muller, J.J., Creed, P.A., Waters, L.E. & Machin, M.A. (2005). The development and preliminary testing of a scale to measure the latent and manifest benefits of employment. *European Journal of Psychological Assessment, 21 (3),* 191-198.

Nordenmark, M. & Strandh, M. (1999) Towards a sociological understanding of mental well-being among the unemployed: The role of economic and psychosocial factors, *Sociology, 33,* 577-597.

Paul, K. & Moser, K. (2006). Incongruence as an explanation for the negative mental health effects of unemployment: Meta-analytic evidence. *Journal of Occupational and Organisational Psychology,* 79, 595-621.

Rogge, B.G., Kuhnert, P. & Kastner, M. (2007). Zeitstruktur, Zeitverwendung und psychisches Wohlbefinden in der Langzeitarbeitslosigkeit [Time structure, time usage, and psychological well-being in long-term unemployment]. *Psychosozial, 109,* 85-103.

Rogge, B.G. (2007). *Understanding mental health in situations of social inequality. A sociological approach.* Unpublished Manuscript.

Schatz, T. (1999). *Temporale Musterpräferenzen – Eine empirische Untersuchung zur Validität und persönlichkeitsspezifischen Relevanz prototypischer realer und latenter idealer temporaler Muster bei Studierenden [Temporal pattern preferences – An empirical investigation regarding the validity and personality-specific relevance of prototypical real and latent ideal temporal pattern amongst students].* Berlin: Logos.

Schatz, T. (2000). Wie sich Einstellungen gegenüber der Zeit auf deren Gestaltung auswirken. In R. Dollase, K. Hammerich, K. & W. Tokarski (Eds.), *Temporale Muster: die ideale Reihenfolge der Tätigkeiten* (pp. 65-74). *[Temporal patterns: The ideal sequence for activities]*. Opladen: Leske + Budrich.

Sheeran, P. & Abraham, C. (1994). Unemployment and self-conception: A symbolic interactionist analysis. *Journal of Community and Applied Social Psychology, 4,* 115-129.

Singh-Manoux, A., Richards, M. & Marmot, M. (2003). Leisure activities and cognitive function in middle age: Evidence from the Whitehall II study. *Journal of Epidemiology and Community Health, 57 (11),* 907-913.

Sorokin, P. & Merton, R. (1937) Social time: A methodological and functional analysis. *American Journal of Sociology, 42,* 615-629.

Starrin, B. & Jönsson, L.R. (2006). The finances-shame model and the relation between unemployment and health. In T. Kieselbach, A.H. Winefield, C. Boyd & S. Anderson (Eds.), *Unemployment and health. International and interdisciplinary perspectives* (pp. 75-97). Bowen Hills: Australian Academic Press.

Stets, J. (2003). Justice, emotion, and identity theory, In P.J. Burke, T.J. Owens, R.T. Serpe & P.A. Thoits (Eds.), *Advances in identity theory and research.* (pp. 105-122). New York: Kluwer-Plenum.

Swinburne, P. (1981). The psychological impact of unemployment on managers and professional staff. *Journal of Occupational Psychology, 54,* 47-64.

Thoits, P.A. (2006). Personal agency in the stress process. *Journal of Health and Social Behavior, 47,* 309-323.

Wacker, A. (1983). Differentielle Verarbeitungsformen von Arbeitslosigkeit [Differential managing of unemployment]. *Prokla, 53,* 77-88.

Wanberg, C.R., Griffiths, R.F. & Gavin, M.B. (1997). Time structure and unemployment: A longitudinal investigation. *Journal of Ocupational and Organisational Psychology, 70,* 75-95.

Wanberg, C.R., Kammeyer-Mueller, J.D. & Shi, K. (2001). Job loss and the experience of unemployment: International research and perspectives. In N. Anderson, D.S. Onies, H.R. Sinangil & C. Viswesvaran (Eds.), *Handbook of Industrial, Work and Organisational Psychology* (pp. 253-269), Volume 2. London: Thousand Oaks.

Wanberg, C.R. & Marchese, M.C. (1994). Heterogeneity in the unemployment experience: A cluster analytic investigation. *Journal of Applied Social Psychology, 24,* 473-488.

Warr, P. (1987). *Work, unemployment and mental health.* Oxford: Clarendon Press.

Waters, L.E. & Moore, K.A. (1999). *Measuring the subjective meaning of leisure activities.* Unpublished manuscript.

Waters, L.E. & Moore, K.A. (2002). Reducing latent deprivation during unemployment: The role of meaningful leisure activity. *Journal of Organisational Psychology, 75,* 15-32.

Waters, L.E. & Muller, J. (2003). Money or time. Comparing the effects of time structure and financial deprivation on the psychological distress of unemployed adults. *Australian Journal of Psychology, 55 (3),* 166-175.

Weber, M. (1986/1920) *Gesammelte Aufsätze zur Religionssoziologie [Collected essays to the sociology of religion].* (8th ed). Tübingen: J. C. B. Mohr.

Winefield, A.H. (2002). Unemployment, underemployment, occupational stress and psychological well-being. *Australian Journal of Management, 27,*137-148.

Winefield, A.H., Tiggemann, M. & Winefield, H.R. (1992). Spare time use and psychological well-being in employed and unemployed young people. *Journal of Occupational and Organisational Psychology, 65,* 307-313.

Informal Labour and Health Effects: Introducing a Research Problem

Simo Mannila

1 What is informal economy?

Informal economy is defined as a system of exchange used outside state-controlled or money-based economic activities (e.g., www.nationmaster.com/encyclopedia/). A supplementary explanation from the same source states that, the informal economy consists of barter, mutual self-help, odd jobs, allotment farming, street trading, and other similar activities. Most of the world's population participate in local informal economies.

Conceptually, we can separate the informal economy from formal labour (Hussmans, 2004; Negrete, 2001). As there is also informal employment in a formal economy, work conditions may not comply with what is required by the law; there are also some sectors of the informal economy that are regulated to a certain extent, for instance, by means of traditional patron-client relations or schemes developed in particular by NGOs for the informal economy. A large share of both the informal economy and informal employment consist of self-employed persons or family enterprises. Here we do not make a strict distinction between the informal economy and informal employment, but we use mainly the former term.

The above source shows also the share of the informal economy in 100 selected countries. The top five countries are Georgia (67%), Bolivia (67%), Panama (64%), Azerbaijan (61%) and Peru (60%). The bottom five countries are Japan (11%), United Kingdom (13%), New Zealand (13%), Netherlands (13%) and China (13%). Additionally, we see that, for instance, Russia is number 17 (46%), Finland number 88 (18%) and Germany number 90 (16%). This information is based on general Internet sources (www.nationmaster.com/graph-T/eco_inf_eco). Although we should take these figures with a certain reservation, since the measurement of the informal economy is very difficult, they indicate how widespread the informal economy is, and in which countries its share is largest: in developing countries and in countries of transition. According to the above data, the weighted average of worldwide informal economy would be 27%. The average size of the informal economy as a percent of official gross national income in 2000 was 41% in developing countries, 38% in transition countries and 18% in OECD countries. These figures were based on a study of 110 countries (Schneider, 2002). The main determinants of the size of the informal economy were the burden of taxation including social benefits combined with government regulative policy.

2 Traditional and new approach to the informal economy (with some Russian examples)

The informal economy is widely criticised for leaving employees working in an informal economy without formal protection. There are, however, new insights into this matter based on the recognition that the informal economy is not withering away. This is reflected in the increased interest of international organisations in the informal economy (e.g., Tacis, 1997; ILO, 2001; OECD, 2003; World Bank, 2003). The controversy between traditional and modern approaches to the informal economy can be described as in Table 1 (see www.wiego.org.).

Table 1. Critical discrepancies between the traditional and modern approaches to informal economy

Traditional	Modern
1. The informal economy is a traditional economy that will wither away with industrial growth	1. The informal economy increases with industrial growth in many developing countries with no signs of withering
2. The productivity of the informal economy is low	2. The informal economy creates employment and is a major source of subsistence for low-income groups contributing to GDP
3. The informal economy and the formal economy exist independently of each other	3. The informal economy and the formal economy are linked
4. The informal economy contains surplus labour for the formal economy	4. The decline of the formal economy amounts to an increase in the informal economy
5. The informal economy means tax evasion and avoidance of employment regulations	5. The informal economy is not equivalent to a shadow economy: a large share of it consists of self-employed persons or non-standard workers
6. Since persons working in the informal economy do not pay taxes they are well-off	6. Average income of persons employed in informal economy are lower than those in formal economy
7. Regulation of the informal economy is unnecessary	7. Clear rules and appropriate legislation are needed to encourage a transfer from the informal to the formal economy and also to regulate the informal economy as much as possible.

The controversy between traditional and modern approaches is quite clear for instance in the analysis of Russian economy, but neither is it without any relevance in more stable economies (OECD, 2004). The Russian transition has meant a decline of the formal economy, which has partly been compensated by the growth in the informal economy. The figures that show a decline of 40% or more in the formal economy takes into account this compensation by the increase of informal economy only to a limited degree (cf. World Bank, 2002). Official Russian statistics usually show much lower figures concerning the share of the informal economy depending on the calculation system (cf. Sinyavskaya & Popova, 2003). In the Russian Federation, Bandyukova (2004) has studied "casual employment", which is generally equivalent to informal self-employment. She states that the share of casual employment is at least as high as that of unemployment. We must also point out that major regional differences most likely exist in the proportion and types of informal economy in the Russian Federation.

Various studies show that informal employment is more common among younger than older persons in Russia. As far as education is concerned, the results are contradictory: educational background is in general linked in a very complex way to status attainment during transition (cf. Evdokimova, 2002).

In Russian informal economy, industry, construction, transport and communications plus trade and catering are overrepresented compared to the formal economy. One third of informal employment is estimated to consist of various services, and approximately one fourth are workers (blue-collar or unskilled), 15% are self-employed in various types of crafts, 14% belong to the managerial class; similarly 14% are various experts and professionals (RLMS 2001 by Sinyavskaya & Popova). The occupational variety shows that informal economy does not always mean working in secondary labour markets or some other disadvantaged position. According to Bandyukova (2004), one third of those casually employed move into permanent jobs, 43% remained trapped in casual jobs while the rest moved to unemployment or out of labour force. Her results showed that casual workers had much smaller monthly earnings, while the hourly wages were higher: transition into a permanent job meant increased earnings as a whole. She comments that casual employment can be good or bad: the former may be true for e.g. pensioners or students. There are also similar results pointing out this ambiguous role of the informal economy and internal variation of informal employment in the European Union (Borghi & Kieselbach, this volume).

The informal economy is harmful for social and health policies in countries of transition for a number of reasons. Firstly, it reduces tax revenues and, thus, the financial basis for social and health policy-making. Unrealistically high payroll and other tax levels combined with weak law enforcement – as stated above by Schneider - have led to a growth of the informal economy. This situation, combined with deficit budgets and soft budget constraints, has lead to unfunded mandates, i.e. a situation where citizens do not receive their legal entitlements (e.g., social benefits, health services). This is harmful for the rule of law, already weakened by the widespread informal economy and may lead to a vicious circle. Secondly, the informal economy distorts the results of household budget surveys and living conditions surveys as well as the operation of the basic safety nets at the level of local administration: anti-poverty policies and targeting are difficult since income becomes hard to measure and we do not always know who is actually poor. This has lead into a situation where a large share of population, in the Russian Federation during 1992-2000, between 21% and 35% (Sinyavskaya & Popova, 2003) lived below the poverty line defined by a minimum consumption basket: one reason for this result is that official survey instruments do not record income from the informal economy or informal employment and thus overestimate poverty. The problem here is that the distortion differs by population groups.

The contribution of the informal economy to the survival strategies of the Russian population is obvious to all social research about countries of transition. The choice between a formal and an informal economy is not a moral one, it is a choice of survival and culture. The participation in the informal economy is often not only the only option available, it is also a rational choice: with the experience of hyperinflation and political instability in mind and having to wrestle with a complicated bureaucracy, the population is suspicious of entrusting a part of their – very modest - income in the form of taxation to the government to be used later for social protection schemes, for example. In national accounts and other statistics, complicated methods have been developed in order to include somehow the contribution of informal economy to the GDP so that the economic progress of the country can be properly assessed. In order to achieve a more accurate picture of poverty, a multitude of indicators have been recommended, although the many-facetted character of poverty is often not well understood in everyday social policy-making (e.g., OECD, 2003; Masakova, 2004). There is also a branch of research which, on the one hand, investigates the deep cultural roots of informal economic relations which cannot be defined solely in terms of economics and which date back to the Soviet era, and on the other hand, surveys these relations critically from the point of view of post-modern society and sometimes emphasising their corrupt character.

Some authors, such as Piirainen (1997) or Birdsall (2000) and Clarke (2000), see involvement in the informal economy as a part of a survival strategy which may be good or bad. For instance, a large share of the informal economy in countries of transition consists of in-kind income from plots of land ("dacha economy"): the yield from these plots of land is mainly consumed by the family, but very often part of it can be sold in the market. Sometimes this micro-business has initiated enterprises. The feasibility of informal employment is in the long run linked with its ability to promote the ability of families and individuals to cope with a new market economy. Informal economy related to migration and, in particular, its illegal forms, has similar roots: it is a survival strategy (Reyneri, 2003). Some authors also point out that involvement in the informal economy is sometimes voluntary and those working in it are not necessarily worse off economically or in terms of other stratification indicators than others (Maleva, 2003; Bandyukova, 2004). Sometimes people are engaged in informal employment solely for additional income. This engagement often entails some form of risk-taking, which is influenced by individual propensity and situational life chances: successful career or business opportunities may be linked with involvement in the informal economy and may also involve some health risks. However, we must bear in mind that a large share of the world population incurs these social and health risks of the informal economy without any real perspective of material or occupational success.

Some writers, such as Ledeneva (1999, 2001) or Yakovlev (1999) have described the criminal character of the informal economy – in this case it can be called the shadow economy. While showing how the informal economy is culturally embedded, they point out the incompatibility of it or at least some of its forms with (post-)modern society. Finally, for instance, Busse (2001) and Lomnitz (2004) view the informal economy as linked to social capital: the fashionable concept of "social capital" may allow us to understand better the development of various social networks in transition: some forms of social capital can be devaluated, creating new forms characterised by trust in modern society.

3 Informal economy and health

Independently of whether we adopt a traditional or modern approach to the informal economy, there remain serious health concerns linked with the informal economy. They can be structured as follows:

- risks related to the insecurity of labour market status and income
- risks related to work load
- risks related to inadequate occupational safety and health or lack thereof
- risks related to culturally induced behavioural factors
- general risks related to poverty and deprivation and indirectly linked with the informal economy.

There does not seem to be very much research into the health impact of the informal economy. On the basis of what was said above, it is clear that the health impact is harmful, but some reservations must be made (cf. Maleva, 2003) and, in many cases, we must inquire into the alternative to involvement in the informal economy and whether there is one.

The issue has a certain resemblance to the problem of unemployment and health. There is evidence that job insecurity (precarious labour market status) is linked with psychological distress, in some cases developing into more significant health problems, but this distress is lower than that linked with unemployment (Gallie et al., 1997). This means that being in the labour force and receiving some of its benefits is usually better than being involuntarily outside it. Here we do not expand on the possibility that some parts of the informal economy may contain acute health risks because of a lack of occupational safety and health.

Research on unemployment and health is relevant here for another reason as well. Most research on the links between unemployment and health shows that the negative impact of unemployment starts while a person who is bound to lose her or his job is still working, i.e. in the phase of anticipation. This impact is not limited solely to those becoming redundant: precariousness in the labour market also affects those who will be able to maintain their jobs. The size of the informal sector and its links with the formal sector permeate the economic culture of society. Finally, there is evidence that mental disorders are more common to informal work than formal employment, but it remains unclear whether informal work is the cause of these disorders or whether involvement in informal work and mental illness may have some causes in common, at times resulting in a vicious circle (Ludermir & Lewis, 2003). This is analogous to a number of studies concerning unemployment and health and their complex mutual relationships (e.g., Jäntti et al., 2000; Nyman, 2002).

4 Policy-making for persons in the informal economy

Various international organisations, in particular the ILO and women´s organisations, have campaigned for increased protection of persons working in the informal economy. The ILO has recognized the informal economy as a non-transitory phenomenon at a high level in 1991where the labour force also needs to be protected in the informal economy. The participation of women's organisations in the issue of informal economy is due to the fact that informal employment is even more typical of women than men in developing countries.

The ILO (2001) strategies include:

- institutional support of occupational safety and health instead of regulating or taxing
- emphasis on improved productivity and income,
- introducing and reforming labour standards and consolidating standards and assistance,
- information and awareness–raising,
- enterprise-based action programmes (application of WISE: work improvement in small enterprises),

- community-based action programmes, and
- programmes for particular target groups (e.g., children, women).

Later this approach has been developed into the decent work agenda, which has contributed to the visibility of informal labour among the policy makers and is widely supported by other international organisations and donors (http://www.ilo.org/global/about-the-ilo/decent-work-agenda/lang--en/index.htm).

5 Conclusions

Taking into account how widespread the informal economy is and how controversial it seems to be from the policy-making perspective and from the point of view of its social impact, the relationship between the informal economy and health deserves to be researched as extensively as the relationship of unemployment and health has been studied in the developed societies during the past 20-30 years.

More people in the world face informal work than unemployment, and there are no signs that their numbers will decrease.

The bulk of research available from unemployment research together with experiences from the consultative work of international organisations can provide some frame of reference as well as hypotheses to this largely virgin area of social research.

References

Adler Lomnitz, L. (2004). *Informal exchange networks in formal systems: A theoretical model.* Retrieved from www.colbud.hu/honesty-trust/lomnitz/pub01.html.

Bandyukova, T. (2004, September). *Casual work as an example of informal sector (Russia 1994-2002).* Paper from EGDI & UNU-WIDER, Conference Unlocking Human Potential: Linking the Informal and Formal Sectors. Helsinki, Finland.

Birdsall, K. (2000). *Covert earning schemes at the Russian workplace: Everyday crime or practices of survival.* Retrieved from www.geog.susx.ac.uk/research/changing_europe/web_hholds.html.

Borghi, V. & Kieselbach, T. (2000). *The submerged economy as a trap and a buffer: Comparative evidence on long-term unemployment and the risk of social exclusion in Southern and Northern Europe* (this volume).

Busse, S. (2001). Strategies of daily life: Social capital and informal economy in Russia. *Sociological Imagination, 38*, 2-3.

Clarke, S. (2000). *Do Russian households have survival strategies.* Retrieved from www.geog.susx.ac.uk/research/changing_europe/web_hholds.html.

Evdokimova, E. (2001). Ot sovetskoy intelligencii k rossiyskomu srednemu klassu. In I. Tranin (Ed.), *Novyie potrebnosti i novyie riski - real'nost'90-kh godov* [New needs and new risks - Russian reality of the 1990s] (pp. 233-244). Saint Petersburg: Norma.

Gallie, D., Marsh, C. & Vogler, C. (Eds.). (1994). *Social change and the experience of unemployment.* Oxford: Oxford University Press.

Hussmans, R. (2004, June). *ILO Guidelines on the measurement of employment in the informal sector and informal employment.* Joint Seminar of the Federal Services of State Statistics, Russian Federation and Central Statistical Office, Republic of Poland. Saint Petersburg, Russia.

ILO (2001). *Promoting gender equality – a resource kit for trade unions.* Retrieved from www.workinfo.com/free/links/Gender/cha_5.html.

Jäntti, M., Martikainen, P. & Valkonen, T. (2000). When the welfare state works: Unemployment and mortality in Finland. In G.A. Cornia & R. Paniccià (Eds.), *The mortality crisis in transitional economies* (pp. 351-369). Oxford: Oxford University Press.

Ledeneva, A.V. (1999). *Russia's economy of favours. Blat, Networking and informal exchange*. Cambridge: Cambridge University Press.

Ledeneva, A.V. (2001). *How Russia really works: Towards an understanding of the informal order*. Retrieved from www.cer.org.uk/n5publicatio/essays.html.

Ludermir, A.B. & Lewis, G. (2003). Informal work and common mental disorders. *Social Psychiatry and Psychiatric Epidemiology, 38*(9), 485-489.

Maleva, T. (2003, red.). *Sredniye klassy v Rossii. Ekonomicheskiye i social'niye strategii*. Moskva: Gendal'f.

Masakova, I. (2004, June). *Current practices in estimating the non-observed economy and the problems of measuring in view of the changes in the national classifications*. Joint seminar of the Federal Services of State Statistics, Russian Federation and Central Statistical Office, Republic of Poland. St. Petersburg, Russia.

National Public Health Institute (2002). *Does unemployment contribute to ill-being: Results from a panel study among adult Finns 1989/90 and 1997*. Publications of the National Public Health Institute A 4. Helsinki: J. Nyman.

Negrete, R. (2001, September). *Case studies on the operation of the concept of "informal employment" as distinct from "informal sector employment"*. Paper from the Expert Group on Informal Sector Statistics (Delhi Group), 5th Meeting, New Delhi, India.

OECD (2004). Informal employment and promoting the transition to a salaried economy. In OECD (Ed.), *OECD Employment Outlook 2004, Chapter 5* (pp. 225-289). Paris: OECD.

OECD (2003). *Izmereniye nenablyudayemoy ekonomiki. Rukovodstvo*. [Measurement of non-observable economy. Handbook.] Moskva: Statistika Rossii.

Piirainen, T. (1997). *Towards a new social order in Russia. Transforming structures and everyday life*. Dartmouth: Aldershot.

Reyneri, E. (2003, February). *Illegal immigration and the underground economy*. Paper presented at The Challenges of Immigration and Integration in the European Union and Australia, Sydney, Australia.

Schneider, F. (2002). *Size and measurement of the informal economy in 110 countries around the world*. Retrieved from www.ecoomics.uni-linz.ac.at.

Sinyavskaya, O. & Popova, D. (2003, June). *Informal employment in Russia: Overview*. Paper from Workforce Development and Skills Mismatch – GPN Methodology Seminar, Independent Institute of Social Policy, Moskva, Russia.

Tacis Task Force. (1997). *Report of the Task Force on the measurement of the non-observed economy*. Luxembourg: Tacis Task Force.

The World Bank. (2003). *The informal economy: Large and growing in most developing countries*. Retrieved from http://rru.wordbank.org/Discussions/Discussion.aspx?id=18.

The World Bank. (2002). *Transition. The first ten years*. Washington D.C.: World Bank.

Yakovlev, A. (1999). Ekonomika "chernogo nala" v Rossii: mekhanizmy, prichiny, posledstviya. In T. Shanin (Ed.), *Neformal'naya ekonomika*, [Informal economy] (pp. 270-271). Moskva: Logos.

Long-term Youth Unemployment in East and West Germany: A Qualitative Analysis of Personal and Situational Factors

Gert Beelmann

Introduction

Ever since the mid-1970s, unemployment among young people has been a continuing problem in Germany. Although the youth unemployment rate has been only about half of the EU-15 average, the existing levels of youth unemployment are of concern on a societal and political level. This concern has lead to the implementation of the "Jugend mit Perspektive" (JUMP) programme (Instant Youth Programme towards Combating Unemployment) by the German government.[1] In fact, soon after coming to power the German chancellor declared that a reduced unemployment rate – especially the youth unemployment rate – will provide evidence for the success of his government's policies. However, after three years only slight changes were observed.

If one looks at the official employment agency statistics on unemployment, one sees a great variation of registered unemployed persons over the last 15 years, with a drastic difference between western and eastern Germany. Figure 1 shows a continuous increase in the 90's with a maximum of 12.2% in 1997. The unemployment rate among young people declined and reached its lowest point of only 10% in 2004, not least due to the change of government in 1998. Due to the Hartz legislation, which combined unemployment assistance with social assistance, unemployment rates rose rapidly but dropped again within three years to their lowest level in 15 years. The unemployment figure among youths in Germany in 2008 was 7.2%. However there was a significant difference between the old states with 5.8% and the new states with 12.8%.

Currently, the increasing East-West discrepancy is of most concern. One explanation might be found in the different success rates of the Jump programme. In the long run, there is a strong possibility that young people in the Western part will profit more than their counterparts from the East because of the prevailing strong economic differences and also because of the different vocational training systems. In the East, vocational training is essentially inter-company training carried out in centres run by organisations which are not part of any company. However, in the Western part, vocational training is a dual system which combines part-time education in a vocational school with on-the-job training.

1 The JUMP programme has been initiated by the Social Democratic/Green government in Germany which replaced the Conservative government in 1998.

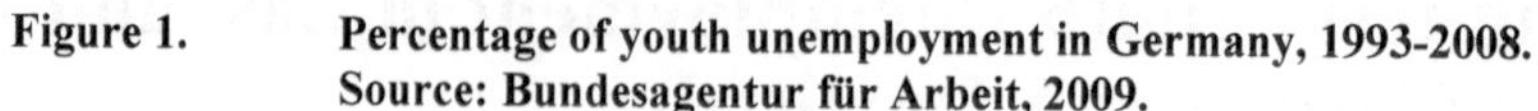

Figure 1. Percentage of youth unemployment in Germany, 1993-2008. Source: Bundesagentur für Arbeit, 2009.

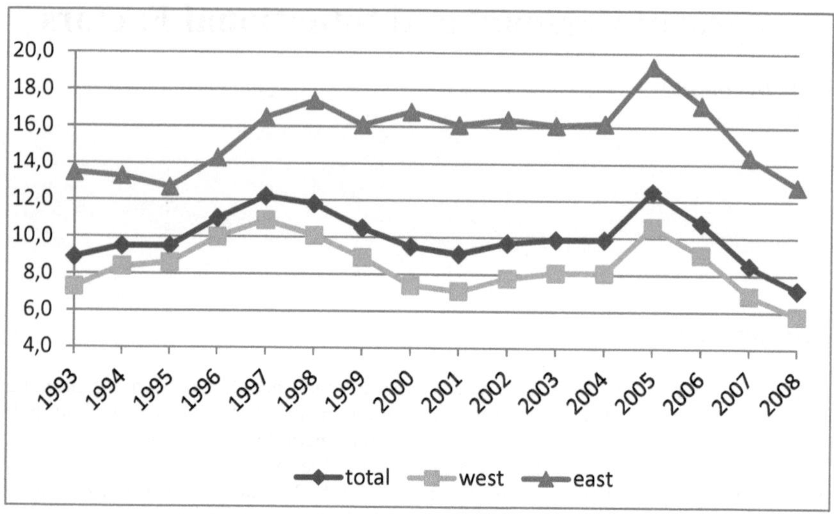

The data of the following article is based on a German study which was part of the European project (1998-2000) "Youth Unemployment and Social Exclusion: Objective Dimensions, Subjective Experiences, and Innovative Institutional Responses in Six European Countries" (YUSEDER), TSER 4th Framework Programme in DG Research (Co-ordinator: Thomas Kieselbach). The main task of the YUSEDER project was to analyse key mechanisms responsible for social exclusion among the young unemployed (Kieselbach & Beelmann, 2000; Kieselbach et al., 2000; Kieselbach et al., 2001).

This article is based on a secondary analysis of these data with the focus on the personal and situational resources (Beelmann, 2003). Specifically, it is concerned with the following general issues: 1) What are the young long-term unemployed really doing without a job, 2) In which way are they acting; and 3) What kind of resources do they find useful for carrying out their aims. This is connected to the discussion about the action-related approach by David Fryer (see Fryer & Payne, 1984; Fryer 1986) within the community of unemployment researchers.

1 Method and sample

Method

For the qualitative study with long-term unemployed youth in the YUSEDER project, we used problem-focused interviews, a method which was developed at the University of Bremen in the 1980s. The idea behind this method is to combine previous approaches within qualitative research. The main focus is on exploring the subjective experiences, the individual courses of action and the interpretations given by the interviewee. As the producer of his or her own reality, the individuals are the experts to interpret and explain their actions and events in their life (Witzel, 2000).

The data analysis is based on descriptions of each case which is structured along the following different aspects: situation in the labour market, economic aspects, institutional aspects, social environment, cultural and spatial aspects, and the daily situation of the individuals, psychosocial strains, the biographical course and socio-demographic variables. A typology is formed in regard to the central dimensions of the research question based on a concept developed by Kluge (2000). This concept brings together several theoretical approaches of qualitative research and also implies the possibilities of quantifying qualitative material.

Access to sample

Altogether 50 interviews were conducted: half of them in the Western and the other half in the Eastern part of Germany. Difficult regions in terms of their economic structures (e.g. structurally weak regions, shipyard problems in the past, etc.) can be found in both areas of Germany. The interviewees were selected based on the following criteria:

They were all:

- German nationals under 25 years of age;
- Long-term unemployed (registered as unemployed for at least the past year).
- In addition, none had participated in any vocational training in the previous year or any other training lasting more than one month.

Sample description

A total of 26 West and 24 East German youths averaging 22 years were interviewed. The youths from the East German survey were slightly younger, unemployed for an average of 33 months and most lived on their own. A significant number had parents who were also unemployed at the time of the survey. The youths from the Western German survey were slightly older on average, employed for about 27 months and most lived with a partner. Only a very small number of these individuals had unemployed parents (see Table 1).

Table 1. Sample description

	West Germany	East Germany
Sample	26	24
- male - female	14 12	9 15
age (in years)	M = 22.7 (SD = 1.19)	M = 22 (SD = 1.33)
length of unemployment (in months)	M = 27 (SD = 13.3)	M = 33 (SD = 15.6)
Unemployment within social context Reported from all: - parents - siblings - friends - partner	 18 4 3 12 4	 20 12 4 13 5

2 Results

Three central dimensions are: possibilities for action, environmental context and psychosocial strain (description below). Each dimension is further split into several levels, creating a multidimensional matrix. For instance, possibilities for action and psychological strain can be divided into three levels: high, average and low. The environmental context can be characterised by enabling/facilitating or impeding.

a) Job-related possibilities for action:
This dimension is defined as the individual's objective, but also subjectively perceived probability of being successfully reintegrated into the job market. Objective chances specifically refer to educational qualification and vocational training, chances that are also dependent on institutional offers and family support. The subjective assessment of objective chances is relevant for deciding the extent to which the objective possibilities can be realised. Therefore, general activity (training, work, leisure) and job searching strategies are also taken into account as well as perceived institutional effectiveness and family support.

b) The environmental context:
The individual and social context will be described by some variables from the vitamin model (Warr, 1987). Here the subjective viewpoint of one's financial security and social network is very important. There is no objective definition of poverty in Germany. But this lack of a definition disregards subjective experiences: How are people able to make ends meet if they realise their own financial limitations and if so, what kind of strains may result from financial insecurity.

Social networks comprise interaction with the family (including partners), close friends and acquaintances. It is important to take the availability of social contacts into account. In addition, it is important to assess the minimal social support network required that would still be able to assist those trying to cope with their unemployment situation. There are also two other variables which are relevant for the operationalisation of the environmental context: the sense of physical security at home and the sense of personal control over one's life (and thus over one's future and occupational perspective).

c) Psychosocial strain:
The dimension psychosocial strain consists of physical, psychological and social factors in regard to the unemployment situation. It therefore considers important aspects such as an individual's health behaviour, suicidal thoughts and self-esteem. This dimension distinguishes between strains not dependent on occupational status.

The second step following the composition of the matrix involved an investigation of empirical regularities amongst different groups of youths. The examination of meaningful relationships between these three dimensions and the characteristics of the various groups represented the third step. The results of the analysis are outlined in the following table representing the matrix (Table 2).

Following the analysis of groups, the next step was to examine their characteristics to identify their action styles. This led to the five different types, which are described by socio-demographic variables and which represent combinations of the three different central dimensions.

Type 1: "active-initiative action style"
A group of 10 youths were interviewed. An equal number were from the Eastern and Western survey groups. Four of these were male and six female. All the youths in this group were well educated and had completed their vocational training. For these youths, unemployment was not as great a threat as for others. They were financially secure and had their family's and friends' social support. Their mental health conditions were excellent. The females in this group were highly motivated to pursue new job opportunities. In the case of the East German survey group, the major barrier in finding a job was that their having completed vocational education via the inter-company training system represented a drawback in the competitive job market.

Various job-related possibilities of action are available to these individuals since they have completed their vocational training. They are active and take the initiative in looking for a job. They also use their time to pursue hobbies or do volunteer work. They are not willing to accept any job, but they try to find a workplace which reflects their individual preferences. These adolescents are interested in specific qualification measures to increase their employability. One problem facing these youths are their vocational training qualifications which are based on the system in the Eastern part of Germany.

At the same time, this group experiences few financial limitations. They also reported to be living in stable social environment with good friends and acquaintances which meant they were largely satisfied with their current living situation. Their level of psychosocial strain was due to the lack of daily structure in their lives. They compensate by spending more time in pursuing leisure activities, to work the occasional job and to continue their job search. In general, these young people are able to pursue their needs and occupational interests. They can rely on support from their social and family environment and possess personal resources (such as past educational and vocational training), personal initiative, self-confidence, and optimism. They have been unemployed for a relatively short period. They have developed alternative strategies for personal enhancement and were optimistic about their future careers. They are young people with an active-initiative action style.

Table 2. Three-dimensional matrix arrangement for groups of youths

Environmental context	Psychosocial strain	Possibilities for action: High	average	low
Impeding (N=18): West: 8 East: 10	low			
	average		Group 4 (N=9): 6 West 3 East	Group 5 (N=9): 2 West 7 East
	high			
Enabling (N=32): West: 18 East: 14	low	Group 1 (N=10): 5 West 5 East	Group 2 (N=12): 8 West 4 East	
	average		Group 3 (N=10): 5 West 5 East	
	high			
Total		(N=10): West: 5 East: 5	(N=31): West: 19 East: 12	(N=9): West: 2 East: 7

Type 2: "inefficient action style"

Twelve youths (eight from the West and four from the East / five male and seven female) belonged to this group. All of them were well educated but most of them did not complete their vocational training. They were unemployed for a shorter period compared to other groups (except Group 1). However, their motivation to look for work was much lower, particularly amongst the male youths and the West German subgroups. Like Group 1, they had definite future goals and were confident about reaching these goals. They received social support from friends but family support was of less relevance for this group as compared to Group 1, as they rated the emotional support from their partners and their circle of friends higher. In the greater scheme of things, they are in a relatively stable financial situation. Although they reported low strains, being unemployed caused them to feel shame.

The main characteristic of this type is the group's deficit in vocational training. Many broke off their vocational training at an early stage. Frustrated about their progress and lacking realistic occupational goals, these adolescents have no occupational orientation. Their individual action depends strongly on external institutions like the employment office. They feel personally responsible for their unemployment and feel they only have the option to become unskilled labourers. How to successfully carry out their occupational ideas is a major issue for many, leading them to question the general usefulness of gainful employment. Their daily life is dominated by various time-consuming activities, thus reducing their actual unemployment to a marginal problem. These young people can be characterised as persons with an inefficient action style.

Type 3: "reflective-sceptical action style"
10 youths (five each from the West and the East / five male and seven female) constituted this group. Here, only half of the group had finished their vocational training. They have been unemployed for a somewhat longer period than the earlier two groups. Although males are more active in searching for jobs, female youths are more involved in various job- related activities. This engagement is, however, again accompanied by feelings of frustrations due to repeated failures.

The young people of this type show higher psychosocial strain in the unemployment situation compared to those youths in Group 1 and 2. The strain of the situation means higher stress, which also results in increased smoking rates and low self esteem. Their daily rhythm is severely affected by unemployment. They experience no structure in their life and even leisure activities are increasingly neglected. As a result, these young people do not use their spare time to consider or work towards their occupational goals. They suffer under the occupational situation, complain about mood swings and are worried about the future. Although these young people are in principle optimistic about their long-term occupational future, they are aware that they will not be able to carry out their specific occupational aims. Despite the difficult labour market, they try to change their situation. That these young people continue to search for job shows that they are still proactive. However, they are well aware of their precarious labour market situation. In consequence, they judge their own chances of getting work very pessimistically due to not having the required occupational qualifications. These adolescents can be described as persons with a "reflective-sceptical action style".

Type 4: "reactive action style"
The interviewees in this group comprised nine youths - six from the West, three from the East. The gender composition was five males and four females. About half of this group had completed vocational training but had been, in general, unemployed for only a short duration. While the entire group lacked institutional or social support, male interviewees faced a higher financial burden compared to the females. All of the female interviewees, though mostly pessimistic and depressed, were still actively searching for

jobs. Only half of the male interviewees were similarly active in their respective job searches.

These young people characteristically retreat from the labour market. They are not that active in looking for a job in comparison to the groups before. Their application activity was determined by the requirements of the employment office. The participation in job-related activities is limited to attending further qualification measures and vocational retraining. These young people doubt that they will still be able to carry out their occupational plans and fear a future dependent on social welfare. Their action is driven by external factors and their spare time is filled with passive and monotonous activities. The adolescents no longer believe to be able to change their life themselves because they do not have sufficient resources to cope with their situation. Neither the family nor friends or acquaintance can support them effectively. The young people criticise the welfare institutions for the lack of guidance and for the unspecific counselling. They have a "reactive action style" because they show only passive behaviour in regard to the changing labour market situation and their living situation in general.

Type 5: "resignative action style"
Nine interviewees - two from the West and seven from the East - six male and three female constituted this group with low educational qualifications and no vocational training. They all had a very much extended unemployment period. They also lack financial security, are burdened with debts and worry about their economic future. All job-related activity is dependent on instructions from state institutions such as the Employment Office.

The main characteristic of these young people is their lack of an educational certificate and vocational training. They could not properly enter the labour market. Without these qualifications it is extremely difficult for them to find a job. The only offers they received from the welfare institutions are further vocational measures. But these measures often lead to so-called career measures which further increase their frustrations. These adolescents have no concrete ideas about their occupational future and no occupational or private perspectives as they have but little influence on their life. They receive little support from the welfare institutions and their social environment. Not surprisingly, this group also suffers from most severe physical and psychological health problems and feels very pessimistic about their future. Their extreme, precarious living situation is associated with indicators of high psychosocial strain such as depression, risky health behaviours and somatic symptoms. As a result, these young people have adopted a "resignative action style" because they feel that their missing competencies will prevent them from being successful in ever finding a job.

3 Intervention options and conclusions

What options and chances - based on existing labour market measures and psychological interventions - are open to those unemployed youths? Most interventions focused on youth unemployment aim to help youths integrate in the labour market as quickly as possible using the most effective means available.

A special case is represented by those unemployed youths who not only have insufficient qualifications but who also experienced failure on the labour market, which prevented them from gaining a permanent foothold in a profession. Many of these youths have been through several apprenticeships. Especially when they have discontinued many vocational programmes, their subsequent chances to obtain work based on their previous training experience and knowledge are slim. What kind of requirements should a group-specific intervention for unemployed youths, as in the described case study, fulfil to be effective?

The following recommendations can be made for those youths who can be categorized "active-initiative action style" individuals: These youths have sufficient resources at their disposal to overcome unemployment at short notice. They do not require an individual intervention; instead, their main problem is how to overcome the threshold between vocational training and obtaining an actual job. Salary subsidies are an instrument based on the practice in the labour market politics. These contributions attempt to entice employers to provide work for trained youths. By subsidizing the salary costs for companies who take on unemployed youths, the entry threshold into the labour market is reduced for these youths.

A central deficit for the surveyed individuals who fall into the category associated with an "inefficient action style" (Type 2) is their lack of vocational qualification. This is the result of youths breaking off their training and unrealistic occupational goals. However, specific intervention approaches can address these shortcomings. To begin with, even when youths do not complete a training program, their professional experience and qualifications to date should still be documented. Only by doing so will their existing knowledge be considered and recognized as already partially qualifying them for work. To achieve this, it would be desirable to have vocationally oriented accompanying measures that provide youths with an early overview of the occupational alternatives and outline a realistic picture of their opportunities. Lengthy phases of unemployment can be avoided only through early interventions of this kind that also work with youths on developing realistic professional goals.

The main problem found amongst youths of the "reflective-sceptical action" style (Type 3) is that they no longer believe that they have any chance carrying out their occupational ambitions. They consider the qualifications they have gained to date as useless, resulting in the belief that their occupational perspectives are very limited. A first intervention strategy would be that these youths complete a form of follow-up qualification that builds on their previous training and experience and leads to a recognized vocational certificate or diploma that is also in demand on the labour market. Concepts such as "Working and Learning" would be appropriate for these youths as well to correct their often negative sense of perspective and to define new professional goals.

Those youths who fall into the fourth category, the "reactive action style", have largely withdrawn from the labour market. In these cases, youths have to be reintroduced to work and qualifying measures. This could be achieved by preparatory measures, which involve youths getting experience in various enterprises and workshops. This will educate them about various available vocations.

Such an approach would also be suitable for those youths who fall into Type 5, associated with a "resignative action style". This group does not only lack school qualifications as well as vocational training experience, but these youths also have behavioural problems, thus, further complicating their situation. Vocational training for this type will only be effective if the training is under socio-pedagogical supervision. Such supervision is important as it provides assistance to the youths in terms of the vocational orientation. But more importantly, it would provide much-needed counselling to youths in terms of their individual problems (which include debt, drug dependence etc.). These youths require a multi-dimensional support system that incorporates social, psychological as well as labour market oriented assistance in order to stabilise their private as well as professional lives over the long-term.

When conceiving interventions, it is important to develop a counselling and supervisory approach that is both supportive and enabling for youths. An intensive interview should stand at the beginning of every intervention to explore the wishes and vocational ambitions of each youth. In addition, such an interview allows for these to be matched with realistic offers and opportunities on the labour market. The next step is the development of a tailored personal development plan that consists of a number of steps. Each step should build on the next and allow the youth to successively move from one step to the next. This type of tailoring approach can, in principle, be applied to all unemployed. However, it is important that such concepts require the support and approval of the employment service and other social welfare institutions.

Another useful strategy is to employ a mediating institution to better integrate the young unemployed into the labour market. Such an organisation would take on the role of a mediator between those enterprises that have apprenticeship positions and wish to hire new staff and those seeking employment, forming a bridge between the two. This establishment would be responsible for two tasks. First, it would assess the qualifications in terms of meeting the necessary standards for jobs, but also consider the personal situation of the youths. Second, it would arrange for the placement of the youths in a suitable company. Both work and vocational training in the company should be accompanied by predetermined, regular meetings to address emerging problems early and thus prevent a potential failure.

One result of this work was the recognition that Germany did not lack effective and innovative strategies. It is especially the least qualified who end up having a particularly difficult time in finding another job. It is important to remember that the various situations of the unemployed youths would be even more challenging if no such policy interventions and the vocational assistance existed.

It can be assumed that several of the surveyed unemployed youths will not succeed in entering the labour market under the current economic conditions. This might apply especially to those youths typed as having a "reactive action style" and "resignative action style". Instead, these youths will end up being dependent on social welfare as well as on the development of new forms of work and employment.

References

Beelmann, G. (2003). *Langzeitarbeitslose Jugendliche in Deutschland. Eine handlungsorientierte Analyse personaler und situativer Faktoren.* [Long-term unemployed youth in Germany. An action related analysis of personal and situative factors] Hamburg: Dr. Kovac Verlag.

Fryer, D. & Payne, R. (1984). Proactive behaviour in unemployment: Findings and implications. *Leisure Studies, 3*, 273-295.

Fryer, D. (1986). Employment deprivation and personal agency during unemployment: A critical discussion of Jahoda's explanation of the psychological effects of unemployment. Social Behavior, 1(3), 3-23.

Kieselbach, T. & Beelmann, G. (2000). Youth unemployment and health in Germany. In Kieselbach T. (Ed.) *Youth unemployment and health. A comparison of six european countries* (pp. 109-136). Leske & Budrich, Opladen.

Kieselbach, T., Beelmann, G., Erdwien, B. Stitzel, A. & Traiser, U. (2000). Youth unemployment and social exclusion in Germany. In T. Kieselbach (Ed.), *Youth unemployment and social Exclusion. A comparison of six European countries* (pp. 131-174). Leske & Budrich, Opladen.

Kieselbach, T., Beelmann, G., Traiser, U. & Meyer, R. (2001). Long-term unemployed youths in Germany: A qualitative study on the risk of social exclusion. In T. Kieselbach, K. van Heeringen, M. La Rosa, L. Lemkow, K. Sokou & B. Starrin (Eds), *Living on the edge – An empirical analysis on long-term youth unemployment and social exclusion in Europe* (pp. 183-241). Opladen: Leske & Budrich.

Kluge, S. (2000). Empirically grounded construction of types and typologies in qualitative social research. *Forum Qualitative Social Research, 1* (1) [Online-Journal]. Available under: http://qualitative-research.net.

Warr, P. (1987). *Work, unemployment and mental health.* Oxford: Oxford University Press.

Witzel, A. (2000). The problem-centred interview. *Forum Qualitative Social Research,* 1 (1) [Online-Journal]. Available under: http://qualitative-research.net.

The Submerged Economy as a Trap and a Buffer: Comparative Evidence on Long-Term Youth Unemployment and the Risk of Social Exclusion in Southern and Northern Europe

Vando Borghi & Thomas Kieselbach

Introduction

The following considerations constitute a summarised account of the YUSEDER research project's results (Kieselbach, 2000a, 2000b; Kieselbach et al., 2001; La Rosa & Kieselbach, 1999) with regard to one specific aspect of youth unemployment: that is, the role played by the submerged economy in regard to the risk of social exclusion of long-term unemployed young people.

Generally speaking, social researchers need to adopt a somewhat cautious approach to the question of informal economy and irregular work, as this phenomenon is one that is difficult to define using fixed theoretical categories (irregular work being often interwoven with, or superimposed on, regular employment), one that escapes official classification (it is after all, by definition, a hidden, underground phenomenon[1]) and simple generalisations. Not only does it vary from one country to another, but its social effects often differ according to the social groups involved, the national welfare benefit system in question, the type of work involved and the different local socio-economic contexts present within the same country (Williams & Windebank, 1998). As a rule, the concept of irregular work refers to the gradual process of separation between work and formal employment that has traditionally characterised capitalist development (Leonard 1998, 3). Those persons who cross this dividing line are classified differently, at least in symbolic terms, with regard to the relationship between work, the economy and society.

If we then look at this phenomenon and its relationship with the problem of social exclusion and unemployment, then what we see are a number of socio-cultural connections, as well as the more obvious economic ones. This multi-dimensional aspect is strongly confirmed by the YUSEDER study, as one could have imagined starting from a position whereby equal weight is given to the objective and subjective nature of em-

1 Moreover, when social policy debate tends to blame those persons who depend upon public welfare, research into the real nature of this phenomenon proves even more complicated: "given the periodic campaigns against supposed abuses of the welfare benefit system in countries such as the UK and the USA, it is very unlikely that people receiving these benefits will declare their involvement in informal employment" (Leonard 1998, p. 67; Mattera, 1985).

ployment, and in this particular case, to the lack of employment (e.g. Paugam, 2000, p. 15; Bourdieu, 1997, p. 241).

In the YUSEDER study we have used the term *irregular employment or work* in order to avoid the misleading connotations inherent in the different terms mentioned above.

1 The submerged economy and long-term youth unemployment: the North-South divide

Research carried out in the six countries taking part in the YUSEDER project, basically confirms the general hypothesis that the submerged economy plays only a limited role in the lives of long-term unemployed young people in *northern European* countries. For example, the submerged economy appears very sporadically in the results of the study conducted in Belgium, and the fear of "severe (financial) sanctions" (Willems, Vanderplasschen & van Heeringen, 2001) is often given as a reason for the limited nature of this phenomenon. The submerged economy seems to be of equally limited importance to the phenomenon of youth unemployment, as is in fact shown by a study conducted in Sweden (Rantakeisu et al., 2001) which, among other things, shows the reluctance of young people to do irregular work for reasons of "civicness" and the lack of social insurance, as well as for those mentioned in relation to the Belgian case. The submerged economy seems to play a greater role in Germany. In some cases, resorting to irregular employment is justified as being the only response to severe economic difficulty, and sometimes as a preventive measure to ensure that the situation does not precipitate towards criminal involvement. The submerged economy is also made use of as it gives people a certain financial independence and maintains their standard of living at an acceptable level. Nevertheless, in the German case the effects of the submerged economy were relatively limited in the case of young unemployed people: this sector of the economy emerges from time to time and they are involved in it in an episodic way, and yet it is never really seen as a realistic alternative to regular employment.

The situation in the three *southern European* countries is very different, however. National estimates of the consistency of the submerged economy in each of the three countries in question show a socio-economic reality that is strongly conditioned by this phenomenon. As far as Greece is concerned, irregular employment is estimated at about 30% of total employment, while according to the Employment Institute of the General Workers Union in Greece, "six out of ten wage-earner workers were unemployed, illegally employed or underemployed" (Sokou et al., 2001). In Italy, the share of "irregular positions of employment" has been at about 37% of total employment (Borghi et al. 2001), with a considerable gap between the central and northern regions (with irregular employment at about 31%) and the southern regions (where it stands at almost 51%). Finally, the Spanish case also confirms the greater presence of the submerged economy in southern regions (Espluga et al., 2001). The Spanish Youth Report indicated that 38% of young people aged between 16 and 29 work without an employment contract, and the majority of them (75%) work in family businesses. This latter statistic, which further underlines the role of family-run businesses and, in more general terms, of small companies, in the submerged economy, is a feature common to all three southern European countries. Other common features included the strong presence of irregular work in farming, the building trade, tourism, manufacturing industry and non-commercial services, and it is more widespread in rural areas than in large towns and cities.

Our field research has confirmed these differences between the northern and the southern European countries involved in the study. Thirty-six cases (24%) of involvement in the submerged economy were discovered among the interviewees from the Centre and North of Europe, many of which being only sporadic and of little importance[2]; in the South of Europe, on the other hand, one hundred and twenty-one cases of long-term unemployed (81%) with experience of irregular work[3] were discovered (out of a total of one hundred and fifty people interviewed in the three countries), albeit of different kinds and with differing results. Once again, the territorial factor is shown to be of considerable importance (although it is not the only one[4]) in understanding the phenomenon of the submerged economy.

Generally speaking, in order to understand the forms taken by the involvement of unemployed youth in the submerged economy in such different contexts, we can use the following typology[5]:

a) *permanent irregular employment:* this category includes those jobs that take up a significant part of the individual's time during the course of the day and the week; these are real, permanent jobs (inasmuch as a job by definition insecure can be considered permanent), of a full-time nature, although not formally recognised as such and thus lacking any kind of social security and welfare protection or trade union rights; this is an employment situation common to a number of different sectors;

b) *irregular seasonal employment*: this category of job is characterised by an intense period of employment (often with longer than normal working hours) for a limited period of time; this kind of employment is most common in tourist areas and in farming;

c) *casual irregular employment*: this category includes all those jobs which people have done, or are currently doing, of a strictly casual nature; it includes mainly casual and often unconnected jobs which last for a very short period, which are in no way associated with the education or training of the person in question and which, as the Spanish researchers have pointed out, cannot be used towards establishing a homogeneous working experience.

These diverse forms of involvement in the submerged economy, together with the nature and quality of the work done, the specific characteristics of the local labour market and social framework, the education and work experience of the unemployed youth, their employment prospects, as well as their social and family networks, all contribute towards the effect produced by irregular work on the risk of social exclusion to which these young people are exposed.

2 For example, in the case of the Swedish study many interviewees said that the irregular jobs they had done were connected to their time as students rather than with any period of unemployment.

3 To be more exact, thirty-eight in Spain, forty-one in Greece and forty-two in Italy.

4 For a further discussion of this point, see Williams & Windebank 1998, pp. 98-111: the authors underline the spatial divisions in informal unemployment using the tentative typology of localities according to the magnitude (high/low levels of informal employment) and the character (autonomous / exploitative work) of irregular employment present in certain areas.

5 This is a classification drawn up and utilised in a similar fashion by both Spanish and Italian researchers, and which also seems to be applicable to the results of the study conducted in Greece.

2 Irregular employment and the risk of social exclusion of unemployed young people

It is clear that the greater presence of irregular employment in southern European countries must be seen from the point of view of the need to soften the financial burden of unemployment on those affected. In those countries where the unemployed receive very little or no financial support at all, there is clearly a greater incentive to involvement in the submerged economy (Leonard 1998, p. 68). Nevertheless, the economic parameter is not the only one with which to measure the consequences of irregular employment. What we have tried to do during the course of this study is to establish whether there are any other factors that may soften hardship, apart from such financial benefits; whether there are any aspects which tend to intensify the problems young unemployed experience; and finally, in which cases, and for what reasons, the first or the second aspects tend to prevail.

The various observations that emerge from the results of the studies conducted in Greece, Italy and Spain confirm the *ambiguous, ambivalent role* played by the involvement of young unemployed people in the submerged economy with respect to the increase in, or limitation of, the risk of social exclusion.

Income earned through irregular work, despite being low, often discontinuous and rarely sufficient for a person to be financially independent, does nonetheless constitute a temporary solution, often a kind of "waiting wage", whilst waiting to find something better, waiting to do military service, waiting for the employer to give you a regular employment contract, and so on. It partially satisfies the individual's financial requirements. Moreover, in some local contexts, these "shadow" jobs may constitute the only available alternative to unemployment, given the unsatisfactory nature of employment measures and the paucity of chances to create regular employment, even short-term, low-paid work and/or work suitable for first-time workers, which otherwise takes the form of irregular employment[6]. However, irregular employment perceived in its role of softening the phenomenon of economic exclusion, shows a more problematic side precisely in those places where it is more commonly found, and where in many cases it constitutes full-time employment for those involved, and thus the source of a discretely substantial income. The studies conducted in the three southern countries, in fact, show just how this employment situation often threatens to subtract valuable resources (time, motivation, etc.) from the search for regular work, and that in the case of those jobs that do not reflect a person's professional training and education or his job aspirations, often leads the individual in question into a professional blind alley.

The studies conducted in the three southern European countries highlight the ambivalent nature of irregular employment also with regard to the risk of *socio-cultural exclusion*. On the one hand, there is the opportunity to develop new social contacts through work, regardless of its irregular nature, and thus to develop a relational network which enables the individual to escape his or her restrictive family circle and to live out and develop social experiences which would otherwise be almost impossible to develop within the family. Such working experience, albeit limited from a number of different points of view, contributes either directly or indirectly to a process whereby life skills are further developed. On the other hand, however, this irregular employment, by its very nature, tends to encourage exploitation, the denial of basic rights, and the creation of humiliating, arbitrary relations, as shown by the experiences reported during the

6 A hypothesis submitted by the Research Report on the Greek situation (Sokou et al., 2001).

interviews, and which drag individuals in the opposite direction to the one we have just described. Furthermore, even when these more extreme situations are absent, the fact that such activity takes place in any case within the family sphere or within a socio-cultural context similar to the individual's original one, and/or consists in the completion of very simple, uninteresting duties, considerably hinders the development and widening of the *life skills* mentioned.

In more general terms, as far as the general risk of *social isolation* is concerned, the following may be said: If the expansion of the submerged economy in these countries[7] does not lead to the blaming and criticism of those individuals directly concerned, and this does not seem to emerge even in northern European countries such as Germany, where such a phenomenon is much less frequent, then it still allows for a partial degree of social inclusion, a reduced form of citizenship for these unemployed young people, the individual and social price of which ought to be seen not so much in relation to the present as to the future.

We can identify two general paths taken by the relationship between irregular employment and the social exclusion of long-term unemployed young people. The first of these sees *irregular employment contributing towards increasing the complexity of an unemployed young person's social relations and broadening (or keeping) the options available to that person.* By softening the economic hardship created by unemployment, irregular work enables an individual to persist in his search for regular employment whilst not binding him in an exclusive manner: it may even render this search more effective through the social contacts created by irregular employment. In fact, in some cases irregular employment increases an individual's professional and social skills through the widening of his range of experiences.

The second path taken is that by which, on the contrary, *irregular work contributes towards considerably reducing this complexity – that is, the range of accessible alternatives – in an almost irreversible manner.* It thus contributes towards rendering socio-cultural weaknesses (as well as economic difficulties) increasingly chronic, thus uniting the poor quality of the work carried out and the limited social contacts it may generate, with the limited cognitive and social resources deriving from the family and social setting.

While in the first case, it is clear that the submerged economy can be seen as a form of protection from the risk of social exclusion, in the second case it leads towards marginalization and thus to an increased degree of exposure to the risk of permanent exclusion. In general, we can see that this second case emerges in situations where the irregular work – as the only available possibility for a young unemployed person in a deprived area – is characterised by its not only economic, but also social, cognitive and cultural poverty: In such a situation, the irregular work reinforces (instead of limiting) the disadvantages of origin.

The submerged economy is also intrinsically contradictory from the point of view of the types of intervention it can generate. On the one hand, it is seen as a social problem, given the wealth it removes from the community, social dumping among workers, and failed socio-economic integration. On the other hand though, as the YUSEDER studies confirm, the submerged economy provides partial socio-economic support to individuals, thus protecting them to a certain degree against the risk of a more radical

7 To be more precise, we really ought to be talking about those regional areas within the southern European countries themselves, characterised by the highest concentration of irregular employment, and to which these observations refer.

form of social exclusion. At the practical level, however, such a contradiction may in fact be resolved by intervening on both sides of the same problem at different times.

In this way, those measures aimed at *clamping down* on the phenomenon of irregular employment can be perceived as *future-orientated*, and aimed at eliminating its structural causes and at encouraging it to gradually surface. This is the guiding principle that emerged from the final section of the YUSEDER study of the three southern European countries. These measures are not usually aimed specifically at young people as such, but are designed to clamp down on tax evasion within the submerged economy, to encourage the above-mentioned "irregular" employment to surface through the offer of economic and tax incentives, and to expand regular job opportunities through job flexibility[8] and self-employment, and so on.

At the same time, awareness of the fact that the effects of such measures take time to emerge legitimises the promotion of local projects and actions within a given social, cultural and economic context, projects which are aimed at *empowerment* and at improvements in the quality of life of those young people for whom, *as things stand*, the irregular labour market represents the only available employment option. Thus, encouragement and support for projects of this kind, where the emphasis is placed more on the overall empowerment of the young people in question than on clamping down on irregular employment, must be seen as "reducing the damage". It should not indirectly legitimise the lack of institutional measures to combat conditions which do not enable these young people to enjoy their full rights as citizens and members of the community.

3 Submerged economy and risk of social exclusion: A case study from Italy

An example of the effects of the relationship between irregular work and risk of social exclusion can be drawn from the Italian empirical research (Borghi et al., 2001, p. 388). The situation of a young long-term unemployed, living in a big city of Southern Italy (Naples) is summarised here. The analysis is based on a multidimensional approach to the concept of social exclusion (Kronauer, 1993), as it was adopted in the YUSEDER research project.

Piero is 25 years old, holds a high school diploma in electronics, and has been unemployed for six years (for ten according to official records). He lives with his mother (unemployed and separated) and his grandfather (who provides for both Piero and his mother) in a village on the outskirts of Naples.

Chronological description of the life course - Piero has always worked in irregular jobs. According to Piero, his military service interrupted a more stable job he was doing at the time (again of an irregular nature), and at the moment he works, unregistered, as a gym instructor. He says that his school diploma is practically useless. Immediately after he left school, he started doing irregular work. This job was interrupted by his being called up for military service, which constituted a turning point for him in the negative sense, both in terms of employment and in terms of his financial situation, as he was forced to use his own savings to supplement the little money he received while in mili-

8 The question of whether or not job flexibility constitutes an effective means of combating irregular employment remains open in discussions on employment policy. The views of some of the experts given in the Spanish study, for example, in fact point out that often the employer-worker relationship within irregular employment is often advantageous for both parties compared with other forms of flexible regular employment.

tary service. This marked the beginning of a precarious working career, made up mainly of casual or seasonal work, and a series of negative experiences: His jobs ranged from the factotum in a holiday village to gym instructor.

Description of the unemployment situation - Although he believes that the irregular jobs have served in some way to provide him with access to the regular labour market, in practice it would seem that the current one does not really offer this chance; he believes he needs to do some further training. Unemployment is seen as a kind of *existential condition.* According to Piero, working irregularly is a common feature of the society he lives in, but it does not enable you to make any progress in terms of the labour market or of your financial situation.

Description of the other social exclusion dimensions - His present job enables him to earn about 315 € a month, and this severely restricts one's options: » *Sometimes I'm even so hard up that I can't afford a cup of coffee, although the thing that gets me down is not the fact that I can't buy a pair of shoes, for example, or a new jacket, it's the fact that it's not normal for a person of my age.*" (I-S24: 153-161). This has also led to Piero getting into debt. He is also envious of those people who do not have his kind of problems, simply because they were born into wealthier families, even though he does not consider himself to be particularly poor: »Q: *Have you ever felt poor? A: Well, I don't know, I mean not really poor as such, because in my opinion poor means someone who's got no money at all...* " (I-S24: 162-166). Irregular work does not enable him to save money, and so the grandfather has to provide for both him and his mother.

Apart from the inadequacy of the preparation he got at school, Piero is also negative about his experiences of government institutions (job centres): » *I live outside the city, and so I get information about the jobs that come up second-hand almost, too late, the information they give is poor, they don't do their job properly, they've got their job, they do it badly, and so they don't give you the chance to have a go at the jobs on offer.*" (I-S24: 308-320). As for the banking system, Piero says he has been literally excluded from it: «*Once I tried to open a current account at the bank, I thought that if I managed to earn something I could put a few bob aside and put the money into an account, save it for some future need .. but they wouldn't let me, because the first thing they asked me was whether I was registered as employed by someone, if I had a regular job, and that was the end of that*» (I-S24: 321-326). Generally speaking, he is rather discouraged by the way the institutions treat him and has little faith in them on the whole.

He has a healthy relationship with his friends, on the other hand, who provide him with material and moral support, despite criticising him for the way he deals with his unemployment. His family both blames him and provides him with support. He is not interested in politics, he keeps himself up-to-date with what is going on from what he sees on the TV. He is a member of sporting associations, does not have any other particular interests, and is not involved in other groups or associations. Nevertheless, he notices a certain difference between himself and his contemporaries, the result of his own socio-economic condition.

He lives in a "small house", albeit equipped with the indispensable facilities, and does not exclude the possibility of being evicted at some point in the future. The area he lives in shows clear signs of degradation and delinquency, and services are poor, there are no real leisure facilities and he has got used to living in the midst of all this, and therefore does not feel that he has any particular problem with it.

Health effects of unemployment - He does not suffer from any serious psychological problems caused by his plight, although he is clearly not happy with the lack of control he exercises over his own life and with the build up of a series of little daily problems, and is frustrated as a result. He partially blames himself for his lot – "*Perhaps I'm not up to it, or I'm not ready for work, or I made a mistake studying what I did.*" (I-S24: 112-129) – although this feeling is tempered by his criticism of the failings of others with respect to the problem of unemployment. He has never thought of, let alone attempted, suicide.

Key mechanisms and categorisation of social exclusion type - The case described here thus seems to conform completely to the criteria by which we identified the high social risk category: the material, and as a result, relational problems within the family, which restrict the support that may be offered; the extremely precarious and poorly paid nature of irregular work; the consequent degree of economic exclusion, with the effects that this has on one's social life in general. The limited resources that may be utilised (the limited use for current educational qualifications, lack of institutional support, degraded environment in which one lives) in fact intensify this risk.

The scheme presented below (Figure 1) summarises the different main aspects of such a case study:

Figure 1. Case study

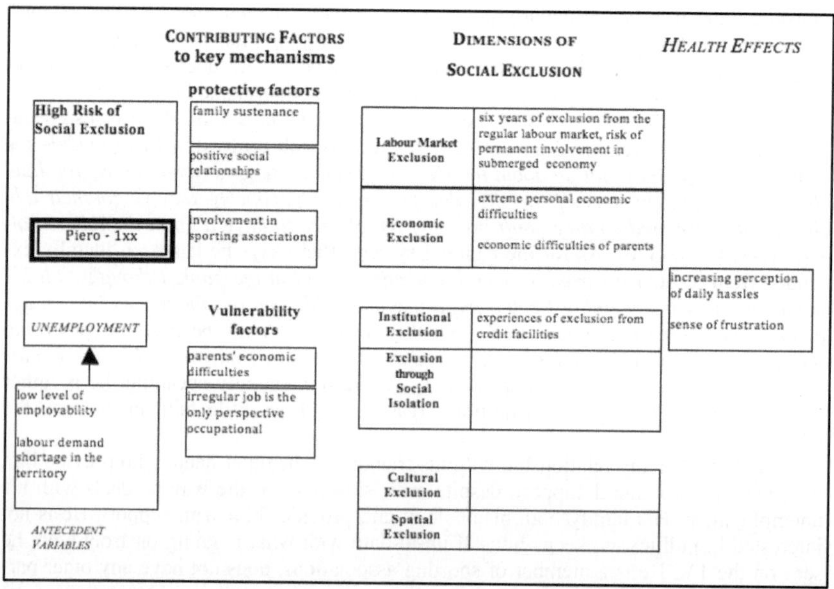

4 Conclusions

Apart from the economic support (constant or otherwise, depending on the individual in question) that the submerged economy may provide to long-term unemployed young people, everything else (on the one hand the alleviation of social isolation, on the other hand involvement in a vicious circle of marginalization; the maintaining of some form of contact with the world of work, but at the same time the worsening of the individual's chances of ever getting a regular job, and so on) may vary considerably. This depends on the kind of job done, the skills and training of the individual, the gap separating the irregular job a person in fact does and the regular one he hopes to find, and so on. With regard to the link between irregular employment and the risk of social exclusion, there is evidence that certain features of the submerged economy make long-term unemployment more bearable from the economic and social points of view. However, it also has to be said that it may lead to the *partial social recognition* of those individuals in question. In denying access to the collective identity normally associated with regular employment, the submerged economy denies individuals the full experience of intersubjective recognition which lies at the root of their very integrity (Honneth, 1995). Among other things, this incomplete recognition, within a socio-political context characterised by an increasing divide between standard employment, accompanied by full social integration, and precarious employment (Castel, 1995), of which irregular employment is the most extreme example, may lead to splits and conflicts among workers themselves[9].

At the same time, irregular employment may also be seen in a positive light: the view that the formal economy constitutes a benchmark, and thus involvement in the informal economy is an indication of failure, may in fact be overturned. As other scholars have also shown (MacDonald, 1994; Bryson & Jacobs, 1992), the very ability of individuals to find some form of work in extremely difficult socio-economic conditions, and to execute this work despite the limitations, insecurity and hardship that the submerged economy implies, highlights the abilities and skills of such subjects in coping with the failure of the economic system and the welfare state in achieving the socio-economic integration of citizens. However, once again we need to bear in mind that such "submerged resources" must be seen as the starting point from which to launch projects aimed at their empowerment and the consolidation of their motivation and initiative[10], rather than as the justification for reducing the support provided to unemployed people.

As we can see, therefore, the ambivalent role played by the submerged economy with regard to the risk of social exclusion makes it impossible to come up with a simple, clear interpretation. Hence the need for the constant monitoring of its evolution, both as

---

9 As Espluga et al. (2001, p. 411) write: "Workers who perform submerged, poorly qualified, occasional jobs in different places have great difficulties relating to stable workers in the companies which offer them submerged jobs. They have even more difficulty in obtaining support in any kind of complaint or claim they may make. Often, the submerged jobs are performed in unsuitable working conditions from the point of view of health, working hours and salary, and the lack of support from stable workers (and from the unions) does not help to resolve the problems. This means that not all submerged jobs lead to an extension of the network of social relations and the most occasional and disparate jobs can increase the experience of loneliness for the young person performing them".

10 Other studies have highlighted the fact that involvement in the submerged economy may contribute towards the subject's perception of his playing an active part in the economic sphere, whereas objectively speaking this involvement in fact exacerbates the subject's exclusion from such a sphere, and thus encourages him to accept his condition (Ditton & Brown, 1981).

a significant phenomenon in its own right, and as an important key to a full understanding of social exclusion, especially in southern Europe.

References

Borghi, V., Chicchi, F. & La Rosa, M. (2001). Empirical analysis of the risk of social exclusion of long-term unemployed young people in Italy. In T. Kieselbach, K. van Heeringen, M. La Rosa, L. Lemkow, K. Sokou & B. Starrin (Eds.), *Living on the edge. An empirical analysis on long-term youth unemployment and social exclusion in Europe* (pp. 319-392) (Psychology of Social Inequality, vol. 11). Opladen: Leske+Budrich.

Bourdieu, P. (1997). *Méditations pascaliennes* [Pacalian meditations]. Paris: Seuil.

Bryson, A. & Jacobs, J. (1992). *Policing the workshy.* Aldershot: Avebury.

Castel, R. (1995). *Les metamorphoses de la question sociale. Une chronique du salariat.* [Metamorphoses of the social question. A Chronic of Employment]. Paris: Fayard.

Ditton, J. & Brown, R. (1981). Why don't they revolt: Invisible income as a neglected dimension of Runciman's relative deprivation thesis. *British Journal of Sociology, 32,* 521-30.

Espluga, J., Baltiérrez, J. & Lemkow, L. (2001). Empirical analysis of the risk of social exclusion of long-term unemployed young people in Spain. In T. Kieselbach, K. van Heeringen, M. La Rosa, L. Lemkow, K. Sokou & B. Starrin (Eds.), *Living on the edge. An empirical analysis on long-term youth unemployment and social exclusion in Europe* (pp. 393-450) (Psychology of Social Inequality, vol. 11). Opladen: Leske+Budrich.

Honneth, A. (1995). *The struggle for recognition.* Cambridge: Policy Press.

Kieselbach, T. (Ed.) in collaboration with K. van Heeringen, M. La Rosa, L. Lemkow, K. Sokou & B. Starrin (2000a). *Youth unemployment and health. A comparison of six European countrie* (Psychology of Social Inequality, vol. 9). Opladen: Leske+Budrich.

Kieselbach, T. (Ed.) in collaboration with K. van Heeringen, M. La Rosa, L. Lemkow, K. Sokou & B. Starrin (2000b). *Youth unemployment and social exclusion. A comparison of six European countries,* Psychology of Social Inequality, vol.10). Opladen: Leske+Budrich.

Kieselbach, T., Heeringen, K. van, Lemkow, L., Sokou, K. & Starrin, B. (Eds.). (2000). *Living on the edge - a comparative study on long-term youth unemployment and social exclusion in Europe.* (Psychology of Social Inequality, vol.11). Opladen: Leske+Budrich.

Kieselbach, T., Beelmann, G., Traiser, U. & Meyer, R. (2001). Empirical analysis of the risk of social exclusion of long-term unemployed young people in Germany. In T. Kieselbach, K. van Heeringen, M. La Rosa, L. Lemkow, K. Sokou & B. Starrin (Eds.), *Living on the edge. An empirical analysis on long-term youth unemployment and social exclusion in Europe* (pp. 183-242) (Psychology of Social Inequality, vol. 11). Opladen: Leske+Budrich.

Kronauer, M. (1998). 'Social exclusion' and 'underclass' – new concepts for the analysis of poverty. In H.-J. Andreß (Ed.), *Empirical poverty research in a comparative perspective* (pp. 51-75). Aldershot: Ashgate.

La Rosa, M. & Kieselbach, T. (Eds.). (1999). *Disoccupazione giovanile ed esclusione sociale. Un approccio interpretativo e primi elementi di analisi* (Youth unemployment and social exclusion. An interpretative approach and first attempts of analysis) (Sociologia del Lavoro - Teorie e Ricerche, vol. 47). Milano: Franco Angeli.

Lemkow, L., Baltierrez, J. & Kieselbach, T. (Eds.). (2001) Atus Juvenil, Salut i Exclusió Social. Recerques, Experiencias i Acciones Institutionals a Espanya [Youth unemployment, health and social exclusion, research, experience and institutional responses in Spain] Barcelona: Autonomous University of Barcelona.

Leonard, M. (1998). *Invisible work, invisible workers. The informal economy in Europe and the US.* London: MacMillan,

MacDonald, R. (1994). Fiddly jobs. Undeclared working and the something for nothing society. *Work, Employment and Society, 4,* 507-30.

Paugam, S. (2000). *Le salarié de la précarité* [The precariously employed]. Paris: Puf.

Rantakeisu, U., Forsberg, E., Kalander-Blomqvist, M., Löfgren, U.B., Johansson, M. & Starrin, B. (2001). Empirical analysis of the risk of social exclusion of long-term unemployed young people in Sweden. In T. Kieselbach, K. van Heeringen, M. La Rosa, L. Lemkow, K. Sokou & B. Starrin

(Eds.), *Living on the edge. An empirical analysis on long-term youth unemployment and social exclusion in Europe* (pp. 77-138). (Psychology of social inequality, vol. 11). Opladen: Leske+Budrich.

Sokou, K., Bayetakou, D. & Papantoniou, V. (2001). Empirical analysis of the risk of social exclusion of long-term unemployed young people in Greece. In T. Kieselbach, K. van Heeringen, M. La Rosa, L. Lemkow, K. Sokou & B. Starrin (Eds.), *Living on the edge. An empirical analysis on long-term youth unemployment and social exclusion in Europe* (pp. 243-318) (Psychology of social inequality, vol. 11).Opladen: Leske+Budrich.

Willems, T., Vanderplasschen, W. & van Heeringen, K. (2001). Empirical analysis of the risk of social exclusion of long-term unemployed young people in Belgium. In T. Kieselbach, K. van Heeringen, M. La Rosa, L. Lemkow, K. Sokou & B. Starrin (Eds.), *Living on the edge. An empirical analysis on long-term youth unemployment and social exclusion in Europe* (pp. 139-182) (Psychology of social inequality, vol. 11). Opladen: Leske+Budrich.

Williams, C.C. & Windebank, J. (1998). *Informal employment in the advanced economies. Implications for work and welfare.* London/New York: Routledge

The Design of Previous Job and Vocational Behaviour during Unemployment

Annika Lantz & Kin Andersson

Introduction

Experiences at work not only have an immediate impact on work-related behaviour, well-being, and health but generalize through socialization at work into attitudes and behaviours away from the workplace (Kohn & Schooler, 1973, 1978, 1982). Low-qualified work, lack of job-training, and a poor work situation have a negative impact on the individual's hope and belief in his or her own possibilities to find other work as well as on job-seeking behaviour (Aronsson et al., 2000; Aronsson & Göransson, 1997; Nutec, 2000; Wirkkala, 2002).

Baethge (2003) argues that efforts to reduce unemployment, especially long-term unemployment, should not begin once an individual is unemployed, but rather at the workplace with preventive measures to enhance the individual's learning possibilities.

The aim of the study is to test, and extend the model of personal initiative presented by Speier and Frese (1997) and its application by giving empirical evidence of a positive relationship between the *design of the previous job* on the one hand, and *personal initiative, self-efficacy, competence efficacy, and mobility-related attitudes and activities in unemployment* on the other hand.

1 Self-efficacy, competence efficacy, and mobility-related attitudes and activities

Job-seeking behaviour is partly influenced by general self-efficacy and task-specific measures of self-efficacy (Eden & Aviram, 1993; Moynihan et al., 2003). Self-efficacy beliefs are the individuals' judgements of their capabilities to organise and execute courses of action required to attain designated types of performance (Bandura, 1986). Self-efficacy determines the initial decision to perform, the effort expended and persistence in the face of obstacles. According to Bandura (1986), self-efficacy is developed throughout an individual's learning history and one important source is enactive mastery; i.e. being able to make decisions, to perform challenging tasks, and to make use of one's competencies. Speier and Frese (1997) view work-related self-efficacy as part of an occupational socialization process. At work, control and work task complexity are crucial for the development of self-efficacy (Speier & Frese, 1997).

Self-efficacy varies in generality. Self-efficacy while job hunting refers to one's confidence in performing tasks that are important in the job search process (Bandura, 1997; Kanfer & Hulin, 1985; Vinokur & Schul, 1997). Competence efficacy refers to the individual's belief in how useful and sought after his/her competencies are on the labour market. Self-efficacy and competence efficacy are related, and affect job-seeking activities (for a review see Moynihan et al., 2003).

Job-seeking activities have been measured by single items as well as in two-dimensional scales including preparatory and active search behaviour among unemployed individuals (Saks & Ashforth, 1999; Wanberg et al., 1999). Mobility-related attitudes and activities refer to attitudes towards mobility, proactive measures to avoid unemployment, and to job-seeking activities during unemployment (Fay & Frese, 2001; Filipczak, 1995; Romaniuk & Snart, 2000).

2 Personal initiative at work and in unemployment

Personal initiative is a behavioural trait, characterized by the individual's taking an active and self-starting approach to work, and going beyond what is formally required in a given job (e.g. Fay & Frese, 2001; Frese et al., 1997; Frese & Zapf, 1994). It is characterized by complying with the organisation's mission, a long-term focus, goal- and action orientation, persistence in the face of barriers, and by taking initiatives and being proactive (Speier & Frese, 1997). Taking initiatives implies that the employee translates externally given tasks into internal tasks through reconstruction, and this process allows employees to define extra-role goals that lie outside role requirements (Fay & Frese, 2001; Frese et al., 1997; Frese et al., 1996).

Personal initiative is linked to initiative at work in terms of managing uncertainty, taking responsibility, being innovative and responding to change and challenge (Bateman & Crant, 1993; Fay & Frese, 2000; Fay & Frese, 2001; Frese et al., 1996; Frese et al., 1997; Frese et al., 1999), and to organisational citizen behaviour (as defined by Organ, 1990). Further, the concept of personal initiative is linked to success in getting a job, the need for achievement, coping with stress, making career plans and carrying them out. It is also linked to interest in self-employment, too (Frese et al. 1997).

3 Job design and its impact on self-efficacy and personal initiative at work

Frese (2001) presents a model of personal initiative at work, in which three approaches to enhancing personal initiative are presented: Environmental support can be increased, skills can be developed, and orientations can be changed. Personality factors are more constant. Environmental support includes control, complexity, social support and support for personal initiative at work. A range of studies show that differences regarding personal initiative at work are related to differences in degrees of control, autonomy and complexity afforded to employees (Fay & Frese, 2001; Frese et al., 1997; Frese et al., 1996).

Speier and Frese (1997) tested a model of the relation between control and complexity and personal initiative with self-efficacy as an intervening variable at work. The results are based on subjective measurements. The results confirm the mediating effect of self-efficacy in the relation between control and complexity and personal initiative (Speier & Frese, 1997).

The predominant method of operationalising control and autonomy is the subjective measurement of the individual's perception of degrees of freedom in planning and carrying out the work tasks whereas complexity is measured in terms of how difficult the job is to perform (see e.g, Breaugh, 1985, 1999; Breaugh & Becker, 1987; Frese et al., 1996; Speier & Frese, 1997). Richter et al. (2000) point out that an imprecise definition of the concepts of control, autonomy and complexity, as well as a lack of clarity in methodology has led to disparate results in the study of work, individual behaviour and health.

4 Work task analysis and a differentiated measure of job design

Within work psychology and action regulation theory, there is a long tradition of work task analysis for the design of a humane work (Hacker, 2001). A humane work gives opportunities to deal with physically and mentally varying tasks that can be predicted, understood, and influenced by the worker (Richter et al., 1999). Action regulation theory provides a broad-ranging list of work task characteristics and aspects of job design that are important for an individual's learning, personal development, motivation and health (Frese & Zapf, 1994; Hacker, 2001; Richter et al., 1999; Ulich, 1990; Ulich & Weber, 1996; Volpert et al., 1983). Some of this knowledge has not previously been taken into account by international research on vocational behaviour. One reason for this is probably that several of the most interesting studies addressing this issue have only been published in German.

Work task analysis by expert assessment is regarded as a more valid tool than subjective measurement, and some variables cannot be shown through interviews or questionnaires (Hacker, 2001; Richter et al., 1999; Volpert et al., 1983). We use work task analysis and this elaborated framework to explore the design of the previous work. We propose that the aspects of job design, which are most predominant in the study of personal initiative, self-efficacy, competence efficacy and mobility-related attitudes and activities are too limited. Although careful work task analysis is a tedious and time-consuming method, it is needed in order to get detailed data that cannot be provided by questionnaires. By providing a richer framework for job design, we might be able to give practitioners better tools for designing jobs, and contribute to creating conditions that have a positive impact on vocational behaviour at work and in eventual unemployment later.

4.1 Dimensions of job design that have impact on personal development and learning

Organisational and technological conditions determining the completeness of work (Completeness). A complete work structure (Frese & Zapf, 1994; Hacker, 2001, 2003) includes preparing, executing, checking and organizing. Goal-setting, planning, decision/action programmes and autonomous decision-making are essential parts of the planning phase. The conditions determining the completion of work are: number and type of job tasks; frequency of performance; scope and content of organisational functions assigned to the employee; required information about the work organisation and the results, and absence of goal conflicts. The completion of work includes aspects previously studied in relation to personal initiative as well, such as control, autonomy and the participation (Frese & Zapf, 1994). Autonomy and control are two separate parts of the work task, complete and distinct from complexity, see below (see Richter et al., 1999).

Demand on cooperation and communication (Demand on cooperation). Work activities are social and intrinsically cooperative. A certain work task can be performed individually, but it is always related to a context in which others are present. Ever since the Hawthorne studies (Mayo, 1933), research has provided substantial evidence that well designed jobs contain opportunities to cooperate within the work context. Richter et al. (1999) showed that demand for cooperation and communication in terms of the type of cooperation, amount of cooperation and content of communication are important aspects of the job's inherent learning opportunities.

Responsibility resulting from tasks (Demand on responsibility). Responsibility resulting from tasks includes the responsibility for the fulfilment of quantity targets, meeting deadlines or for the adherence to quality parameters within the individual's own work tasks. Responsibility for material assets, information, cost development, and for the safety and health of fellow employees is conducive to learning. Group responsibility exists when sub-tasks regarding preparation, organisation and control have to be completed collectively to fulfil a certain task and if the work done is balanced, evaluated and remunerated with regard to the group (Richter et al., 1999).

Required cognitive operations (Cognitive demand). A well-designed job gives opportunities to deal with demanding tasks; i.e. tasks that are mentally stimulating and require creativity at least some of the time (Hackman & Oldham, 1980; Fay & Frese, 2001; Hacker, 2001; Volpert et al., 1983). Among action regulation theorists, there is a long tradition of analyzing work task complexity as the degree of demand on cognition, in terms of required cognitive operations for planning, decision-making, and problem solving (Hacker, 2001; Volpert et al., 1983). Cognitive regulation requirements are measured through a careful analysis of the relation between work activity and goal for that specific activity. The scale's lowest point describes a work task that requires only sensory motor regulation; that is, it is purely routine, while at the other end of the scale the work task requires the establishment of new working processes and puts challenging demands on planning, decision-making, and problem-solving capacity.

Qualification and learning opportunities (Learning opportunities). Qualification and learning opportunities refer to maintaining and using acquired skills and to continuous learning. Maintenance of skills concerns formal training and previous work experience. Enlargement of qualifications and abilities are related to the extent the work tasks require continuous learning (Richter et al., 1999).

5 Research model

The hypothetical model employed for the current study is presented in Figure 1. The model is an extension of the model by Speier and Frese (1997) in which self-efficacy is a mediating variable between control and complexity and personal initiative at work. In the hypothetical model, self-efficacy is conceptualized as both general self-efficacy and competence efficacy. Vocational behaviour in terms of mobility-related attitudes and activities during unemployment is an effect of personal initiative.

Figure 1. A hypothetical model of the causal relation between Job Design in five dimensions, Self-Efficacy, Competence Efficacy, Personal Initiative, and Mobility-Related Attitudes and Activities

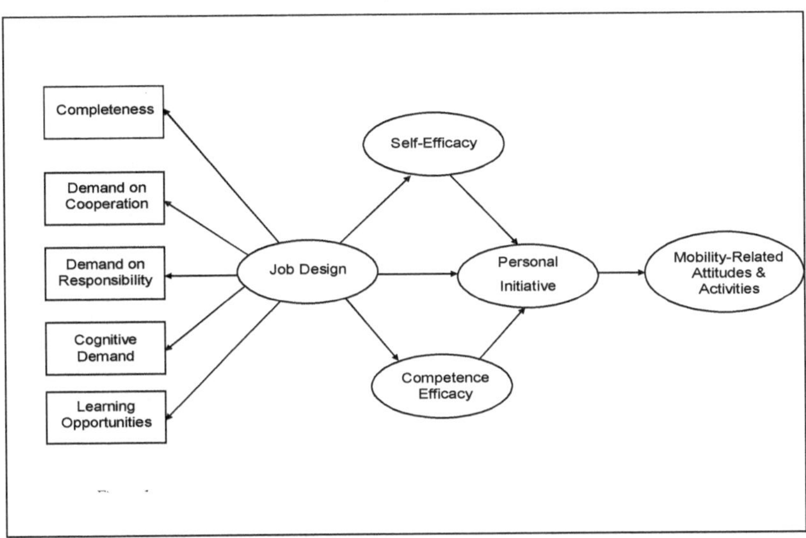

The results are based on a detailed work task analysis of the previous position held and self-reported data on behaviour and attitudes during unemployment by individuals who had been made redundant through the downsizing of a manufacturing plant that was part of a large multi-national company in Sweden. A major restructuring was initiated at a time when Swedish production plants had all been affected by an outsourcing trend. It was anticipated that the company under study would be affected as well. There were earlier media reports of redundancies within the region and elsewhere, and the labour unions were involved in various negotiations. Finally, about half the production line was closed and staff were made redundant after more than a year of uncertainty and negotiations.

Work task analyses of all different work tasks in the various jobs were carried out while the production was still running. The data was analysed by two independent experts. All former employees made redundant were asked to fill out the questionnaire three to six months after being dismissed. At that point, they were all unemployed.

5.1 Sample

A total of sixteen different jobs were analyzed. Participating were 201 (out of in total 217) former employees at the manufacturing plant that had carried out the jobs under study. The response rate was 93 %. It was possible to carry out work task analyses for 176 of these 201 employees.

Measures

During the last thirty years, a valid system for work task analysis has been developed as a tool for human-centred job design in Germany (Frese & Zapf, 1994; Hacker, 2001; Richter et al., 1999; Ulich, 1990; Ulich & Weber, 1996). The REBA instrument (in German: Rechnergestütztes Dialogverfahren zur psychologischen Bewertung von Arbeitsinhalten) is a semi-standardised toolkit for work analysis. It is intended for the analysis, evaluation and design of work content and job design with the objective of increased effectiveness, optimal mental load, health and personal development at work. The latter is understood as maintenance and development of qualifications and abilities, maintenance and stimulation of work motivation, and maintenance and strengthening of physical and mental health (Richter et al., 1999).

The starting point for the work task analysis is the job and the different work tasks it comprises. Data is based on observations of staff carrying out each specific work task and on complementary interviews with staff. After gathering background information, such as documents describing equipment used, the work structure and the workflow of each job were studied. The work structure was described in terms of the number of work elements or tasks. Once the workflow had been described, the time needed for each work task was noted. Each work task was measured according to 22 different scales and these were grouped into five dimensions (see description above). Each task, as well as the job as a whole, received a profile based on the values of the scales in each of the dimensions as well as an overall value of the 22 ordinal scales.

Observer reliability was attained by using a guide handbook for the REBA instrument, supervision by experts, and by using two independent observers for the analysis. When discrepancies in evaluation between the observers were found, these work tasks were analysed further. The initial inter-rater reliability was $r = .96$.

Cognitive Demand was measured with one score for the principal task, as well as the lowest and highest demands for the work tasks in the job as a whole. The correlations between the three measures were high and statistically significant ($p < .01$), and internal consistency showed $\alpha = .96$, hence only one score based on the score for the principal task was used in the analysis.

The Questionnaire

Personal initiative was measured with a set of nine items developed by Frese and coworkers (Frese et al., 1997; Frese et al., 1996) with instructions to answer the questions in the perspective of taking initiatives in the current unemployment situation. *Self-efficacy* was measured with nine items (Koskinen et al., 1999; Schwarzer & Jerusalem, 1993) with instructions to answer the questions with regard to the current unemployment situation. *Competence efficacy* was measured with six items partly constructed for this study, and an exploratory study was performed to test the items' suitability (Fogde & Lundqvist Medén, 2002). *Mobility-related attitudes and activities* were measured with 12 items, partly constructed for this study, and partly by Wirkkala (2000). To further test the reliability, an exploratory study was conducted in a company not part of the study (Fogde & Lundqvist Medén, 2002). Table 1a/b shows M, SD, and intercorrelations for the main variables. Based on these questions, an attempt was made to arrive at multi-dimensional measures founded on the general theoretical model illustrated in Figure 1.

Table 1a. Correlations among the main variables, means, standard deviations and Cronbach alpha (α) (N = 176)

	1	2	3	4	5	6	7	8	9	10
1. Personal Initiative	–									
2. Self-efficacy	.28**	–								
3. Competence Efficacy	.43**	.31**	–							
4. Mobility related Activities & Attitudes[a]	.22**	.12	.17*	–						
5. Completeness (13 item)[a]	.21**	.12	.20**	.03	–					
6. Cooperation (3 item)[a]	.18**	.10	.13*	-.01	.74**	–				
7. Responsibility (2 item)[a]	.17*	.10	.24**	-.01	.62**	.77**	–			
8. Cognitive de mand (2 item)[a]	.16*	.12	.13*	.07	.73**	.90**	.72**	–		
9. Learning (2 item)[a]	.12	.11	.20**	.08	56**	.79**	.74**	.88**	–	
10. Job Design *total* (22 item)[a]	.21**	.13*	.21**	.03	91**	92**	.82**	.91**	.82**	–

a = built on standardized values.
*= $p > .05$, ** = $p > .01$, one tailed.

Table 1b. Means, standard deviations and Cronbach alpha (α) (N = 176)

	M	*SD*	*α*
1. Personal Initiative	5.06	.90	.87
2. Self-efficacy	5.09	.65	.63
3. Competence Efficacy	3.49	.95	.73
4. Mobility related Activities & Attitudes[a]	.01	.49	.63
5. Completeness (13 item)[a]	.0	.49	.72
6. Cooperation (3 item)[a]	.0	.90	.87
7. Responsibility (2 item)[a]	.0	.87	.64
8. Cognitive demand (2 item)[a]	.0	.94	.81
9. Learning (2 item)[a]	.0	.96	.90
10. Job Design *total* (22 item)[a]	.0	.59	.91

6 Research results

At the production plant, there were sixteen different jobs. They were each carefully analysed and evaluated. In terms of Completeness of Work, the results show that the jobs were all designed in such a way that the worker has some possibility to prepare, execute, check and organise his or her own work. Demand on Cooperation, Demand on

Responsibility, Cognitive Demand and Learning Opportunities were overall at the lower end of the scales. The jobs were mostly of routine character.

The hypothesized model (Figure 1) was tested using LISREL 8.30 (Jöreskog & Sörbom, 1993) and employing maximum likelihood estimation on the covariance matrix. To allow a proper examination of the hypothetical model two sets of analyses were carried out, one using the overall measure of job design (see Figure 2) and one with the five work task dimensions kept separate for each of the two data sets (see Figure 3). As the reliability estimates of the manifest variables affect the parameters in the model, the error variances of the manifest variables were calculated using reliability estimates (see Jöreskog & Sörbom, 1993, pp. 37-38). This procedure allows an analysis of the linear structural relations among the latent rather than the manifest variables. To simplify the presented models, the manifest variables are not depicted in the figures. Further, in all the tested models the five work task dimensions of job design were allowed to correlate as they were correlated in the present sample.

Model fit was determined by using χ^2 tests first, but as sample size affects the χ^2 value, the Root Mean Square Error of Approximation (RMSEA $\leq .06$ indicating good fit) was used as well.

The final causal model (Figure 2), based on personal work experience (REBA variables) confirmed all but one (between Job Design and Personal Initiative) of the hypothesized paths in Figure 1. Self-Efficacy and Competence Efficacy mediate the relationship between Job Design and Personal Initiative but there is no direct path between them. All path coefficients (partial regression) are statistically significant at $p < .05$. Self-Efficacy is correlated with Competence Efficacy at a statistically very significant level $p < .001$). This model showed a very good fit to the data, χ^2 (4) = 3.04, p = .55, RMSEA = 0.00.

In Figure 3 on the next page (N =176), all five dimensions of Job Design are examined separately. The path from Job Design to Self-Efficacy is marginally significant ($p = .057$) whereas all other path coefficients show statistical significance ($p < .05$). All five dimensions of Job Design provide a significant contribution. Self-Efficacy is correlated with Competence Efficacy ($r = .28$, $p < .001$) and the five dimensions of Job Design are correlated in pairs. The model shows a good fit to the data, $\chi^2(20) = 23.31$, p = .27, RMSEA = .03.

The aim of the study was to test and extend the model of personal initiative presented by Speier and Frese (1997) and its application by data concerning the relationship between the design of the previous job on the one hand, and personal initiative, self-efficacy, competence efficacy, and mobility-related attitudes and activities during unemployment on the other hand. Job design was operationalised as a) organisational and technological conditions determining the completeness of work, b) demand on cooperation and communication, c) responsibility resulting from tasks, d) required cognitive operations, and e) qualification and learning opportunities. Some of these dimensions were to our knowledge rather new for the research into personal initiative and vocational behaviour during unemployment.

The results are based on detailed work task analysis of the previous work and self-reported data on behaviour and attitudes during unemployment among former employees (n = 176) who were made redundant through downsizing at a large Swedish manufacturing plant.

Figure 2. Causal model of *personal experience* (N = 176) of Job Design, Personal Initiative, and Mobility-Related Attitudes and Activities through Self-Efficacy and Competence Efficacy.

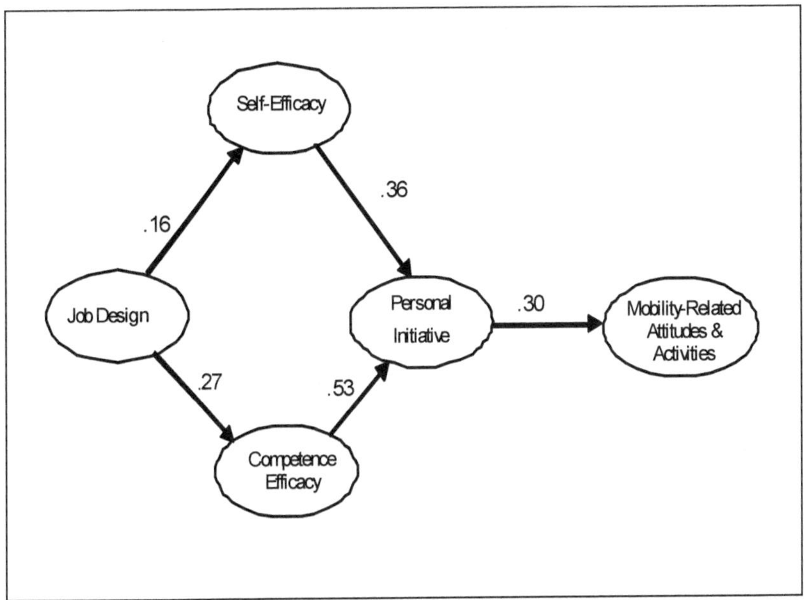

The background variables (sex, age, marital status, having children at home and education) when kept separate provide no significant contribution to the variance in Personal Initiative in a regression analyses. The background variables as one measurement showed $R^2 = .073$, $p = .01$.

The results confirm previous results on the relations between job design, and self-efficacy and personal initiative (Fay & Frese, 2001; Frese et al., 1997; Frese et al., 1996). However, during unemployment self-efficacy and competence efficacy have more impact on personal initiative than job design. The job design of previous work was important for personal initiative during unemployment, with regard to how job design enhances self-efficacy and competence efficacy as mediating variables. The results suggest that the model (1997) of personal initiative by Speier and Frese can be extended beyond the work situation.

Since personal initiative is of importance in most jobs, as well as during unemployment, it is imperative to know the dimensions of job design that increase or hinder the development of personal initiative. Detailed work task analysis is worthwhile since it captures aspects previously not accounted for, such as aspects of adequacy, demand

Figure 3. Causal model of *personal experience* (N = 176) of job *in five dimensions*, Personal Initiative, and Mobility-Related Attitudes and Activities through Self-Efficacy and Competence Efficacy.

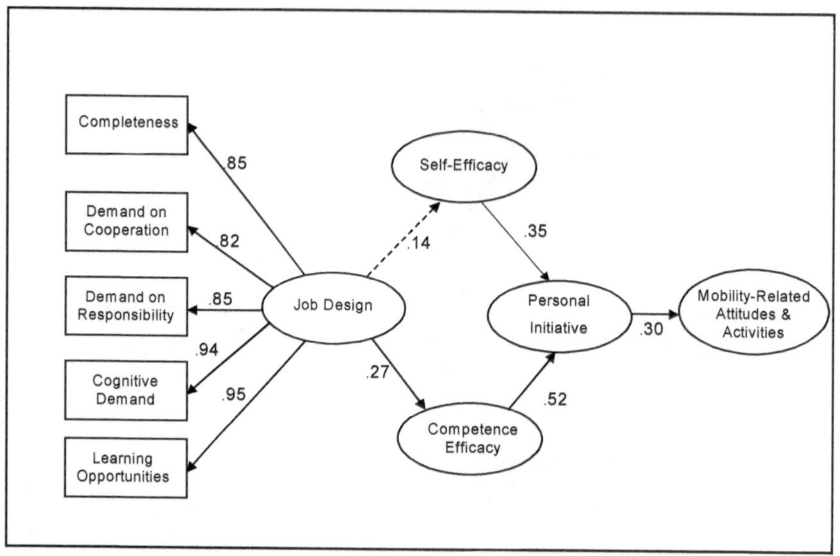

on cooperation and on responsibility, cognitive demand as action regulation, and learning opportunities. Some of these aspects cannot be examined with questionnaires. Human-centred job design (Richter et al., 1999) is important not only for the individual at work but also for the individual's coping with eventual unemployment and for labour market mobility in general. The results are supported by other findings as well. Saks and Ashforth (1999), Wanberg et al. (1999) and Moynihan et al. (2003) showed that self-efficacy is related to job search behaviour.

It is somewhat surprising that personal background has little impact on personal initiative in this study. However, these results are supported in an overview of the literature (Baethge, 2003). Early socialization is in adulthood related to initiative-taking and life-long learning, but socialization at work has a great impact on behaviour and attitudes as well (Kohn & Schooler, 1973, 1978, 1982).

7 Some limitations of the study

One major limitation is that the research group in general had jobs that were low-skilled and routine which did not give much input for personal development or learning. It is likely that a research group with larger differences in job design, especially cognitive demand, would show stronger relations between the design of previous work and personal initiative in unemployment. However, this does not affect the presented results; on the contrary, it supports the fact that the relations shown are valid. Employees with low-skilled jobs are also the most vulnerable group in the restructuring of the labour market (von Otter, 2003) and from that perspective it is of importance to study how job design affects these groups' mobility when unemployed.

The research group was not large and only 16 jobs were analysed. The research design makes it difficult to utilize large samples. The measurement of self-efficacy and mobility-related attitudes and activities should be improved. Although their reliability is higher than .60 and, therefore sufficient (Nunnally, 1987), the reliability is on the lower end of acceptability.

8 Implications for future research

An important task for future research is to replicate the study in a setting where the variance in job design is large. Further, this study raises the question as to what characterizes those individuals who are vocationally more mobile than others, with regard to personal background, work experiences, competencies and life-situation.

Research on the predictors of mobility and employment outcomes can provide important information for career guidance counsellors. However, the results also should be of importance for job design. Much can be done to design jobs that enhance productivity, long-term learning, initiative taking and vocational mobility. This mobility, under certain conditions, is likely to have a positive impact on individual learning and development and on innovation at the organisational level (Brown et al., 2001; Michie & Sheehan, 2003).

The relationship between learning possibilities and personal development has a socio-structural dimension. The concept of humane job design needs to be implemented to improve low-skilled jobs. Poor working conditions will affect not only the individuals' well-being at work, but it will also increase differences between various labour market and social segments. Socialization at work can be a second chance for those who are underprivileged in terms of education.

Acknowledgements

The authors would like to thank Bo Ekehammar, Nazar Akrami, and Anna-Christina Blomkvist, for valuable help with the LISREL statistics, and Peter Friedrich and Peter Richter for sharing their expertise in work task analysis.

References

Aronsson G., Dallner, M. & Gustafsson, K., (2000). Yrkes- och arbetsplatsinlåsning. En empirisk studie av omfattning och hälsokonsekvenser [Locking-in, in one's job and at the workplace. An empirical study of prevalence and health consequences]. *Arbete och Hälsa, 5,* 1-25.

Aronsson, G. & Göransson, S. (1997). Fasta anställningen men inte det önskade jobbet [Permanent employment but not the desired position]. *Arbetsmarknad & Arbetsliv, 3,* 193-205.

Baethge, M., (2003). Lebenslanges Lernen und Arbeit. [Life-long learning and work]. *SOFI-Mitteilungen, 31,* 91-104.

Bandura, A. (1986). *Social foundations of thought and action: A social cognitive theory.* Englewood Cliffs, NJ: Prentice-Hall.

Bandura, A. (1997). *Self-efficacy: The exercise of control.* New York: W. H. Freeman.

Bateman, T. S. & Crant, J. M. (1993). The pro-active component of organisational behaviour: A measure and correlates. *Journal of Organisational Behaviour, 14,* 103-118.

Breaugh, J.A. (1985). The measurement of work autonomy. *Human Relations, 38,* 551-570.

Breaugh, J.A. (1999). Further investigation of the work autonomy scales: Two studies. *Journal of Business and Psychology, 13,* 357-373.

Breaugh, J.A. & Becker, A.S. (1987). Further examination of the work autonomy scales. Three studies. *Human Relations, 40,* 381-400.

Brown, P., Green, A. & Lauder, H. (2001). *High skills. Globalization, competitiveness, and skill formation.* Oxford University Press.

Eden, D. & Aviram, A. (1993). Self-efficacy training to speed reemployment: Helping people to help themselves. *Journal of Applied Psychology, 78,* 352-360.

Fay, D. & Frese, M. (2000). Conservative at work: Less prepared for future work demands? *Journal of Applied Social Psychology, 30,* 171-195.

Fay, D. & Frese, M. (2001). The concept of personal initiative: An overview of validity studies. *Human Performance, 14,* 97-124.

Filipczak, B. (1995). You're on your own: Training, employability, and the new employment contract. *Training, 1,* 29-34.

Fogde, S. & Lundqvist Medén, S. (2002). *Motivation till arbete och mobilitet: Ett steg i konstruktionen av mätinstrument [Work and mobility motivation: A step toward constructing a measurement instrument].* Unpublished manuscript, Eskilstuna, Sweden: Mälardalens högskola, Department of Social Science.

Frese, M. (2001). Dynamic self-reliance: An important concept for work in the twenty-first century. In C.L. Cooper & S.E. Jackson (Eds.), *Creating tomorrow's organisations* (pp. 399-416). Chichester, England: Wiley.

Frese, M., Fay, D., Hilburger, T., Leng, K. & Tag, A. (1997). The concept of personal initiative: Operationalization, reliability and validity in two German samples. *Journal of Occupational and Organisational Psychology, 70,* 139-161.

Frese, M., Kring, W., Soose, A. & Zempel, J. (1996). Personal initiative at work: Differences between East and West Germany. *Academy of Management Journal, 1,* 37-63.

Frese, M., Teng, E. & Wijnen, C.J.D. (1999). Helping to improve suggestion systems: Predictors of making suggestions in companies. *Journal of Organisational Behaviour, 20,* 1139-1155.

Frese, M. & Zapf, D. (1994). Action as the core of work psychology: A German approach. In H.C. Triandis, M.D. Dunette, & M.H. Leatta (Eds.), *Handbook of industrial and organisational psychology* (2nd ed.) (pp. 271-340). Chicago: Consulting Psychologist Press.

Hacker, W. (2001). Activity theory. In N.J. Smelser & P.B. Baltes (Eds.), *International encyclopedia of the social and behavioral sciences* (pp. 58-62). Amsterdam: Elsevier.

Hackman, J.R. & Oldham, G.R. (1976). Motivation through the design of work: Test of a theory. *Organisational Behaviour and Human Performance, 16,* 250-279.

Hackman, J.R. & Oldham, G.R. (1980). *Work redesign.* Reading, MA: Addison-Wesley.

Jöreskog, K. & Sörbom, D. (1993). *LISREL 8: Structural equation models with SIMPLIS command language.* Chicago, IL: Scientific Software International.

Kanfer, R. & Hulin, C.L. (1985). Individual differences in successful job searches following lay-off. *Personnel Psychology, 38,* 835-847.

Kohn, M.L. & Schooler, C. (1973). Occupational experience and psychological functioning: An assessment of reciprocal effects. *American Sociological Review, 38,* 97-118.

Kohn, M.L. & Schooler, C. (1978). The reciprocal effects of the substantive complexity of work and intellectual flexibility: A longitudinal assessment. *American Journal of Sociology, 84,* 24-52.

Kohn, M.L. & Schooler, C. (1982). Job conditions and personality: A longitudinal assessment of their reciprocal effects. *American Journal of Sociology, 87*(6), 1257-1283.

Koskinen-Hagman, M., Schwarzer, R. & Jerusalem, M. (1999). Swedish version of the general self-efficacy scale. Accessed online August 9, 2008: http://userpage.fu-berlin.de/~health/swedish.htm

Mayo, E. (1933). *The human problems of an industrial civilization.* Cambridge, MA: Harvard Graduate School of Business Administration.

Michie, J., & Sheehan, M. (2003). Labour market deregulation, flexibility, and innovation. *Cambridge Journal of Economics, 27,* 123-143.

Moynihan, L., Roehling, M., LePine, M. & Boswell, W. (2003). A longitudinal study of the relationships among job search self-efficacy, job interviews, and employment outcomes. *Journal of Business and Psychology, 18,* 207-233.

Nunnally. J.C. (1978). *Psychometric theory* (2nd ed.) New York: McGraw-Hill.

Nutec (2000). *Arbetskraftens rörlighet – ett smörjmedel för tillväxt [Workforce mobility: A catalyst for growth]* (Report No.15). Stockholm: Nutec.

Organ, D.W. (1990). The motivational basis of organisational citizenship. *Research in Organisational Behavior, 12,* 43-72.

von Otter, C. (2003). *Låsningar och lösningar i svenskt arbetsliv. Slutsatser från en trendanalys [Deadlocks and solutions in Swedish work life. Conclusions from a trend analysis].* Stockholm: National Institute of Working Life.

Richter, P., Hemman, E., Merboth, H., Fritz, S., Hänsgen, C. & Rudolf, M. (2000). Das Erleben von Arbeitsintensität und Tätigkeitsspielraum – Entwicklung und Validierung eines Fragebogens zur orientierenden Analyse (FIT) [The experience of work intensity and autonomy – A development of, and validation of a questionnaire]. *Zeitschrift für Arbeits und Organisationspsychologie, 3,* 129-139.

Richter, P., Hemman, E. & Pohlant, A. (1999). Objective task analysis and the prediction of mental workload: Results of the application on an action oriented software tool (REBA). In M. Wiethoff & F.R.H. Ziljstra (Eds.). *New approaches for modern problems in work psychology* (pp. 67-76). Series: WORC report 99.10.001. Tillburg University, The Netherlands.

Romaniuk, K. & Snart, F. (2000). Enhancing employability: The role of a prior learning assessment and portfolios. *Journal of Workplace Learning: Employee Counselling Today, 12,* 29-34.

Saks, A. & Ashforth, B. (1999). Effects of individual differences and job search behaviors on the employment status of recent university graduates. *Journal of Vocational Behaviour, 54,* 335-349.

Schwarzer, R. & Jerusalem, M. (1993). The general perceived self-efficacy scale. Accessed online August 9, 2008: http://userpage.fu-berlin.de/health/selfscal.htm

Speier, C. & Frese, M. (1997). Generalized self-efficacy as a mediator and moderator between control and complexity at work and personal initiative: A longitudinal field study in East Germany. *Human Performance, 10,* 171-192.

Ulich, E. (1990). *Arbeitspsychologie [Work psychology].* Stuttgart, Germany: Poeschel.

Ulich, E. & Weber, W.G. (1996). Dimensions, criteria and evaluation of work group autonomy. In M.A. West (Ed.). *Handbook of work group psychology* (pp. 247-282). Chichester, England: Wiley.

Vinokur, A.D. & Schul, Y. (1997). Mastery and inoculation against setbacks as active ingredients in the JOBS intervention for the unemployed. *Journal of Consulting and Clinical Psychology, 65,* 867-877.

Volpert, W., Österreich, R., Gablenz-Kolokovic, S. & Resch, M. (1983). *Verfahren zur Ermittlung von Regulationserfordernissen in der Arbeitstätigkeit (VERA) [A method to assess demand on regulation in work]. Handbuch und Manual.* Cologne; Germany: TÜV Rheinland.

Wanberg, C., Kanfer, R. & Rotundo, M. (1999). Unemployed individuals: Motives, job-search competencies, and job-search constraints as predictors of job seeking and reemployment. *Journal of Applied Psychology, 6,* 897-910.

Wirkkala, L. (2002). Att vilja men inte kunna byta jobb [Willing but unable to change jobs.] (Report No. 31). Stockholm: TCO.

The Re-Structuring of Occupational Life and Mental Health in Insecure Employment

Livia de Oliveira Borges & Janine Maranhão

Introduction

The *Banco do Estado do Rio Grande do Norte S/A (BANDERN)* was an economically mixed enterprise linked to the Northern Brazilian state of Rio Grande do Norte and was active mainly in its own state where it had agencies in many cities and was also active in a few capitals of other states, as well as in Brasília (the capital of Brazil).

During the 1980's, and particularly during the first half of this decade, the bank along with other state banks in Brazil were economically stimulated by high inflation, economic politics adopted by the federal government and, by electoral interests. As a consequence, these banks continued with their expansion plans despite financial difficulties. In the case of BANDERN, many agencies were opened both within the state of Rio Grande do Norte and in other parts of the country. By the end of the 1980's, the financial health of state banks had become a cause for serious concern, to both regulators such as the Central Bank and the population in general.

In 1990, the federal government decided to close down BANDERN along with other banks in the country. The reasons for this decision have, up to now, still not been clarified. Other banks, which were affected by the same decision, were later re-opened. In the case of BANDERN, it was stated in the local newspapers that there were sufficient resources for its re-opening. The public of Natal (the capital of Rio Grande do Norte) became mobilized over this matter, and some sectors made claims to state and federal government for a full return of BANDERN to its former activities. The union of bank workers negotiated in order to retain all the workers' rights and benefits, as declared in the state constitution. The objective was to create a more modern and simpler bank, although today the bank, as such, no longer exists, leaving in its place only an office to conclude the closing down process.

The final decision to close the Bank was taken in 1998, but before this date, the employees had to endure a prolonged period of uncertainty starting in 1990, when the Federal Government first announced the closure. At the beginning of the closure process in 1998, the bank tried to carry out its obligations to its workers, although this did not satisfy many of those working in the bank and many of them continued making legal claims. As a consequence of the closure, there followed a long period of uncertain employment, as described in a previous study (Borges, 2003).

Today the employees are still experiencing many difficulties. In the beginning, the worker and social mobilization stimulated a search for a collective solution. However, later differences among the workers themselves and the different implications resulting from the closure ended in the adoption of a more individualistic approach and subsequent fragmentation into smaller groups. Approximately two years after the first announcement of the bank closure, negotiations for re-opening the bank were well under way and the employees were given the choice between: (1) participating in a selection process to continue as employees with the bank, (2) being relocated by direct management of the state (DMS) or by the "Tribunal de Contas do Estado (TCE)", (3) resigning from the job or (4) taking early retirement. The workers had serious doubts about both the transparency and the technical quality of the selection process. But in spite of these doubts, those workers who were selected felt like winners. But as the bank was never re-opened, these same employees were, in reality, harshly dismissed by the bank later. Those continuing at the closure office are now facing an ambiguous situation. If they work efficiently their work will finish sooner, ironically curtailing their own period of employment. Furthermore, their status as employees at DMS is ambiguous because although they are working at DMS, their legal status is that of a bank employee. Employment was guaranteed but salary and benefits were not. They did not have any realistic possibilities of a professional career. The case of those employees who were allocated to TCE is different, as they had an opportunity to change their job and to receive a wage plan. Each one of the ex-employees of the bank obtained a permanent position.

In summary, these different cases implied a number of distinct consequences for the ex-employees of BANDERN – loss of job for some, relocation for others and early retirement for the remainder – which has created a critical situation in a person's life.

A previous study on this subject (Borges, 2003) covering the period from March 1998 to March 2000 was a preliminary exploration of the subject. In part, the study consisted of a content analysis of interviews that were conducted with ex-employees of BANDERN. Such analysis permitted the identification of three levels of occupational life re-structuring: re-structured, partially re-structured and not- re-structured. This classification was based on the following dimensions: (1) being employed or not; (2) level of satisfaction with the present situation; (3) recovery to a socioeconomic standard of living; (4) building new life perspectives; (5) calmness in related experiences at BANDERN; and (6) a balance between work life, personal life and family life.

It was observed that the level of occupational life re-structuring of these ex-bank employees varied with the level of education and with the type of job (administrative-bureaucratic, managerial, or technical). In the period when the banks in general plan a reduction of personnel, returning to become a bank employee again is not very common. The high level of education and the managerial experience, however, can support the reinsertion into other economic sectors. Banking experience, however, shows very little transfer to other economic sectors. Other results with same importance were found, highlighting, among others: (1) a greater association between income and the socioeconomic standard of the family; and (2) the presence of depressive feelings (for example, melancholy, sadness, lack of enthusiasm, reduction of expectation, and lack of belief in occupational improvement).

Regarding life with long term unsecure employment, there is no theoretical background approach relating this to unemployment, although it seems plausible that reactions to it are similar to those felt in unemployment, on which there is a long tradition of empirical research (for example: Álvaro & Paez, 1996; Álvaro-Estramiana, Torregrosa & Garrido-Luque, 1992; Banks et al., 1980; Borges & Argolo, 2002; Sarriera, Schwarcz & Câmara, 1996). All these studies are based on the notion that mental health is synonymous with well-being and as a complex process it is closely associated with other factors. As a consequence, attention was focused on small mental health alterations and/or behaviour changes, using GHQ-12 (General Health Questionnaire, Goldberg's 12-item version, 1972). An adapted version was developed in Natal (Borges & Argolo, 2002). The study used a sample of unemployed bank employees and health professionals, and showed that levels of mental health deterioration in the unemployed bank employees were higher than in the health professionals, possibly indicating that the GHQ-12 is more sensitive to the psychological problems of bank employees. Furthermore, the technical literature (for example: Álvaro-Estramiana, 1992; Banks et al., 1980; García-Rodríguez, 1993; Garrido, 1996) has consolidated the use of GHQ-12 in studies concerning the effects of unemployment and has established comparisons among different occupations (Borges & Argolo, 2002; Sarriera et al., 1996). This observation is also supported by the results of a previous study using ex-employees of BANDERN in which depressive feelings were indicated. The validation study in Natal corroborated the consistency of GHQ-12, permitting the estimation of scores in two factors: self-efficacy reduction and emotional and depressive tension. Jahoda (1987) also showed the stress effects and the lack of enthusiasm in those experiencing long-term unemployment.

On this basis it was assumed that the results would be more consistent if occupational life re-structuring levels of the ex-employees were assessed more directly and systematically. The results of the previous study about these ex-employees allowed for the elaboration of a structured questionnaire and also indicated the importance of assessing mental health states from small mental health changes. To achieve this, the present study had the following objectives: (1) to study the association between levels of occupational life re-structuring and the mental health of the ex-employees of BANDERN and (2) to explore the antecedent variables of re-structuring, comparing different groups of ex-employees in accordance with the level of occupational life re-structuring.

1 Method

Field research was carried out adopting the systemic-cognitive approach and assuming, as its starting point, the perception of BANDERN'S ex-employees about the re-structuring of their occupational lives.

1.1 Sample

The sample for this study consisted of ex-employees of BANDERN (approximately 1,500 persons), of which 625 were relocated to DMS and to TCE, while 350 were selected to continue working in the bank and others chose retirement or resigned from their job. As the bank did not re-open again, of the 350 employees selected to continue working in the bank, only 30 continued working at the closure office. The others were dismissed as well.

After data collection, a sample of 105 persons was formed. Of these, 86 were ex-employees who had a job with the state government: 50 in DMS, 11 in the TCE and 25 in the closure office. Among this last group of 25 persons, there are two already retired employees and two participants who have been officially relocated to DMS.

The study tried to include ex-employees of BANDERN working in both the private and the public sector, but there were difficulties in finding such subjects and, together with limited resources, this resulted in the use of only a small number of questionnaires for these sectors. As a consequence, it was possible to obtain 14 questionnaires from individuals working in the private sector (employees, managers and self-employed), 2 questionnaires from retired persons, 2 from unemployed and 1 from an employee in the public sector.

Because there were only four people outside the labour market (unemployed or retired) in the sample, these were excluded from the analysis, which means that the analyses used a total sample of 101 participants.

1.2 Instruments

Based on the use of content analysis from the previous study, on assessing the level of occupational life re-structuring, a questionnaire was elaborated with 21 items, asking about: their last job at BANDERN; their decisions about what to do after the announcement of the closure; present day job situation; evaluation of their present salary; evaluation of the assistance received; leisure activities; time spent in believing that the bank would re-open; evaluation of their present job; expectations of professional improvement and the emotional reaction to the information of the closure. Each question contained alternative answers, from which the participants were asked to choose the most appropriate ones.

Furthermore, the study also contained questions about social support (Argolo, 2001); the GHQ-12 (Goldberg, 1972) in the adapted version (Borges & Argolo, 2002) and a socio-demographic part.

1.3 Collection of data

Interviewing the BANDERN ex-employees about the closure was difficult because of the emotional stress they experienced. The problem was even greater, as the present research intended to approach the same people for the second time. To minimize this problem, the results of the first study were published widely in the media with the intention of informing the public.

The questionnaires were handed out individually and each participant received personal instruction. Some participants filled in the form immediately, but most of them returned it some days later.

1.4 Analysis of data

Answers to specific questionnaires about the experience of ex-employees at BANDERN were coded. For some questions, the answers corresponded to a scale from 1 to 3 or from 1 to 4 (for example: evaluation of salary, expectation for occupational improvement, etc.) and in others, to a nominal scale (type of job at the bank, occupational situation, etc.). Answers to the questions about social support were not coded, because they were answered attributing number symbols.

The occupational life re-structuring variable was assessed in two ways: (1) considering the different styles (cluster analysis); and (2) considering the levels of re-structuring (coefficient of intensity). In both evaluations, the dimensions used for occupational life re-structuring came from a previous study, except when reference was made to being employed or not, because all the participants in the sample of the present study are employed. Because of this characteristic, the expected variance for occupational life re-structuring levels became smaller than in the previous sample.

For the remaining criteria, the following procedures were adopted:

- *Satisfaction with present situation* was estimated by arithmetic mean of answers to the three questions, of which one is an indication of present job satisfaction; the second is an evaluation of the actual salary; and the third is an evaluation of assistance or benefits.
- *Recovery of socioeconomic standard of living* was estimated by weighted mean of answers to three questions: the first is referred to the participants' evaluation of their economic situation; a second is an evaluation of the actual salary comparing it to the previous employment; and the last one evaluates the present level of assistance and comparing it to that before the closure of the bank.
- *Diversity of leisure activities* was estimated using a frequency rating of leisure activities, such as trips, going to bars/restaurants, attending cultural activities, going to the beach, shopping and leisure at home. Answers were rated on a scale from 1 to 4.
- Different styles of re-structuring were identified using cluster analysis of the scores above. To assess the level of occupational life re-structuring, it was observed that each criterion was evaluated using scales with different variance. For this, each distribution was evaluated, splitting them into three levels with scores of 0, 0.5, or 1. The sum of all the scores for each participant was taken as the level of occupational life re-structuring. This procedure uses the same dimensions as those in previous studies, although it is different because it is more focused on the participants' own statements about themselves and less on evaluations made by the researchers. With a structured questionnaire the answers were direct and not categories deduced from interviews by researchers.

The answers to GHQ-12 were coded using a Likert scale from 0 to 3. The scores in the factors were calculated for means of attributed values for each item.

2. Results

2.1 Re-structuring of occupational life

The corresponding scores for occupational life re-structuring were estimated. The descriptive statistics are shown in Table 1. Concerning levels of work satisfaction (scale from 1 to 4), it was observed that the mean was 3.01 and the median was 3.00. The curve shows a higher concentration on the right, indicating a tendency towards a high satisfaction.

Table 1. Descriptive statistics for dimensions of re-structuring

Statistics		Occupational satisfaction	Life standard	Life perspectives	Leisure diversity
Mean		3.0	1.2	2.2	2.3
Standard deviation		0.7	0.5	0.9	0.6
Minimum		1.3	0.3	1.0	1.0
Maximum		4.0	3.0	4.0	3.6
Quartiles	25	2.7	0.8	1.5	1.8
	50	3.0	1.1	2.0	2.3
	75	3.5	1.4	2.5	2.8

With respect to the recovery of socioeconomic standard (scale from 0 to 3), the mean was 1.2 and the median was 1.1. Therefore, there is a large dispersion of high scores, and a large concentration of low scores. This observation is corroborated by distribution of ex-employees in accordance with the salary intervals, because such distribution is decreasing: 44 ex-employees of the sample (41.9%) received up to 3 minimum salaries[1]; 25 of them (23.8%) received between 3 - 5 minimum salaries; 22 of them (21.0%) received from 5 - 10 minimum salaries; and only 8 persons (7.6%) received more than 10 minimum salaries.

These last results contrast with the assessment of satisfaction with the present occupational situation, because their scores are inversely correlated ($r = -0.57$; $p < 0.001$). This contradiction is likely apparent and this really reveals that answers about levels of satisfaction with the present situation have a compensatory psychological function with an unpleasant situation.

For occupational life perspectives, the mean is 2.19 and a median is 2. But it should be noted that the first quartile of the distribution is 1.5, when the smallest score is 1. This indicates the existence of a strong concentration of very small scores. Besides this observation, such scores are inversely correlated with work satisfaction ($r = -0.50$; $p < 0.001$), what corroborates with the hypothesis already referred to: high satisfaction as a compensatory conformism.

On the calmness of related experience it was observed that the mode of distribution indicates that participants are feeling sad, but are not in shock (35.2%). This type of reaction is followed by another one which indicates that participants tend to avoid hearing about the closure of BANDERN. Finally, there are two other answers – "paid attention", and "without emotional reaction" – they are represented by the same proportion of the sample (18.1%). Then, there is a predominant reaction of affection to hear others speaking about BANDERN.

---

1 The Brazilian government regulates the payment of employees, establishing a minimum salary which an enterprise must pay to their employees (at the time of the study the monthly minimum salary was the equivalent of 68 € for a week with 44 hours).

Concerning diversity in leisure activities (scale from 1 to 4), it was observed that the participants have a mean of 2.3. The quartiles indicate that 50.0% of the sample obtained a score above 3.0, which indicates a great variety of leisure activities among the participants. There is enough dispersion in these answers, despite a half of the sample obtaining means between 3 and 4 on the scale. It is also observed that, in the same direction of the other indicators, the diversity of leisure activity is inversely correlated with occupational satisfaction ($r = -0.29$; $p = 0.004$).

Cluster analysis (three group solution) was calculated (Table 2) to identify how the participants are grouped in accordance with the five dimensions of life restructuring. The first group, 40.0% of the sample, is lower in occupational satisfaction (≅2.8), has a slightly higher life standard (≅1.4), a slightly higher occupational perspective (≅2.4) and the highest mean for leisure diversity (≅2.5) among the three groups. This group tends to emotional reactions with reference to the bank as "attention without emotional alteration". In summary, the group is characterized by a style of moderated re-structuring by rational control and with a trend to use leisure time as a compensatory solution.

Table 2. Clusters of occupational life re-structuring styles

Dimensions	Clusters		
	First	Second	Third
Occupational satisfaction	2.8	3.0	2.4
Life standard	1.4	1.0	1.4
Life perspectives	4.0	2.0	2.0
Leisure diversity	2.5	2.1	2.3
Occupational perspective	2.4	1.7	3.0

The second group, 43.0% of the sample, shows the highest levels of occupational satisfaction compared to the other groups, but this result goes beyond the smaller scores for living standards and for occupational perspectives. This group tends to react to the closure of BANDERN with sadness and its leisure diversity (≅ 2.1) is smaller than in the general sample. This group, therefore, tends to attempt a type of re-structuring through compensatory conformism.

The third group, 17% of the sample, has the lowest scores for occupational satisfaction, slightly higher than average for living standards and the best evaluation for occupational perspectives as well as an average diversity of leisure activities. This group demonstrates sadness when talking about their experience of the closure. The group as a whole tends to present styles of life re-structuring based on worry but retained their belief in occupational improvements and/or in overcoming difficulties.

Exploring the associations for each style of re-structuring with sociodemographic sample characteristics, it was found among participants with superior positions that those in specialized technical jobs or in managerial jobs in the bank tended to be the ones with the third style of re-structuring ($\chi^2 = 4{,}7$; p=0,05).

The levels of re-structuring were calculated in accordance with the description already presented. The results of this procedure indicate that the level of re-structuring varies from 0.50 to 4, with a mean of 2.3, and a median of 2.5. In Figure 1, it can be seen that there is little incidence of extreme scores. This is probably a consequence of the sample composition, as explained above. Corroborating this observation, when the eliminated participants (four persons who are no longer in the labour market) are examined, then it is seen that only one unemployed person has a score of zero.

Figure 1. Occupational life re-structuring: sample distribution

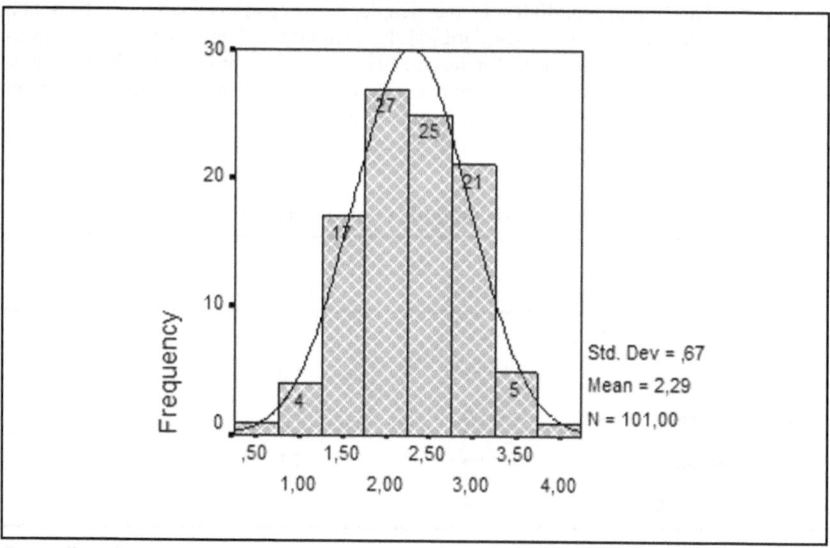

Re-coding the scores for re-structuring levels in the three intervals (up to 1.99, from 2.0 to 2.99 and from 3.0 upwards), it was found that 22.8% of the participants are non-structured, 50.5% of them are partially structured, and 26.7% of them are structured. There is, therefore, a concentration of scores above the mean. In the first study it was found that 29.5% of participants were non-structured, 29.5% were partially structured and 40.9% were structured. In the present sample, a smaller dispersion of the scores was found. It is possible that such differences are due to changes in the assessment criteria for re-structuring. The fact that this sample has proportionally more participants in groups with connections to the state government (relocation to DMS, TCE, and closure office of BANDERN) had already made it more predictable that the variance would be smaller. However, the proportional reduction in the number of structured participants should be considered in light of the fact that data for the second research were collected when the closing down process of the bank was already in an advanced stage. This suggests that the participant's belief in the bank re-opening was in decline. This was corroborated by small scores regarding occupational improvement and by additional observations during data collection. Indeed, many participants showed emotional stress simply on being contacted and asked to fill in the questionnaire.

When the variance is small, it is possible to predict that the results would also indicate a smaller association between levels of re-structuring and other variables. When the means of the three groups are compared with those ex-bank employees relocated to DMS, TCE, and the closure office, the analysis of variance indicates significant differences ($F = 4.98$; $p = 0.009$), with means of 2.68 in the case of TCE, 2.35 for DMS, and 1.98 for the closure office. T-test results emphasize the small mean for those persons who continue working in the closure office ($t = 2.28$; $p = 0.02$).

If the sample is split into two groups in accordance with educational achievement, using education up to secondary level education as the cut off point, the results ($t = 2.21$; $p = 0.03$) indicate a significant difference for levels of occupational re-structuring between the two groups. The result shows that occupational re-structuring increases with the educational level.

2.2 Participants' mental health

Table 3 shows the GHQ scores. The first column gives the general coefficient. The mean of 12.8 has an intermediary value between results for unemployed and employees in international studies (for example: Álvaro-Estramiana, 1992). When the results are examined by arithmetic mean of answers to the 12 items (second column of Table 3), it can be seen that this mean is exactly equal to the mean which was obtained for the unemployed in the previous study about the validation of GHQ-12 (Borges & Argolo, 2002). The means of the specific factors – reduction of self-efficacy, and emotional tension and depression – are 1.0, and 1.3 respectively. These means follow the results of the quoted local research for the unemployed. Scores in the second factor are higher than in the first study ($t = 5.3$; $p < 0.001$), corroborating what was observed in interviews of the first study about ex-employees of BANDERN regarding depressive tendencies.

Comparing the scores of the three groups by an analysis of variance, with respect to the styles of occupational life re-structuring significant differences were found ($p < 0.002$). In all the GHQ-12 scores, the highest means (Table 4) are for the re-structuring style named compensatory conformism (second cluster), followed by the group with moderate re-structuring by rational control and, finally, the group which presents the style of belief in the occupational improvement with the lowest values. Applying the t-test, using the general scores of GHQ-12 (calculated by means), their results show that a real difference exists in the second cluster (compensatory conformism) which has the highest mean (t=4.42; p<0.001). This pattern is repeated in all the GHQ-12 scores.

Table 3. GHQ-12 scores, self-efficacy, emotional tension and depression

		General Score (sum)	General score (mean)	Self-efficacy reduction	Emotional tension and depression
Cases	valid	101	101	101	101
	missing	0	0	0	0
Mean		12.8	1.1	1.0	1.3
Median		11.0	0.9	0.9	1.3
Standard deviation		7.4	0.6	0.6	0.9
Minimum score		0.0	0.0	0.0	0.0
Maximum score		32.0	2.7	2.6	3.0
Quartiles	25	7.5	0.6	0.6	0.7
	50	11.0	0.9	0.9	1.3
	75	18.0	1.5	1.4	2.0

Taking the general score of GHQ-12 as the dependent variable (DV), stepwise regression analyses were used to identify variables that are able to predict the variance from DV. The model explains 19.0% of the variance through the use of two variables: perceived social support since the closing down notification ($\beta = 0.32$), and belonging to the group of compensatory conformism style ($\beta = 0.25$). These results mean that: (1) the participant who perceives more social support tends to exhibit less psychological deteriorations, and (2) the fact of being included in the group of compensatory conformism increases the probability of exhibiting the previously mentioned psychological deteriorations.

If re-structuring styles, which were used as antecedent variables in the regression equation, are substituted by component variables (criteria), the regression equation is able to explain 28.0% of variance, but variables effectively added (with significantly statistical level) in the equation are only the perspective of occupational increases ($\beta = -0.27$), and the occupational life satisfaction ($\beta = 0.22$), maintaining the perceived social support ($\beta = 0.26$) in it. When the two new variables were introduced into the equation, they showed an inverse tendency, meaning the greater the perceived perspective of occupational increases the less they tended to show psychological alterations while, on the other hand, the more they declared themselves satisfied, the more psychological alterations they exhibited.

Taking the level of re-structuring as the dependent variable, the best regression analysis explains 35.0% of the variance and subsequent independent variables predict significantly the values from the dependent value: to adopt a re-structuring style of compensatory conformism ($\beta = -0.23$), to be relocated to intervention schemes of the state government ($\beta = 0.39$), the amount of work experience ($\beta = 0.26$), and GHQ-12 scores ($\beta = -0.27$). These results indicate that the probability of having high occupational life re-structuring scores increases among those people who do not dispose of the

Table 4. Descriptive statistics of GHQ-12 scores for re-structuring style clusters

		N	Mean	Standard deviation	Minimum	Maximum
General score (sum)	Moderated structure of rational control	31	10.5	5.1	0.0	20.0
	Compensatory resignation	35	15.1	8.0	2.0	32.0
	Belief in professional improvement	11	8.1	6.3	1.0	21.0
	Total	77	12.2	7.2	0.0	32.0
General score (mean)	Moderated structure of rational control	31	0.9	0.4	0.0	1.7
	Compensatory resignation	35	1.3	0.7	0.2	2.7
	Belief in professional improvement	11	0.7	0.5	0.1	1.8
	Total	77	1.0	0.6	0.0	2.7
Self-Efficacy reduction	Moderated structure of rational control	31	0.8	0.4	0.0	1.7
	Compensatory resignation	35	1.2	0.6	0.2	2.6
	Belief in professional improvement	11	0.7	0.5	0.1	1.8
	Total	77	1.0	0.6	0.0	2.6
Emotional tension and depression	Moderated structure of rational control	31	1.1	0.7	0.0	2.7
	Compensatory resignation	35	1.4	0.9	0.0	3.0
	Belief in professional improvement	11	0.7	0.7	0.0	2.0
	Total	77	1.2	0.8	0.0	3.0

re-structuring style of compensatory conformism, are relocated to state government jobs (as opposed to working in the bank closure office), have more work experience and have less psychological alterations.

3 Discussion of results and conclusion

The study allows for a more exact assessment of the levels of the restructuring of the participants' occupational life. It furthermore identifies different styles of occupational life re-structuring and strengthens the exploration of variables associated with the phenomenon of long-term insecure employment.

The results indicate that the long period of uncertainty experienced by the participants continues to affect the re-structuring of their occupational life, and these effects have probably been intensified by persistent lower standards of living than those experienced during their employment in the bank and by a persistent lack of perspectives for regaining such standards. The majority tried to maintain psychological balance by compensatory conformism, stating work satisfaction as a form of denying reality. This situation has most likely been generated by occupational expectations not having been fulfilled over a long period of time, and adapting in a resigned way as a strategy of coping with feelings of frustration. Re-structuring levels and re-structuring styles are associated with the participants' educational achievements. The methodological approach applied in this study shows that this relationship is stronger than in the previous research, because there were only significant differences for participants with a high level of education.

There are differences in regard to the re-structuring style in relation to the different working conditions in DMS, TCE, and the bank closure office, which clearly connect the quality of employment and specific working conditions with different levels of restructuring. Employees working in TCE have the most structured level with less frequent compensatory conformism style. These employees, besides having stable employment, are included in career plans with concrete perspectives for occupational improvements. This group is followed by employees at DMS, who have stable employment, but have ambiguous roles and are not considered official state civil servants. The most non-structured group belongs to those people who are still working in the bank closure office and who show a compensatory conformist style. These people are in temporary employment with the prospect of potential unemployment. The retired ex-employees working in the closure office have scores which differ from their group in general (workers of closure office), which suggests that temporal employment may contribute to a style of non-restructuring. Beyond this fact, it is important to point out that the use of such coping styles would also be revealed in the participants' day-to-day talking.

The association between the type of previous job and the level of re-structuring is weaker in this study compared to the previous one. This phenomenon probably occurred as a result of the composition of the present sample, which has proportionally more participants in state employment.

More relevant, however, was the observation that the average GHQ-12 scores follow the standard for unemployment found in the other study. The increased levels of tension and depression found in those in the non-structured or partially structured group indicates that it is difficult for them to build new perspectives for themselves. This indicates the need for social support in order to overcome this difficult situation, a fact confirmed by the result that the perception of social support predicts the GHQ-12 scores. The same is true for the compensatory conformism style. This means that the coping strategy adopted by this group is not effective in maintaining their own health or well-being.

Why is this coping style not effective? If there is variance of mental deterioration levels for different styles of re-structuring and also for types of work (TCE, DMS, and closure office), it means that there is an inherent problem with the working environment that is more endemic. If this is the case it needs to be resolved on this level as well, requiring both institutional and collective solutions. The use of the compensatory conformity style by individuals represents an attempt at re-gaining control of the situation. This attempt, however, left the origins of the problem intact but it is able to generate temporarily relief. The persistent use of this strategy will gradually affect the persons' health because it can become impossible to mobilize the search for effective solutions to the problems. A better solution requires, inversely, a collective solution. This understanding was supported by specializing on a strategy of coping (for example: Gil-Monte & Peiró, 1997; Locke & Taylor, 1990; Maslasch & Leiter, 1999; Maslasch, Schaufeli & Leiter, 2001), emphasizing that Maslasch & Leiter (1999) paid attention to the emotional endemic stress and its origin in the work environment, requiring organisational action.

With regression analysis, when the general GHQ-12 scores (indicator of the level of mental deterioration) are taken as the dependent variable, and using specific dimensions of re-structuring as the independent variable and not the clusters (styles) or the re-structuring level, the evaluation for occupational improvements and the declared satisfaction offer a significant prediction of those scores: The first variable in the association is inversely proportional, while the second variable is directly proportional. These results can indicate a way in which institutional action may become involved. These must be characterized by supporting ex-employees in their efforts to build more positive perspectives and by developing a more realistic analysis of their situation.

During the description of the results, analyses of other variables such as religious beliefs, the frequency of church attendance, sex, etc. were omitted. This was because the statistical tests did not indicate any significant differences. These results, however, strengthen the relationship between life re-structuring problems and mental health deterioration and employment, thereby weakening the link between such deterioration and individual traits. But it is important to note that the possible association between re-structuring levels and/or mental health and working in the private or public sector were not explored because the composition of the sample did not allow this.

The analysis shows that mental health deterioration is associated with occupational life re-structuring and it implies the need for action to be taken in supporting ex-employees in overcoming a difficult situation. These are important considerations for employment policies as well as public health policies. The evident contextual determinants of this phenomenon of mental deterioration reveal that the closure procedures of the bank were executed without any serious consideration of its human aspects. Such lack of consideration can generate associated health damages to individuals and to society as a whole, reducing the value of the State and, simultaneously, of dedication to work.

The situation of the ex-employees of the bank requires immediate action to reduce its negative consequences, and the following actions were recommended: (1) to regulate the situation of ex-employees in state government institutions by removing their role ambiguity; (2) to offer stable conditions and stimulus to those working in the closure office; (3) to develop a program of support for bank ex-employees, helping them to use coping strategies more constructively and to change their perceptions about institutional support; (4) to stimulate a program of qualification, emphasizing the development of new competencies and the transfer of experiences; (5) to consider the ex-employees in DMS and in the closure office as a group that needs consistent support, without neglecting other groups of unemployed which were not represented in this research that may be in need of social support as well.

Acknowledgements

The author would like to thank Roy Robinson for his patience in improving the language of the present text.

References

Alexandre, M.L.O. (1997). *Análise da Gestão Bancária na Perspectiva da Administração Estratégica: O Caso do Banco do Estado do Rio Grande do Norte S. A. De 1980 a 1990.* Master's dissertation. Universidade Federal do Rio Grande do Norte, Natal.

Álvaro-Estramiana, J.L. (1992). *Desempleo y Bienestar Psicológico.* Madrid: Siglo Veintiuno de España Editores.

Álvaro, J.L., & Paez, D. (1996). Psicología Social de la salud mental. In J. L. Álvaro, A. Garrido & J. R. Torregrosa (Eds.), *Psicología Social Aplicada* (pp. 381-406). Madrid: McGraw-Hill/Interamericana de España.

Álvaro-Estramiana, J.L., Torregrosa, J.R., & Garrido-Luque, A. (1992). *Influencias Sociales y Psicológicas en la Salud Mental.* Madrid: Siglo Veintiuno Editores.

Argolo, J.C.T. (2001). O Impacto do Desemprego Sobre o Bem-Estar Psicológico dos Desempregados da Cidade do Natal. Master's Dissertation. Universidade Federal do Rio Grande do Norte, Natal.

Argolo, J.C.T. & Araújo, A.D. (2002). O impacto do desemprego sobre o bem-estar psicológico dos desempregados da cidade do Natal [Complete text]. *Annals of Encontro da Associação Nacional dos Programas de Pós-Graduação em Administração 2002* (Published in CD-ROM). Salvador: ANPAD.

Banks, M.H., Clegg, C.W., Jackson, P.R., Kemp, N.J., Stafford, E.M., & Wall, T.D. (1980). The use of the General Health Questionnaire as an indicator of mental health in occupational studies. *Journal of Occupational Psychology, 53*(3), 187-194.

Bastos, A.V.B. (1994). *Comprometimento no Trabalho: a Estrutura dos Vínculos do Trabalhador com a Organização, a carreira e o Sindicato.* Doctorate thesis. Universidade de Brasília, Brasília.

Bastos, A.V., Braga, S.C., Torres, L., Gomes, R.O., & Nunes, K. (1995). Significado do trabalho: analisando tendências de mudança entre estudantes e profissionais de administração e processamento de dados [Abstract]. In Sociedade Brasileira de Psicologia (Ed.) *Resumos de comunicações científicas. XXV Reunião Anual de Psicologia* (p. 197). Ribeirão Preto: SBP.

Bastos, A.V.B., Pinho, A. P.M., & Costa, C.A . (1995). Significado do trabalho: Um estudo entre trabalhadores em organizações formais. *Revista de Administração de Empresas,* 35(6), 20-29.

Borges, L.O. (2003). *Extended experience of undefined employee status.* Paper presented to the 27th International Congress on Occupational Health (ICOH), Iguassu, Brazil.

Borges, L.O., & Argolo, J.C.T. (2002). Validação e adaptação de uma escala de bem-estar psicológico para uso em estudos ocupacionais. *Revista de Avaliação Psicológica,* 1(1), 17-27.

Borges, L.O., & Tamayo, A. (2001). A estrutura cognitiva do significado do trabalho. *Revista Psicologia Organizações e Trabalho, 2*(2), 11-44.

Garrido, A. (1996). Psicología social del desempleo. In J.L. Álvaro, A. Garrido & J.R. Torregrosa (Eds.), *Psicología Social Aplicada* (pp. 122-154). Madrid: McGraw-Hill.

García-Rodríguez, Y. (1993). Principales medidas en psicología del desempleo. In Y. García-Rodríguez (Ed.), *Desempleo: Alteraciones Psicológicas* (pp. 55-94). Valencia: Promolibro.

Gil-Monte, P., & Peiró, J.M. (1997). *Desgaste psíquico en el trabajo: El síndrome de quermarse.* Madrid: Editorial Síntesis.

Goldberg, D.P. (1972). *The detection of psychiatric illness by questionnaire.* London: Oxford University Press.

Jahoda, M. (1987). *Empleo y desempleo: Un análisis sociopsicológico.* Madrid: Ediciones Morata.

Kohn, M.L., & Schooler, C. (1983). *Work and personality.* New Jersey: Ablex Publishing Corporation.

Locke, E., & Taylor, M.S. (1990). Stress, coping, and the meaning of work. In A.P. Brief & W.R. Nord (Eds.), *Meaning of occupational work: A collection of essays* (pp. 135-170). Massachusetts/Toronto: Lexington.

Maslach, C. (1994). Stress, burnout, and workaholism. In R.W. Thoreson, R.R. Kilburg & P.E. Nathan (Eds.), *Professionals in distress* (pp. 53-75). Washington: APA.

Maslach, C., & Leiter, M.P. (1999). *Trabalho: Fonte de prazer ou desgaste? Guia para vencer o estresse na empresa.* Campinas: Papirus.

Maslach, C., Schaufeli, W.B., & Leiter, M.P. (2001). Job burnout. *Annual Review of Psychology, 52,* 397-422.

Sarriera, J.C., Schwarcz, C., & Câmara, S.G. (1996). Bem-estar psicológico: Análise fatorial da escala de Goldberg (QSH-12) numa amostra de jovens. *Psicologia: Reflexão e Crítica,* 20(9), 293-306

Quality of Life of the Long-Term Unemployed[1]

Božena Buchtová

Introduction

Initially, quality of life used to be examined from the perspective of health and illness. Strauss' monograph (1975) dealing with the quality of life of the chronically ill and aged people was one of the first studies dealing with this subject. Methodologically, scholars concentrated in particular on the effectiveness of healing methods for patients' quality of life and tried to establish intervention approaches (Bergsma & Engel, 1998; McGee et al., 1991; O´Boyle & McGee, 1992; Browne et al., 1994; Browne et al., 1997). It was only later that a broader human life perspective was taken into account when studying the quality of life (Emmons & Diener, 1985; Ryff & Keys, 1995; Oishi, 1999; Dzúrová & Dragomirecká, 2000; Hnilica, 2000). It became apparent though that quality of life was not determined solely by a set of identifiable external factors, but also – and substantially – by the individual perception of life's meaningfulness (Zika & Chamberlain, 1987; Frankl, 1994; Thompson & Janigian, 2000; Halama, 2000; Balcar, 1995c; Machovec, 1967; Šmajs & Krob, 2003). The findings and experience acquired from measuring the quality of life have shown the following:

1. Rather than a system of values set and evaluated externally, an individual's own perception of the priorities concerning quality of life is relevant for evaluating his or her quality of life.
2. Various quality of life dimensions have different levels of importance for every individual.
3. The importance of the quality of life dimensions changes during an individual's life span as a result of his or her going through different stages of life and facing various situations.
4. A personal approach to the quality of life is closely related to expressing satisfaction with achieving goals and fulfilling plans.

1 The study was completed with a grant aid from GAČR Reg. No. 406/02/1562 and grant aid from ÖSI, branch Brno in 2001 and 2002 Statistical data processing: Dr. Tomáš Urbánek , Psychologický ústav ČSAV Brno Technical assistance: Viktor Kulhavý, student of the Faculty of Economics and Administration of the Masaryk University in Brno.

Our approach derives from the above-mentioned findings as well as from the approach to the quality of life issue by the Irish psychologists O´Boyle, McGee and the Swiss doctor Joyce (1994, p. 160): "Quality of life should be defined individually depending on how the person has determined it". Their method called SEIQoL (Schedule for the Evaluation of Individual Quality of Life)[2] is currently one of the most frequently employed methods of evaluation. As the sources available suggest, we are unique in using SEIQoL for examining the quality of life of the long-term unemployed.

Our study was based on the following assumptions:

- Loss of work is a serious milestone in an individual's life with significant effects on his/her quality of life and its evaluation.
- Long-term unemployment has negative effects on the overall quality of life as well as on the composition and importance of the different aspects of life.
- For the long-term unemployed their quality of life is strongly influenced by age, sex, education and duration of unemployment.
- Long-term unemployment negatively impacts the meaningfulness of life for the afflicted individuals.

1 Data and methodology

The study took place in 2001 and 2002 with the participation of 1,957 respondents, who were divided into two sub-sets:
1. The unemployed (N= 966; 558 women and 408 men, average age 34.5 years, aged between 17 and 65. 6% of the respondents had primary education, 19% had completed apprenticeship, 60% had secondary education, 2% had attended colleges and 13% had university degrees. On average, they had been unemployed for 19.6 months).
2. The employed (N=949; 528 women and 421 men, average age 35.7, aged between 19 and 73. 2.7% of the respondents had primary education, 6.3% had completed apprenticeship, 59% had secondary education, 2% had attended colleges and 30% had university degrees).

The respondents participated in the study on a voluntary basis. They were interviewed by trained researchers throughout the Czech Republic.

We used the SEIQoL method to examine the quality of life of the unemployed. The concept published by O'Boyle, McGee and Joyce in 1994 in the Advances in Medical Psychology (5, p. 159-180) is based on a subjective individual evaluation of the quality of life. In a structured interview the individual freely contemplates his/her value system without any previously set dimensions. The individual evaluates which aspect of life he/she currently considers to be the most important one and to which of the five aspects of life he/she attributes the greatest importance. For a better understanding of the statement and for an adequate interpretation, the respondents refer to the chosen aspects of life in a descriptive manner rather than just in key words; they use free asso-

2 A Czech translation of the SEIQoL method was published in the Czech Republic by J. Křivohlavý (2001, 2002)

ciations and explain what each aspect means to them. The aspects of life considered to be substantial in their current status are assessed in terms of relevance, i.e. their relative importance for the particular individual. Subsequently, the individual contemplates on his satisfaction with the given life aspect and whether he or she manages to meet the demands and objectives of the current life situation. For the purpose of the data evaluation, the importance of each aspect of life (given in per cent from 0 to 100) is multiplied by the level of satisfaction (also given in per cent from 0 to 100 where 0 per cent represents the lowest level of satisfaction and 100 per cent represents complete satisfaction). The resulting quality of life score is a total sum of all the five products divided by 100 resulting in a value between 0 and 100. The obtained quality of life profiles and the final quality of life scores for the individuals and groups are accompanied by data reflecting the meaningfulness of life in their current stage of life. The individual assessment of the meaningfulness of life is marked by a cross on a sloping line (rising from left to right at a 45° angle) where the bottom is entitled "life is absolutely meaningless" and the top "life is truly meaningful".

The data were analysed using descriptive statistical methods, correlation analysis (Pearson's correlation coefficient, Spearman's rank correlation) as well as the t-test, the Kruskal-Wallis test, the median test and the ANOVA method.

2 Research outcomes, discussion

For each group, we processed the following:

a) sequence of life aspects arranged in the order of importance for the respondent's life and for establishing an individual's *quality of life* (QL) *profile*,
b) sequence of the separate life aspects evaluated by the respondents *based on their satisfaction levels* (in a scale from 0 to 100) with the particular aspects.

The following tables give averages and standard deviations for the different groups of respondents concerning the identified quality of life aspects and their satisfaction with these aspects. Apart from the five most important life aspects we specified also other life aspects for each group that were referred to as being less important.

The same life aspects are important for the quality of life of the employed and the unemployed respondents, in the following order – family, health, work, peace of mind and interpersonal relations. We identified only one statistically significant difference between these two groups: the value of the family, which was a more important aspect in the life of the employed than in the life of the unemployed, with the employed also being more satisfied with this particular life aspect. While the employed showed significantly higher satisfaction with work and health, the unemployed reported hobbies and interests as the life aspects providing them with more satisfaction in their current status.

Family
We can conclude that family has almost an identical meaning for both the employed and the unemployed respondents. 1. Family as a symbol of understanding, satisfaction and fellowship, 2. Family as a psychological support (a symbol of security, background and safety), 3. Family as the most important value in life (to have someone to live for), 4.

Family as a background where children can be brought up, 5. Family as giving meaning to life (self-fulfilment).

However, obvious differences were apparent between both groups in the relative frequency of the family meanings and their ranking by the frequency of reports. Among the unemployed respondents family was primarily associated with psychological support during unemployment. In their families the unemployed reclaim psychological balance, which was previously shattered by the involuntary job loss and even more by the unsuccessful attempts to find a job. Family often restores mental balance of the unemployed. The employed respondents associate family with the peace of mind, harmony, coherence, and understanding among the family members, and only then it comes as a place of security, safety and backup. This seemed to confirm what we had learned during our previous research: long-term unemployment is a test of the interpersonal relations in a family and is better coped with by those who have close people to rely on and to speak with openly about their situation (Buchtová et al., 2002, pp.107f.).

Table 1. The employed. Quality of life (QL) profile and the respective level of satisfaction

Quality of life

	Aspect of life (cue)	Average QL	Standard deviation
1.	Family	**27.44**	17.01
2.	Health	20.79	15.29
3.	Work	9.33	13.4
4.	Peace of mind	6.32	9.02
5.	Interpersonal relations	5.67	8.42
6.	Hobbies	4.64	7.49
7.	Personal growth	4.52	7.91
8.	Money	**1.07**	4.74
9.	Housing	0.28	2.26

Satisfaction

	Aspect of life (cue)	Average satisfaction	Standard deviation
1.	Family	**64.51**	33.30
2.	Health	**58.41**	33.75
3.	Work	**30.13**	37.80
4.	Interpersonal relations	26.96	35.41
5.	Peace of mind	**26.60**	35.06
6.	Hobbies	24.57	34.07
7.	Personal growth	18.41	29.48
8.	Money	3.41	13.72
9.	Housing	0.87	7.41

N=949; significance levels $^*p \leq 0.05$. $^{**}p \leq 0.01$

Table 2. The unemployed. Quality of life (QL) profile and the respective level of satisfaction

Quality of life

	Aspect of life (cue)	Average QL	Standard deviation
1.	Family	25.28	17.04
2.	Health	20.12	15.11
3.	Work	9.70	14.36
4.	Peace of mind	5.70	8.88
5.	Interpersonal relations	5.65	8.47
6.	Hobbies	5.50	8.36
7.	Personal growth	5.06	9.21
8.	Money	2.13	8.14
9.	Housing	0.44	3.18

Satisfaction

	Aspect of life (cue)	Average satisfaction	Standard deviation
1.	Family	58.03	35.14
2.	Health	53.18	34.23
3.	Hobbies	28.81	36.77
4.	Interpersonal relations	25.71	34.35
5.	Peace of mind	20.54	30.44
6.	Personal growth	16.82	27.83
7.	Work	8.60	18.86
8.	Money	3.09	11.90
9.	Housing	1.16	8.50

N=966; significance levels $*p \leq 0.05$. $**p \leq 0.01$; Given the numbers of respondents involved, differences between groups could only be statistically tested between the employed and the unemployed. Where statistically significant differences were identified, the average figures are given in bold.

In addition, we identified differences between the employed and the unemployed in the frequency of reporting family as something that gives meaning to life and provides personal self-fulfilment. By losing their jobs, people experience a shift and transformation of their vital energy from work to family, which gives an alternative arrangement of the social roles. Especially for unemployed women, family is an alternative working field providing self-fulfilment and easing the burden of unemployment (Buchtová et al., 2002, p.100). The psychosocial burden on unemployed men responsible for the livelihood of the family is much heavier than that on unemployed women.

Health

Health was reported to be one of the primary life qualities by all the four groups of respondents. Semantically, health was most frequently associated with the following:

1. the highest value in life, 2. a value people become aware of only after they lose it (increased care for health as a result of illness or injury), 3. a prerequisite for achieving lasting employment (a prerequisite for productive life), 4. healthy lifestyle (healthy food, physical exercise – care for physical health), 5. a source of physical and mental comfort (with a focus on harmony between mental and physical health resulting in satisfaction with life), 6. a guarantee of self-reliance (independence from others, in particular

in old age, not being a burden on others), 7. care for the health of family members and close people.

Health is a supreme value from which a number of other fulfilments stem in the quality of life of both the employed and the unemployed respondents. It is also seen (as it was frequently reported) as a prerequisite for getting and retaining a job. Health is the most valued asset in the actual labour market. The chance of people with disabilities to find jobs is decreasing as a result of the strong focus on the productivity of work and performance. In general, the period of time they are registered at job centres exceeds that of the healthy individuals many times.

Several researchers describe in their studies the connection between unemployment and a decline in health. In our studies as well (e.g., Buchtová, 1992, 1999, 2000), more than a half of the long-term unemployed repeatedly reported subjective symptoms of neurotic complaints such as anxiety, unease, irritation, headache, insomnia, exhaustion. After losing the job, both men and women experienced a worsening of existing health problems – hypertension, stomach ulcers, heart disease, spinal cord problems, asthma, etc. Many people in the Czech labour market live in fear of job loss due to being employed for a limited period of time or as they observe the constantly increasing unemployment figures in many regions. Foreign investors recently resorted to large-scale layoffs as the Czech labour force has become less profitable. Apart from having an impact on those who have lost their jobs, unemployment also influences the behaviour and health of the employed. They either experience anxiety and strain from the anticipated job loss or they have to work under deteriorated conditions. It is obvious that the quality of peoples' emotional comfort represented by health is influenced by the changed economic climate regardless of whether the individual directly experiences the unpleasant job-related events.

Work

Work was ranked third by the respondents considering their quality of life. Work was most frequently related to the following needs: 1. self-fulfilment (employing one's abilities, knowledge and skills), 2. financial independence (material support for the family, means of independence), 3. security in life (a certain future), 4. orderly life (daily programme, time spending, everyday routine), 5. social background (interpersonal relations in the workplace, friendship, celebrations, common eating facilities), 6. emotional response, emotional appreciation (need for success, appreciation, acknowledgment).

A statistically significant difference between both groups was identified in their satisfaction with work. While the employed respondents reported remarkable satisfaction, the unemployed tended to fill in this life aspect with hobbies and interests.

A generation gap is identifiable in the responses of the unemployed respondents. Young people believe they will find a job soon, they strive for self-fulfilment and have plans for the future, while older individuals place the basic needs of their families first. Men more often reported age discrimination in the labour market. Repeated failure to find a job can lead to depression, feeling inferior, losing confidence. In family life, the unemployed feel like "parasites", men believe unemployment is the reason for their inability to satisfy the basic needs of their families. It is obvious that the importance of labour in human life changes, which has an effect on experiencing and coping with the loss of work.

The employed respondents showed a significant satisfaction with work as well as aspirations to get even a better job facilitating their further striving for better education (in particular as computer literacy and language skills are concerned). Work is conceived as an important place in human community, as a place of fellowship with colleagues, as a second "family".

The following part of our research was focused on comparing the average quality of life among the different groups (see Table 3).

Table 3. Quality of life for different groups (in per cent)

	QL average	Standard deviation	Number of respondents
Unemployed	60.3	20.0	966
Employed	70.6	15.8	949
Total			1,915

The individual evaluation of the quality of life is higher among the employed. It is apparent that a loss of employment is a milestone in human life, it has a significant effect on the quality of life and impacts mostly negatively other life aspects and personal goals.

Quality of life aspects and life satisfaction by age and gender

We have confirmed the assumed relationships between the quality of life aspects, satisfaction with them and the age and gender. The results of the statistical analysis confirmed major links between the variables in question.

Among the *unemployed respondents* the importance of health (0.222**), family (0.129**) and finances (0.071*) increases with age, while the importance of personal improvement decreases as well as satisfaction with it (-0.168**); the same holds true for interpersonal relations (-0.097**) and peace of mind (-0.079*). In the same group of respondents, satisfaction with family proved to be statistically significant (0.080*) as well as housing (0.081*).

Among the *employed respondents* age was positively associated with the values of family (0.157**) and peace of mind (0.073*) and also with satisfaction with both. A negative correlation was found in the importance of and satisfaction with personal improvement (-0.134**; -0.129**).

The results show that in terms of gender, both the unemployed and the employed women are more satisfied with family and health life aspects than both groups of men.

In the case of both groups, the employed and the unemployed men, there is a positive correlation with leisure, which is not observed among women.

Employed women tend to be more satisfied with their interpersonal relations (31.25**) than employed men (21.67). Among unemployed men and women no significant differences in satisfaction with this variable were found.

Both unemployed and employed women tend to be more satisfied with their personal improvement and peace of mind than both groups of men.

We tested the differences between employed and unemployed men and women in preferences of the life aspects and the satisfaction with them using the Pearson's correlations and testing the individual variables for the different groups (N=1915; unemployed =966, employed =949; significance level $*p \leq 0.05$; $**p \leq 0.01$).

The data suggest that from the gender perspective family is a more important life aspect for women, both for unemployed and employed women compared with men (-0.239**; 0.120**) and women tend to be more satisfied with this than men.

The value of health and interpersonal relations is considered to be more important by employed women than by employed men (-0.112**; -0.088**) and the employed women tend to be more satisfied with these aspects (-0.116; - 0.134**). Between unemployed men and women gender differences were not confirmed.

The value of work and leisure is statistically more significant for the employed (0.121**; **0.167**) and the unemployed men (0.117**; 0.194**) than for women. Both the employed and the unemployed men find more satisfaction in their hobbies (0.129**; 0.170**) than both groups of women.

Satisfaction with personal improvement and satisfaction with peace of mind is statistically more significant among both groups of women (unemployed and employed) than among men.

Personal improvement is statistically more significant for unemployed women (-.077*) than for unemployed men. Employed women are apparently more satisfied with their interpersonal relations than employed men. Differences between genders in this value were not confirmed between unemployed men and unemployed women.

For other life aspects, no statistically significant differences between men and women were discovered.

Quality of life aspect and life satisfaction according to educational level

We analysed the relationships between the quality of life aspects, satisfaction with them and the achieved level of education using the Spearman's rank correlation. The results of the statistical analysis revealed significant relationships (N=1,915; unemployed =966, employed =949; significance level $*p \leq 0.05$; $**p \leq 0.01$).

The results suggest that the higher the level of education in the unemployed group, the lower the importance of hobbies (-0.089**), value of money (-0.084**) and work (-0.076*), while the importance of personal improvement increases with education (0.094**), as well as peace of mind (0.079*).

In the employed group, the importance of money in their quality of life decreases with the achieved level of education (-0.113**), while the importance of personal improvement increases (0.107**).

More educated unemployed people showed a significant increase in satisfaction with their personal improvement (0.118**) and peace of mind (0.085**) (they cope better with the loss of work) as well as with health (0.065*). On the other hand, their satisfaction with money decreases (-0.095**) along with the satisfaction with hobbies (-0.077*) and housing (-0.064*).

Among the employed people satisfaction with their personal improvement increases with education (0.114), while satisfaction with money decreases (-0.113**).

Quality of life aspects and satisfaction with them according to unemployment duration

We tested the expected relationships between the quality of life aspects, the satisfaction with them and the duration of the unemployment using the Pearson's correlation coefficient. The data suggest that the longer a person is unemployed, the more importance he or she attributes to his or her family (0.081*). At the same time, satisfaction with work decreases (-0.067*).

Relationship between the duration of unemployment and education

Having divided the duration of the unemployment into 5 categories (up to 0.5 year, 0.5 to 1 year, 1 year to 1.5 years, 1.5 to 2 years and more), we calculated the Spearman's correlation of duration of the unemployment with education. The correlation was -0.226, significant at 1%. Our assumption that *the higher the level of achieved education, the shorter the duration of unemployment* was confirmed.

These results in Table 4 show the different frequencies and percentages. Shown *in bold* are the figures that are more frequent, in a statistically significant manner, than we would assume. *Bold italic* refers to figures significantly less frequent.

Table 4. Duration of unemployment by education

Education	up to ½ year	½ to 1 year	1 to 1½ year	1½ to 2 years	over 2 years	Total
Primary	***10***	***8***	***3***	9	**38**	68
Apprenticeship	***58***	62	46	27	**64**	257
Secondary	145	**137**	81	50	**72**	485
FE college	5	5	3	1	2	16
University	**54**	27	25	***6***	***15***	127

Note: The relationship between duration of unemployment and the level of education could be examined only in 953 respondents, in thirteen of the total 966 respondents education was not specified unambiguously.

The relationship between the level of education and the duration of unemployment is, similar to the Spearman's correlation above, statistically significant: χ^2=94.309, df=16, p=0.000

As Table 4 suggests, the most frequent combination consists of unemployed people with a secondary education who were jobless for less than half a year (145 persons) and ½ to 1 year (137 persons).

Significant relationships in the individual categories: People with lower education are significantly more frequently unemployed for a period exceeding 2 years (primary education and apprenticeship), people with higher education for a period up to 1 year (secondary and university education) – see the figures in *bold*. In contrast, combinations of short unemployment and low education as well as long-lasting unemployment and higher education are significantly less frequent – see the figures in *bold italics*.

Quality of life aspects for men and women and life satisfaction

We further examined the question whether the average QL figures differ for the different life aspects and the satisfaction with them between men and women (regardless of their employment status) and between the employed and the unemployed men and women. We used the ANOVA method. Respondent group sizes: 1,106 women, 851 men, 966 unemployed, 949 employed.

Family. Results of statistical analysis show that the average QL figures differ in the *family* aspect for men and women regardless of their employment status. Specifically (see the table containing the descriptive statistics), women show an average of 29.06 and men 22.92. The average QL figure for the *family* aspect is higher for women than for men.

While the average QL for the *family* aspect is roughly the same for the unemployed (28.82) and the employed (29.31) women, among men – whose QL aspect for *family* is generally lower than in the case of women – there is a great difference between the unemployed (20.59) and the employed (25.2). Among the employed men, the *family* aspects show a significantly higher value than among the unemployed men.

Among the employed men and women, the average satisfaction with *family* is higher (64.64) than among unemployed men and women (58.15), while women tend to be more satisfied with their families (66.82) than men (54.21) in the whole sample.

In the average satisfaction with *family* there is no significant difference between the employed (67.99) and the unemployed women (65.69), although among men certain significant differences were discovered. Unemployed men are significantly less satisfied (47.88) with their families than employed men (60.4).

Health. As far as the quality of life is concerned, the average figure for the *health* aspect is higher for women (21.43) – both employed and unemployed – than for both groups of men (19.27).

In the case of both employed men and women the average satisfaction with health is higher (58.54) than among unemployed men and women (53.29). Women in general are more satisfied with their health than men (51.48) (59.25). The highest average figure for the satisfaction with *health* is among employed women (62.00).

Work. The average quality of life data for *work* is in general significantly higher for men (11.40) then for women (8.11.

As anticipated, satisfaction with *work* is higher among employed men and women (30.19), than among unemployed men and women (8.62).

Peace of mind. No significant differences between employed and unemployed and between men and women regarding quality of life for the *peace of mind* aspect were discovered.

The average satisfaction with *peace of mind* is statistically more significant among women (25.58) than among men (20.99). Employed men and women are more satisfied with it (26.66) than the unemployed (20.58). Employed women show the highest average satisfaction with the *peace of mind* aspect (29.08).

Interpersonal relations. Women tend to be more satisfied with the *interpersonal relations* aspect (28.85) than men (23.15) across the entire set of respondents. The average quality of life for the *interpersonal relations* aspect is higher among women (6.11) than among men (5.1) regardless of their employment status.

While the average satisfaction with the *interpersonal relations* aspect is roughly the same for both the employed and the unemployed men (21.67; 24.66), for women, who generally show a higher average value in the *interpersonal relations*, there is a significant difference between the employed (31.25) and the unemployed (26.58) ones. Employed women are more satisfied with their interpersonal relations than unemployed women.

Hobbies and interests. The average quality of life for the *hobbies* and *interests* aspect is higher among men (6.73) than among women (3.84) and is higher among all the unemployed men and women (5.52) than among the employed men and women (both 4.65). Among the unemployed men and women you can see a higher satisfaction with this aspect (28.86) than among the employed men and women (24.62). The highest average quality of life figure for the *hobbies* and *interests* aspect appears for the unemployed men (7.41) who also express the greatest satisfaction with it (36.15).

Personal improvement. The average quality of life figure for the *Personal improvement* aspect is more relevant for women (5.3) than for men (4.15). Across the entire set of respondents, women are more satisfied with their personal improvement (20.00) than men (14.55).

Money. As the table shows, the average quality of life figure for the *money* aspect is higher for men (1.97) than for women (1.33). An obvious difference also exists between the unemployed (2.13) and the employed (1.07) men and women. The average figure for the m*oney* aspect is most significant among unemployed men (2.65).

The ANOVA results did not provide statistically relevant differences between the average satisfaction figures for the *money* aspect between men and women and between the employed and the unemployed.

Housing. The average quality of life and satisfaction figures for the *housing* aspect did not confirm significant differences either between the employed and the unemployed or between men and women.

Meaningfulness of life and long-term unemployment

Our study on the quality of life of the long-term unemployed included an evaluation of the meaningfulness of their lives. First we processed an analysis of the average meaningfulness figures reported by the different respondent groups.

Table 5. Analysis of the meaningfulness of life in the different groups (in per cent)

	Average (in %)	Number of respondents	Standard deviation
Unemployed	65.261	966	22.969
Employed	81.270	949	16.137

The average figures indicate remarkable differences between the unemployed (65.261) and the employed (81.270) respondents in their perception of their lives as meaningful.

We checked the differences in meaningfulness of life between the employed and the unemployed respondents using the t-test. The difference between the meaningfulness figures for the employed and the unemployed men and women groups is statistically significant. Work brings meaning and order to human life. No alternative activity is a fair replacement for a lost job.[3]

We further examined as to whether the meaningfulness of life evaluation is influenced by the duration of unemployment, gender and age and we also examined the differences between the employed and the unemployed.

3 For more details refer to Smajs, 2002, Machovec, 1967

Table 6. Pearson's correlations between the meaningfulness of life evaluation and duration of unemployment, gender and age

	Unemployed	Employed
Duration of unemployment (in months)	-0.111**	.
Gender	-0.158**	0.038
Age	-0.130**	0.056

N=1915; unemployed 966, employed 949; significance level *p ≤ 0.05; **p ≤ 0.01

As the data in Table 6 show, relationships are statistically significant for the unemployed group where the meaningfulness of life evaluation slightly decreases with the increasing duration of unemployment (-0.111**). The same holds true for men (-0.158**) and for older people (-0.130**). The results suggest that older men who are unemployed for a long time represent a risk group suffering worst from the loss of employment.

3 Conclusions

Our study has not only clarified our understanding of how job loss is experienced by the unemployed, but also our understanding of work and its role in contemporary life. The notion of the "Quality of Life" arose in the early 1970s as a sociological term initially referring to positive changes in human life by the social and scientific progress. Later, even the negative effects of illness and old age were incorporated into the concept, and one of our conclusions is that the notion will need further theoretical refinement and a more precise definition.

The classic idea of the Age of Enlightenment that human life can be improved through culture was built on the confidence that although the human nature has a mysterious biological element, the elastic social (socio-cultural) element is determinative and more important. It was believed that man shapes his own nature about as much as he creates an artificial cultural environment. Yet this optimistic understanding of man and his human nature is supported neither by the biological sciences nor by human and social life itself. In addition, it must be considered illusory that human work could be completely replaced by machine work using production and other technology. A return to employing human forces in a well-considered and productive way is not a step back, but rather a prospective solution of the existing stalemated unemployment situation in highly developed countries. So far, however, technical progress in combination with the market and relentless demand for profit steal jobs from thousands and tens of thousands of people every day.

Those who lose paid work through no fault of their own bear a load similar to that of suffering from a long-term illness. This is why we tried to employ a method in our study which had been used to measure changes in quality of life before and after interventions with ill and old people.

We chose the evaluation of individual quality of life in unemployed and employed people because the individual feelings when losing a job are rather resistant to a purely objective perspective. Furthermore, the Schedule for the Evaluation of Individual Quality of Life gives the individual a possibility to specify those areas of life he or she considers to be the most important in their life and to give a relative weight regarding their importance. SEIQoL is a method which, unlike traditional approaches, takes into account mainly the individual perspective.

Although we are well aware that we have to be careful in drawing conclusions from our study, we are confident that it has lead to several substantial findings:

Paid work in our young liberal market economy, which did not experience unemployment and its effects for two generations, will be increasingly valued in our lives.

Family is not a waning category with just religious, ethical and educational relevance, but rather a category still important from biological, social and existential perspectives.

Even the transformed lifestyles due to increased consumption, travel and massive expansion of consumer products such as cars, televisions and computers, cannot make up for or litigate the loss of the beneficial effect of work on human satisfaction and health.

References

Balcar, K. (1995). Životní smysluplnost, duševní pohoda a zdraví [Meaningful life, peace of mind and health]. *Československá Psychologie, 39* (6), 420-424.

Bergsma, J. & Engel, G. L. (1988). Quality of life: does measurement help? *Health Policy, l0*, 267-279.

Buchtová, B. (1992). *K psychologické dimenzi nezaměstnanosti* [On the psychological dimension of unemployment]. Brno: FF MU.

Buchtová, B. (1994). Vývoj a analýza psychologických výzkumů nezaměstnanosti [The development and an analysis of the psychological examination of unemployment]. *Československá Psychologie, 38* (2), 119-130.

Buchtová, B. (1995). K psychologické dimenzi nezaměstnanosti [On the psychological dimension of unemployment]. *Psychologie v Ekonomické Praxi, 30* (2), 1-17.

Buchtová , B.(1998). Sebereflexe dlouhodobé ztráty zaměstnání [Self-reflection in long-term unemployment]. *Sociální Politika, 24*, 12-14.

Buchtová, B. (1999a). Podání ruky na odchodnou [Severance handshake]. *Ekonom, 63* (39), 66-67.

Buchtová, B. (1999b). Nezaměstnanost je jako nevyléčitelná nemoc [Unemployment as an incurable illness]. *Psychologie Dnes, 5* (5), 8-11.

Buchtová, B. (2000). Nezaměstnanost a zdraví [Unemployment and health]. *Psychologie Dnes, 6* (5), 24-26.

Buchtová, B. (Ed.) (2000). Psychologické a medicínské aspekty nezaměstnanosti (*Psychological and medical aspects of unemployment*). Brno: ESF MU.

Buchtová, B. (2000). Sociální psychologie nezaměstnanosti [Social psychology of unemployment]. In: J. Výrost & I. Slaměník (Eds.), *Aplikovaná sociální psychologie II: Člověk v sociálním kontextu* (pp. 81-108). Prague: Grada.

Buchtová, B. (Ed.). (2002). *Nezaměstnanost. Psychologický, ekonomický a sociální problem* [Unemployment. A psychological, economic and social issue]. Prague: Grada.

Browne, J. P., C. A. O'Boyle, H. M. McGee, C. R. B. Joyce, N. J. McDonald, K. O'Malley, and B. Hiltbrunner. (1994). *Individual quality of life in the healthy elderly*. Quality of life Research 3, no. 4, p. 235-244.

Browne, J. P., C. A. O'Boyle, H. M. McGee, N. J. McDonald, and C. R. B. Joyce (1997). Development of a direct weighting procedure for quality of life domains. Quality of life Research 6, no. 4, p. 301-309. Dzúrová, D. & Dragomirecká, E. (2000). *Quality of life in the Czech Republic*. Acta Universitatis Caroline, 1, 103-116.

Emmons, R. A. & Diener, E. (1985). Personality correlates of subjective well-being. *Personality and Social Psychology Bulletin, 11*, 89-97.

Frankl, V. E. (1994). *Vůle ke smyslu* [Will to meaning]. Brno: Cesta.

Halama, P. (2000). Teoretické a metodologické prístupy k problematike zmyslu života [Theoretical and methodological approaches to the issue of meaningfulness of life]. *Československá Psychologie, 44* (3), 216-236.

Hnilica, K. (2000). Konflikt hodnot a kvalita života [Conflict of values and quality of life]. *Československá Psychologie,e 44* (5), 385-403.

Křivohlavý, J. (2001). *Psychologie zdraví* [Psychology of health]. Prague: Portál.

Křivohlavý, J. (2002). *Psychologie nemoci* [Psychology of illness]. Prague: Grada.

Machovec, M. (1967). *O smyslu lidského života* [On the meaning of human life]. Prague: Svoboda.

McGee, H. M. et al. (1991). Assessing the quality of life of the individual: The SEIQoL with a healthy and gastroenterology unit population. *Psychological Medicine, 21*, 749-759.

O´Boyle, C. A. & McGee, H. (1992). Individual quality of life in patients undergoing hip replacement. *Lancet, 33*, 1088-1091.

O´Boyle, C. A., McGee, H. & Joyce, C. R. (1994). Quality of life: Assessing the individual. *Advances in Medical Sociology, 5*, 159-180.

Oishi, S. et al. (1999). Value as a moderator in subjective well-being. *Journal of Personality, 67*, 157-184.

Ryff, C. D. & Keys, C. L. (1995). The structure of psychological well-being revisited. *Journal of Personality and Social Psychology, 69*, 719-727.

Strauss, A. L. (1975). *Chronic illness and the quality of life*. St. Louis: Mosby.

Šmajs, J. (2002) Práce – téma k zamyšlení [Work – a theme to think of]. In B. Buchtová (Ed.), *Nezaměstnanost. Psychologický, ekonomický a sociální problem* (pp. 9-13). Prague: Grada.

Šmajs, J. & Krob, J. (2003). *Evoluční ontologie* [Evolutionary ontology]. Brno: Masarykova universita.

Thompson, S. C. & Janigian, A. S. (1998). Life schemes: A framework for understanding the search for meaning. *Journal of Social and Clinical Psychology, 7* (2), 260-280

Ziga, S. & Chamberlain, K. (1987). Relation of hassles and personality to subjective well-being. *Journal of Personality and Social Psychology, 53* (1), 155-162.

2. UNEMPLOYMENT AND HEALTH

Unemployment and Health in East Germany: The Saxony Longitudinal Study

Hendrik Berth, Peter Förster, Ellen Hämmerling, Elmar Brähler, Markus Zenger & Yve Stöbel-Richter

Introduction

For more than 100 years, unemployment has been one of the most frequently examined social phenomena in psychology and social science. Most of the available meta-analyses and reviews covering unemployment and mental health conclude: Unemployment distinctively reduces mental health (e. g. Murphy & Athanasou, 1999; Feather, 1990; McKee-Ryan, Song, Wanberg & Kinicki, 2005, Kieselbach, Winefield, Boyd & Anderson, 2006 or Winefield, 2002). This relates to almost all aspects of the human psyche, ranging from psychological disorders as anxiety or depression to quality of life and well-being, to alcohol and drug abuse. Meta-analyses (e. g. Murphy & Athanasou, 1999) showed that psychological consequences of unemployment are often more serious than somatic ones.

In Germany, psychological research of unemployment boomed in the early 1990s due to the consequences of reunification (cf. Kieselbach & Voigt, 1994). The reorganisation of the former East German economy brought about the closing of many state enterprises, thus resulting in a massive reduction of jobs. The consequent unemployment figures are much higher than in West Germany; in spite of considerable political efforts they have remained the social reality. In May 2008, 3,283,279 people were unemployed in Germany, according to the official statistics of the German Federal Labour Office. This corresponds to a rate of 8.8%. The differences between the newly-formed German states (1,143,387 persons, 15.0%) and the old West German states (2,139,892 persons, 7.2%) are significant (http://www.pub.arbeitsamt.de/hst/services/statistik/000000/html/start/monat/aktuell.pdf).

Since 1990 almost all citizens of the newly-formed German states have had experience with unemployment in the form of their own unemployment or within the family circle, their circle of friends or acquaintances. Unemployment in East Germany is a mass phenomenon with serious problems (e.g. Bormann, 2006). Hence, even though the German reunification occurred twenty years ago, it is still important to examine the health-related consequences of unemployment in the newly-formed German states. For this purpose the study presents selected data from a longitudinal study.

1 Research methods

1.1 Sample

The Saxony Longitudinal Study ("Sächsische Längsschnittstudie", cf. Berth, Förster, Brähler & Stöbel-Richter, 2007; http://www.wiedervereinigung.de/sls/) was launched in 1987 in the former GDR. A sample (*N* = 1.281) of then 14-year-old students was selected as a representative group for the East German cohort of 1973; they were interviewed repeatedly until the spring of 1989. In the spring of 1989, the third poll (n = 587) of these participants consented to take part in further surveys. It has been possible to continue the study following German reunification until today. In 2006, the twentieth survey and in 2007, the twenty-first survey were conducted. The main focus of the study dealt with political and social questions; for example those relating to long-term socialization in the GDR, experiencing German reunification, and changes in living conditions. Starting in 2002, an additional focus of the study has been an examination of the consequences of unemployment (cf. e. g. Berth, Förster & Brähler, 2005; Berth, Förster, Stöbel-Richter, Balck & Brähler, 2006).

The 383 respondents of the twenty-first survey (2007) of the Saxony Longitudinal Study were aged 34 years on average. All participants who took part in the first survey in 1987 were in grade 8 at that time, i.e. the sample was homogeneous in age. Of these participants, 54% were female. The response rate, referring to 587 persons who had agreed to further participate in 1989, amounted to 65%. Most of the interviewees finished their vocational or professional education; only 2 % did not have a completed vocational training. More than 85% lived in a relationship, 45% were married, and 67% had children. More information about the participants in the last surveys of the study is provided in Table 1.

1.2 Questionnaires

A great deal of information was collected on the experiences of reunification as well as the transformation of East Germany. Additionally, the experience of unemployment ("never", "one time", "repeatedly") and cumulative unemployment (in months) have been studied since 1996. Besides some standard instruments (among others HADS-D, GBB-24, SCL-9) some scales developed by the Saxony Longitudinal Study were also employed. The D-Score, comprising four items, is used in order to measure the general psychological burden of an individual, covering, for example, feelings of despondence and dejection or fear of the future. The G-Score investigates some common somatic symptoms (stomach-aches, heart trouble, nervousness, sleeplessness). In both of these screening instruments, higher scores signify a greater burden. The study questions focused also on social and family relations, family planning and the desire to have children.

The HADS-D (Hospital Anxiety and Depression Scale, Herrmann, Buss & Snaith, 1995) is an internationally-employed self-rating questionnaire for anxiety and depression in adults. It comprises 14 items. Its validity was demonstrated through several studies; current, representative reference values are available. The GBB-24, a short version of the "Giessener Beschwerdebogen" GBB (Brähler & Scheer, 1995), is one of

Table 1. Selected characteristics of the participants of the Saxony Longitudinal Study from 2002 until 2007

Survey (year)	**16 (2002)**	**17 (2003)**	**18 (2004)**	**19 (2005)**	**20 (2006)**	**21 (2007)**
Participants (N =)	423	419	414	385	387	383
Response rate (%)	72.1	71.4	70.5	65.5	65.9	65.2
Age (M)	29.0	30.1	31.1	32.1	33.2	34.2
Sex (female, N =, %)	221 (52.6)	227 (54.2)	222 (53.6)	205 (53.4)	211 (54.5)	207 (54.2)
Occupation (N =, %)						
In training	17 (4.0)	12 (2.9)	8 (1.9)	7 (1.8)	4 (1.0)	4 (1.0)
Worker	96 (22.9)	87 (20.8)	77 (18.6)	79 (20.6)	70 (18.1)	75 (19.8)
Employee	181 (43.1)	167 (40.0)	178 (43.0)	150 (39.1)	164 (42.5)	172 (45.5)
Self-employed	25 (6.0)	28 (6.7)	36 (8.7)	36 (9.4)	40 (10.4)	40 (10.6)
At home	50 (11.9)	53 (12.7)	44 (10.6)	42 (10.9)	37 (9.6)	28 (7.4)
Unemployed	22 (5.2)	38 (9.1)	40 (9.7)	42 (10.9)	42 (10.9)	22 (5.8)
Other	29 (6.9)	32 (7.7)	31 (7.5)	28 (7.3)	19 (7.5)	37 (9.8)
Marital status (N =, %)						
single	279 (66.0)	261 (65.3)	239 (57.7)	209 (54.3)	193 (49.9)	192 (50.1)
married	129 (30.7)	149 (35.6)	162 (39.2)	161 (41.9)	174 (45.8)	170 (45.1)
divorced	12 (2.9)	8 (1.9)	12 (2.9)	14 (3.6)	19 (5.0)	15 (4.0)
Children (N =, %)						
yes	182 (43.3)	211 (50.4)	235 (57.0)	235 (61.4)	244 (64.6)	252 (67.2)
no	238 (56.7)	207 (49.5)	177 (43.0)	148 (38.6)	134 (35.4)	123 (32.8)

Note: missing to 100 %: no statement

the most frequently used instruments in Germany for studying subjective physical complaints. Its 24 items cover such physical complaints as "feeling of weakness" or "pain in nape or shoulder". The questions can be grouped into four scales: exhaustion, gastric complaints, pain in the limbs, and heart trouble. All items are used in order to compile a global score ("feeling of discomfort). The SCL-9 (Klaghofer & Brähler, 2001) is a short version of the internationally employed Symptom-Checklist-90-R. From each of the nine scales of the SCL-90-R, one item is used. Applying this efficient instrument, the psychological feeling of discomfort or the global distress of a person can be quantified.

2 Results

2.1 Experiences of unemployment in young East Germans 1996 - 2007

Figure 1 displays the participants′ experiences with unemployment from 1996 to 2007. In 1996, the participants, then 23 years old, were asked about unemployment for the first time. Even at that time 50% of the sample was affected by unemployment. Until 2007 a total of 32% of the remaining sample had been unemployed once and 40% had been repeatedly unemployed. On average the respondents were unemployed for a total of 17.3 months until 2007. Distinct statistically significant differences were found by gender: the cumulative unemployment of women (20.7 months) was longer than that of men (13.5 months).

Figure 1. Proportion of participants with one-time or repeated unemployment, 1996 - 2007 (%)

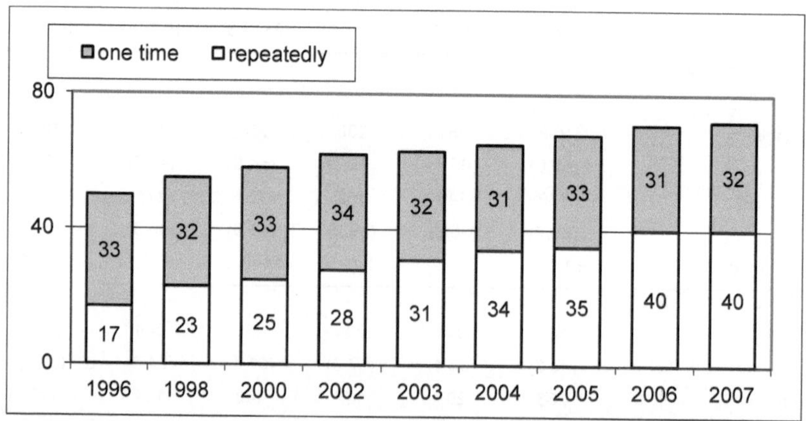

2.2 Unemployment and physical complaints

Table 2 on the next page summarizes the results of the GBB-24 from 2002 until 2006.

Table 2. Frequency of unemployment and physical complaints, 2002 - 2006 (GBB-24, *M, SD*, Oneway Anova)

	Experiences of unemployment								
	2002			2004			2006		
Scale	never	once	re-pea-tedly	never	once	re-pea-tedly	never	once	re-pea-tedly
Exhaus-tion	5.64	5.88	6.85	5.08	5.53	6.31	5.06	4.79	6.03
	(4.52)	(4.12)	(4.80	(4.24)	(4.36)	(4.61)	(4.35)	(4.54)	(5.04)
	$F=2.64$. $p>0.05$			$F=2.81$. $p>0.05$			$F=2.66$. $p>0.05$		
Sto-mach-ache	2.68	2.65	3.32	2.60	3.08	3.36	2.64	2.32	3.14
	(3.08)	(6.88	(3.24)	(2.91)	(3.42)	(3.19)	(3.40)	(2.95)	(3.08)
	$F=2.03$. $p>0.05$			$F=2.04$. $p>0.05$			$F=2.36$. $p>0.05$		
Pain in the limbs	6.60	6.88	7.66	6.07	6.23	7.78	6.29	6.04	7.64
	(4.23)	(4.27	(4.35)	(4.29)	(4.40)	(5.15)	(4.76)	(4.69)	(4.95)
	$F=2.13$. $p>0.05$			$F=5.72$. $p<0.05$			$F=4.36$. $p<0.05$		
Heart trouble	1.89	2.19	2.57	1.63	1.81	2.59	1.54	1.62	2.18
	(2.52)	(2.95	(3.31)	(2.29)	(2.79)	(3.45)	(2.28)	(2.36)	(3.27)
	$F=1.84$. $p>0.05$			$F=4.26$. $p<0.05$			$F=2.29$. $p>0.05$		
Feeling of discom fort	16.81	17.61	20.39	15.38	16.65	20.04	15.54	14.77	19.00
	(11.74)	(11.28)	(11.9	(11.16)	(11.82)	(13.15)	(12.03)	(11.89)	(13.22)
	$F=3.46$. $p<0.05$			$F=5.52$. $p<0.05$			$F=4.47$. $p<0.05$		

In the 2002 data, there was only one significant difference for the global score of "feeling of discomfort"; it was typical of cumulatively unemployed respondents. In the data for 2004 and 2006, "feeling of discomfort" and "pain in the limbs" were in addition higher for these participants compared to the sample who were never unemployed or unemployed only once. A further difference can be found for "heart trouble" in 2004. For all surveys, this exemplifies an overall higher nonspecific burden for participants who had repeatedly been unemployed, whereas interviewees who had never been unemployed or unemployed only once did not differ with regard to their physical complaints.

The G-Score (data not shown, see Berth et al., 2007) records the degree of burden through physical/psychosomatic complaints, and it was applied for the first time in 1996. The data shows a homogenous pattern: Those who were repeatedly unemployed suffered from physical problems in all surveys employing this instrument. According to the results in the GBB-24, there were no differences with respect to physical burden among those participants who were unemployed only once or never at all.

2.3 Unemployment and mental health

Mental health was studied by anxiety and depression scores (HADS-D), the global score of psychological distress (SCL-9, see Table 3) as well as the value of the short questionnaire D-Score.

Table 3. Frequency of unemployment and anxiety, depression (HADS-D) as well as psychological distress (SCL-9), 2002 - 2006 (*M, SD*, Oneway Anova)

	Experiences of unemployment								
	2002			2004			2006		
Scale	never	once	re-peatedly	never	once	re-peatedly	never	once	re-peatedly
HADS-D Anxiety	6.24 (3.16)	6.13 (3.00)	7.44 (3.32)	5.53 (3.57)	5.67 (3.10)	6.30 (3.74)	5.36 (2.29)	5.16 (3.50)	5.81 (3.52)
	F=6.74. *p*<0.01			*F*=1.93. *p*>0.05			*F*=1.28. p>0.05		
HADS-D Depression	3.47 (2.86)	3.67 (2.69)	4.98 (3.00)	3.75 (3.10)	4.21 (3.16)	5.58 (3.66)	3.61 (2.89)	3.83 (3.49)	4.69 (3.62)
	F=10.68. *p*<0.001			*F*=11.38. *p*<0.01			*F*=42.8. *p*<0.05		
SCL-9 psychological distress	6.91 (6.67)	6.27 (5.39)	8.26 (6.89)	6.32 (5.62)	6.62 (5.18)	7.73 (6.20)	5.92 (5.17)	6.82 (6.86)	7.57 (6.08)
	F=3.29. *p*<0.05			*F*=2.37. *p*<0.1			*F*=2.38. *p*<0.1		

Concerning the HADS-D-scale for anxiety, in 2002 there was a statistically significant difference depending on the experience of unemployment, but there was no such difference in 2004 and 2006. Participants with cumulative unemployment reported more symptoms of anxiety in 2002. The scores for depression (HADS-D) showed a variable pattern in the survey years 2002, 2004 and 2006. Participants who had repeatedly been unemployed were, however, more depressive in each survey, and respondents who had never been affected by unemployment reported the lowest depression scores in all surveys.

The highest scores of general psychological distress (SCL-9) were always reported by participants who had repeatedly experienced unemployment. We see that their psychological distress was most distinct. However, these differences reach the level of significance ($p<0.05$) only in 2002. In 2004 and 2006 there are only marginal differences ($p<0.1$).

There was a considerably clearer pattern in all surveys by D-Score (data not shown, see Berth et al., 2007). From 1996 (12th survey) until 2007 (21st survey) for the SCL-9 the general psychological distress for each point in time was significantly related to the experience of unemployment. The highest psychological distress was, again, shown by participants who repeatedly experienced unemployment, followed by respondents who were unemployed only once. Participants who were never affected by unemployment always reported the lowest scores for psychological burden.

3.4 Unemployment and family planning

Significant delays in family formation exist for persons who have experienced unemployment. The participants with cumulative unemployment were less often married than those without any experience of unemployment. Thus, the share of married men differs according to whether they were affected by unemployment. Of all cases included in the surveys since 1998, 37% of the men and 58% of the women were married by the age of 33. Ten percent of the married men were unemployed several times and 14 % never or only once. Of the married women, 19% were unemployed on several occasions, 15% once and 24% never. The persons who reported cumulative unemployment appeared to marry later in life than those who were never unemployed. With respect to the number of children until 2006, no significant correlation with unemployment could be found. However, respondents who had never been unemployed tended to have more children than others.

Clustering the sample according to the cumulated duration of unemployment, we found that those who had not been unemployed more often stated 3+ children as an ideal. In contrast, participants who were among the long-term unemployed (12+ months) considered “no children” as an ideal. There was a statistically significant difference in the optimal number of children by unemployment. Figure 2 shows the changes regarding the ideal number of children for women between 2003 and 2006. Even though the results are not significant for each combination, there seems to be a relationship between the two factors in the longer term among female respondents. Nevertheless, as there was no statistical significance by current employment vs. unemployment, the impact is caused by cumulative unemployment.

Figure 2. **Ideal number of children in relationship to the cumulated time of unemployment for women, from 2003 until 2006 (*M*)**

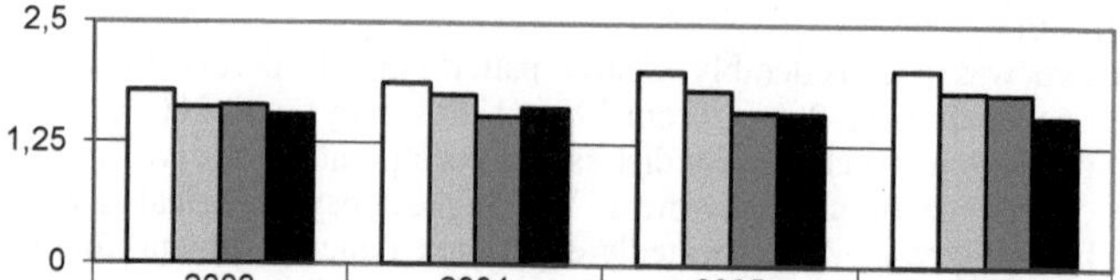

	2003	2004	2005	2006
never	1,78	1,87	2,00	2,03
1-4 months	1,61	1,75	1,81	1,81
5-11 months	1,63	1,53	1,59	1,79
12+ months	1,54	1,62	1,58	1,55

2.5 Predictors of unemployment

As shown above, the Saxony Longitudinal Study also shows cumulative unemployment to be linked with higher physical and psychological burden but the causal direction of this correlation remains unaccounted for. Does unemployment make you sick or do the sick become unemployed more easily? Either hypothesis (shift and drift) can be proven by research (e. g. Kivimäki, Elovainio, Kokko, Pulkkinen, Kortteinen & Tuomikoski, 2003).

Since prevention becomes increasingly important in work life, the study of factors making people vulnerable for unemployment and its consequences takes on greater significance. Twenty-one years ago the goal of the Saxony Longitudinal Study was not to focus on unemployment and health. Thus a systematic test of various models on the basis of the given data is not possible. Nevertheless, some data from earlier surveys can be utilized.

An analysis of the predictive value of the respondents' school performance (cf. Berth, Förster, Balck, Brähler & Stöbel-Richter, 2008) may illustrate the importance of education for one's professional career and risk of unemployment. In the former GDR, most students went to a uniform kind of comprehensive school until grade 10. Only few students continued up to grade 12, which provided qualifications for higher education, while other students continued with vocational training. A statistically significant correlation of the tenth grade performance (1989) with the subsequent experience of unemployment was found. Seventeen years later, the former weak students suffered from unemployment more often and for longer periods of time than the others. The dependency is similar to that between further education and unemployment: If the person did not have a university degree the chance of becoming unemployed increased by 77 %. Even though women were unemployed for longer periods of time than men, gender was not a significant predictor for unemployment. The same applied to several other potential predictors such as having children or civil status.

Concerning psychological burden earlier in life and unemployment, a link was found (Figure 3, see also Berth et al., 2006).

Figure 3. Experiences of unemployment 2007 and psychological burden (D-Score, *M*) 1991, 1996 and 2007 (*n* = 169)

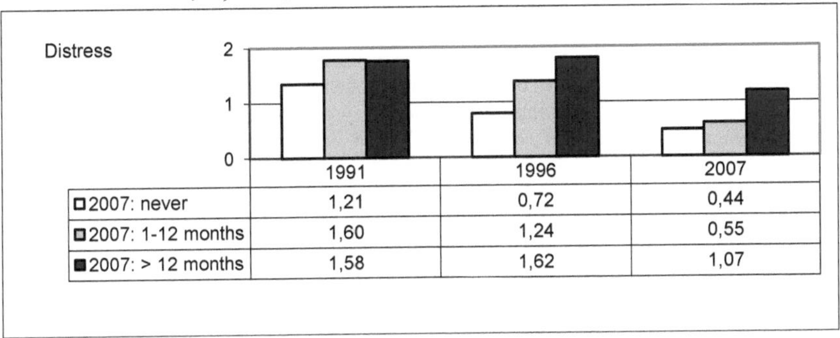

	1991	1996	2007
2007: never	1,21	0,72	0,44
2007: 1-12 months	1,60	1,24	0,55
2007: > 12 months	1,58	1,62	1,07

Figure 3 shows that the psychological burden depended on cumulative unemployment 2007 (three categories: never, 1-12 months, > 12 months) for all participants ($n = 169$) who took part in the Saxony Longitudinal Study in 1991, 1996 and 2007. Firstly, the figure illustrates the already described relation: Respondents who had been cumulatively unemployed until 2007 were psychologically more burdened than others. Yet, it also shows that those who were unemployed longer in 2007 (at the age of approx. 34) already had been more psychologically burdened in 1996 and in 1991. In 1991, the participants were about 18 years old. None of them had been unemployed previously. Those who demonstrated higher psychological burden in 1991 (by D-Score) had a risk of becoming unemployed in 2007 which was higher by a factor of 2.79 (sex-adjusted odds ratio).

3 Discussion

Labour markets in many European countries have changed crucially during recent years. Jobs requiring high education as well as service-related jobs are increasing. Nevertheless, unemployment remains a social, political and health-related topic, especially in many East European countries in transition as well as in the former GDR, presently Eastern Germany.

Unemployment is related to several physical complaints for many people concerned, such as higher risks of chronic bronchitis, back pain, vertigo, hypertension or bronchial asthma (Lange & Lampert, 2005), but there is still relatively little research on the somatic consequences of unemployment (Jin, Shah & Svoboda, 1995). One reason for this may be that psychological distress in unemployed people is often more serious than physical distress. Psychological impact mostly relates to such problems as depression, symptoms of anxiety, psychological distress, psychosomatic symptoms, satisfaction with life, and emotional well-being. The results of this study indicate that unemployment leads to a higher degree of physical complaints, anxiety, depression and a generally increased level of distress. In the course of this, those with cumulative unemployment often described several conditions. In contrast, those who had been unemployed only once or for a short period of time, respectively, did not differ significantly from those who had never been unemployed. These (in-) group differences were consistently found in all surveys of the study. The study proved that increased psychological burden at the age of about 18 was associated with a significantly higher risk of unemployment 16 years later. Further analyses demonstrated, for example, a negative impact of unemployment on the quality of life (Berth, Förster & Brähler, 2005).

The consequence of unemployment on the psyche varies depending on several characteristics. Meta-analyses (e.g. Paul, Hassel & Moser, 2006) show that men, younger persons, persons with lower social or professional status, and people affected by long-term unemployment suffer more from unemployment. However, these consequences are also mediated by personality traits such as self-efficacy (e.g. Berth, Förster, Balck, Brähler & Stöbel-Richter, 2005) or coping (e.g. Christensen, Schmidt, Kriegbaum, Hougaard & Holstein, 2006).

In Germany, unemployment has decreased significantly during the last two years (2006-07). However, as already stated, considerable differences between East and West Germany continue to exist since German reunification. In the former GDR, unemployment rates are still twice as high as in West Germany. The present study could ascertain that the majority of East German respondents (70 %) had experienced unemployment once or repeatedly by the age of 34. On average, women were unemployed for longer periods of time. Well-qualified persons - only 2% of the respondents had not completed vocational training - also have a hard time finding a job. Due to the better economic situation, 25 % of the current study participants are now living in West Germany (see Berth, Förster & Brähler, 2004).

Due to the fact that the participants have reached their middle age, sociological and social-psychological aspects of family planning were one of the focuses of the Saxony Longitudinal Study. We were able to show that unemployment also has effects on civil status and parenthood: There was a time lag for starting a family as well as reduced wishes concerning the number of children as well as a reduced number of children actually born.

Our participants were socialized until the age of 16/17 in the former GDR and completed their schooling there. They witnessed the social changes following German reunification during their vocational training, and they could profit from the advantages of the new social system. Nevertheless, they also often had to face disadvantages in terms of unemployment. We should emphasise that the sample consists of a young, age-homogenous, well-educated and mobile cohort. We are planning to continue the panel study with a special focus on unemployment because there is a risk that the experience of unemployment may have psychologically negative consequences many years after re-employment (Lucas, Clark, Georgellis & Diener, 2004).

Since the Saxony Longitudinal Study was initially not designed to measure the impact of unemployment on health, data was not always gathered for all of those indicators which would have been interesting and important to explore. In unemployment research it is important to consider many indicators because the relation between unemployment and health is multifactorially conditioned (cf. e.g. Beland, Birch & Stoddart, 2002). Furthermore, the sample sizes of the single surveys vary considerably, which limits the possibility of longitudinal analyses.

Policy and decision makers are not sufficiently aware of the negative consequences of unemployment on health (Hammarström & Janlert, 2005). The results of the Saxony Longitudinal Study emphasize that unemployment should be seen as a passing, normal phenomenon which can be part of each person's professional biography, especially in the currently transforming markets, with their requirements of constantly changing professional biographies as well as high mobility and flexibility. Accordingly, society should change its way of dealing with unemployment. The data of the Saxony Longitudinal Study, for instance, emphasizes the importance of adequate health care for unemployed people. Despite the enormous bulk of research on the topic of unemployment and health, there is still a lack of studies examining interventions in health promotion for the unemployed.

Acknowledgements

We would like to express our sincere thanks to the Otto-Brenner-Foundation, Frankfurt am Main (Germany), and to the Rosa-Luxemburg-Foundation, Berlin (Germany), for their generous support of the Saxony Longitudinal Study.

References

Beland, F., Birch, S. & Stoddart, G. (2002). Unemployment and health: Contextual-level influences on the production of health in populations. *Social Science and Medicine, 55*, 2033-2052.

Berth, H., Förster, P. & Brähler, E. (2004). Psychosoziale Folgen einer Migration aus den neuen in die alten Bundesländer. Ergebnisse einer Längsschnittstudie [Psychosocial consequences of migrating from the newly-formed East German states to the old West German states. Results from a longitudinal study]. *Psychosozial, 27*, 81-95.

Berth, H., Förster, P. & Brähler, E. (2005). Arbeitslosigkeit, Arbeitsplatzunsicherheit und Lebenszufriedenheit. Ergebnisse einer Studie bei jungen Erwachsenen in den neuen Bundesländern [Unemployment, job insecurity and satisfaction with life. Results from a study with young adults in the newly-formed German states]. *Sozial- und Präventivmedizin, 50*, 361-369.

Berth, H., Förster, P., Balck, F., Brähler, E. & Stöbel-Richter, Y. (2005). Arbeitslosigkeit, Selbstwirksamkeitserwartung, Beschwerdeerleben. Ergebnisse einer Studie bei jungen Erwachsenen [Unemployment, expected self-efficacy, experience of burden. Results from a study with young adults]. *Zeitschrift für Klinische Psychologie, Psychiatrie und Psychotherapie, 53*, 328-341.

Berth, H., Förster, P., Balck, F., Brähler, E. & Stöbel-Richter, Y. (2008). Schulnoten, Berufsbiographie und Arbeitslosigkeit. Ergebnisse der Sächsischen Längsschnittstudie [Grades, biography and unemployment. Results from the Saxony Longitudinal Study]. In P. Genkova (Ed.), *Erfolgreich dank Schlüsselqualifikationen? Heimliche Lehrpläne und Basiskompetenzen im Zeichen der Globalisierung* (pp 265-267). Lengerich: Pabst.

Berth, H., Förster, P., Brähler, E. & Stöbel-Richter, Y. (2007). *Einheitslust und Einheitsfrust. Junge Ostdeutsche auf dem Weg vom DDR- zum Bundesbürger* [Pros and cons of the German reunification. Young East Germans on their way to becoming a German citizen]. Gießen: Psychosozial-Verlag.

Berth, H., Förster, P., Stöbel-Richter, Y., Balck, F. & Brähler, E. (2006). Arbeitslosigkeit und psychische Belastung. Ergebnisse einer Längsschnittstudie 1991 bis 2004 [Unemployment and psychological burden. Results from a longitudinal study, 1991 – 2004]. *Zeitschrift für Medizinische Psychologie, 15*, 111-116.

Bormann, C. (2006). Gesundheitliche Konsequenzen von Arbeitslosigkeit in den alten und neuen Ländern in der Gender-Perspektive [Health-related consequences of unemployment in the newly-formed East German states and the old West German states from the perspective of gender]. In A. Hollederer & H. Brand (Eds.), *Arbeitslosigkeit, Gesundheit und Krankheit* (pp. 85-96). Bern: Huber.

Brähler, E. & Scheer, J.W. (1995). *Gießener Beschwerdebogen (GBB)* [Gießen complaints questionnaire], 2nd rev. edition. Göttingen: Hogrefe.

Christensen, U., Schmidt, L., Kriegbaum, M., Hougaard, C.O. & Holstein, B.E. (2006). Coping with unemployment: Does educational attainment make any difference? *Scandinavian Journal of Public Health, 34*, 363-370.

Feather, N.T. (1990). *The psychological impact of unemployment.* New York: Springer.

Hammarstrom, A. & Janlert, U. (2005). An agenda for unemployment research: A challenge for public health. *International Journal of Health Services, 35*, 765-777.

Herrmann, C., Buss, U. & Snaith, R. P. (1995). *Hospital Anxiety and Depression Scale - Deutsche Version. Ein Fragebogen zur Erfassung von Angst und Depressivität in der somatischen Medizin* [Hospital Anxiety and Depression Scale – German Version. A questionnaire for detecting anxiety and depression in somatic medicine]. Bern: Huber.

Jin, R.L., Shah, C.P. & Svoboda, T.J. (1995). The impact of unemployment on health: A review of the evidence. *Canadian Medical Association Journal, 153*, 529-540.

Kieselbach, T. & Voigt, P. (Eds.). (1994). *Systemumbruch, Arbeitslosigkeit und individuelle Bewältigung in der Ex-DDR* [Change of society, unemployment and individual coping in the former GDR]. Weinheim: Deutscher Studienverlag.

Kieselbach, T., Winefield, A.H., Boyd, C. & Anderson, S. (Eds.) (2006). *Unemployment and health. International and interdisciplinary perspectives.* Bowen Hills: Australian Academic Press.

Kivimäki, M., Elovainio, M., Kokko, K., Pulkkinen, L., Kortteinen, M. & Tuomikoski, H. (2003). Hostility, unemployment and health status: Testing three theoretical models. *Social Science and Medicine, 56*, 2139-2152.

Klaghofer, R. & Brähler, E. (2001). Konstruktion und teststatistische Prüfung einer Kurzform der SCL-90-R [Construction and test-specific evaluation of the short version of the SCL-90-R]. *Zeitschrift für Klinische Psychologie, Psychiatrie und Psychotherapie, 49*, 115-124.

Lange, C. & Lampert, T. (2005). Die Gesundheit arbeitsloser Frauen und Männer. Erste Auswertungen des telefonischen Gesundheitssurveys 2003 [Health of unemployed women and men. First results of the telephone interview survey on health in 2003]. *Bundesgesundheitsblatt, 48*, 1256-1264.

Lucas, R.E., Clark, A.E., Georgellis, Y. & Diener, E. (2004). Unemployment alters the set point for life satisfaction. *Psychological Science, 15*, 8-13.

McKee-Ryan, F.M., Song, Z., Wanberg, C.R. & Kinicki, A.J. (2005). Psychological and physical well-being during unemployment: A meta-analytic study. *Journal of Applied Psychology, 90*, 53-76.

Murphy, G.C. & Athanasou, J.A. (1999). The effect of unemployment on mental health. *Journal of Occupational and Organisational Psychology, 72*, 83-99.

Paul, K.I., Hassel, A. & Moser, K. (2006). Die Auswirkungen von Arbeitslosigkeit auf die psychische Gesundheit: Befunde einer quantitativen Forschungsintegration [Consequences of unemployment on psychological health: Results of a quantitative research integration]. In A. Hollederer & H. Brand (Eds.), *Arbeitslosigkeit, Gesundheit und Krankheit* (pp. 35-52). Bern: Huber.

Winefield, A.H. (2002). The psychology of unemployment. In C.v. Hofsten & L. Baeckman (Eds.), *Psychology at the turn of the millennium*, vol. 2: Social developmental, and clinical perspectives (pp. 393-408). Florence: Taylor & Francis.

Association of Employment Status and Hypertension in Young Males

Yücel Demiral, Ahmet Soysal, Reyhan Uçku, Gazanfer Aksakoğlu, Dilek Soysal & Mehmet Köseoğlu

Introduction

Several studies support the view that psychosocial factors contribute to differences in cardiovascular risk factors, such as blood pressure (Hammarstrom, 1998; Henriksson, 2003; James, 1984; Janlert, 1992). Unemployment as a psychosocial risk factor has been shown to be associated with an increase in blood pressure in most industrialized countries (Janlert, 2003; Kasl, 1980). Furthermore, positive relationships were observed between being in a lower social class and having an increased prevalence of hypertension, engaging in less physical activity, and having a high cigarette consumption (Kunst, 1999). Social class position and educational level may both determine blood pressure independently or in combination with employment status. It must be noted however, that these results have not been completely consistent. The relative importance of the various types of psychosocial factors on cardiovascular disease (CVD) has not been fully discerned, especially the nature and degree of its relevance in different countries and cultures. Henriksson et al. showed that associations between unemployment and CVD risk factors diminished as unemployment rates increased. A higher unemployment rate within the general population may therefore act to obscure the effects of unemployment on cardiovascular diseases (Mattiasson, 1990). Therefore, in regard to this finding, the relationship between unemployment and health may be different in developing countries than in those that are developed. On the other hand, the studies that contribute to the current knowledge on association between unemployment and blood pressure were mostly conducted in the developed countries. The aim of this study was to explore the association of employment status and blood pressure levels in men in a large urban area of Turkey, a developing country.

1 Methods

Study population
This study was originally conducted to determine the prevalence of hypertension in young adults in the Konak Health District, Izmir, Turkey. There were 302,546 people aged between 20 and 39 in the Konak Health District. The sample size was calculated as 1076 people, since an expected hypertension prevalence was 7.0% with a 2.0% of error within a 99% CI. Stratified and cluster sampling methods were used to select the par-

ticipants. There were 45 Primary Health Centres in the Konak Health District. Primary Health Centres were scored according to the socioeconomic status of each health centre's region. Five strata, based on socioeconomic status, were determined to be: very poor, poor, fair, good, and very good. Every person aged between 20 and 39 years old on a particular street that had been previously selected, was notified of the study and invited to the health centre two days prior to the survey. Since most of the women in the study sample were housewives, and therefore not considered a part of the workforce, the analyses were carried out using only males. There were 509 male in the sample and 318 participated in the study. There were four men that were excluded from the study because of missing data for the employment situation. The study was completed with 314 males, which resulted in a response rate of 62%.

Variables and measurements

Data was obtained using a questionnaire which included socio-demographic variables such as age, education, marital status, number of children, as well as occupation, perceived economic status, smoking habit, current drug therapy and personal and family history of CVD. Blood pressure was measured after a 15 minute resting period by trained nurses twice, with 5 minute intervals between each reading. Participants were classified as hypertensive if the mean of both standardized measurements exceeded 140 mmHg systolic and/or 90 mmHg diastolic. Participants that stated being on antihypertensive medications were also regarded as hypertensive. Body mass indices (BMI) were calculated according to formulae "weight in kg/square of height in meters". Obesity was defined as any BMI of 25 or greater. Family history of premature CVD was defined as myocardial infarction or the sudden death in a first degree male relative <55 years old, and in a first degree female relative <65 years old. The length of education was dichotomized into two groups: those educated for 8 or fewer years and those educated for 9 or more years. Employment status was defined as unemployed, regularly employed and precariously employed (e.g., street vendor etc.). Precarious workers were later added to the unemployed group to dichotomize the employment status.

Statistical analyses

The point prevalence rate of hypertension was calculated. The t-test was used in the comparison of means (SD) and the chi-square test was used for the comparison of proportions. Logistic regression was used in the analysis of multiple variables. The crude and adjusted prevalence of high blood pressure in association with employment status were both determined. Adjustments were made for age, education, smoking habit and family history. Results were expressed with a 95% CI with a p value of <0.05 to be significant. Data were analyzed using SPSS 11.0.

Results

The mean age of the study population was 33.6±6.0 years. The mean age of the unemployed/precariously employed group (32±6) was younger than that of the regularly employed group (34±6), but this difference was not significant (p=0.09). There were 70 unemployed/precariously employed men (22%), and 244 regularly employed men (78%) (Table 1). Most of the study group had completed primary school (n=222, 71%). Fifty three percent of the males rated their economic position as fair, 33% as bad, and 15% as good. The prevalence of hypertension was 17% (n=53) in the study group. Among those with hypertension, 17 were already using antihypertensive drugs. Within the study population, 54% were currently smokers and the family history for CHD (coronary heart disease) was 44%. There were 13 (4%) men with diabetes mellitus (Table 1).

Table 1. Descriptive characteristics of the study population

	N = 314	%
Employment status		
Regularly employed	244	77.8
Unemployed/precariously employed	70	22.2
Education		
Primary	222	70.7
Higher	92	29.3
Perceived economic status		
Bad	102	32.5
Fair	166	52.9
Good	46	14.6
BMI		
<25	152	48.4
≥25	162	51.6
Smoking		
Current smokers	169	54.0
Non-smokers	145	46.0
Family history of CHD		
Yes	137	43.7
No	177	56.3
Diabetes		
Yes	13	4.1
No	301	95.9
Hypertension		
Yes	53	16.7
No	265	83.3

Table 2 shows the distribution of independent variables among both the regularly employed and unemployed/precariously employed groups. The unemployed/precariously employed group was found to have both a significantly lower level of education and a significantly worse economic situation in comparison to the regularly employed group. A family history of CHD was significantly higher among the regularly employed group (p=0.038). Unemployed/precariously employed men were significantly more likely to be smokers than the regularly employed ones (p=0.039). Hypertension was found to be significantly higher among the regularly employed (20.1%) group than among the unemployed/precariously employed (5.7%) group (p=0.008). Systolic blood pressure was found to be significantly higher in the regularly employed group (p=0.02) and diastolic blood pressure was also higher in regularly employed group, but this difference was not significant (p=0.06). There was no significant difference between the regularly employed and unemployed/precariously employed groups for both diabetes mellitus and BMI (p=0.740 and p=0.128, respectively).

Table 2. Distribution of independent variables according to employment status

	Regularly employed (N=244) %*	Unemployed/ precariously employed (N=70) %*	p
Low education	67.6	81.4	0.037
Bad economic status	21.3	71.4	<0.001
Obesity (BMI≥25)	54.1	42.9	0.128
Smokers	50.8	65.7	0.039
Family history of CHD	46.3	31.4	0.038
Diabetes	4.5	2.9	0.740
Hypertension	20.1	5.7	0.008

* Column percentages

The regularly employed group had a 4.1 fold (CI: 1.4-11.9) greater risk for hypertension in comparison to the unemployed/precariously-employed group (Table 3). Hypertension was also found to be 3.2 fold higher (CI: 1.4-7.0) in men with a better perceived economic status than in men with a worse perceived economic status. There was no significant excess risk for hypertension according to education level. After age, education, smoking habit and family history were adjusted, the association between employment status and hypertension persisted (OR: 3.1 (CI: 1.0-9.1)) (Table 3).

2 Discussion

This study showed that hypertension was more common in regularly employed men than among those who were unemployed/precariously employed. Hypertension was also found to be higher among those men who perceived their economic status as good-fair than among those who perceived their economic status as bad.

These findings on the association between employment status and blood pressure were inconsistent with the results from industrialized countries (Hammarstrom, 1988; Kasl, 1980; Morrell, 1998; Yarnell, 2005). The "modernization" process, which includes urbanization, smoking, a fatty diet, and a stressful work environment, has been used to explain the differences observed among the social gradients in regard to coronary heart disease rate. It has been suggested that during the early phase of industrialization, the risk factors for coronary heart disease, which include hypertension, are increased among the higher occupational class (Janlert, 2003). As the living standards improved, a positive gradient (e.g., the higher the class, the higher the risk) was observed. It has been explained that members of the lower socio-economic classes were too poor to be able to engage in risky health behaviours, such as tobacco smoking, animal fat intake or a sedentary lifestyle (Kunst, 1999). Our results suggested that is the current trend in Turkey.

Table 3. **Crude odds ratios for the independent variables for hypertension risks (95% CI) and adjusted odds ratio for the employment status**

	Hypertension %	OR (% 95 CI)	p
Age			
20-29	6.0	4.1 (1.6-10.9)	0.003
30-39	20.9		
Education level			
Primary or less (n=222)	15.8	0.8 (0.4-1.4)	0.514
High (n=92)	19.6		
Economic status			
Bad (n=102)	7.8	0.3 (0.1-0.7)	0.005
Good-fair (n=212)	21.2		
Smoking			
Yes	12.4	0.5 (0.3-0.9)	0.030
No	22.2		
Family history of CHD			
Yes	23.7	2.3 (1.3-4.3)	0.008
No	11.7		
Diabetes			
Yes	46.2	4.6 (1.5-14.4)	0.012
No	15.6		
BMI			
<25	9.2	3.1 (1.6-6.0)	0.012
≥25	24.1		
Employment status			
Unemployed/precariously employed (n=70)	5.7	4.1 (1.4-11.9)	0.008
Regularly employed (n=244)	20.1		
Employment status †			
Unemployed/precariously employed (n=70)	5.7	3.1 (1.0-9.1)	0.043
Regularly employed (n=244)	20.1		

†Adjusted for age, BMI, smoking habits, and family history

It has also been suggested that unemployment rates in a given country could play a role in the relationship between unemployment and the prevalence of coronary heart disease risk factors. Henriksson et al. (2003) showed that the association between unemployment and cardiovascular risk factors diminished as the unemployment rate increased. Turkey has an official unemployment rate of 11%, but research findings indicate that it may be as high as 25% in urban areas. The high unemployment rates in Turkey may therefore diminish the association between unemployment status and high blood pressure that is observed in developed countries. Social support, including relationships with relatives might play a role as protector to negative effects of unemployment in Turkey.

This study has some important limitations, which must be noted. First, this study was originally performed to determine the prevalence of cardiac risk factors among young people, aged 20 to 39 years old, in an urban area in Turkey. Therefore, there wasn't any data available on the duration of the period of unemployment. Thus, the study could not provide information on a dose–response relationship. Second, since the women were excluded from the study, it was not possible to interpret results for the whole population.

Larger population surveys, which would include women, might provide more conclusive evidence about the relationship between unemployment and blood pressure in developing countries. It is important to note that follow-up studies are needed to further explore the causal relationship between unemployment, as well as other social determinants, and the coronary risk factors in developing countries.

This study therefore indicates that there could be possible discrepancies between developed and developing countries with regard to the nature of the relationship between employment status and blood pressure.

References

Hammarstrom, A., Janlert, U. & Theorell, T. (1988). Youth unemployment and ill health: Results from a 2-year follow-up study. *Social Science and Medicine*, *26*(10), 1025-1033.

Henriksson, K.M., Lindblad, U., Agren, B., Nilsson-Ehle, P. & Rastam, L. (2003). Associations between unemployment and cardiovascular risk factors varies with the unemployment rate: The cardiovascular risk factor study in southern Sweden (CRISS). *Scandinavian Journal of Public Health, 31*(4), 305-311.

James, S.A., LaCroix, A.Z., Kleinbaum, D.G. & Strogatz, D.S. (1984). John Henryism and blood pressure differences among black men. II. The role of occupational stressors. *Journal of Behavioral Medicine*, *7*(3), 259-275.

Janlert, U. (1992). Unemployment and blood pressure in Swedish building labourers. *Journal of Internal Medicine, 231*(3), 241-246.

Janlert, U. & Holmgren, L. (2003). Psychosocial factors in the Northern Sweden MONICA project. *Scandinavian Journal of Public Health, 61*, 38-42.

Kasl, S.V. & Cobb, S. (1980). The experience of losing a job: Some effects on cardiovascular functioning. *Psychotherapy and Psychosomatics, 34*(2-3), 88-109.

Kunst, A.E., Groenhof, F., Andersen, O., Borgan, J.K., Costa, G., Desplanques, G. et al. (1999). Occupational class and ischemic heart disease mortality in the United States and 11 European countries. *American Journal of Public Health, 89*(1), 47-53.

Mattiasson, I., Lindgarde, F., Nilsson, J.A. & Theorell, T. (1990). Threat of unemployment and cardiovascular risk factors: Longitudinal study of quality of sleep and serum cholesterol concentrations in men threatened with redundancy. *British Medical Journal, 301*(6750), 461-466.

Morrell, S.L., Taylor, R.J. & Kerr, C.B. (1998). Jobless. Unemployment and young people's health. *Medical Journal of Australia, 168* (5), 236-240.

Yarnell, J., Yu, S., McCrum, E., Arveiler, D., Haas, B., Dallongeville, J. et al. (2005). Education, socioeconomic and lifestyle factors, and risk of coronary heart disease: The PRIME Study. *International Journal of Epidemiology*, *34*(2), 268-275

Studies on Unemployment and its Adverse Health Effects in China

Zhijun Zhou, Qiang-en Wu & Dong Chen[1]

1 Unemployment in China

As China has been changing from a communist economy to one based on "socialism with Chinese characteristics" during the past two decades, more and more workers have lost their jobs in state enterprises. Unemployment is becoming a more serious problem, especially as the country enters the globalized economy and prepares itself for full integration into the international trading system WTO.

According to the Chinese Ministry of Labour and Social Security, there were more than 8 million registered unemployed workers in Chinese urban settlements, and the registered unemployment rate was 4.2% at the end of September 2004. Mr. Meng-kui WANG, the director of the Development Research Center of the State Council (DRC), pointed out in his keynote speech at the China Development Forum, Beijing, June 2003 that the urban unemployment rate, including various forms of unemployment as defined by the ILO, is probably 8-10% at present based on the results of different studies.

It is difficult to measure the unemployment in China exactly. The unemployment figure published by the Chinese government is, in effect, the registered urban unemployment rate. Some laid-off workers are still included in the workforce of some state-owned enterprises. Some workers, who have found a new job, may still be registered as unemployed so that they can be covered by unemployment insurance or other benefits. To calculate the unemployment rate in a realistic way, jobseekers that have not registered themselves should also be taken into the calculation and informal employment should be taken into account in the measurement. It is generally known that the unemployment rate in the south and the east of China is lower than that in the north and in the west because of a significant difference in economic development. We may state that the rate of rural unemployment is still a mystery in China. In Shanghai, the unemployment rate is relatively low, since the economy has been developing very favourably, and the government tries to keep it below 5%.

1 WHO Collaborating Center for Occupational Health (Shanghai), Department of Occupational Health, School of Public Health, Fudan University, Shanghai 200032, China, Email: zjzhou@shmu.edu.cn

2 Effects of unemployment on health status

In China there are few studies concerning the effects of unemployment on health status. The works were conducted and reported only by several social or psychological scientists, not by medical scientists, who pay more attention to the problems of occupational safety and health. Fortunately, the situation is beginning to change. Both the Chinese government and the scientific community have recently started to pay increasing attention to new problems such as unemployment.

As a grave stressor, unemployment has a negative influence on the jobless person's mental and physical health. Research in China indicates that unemployment could cause a series of psychological problems including anxiety, depression, suicide, substance abuse, addiction, etc. This article describes the effects of unemployment on Chinese people's health based on the data published in Chinese medical journals.

3 Unemployment and mental health

The mental health consequences of unemployment have been studied in China with such questionnaires as Symptom Checklist-90 (SCL-90), Cornell Medical Index (CMI), Self-esteem Scale (SES) and Index of Well-being (IOWB). For studying the potential effects on social life, Social Support Questionnaire (SSQ), Thought Control Questionnaire (TCQ) and Impact of Event Scale (IES) questionnaires have been used. Chinese studies show similar results concerning the psychological costs of job stress and job insecurity found during unemployment as those found in international literature. Most studies draw their conclusions from comparisons between the mean symptom scores of groups with different employment statuses, e.g. people matched on a range of variables but differing in employment status.

For example, Xu and her colleagues (2001□interviewed 675 laid-off and 669 employed workers with SCL-90, SES, and IOWB in the city of Changsha, Hunan Province. About two-thirds (63.6%) of laid-off workers showed mental distress with varied degrees. The most typical psychological symptoms among unemployed persons were somatic symptoms, obsession-compulsion, anxiety, depression and interpersonal oversensitivity. Table 1 shows a statistically significant difference in mental health, defined by a number of indicators, between the laid-off and employed workers studied (Xu, Xiao & Chen, 2001).

Liu and his colleagues (2000) used the SCL-90 inventory to survey 628 unemployed workers in Nanchong City of Sichuan Province. It was found that the psychological health status deteriorated significantly among those who were unemployed a longer time. There was an inverse correlation between the SCL-90 scores and the unemployment time period for those who were unemployed for less than six months (Liu & Tan, 2000). Shi and Tao (2000) examined 31 male and 37 female laid-off people with MMPI in Xi'an City of Shangxi Province using a matched control group. The results showed a higher prevalence of depression, sexual indifference, hypochondria amongst unemployed persons, plus somatic problems among female respondents (Shi & Tao, 2000). Zhan and his colleagues studied the mental health of unemployed persons and compared the results with the available reference values in China for the whole pop-

Table 1. Comparison of SCL-90 score between unemployed and employed persons

Factor	*Unemployment*			*Employment*		
	Male (314)	Female (361)	*Incidence (%)*	Male (312)	*Female (357)*	*Inciden-ce (%)*
Total	70.64±62.3	81.71±63.8^△^	63.6	56.98±47.4	55.05±41.	23.6
Somatisation	0.85±0.81	0.97±0.823 △	11.0	0.67±0.68	0.62±0.54	4.0
Compulsive	0.93±0.76	1.05±0.753 △	9.0	0.82±0.64	0.79±0.56	4.2
Inter-personal Sensitivity	0.78±0.733	0.91±0.753 △	9.0	0.69±0.60	0.69±0.55	2.7
Depression	0.81±0.73	0.98±0.753 △	9.4	0.67±0.60	0.66±0.57	3.6
Anxiety	0.77±0.80	0.91±0.823 △	9.5	0.58±0.59	0.57±0.52	1.6
Hostility	0.81±0.82	0.90±0.833 △	7.5	0.61±0.61	0.62±0.56	2.5
Horror	0.52±0.70	0.69±0.783 △	6.8	0.36±0.51	0.38±0.45	1.2
Obstinacy	0.79±0.72	0.82±0.753 △	5.9	0.66±0.63	0.60±0.55	2.5
Psychosis	0.65±0.67	0.73±0.723 △	4.4	0.52±0.54	0.46±0.45	1.3
Other	0.84±0.73	0.91±0.723 △	-	0.67±0.56	0.65±0.5	-

△ The difference between unemployed and employed groups is statistically significant ($p < .01$).

ulation. Somatic symptoms, compulsive, inter-personal oversensitivity, depression, hostility, horror, obstinacy, and psychotic symptoms were compared. A significant difference in mental health in disfavour of the unemployed persons was found (Zhang, Guo & Yang, 1999; Shao, 1999; Chen, 1999; Li, He, Lin et al., 2000; Chen, Zhao, Wang, 1999; Wang, 2002).

Wang's research showed that the happiness index, total sensibility index, and the respondents′ satisfaction to their life quality were usually lower among those unemployed than among the controls. The self-respect scores for the unemployed persons have also decreased dramatically. There were multi-directional changes among the different age groups (Chen, Zhang & Li, 2004; (see Table 2) Wang, Zhang & Ti, 2000 (see Table 3)) . Wang studied the stress and psychological status of laid-off workers in state-owned factories in China in 2002 also with Cawte Stress questionnaire.

Table 2. Cawte Stress Questionnaire Scales of laid-off workers (Chen, Zhang & Li, 2004)

Group	Stress	N	Psychological	N	Physiological	N
Unemployment	6.25 ±4.01	277	3.82 ±2.42	283	3.43 ±2.38	285
Employment	5.52 ±3.72	263	2.91 ±2.71	268	2.96 ±2.12	264
Z-value	2.21*		4.41**		2.47*	

$^{*}P < 0.05$, $^{**}P < 0.01$, $^{***}P < 0.001$

He found a significant difference in stress as measured by the psychological and physiological scores between the unemployed and the employed persons. Depression, anxiety, irritated state of mind, mental strain and oversensitivity of laid-off workers were studied by CMI mental health questionnaires in another study by Wang. The anxiety of the unemployed persons was significantly higher than that of the employed persons (Wang, Zhang & Ti, 2000).

Table 3. CMI mental health questionnaires of laid-off workers (Wang, Zhang & Ti, 2000)

Item	Unemployment			Employment			
	X	S	Z	X	S	Z	Z
Depression	1.59	1.84	292	1.68	1.34	268	0.64
Anxiety	1.22	1.37	285	2.04	1.99	266	5.47***
Irritation	1.34	1.41	287	1.35	1.25	268	0.09
Strain	2.20	2.33	287	2.16	1.80	267	0.23
Sensitivity	2.17	2.15	286	2.13	1.81	268	0.24

4 Unemployment and health - mediation through low income

There is remarkable positive correlation between employment and somatic health in the Chinese data. A person with a high income can obtain better health service, good nutrition and clean drinking water, and the reduction of income due to joblessness can affect health. There are several reasons for this negative impact of unemployment mediated by low income. (1) Low income may expose people to a harmful working environment, since they may have to involve themselves in various activities of informal employment with elevated occupational risks. (2) Low income increases unhealthy behaviour and unhealthy habits. (3) Low income causes psychological pressure and tension, and may thus also cause negative physiological reactions. (4) Low income is linked with inadequate health care: a lack of income determines help-seeking behaviour, and the high expected absolute and relative costs reduce health care demand (Yan, Uhlemann, Weng et al., 2001). The link between low income and inadequate health care is related directly to the system of Chinese health insurance and how it operates.

Health status and the use of medical services were studied by Zhang and his colleagues (2003) in a survey comparing laid-off and the employed workers. The results showed that the prevalence of illness among those who had been unemployed two weeks was higher than among those who remained employed, while the shares of those attending a physician's services and actual hospitalisation were lower among those unemployed. According to the results, the unemployed persons have more health problems but they use less medical services than the employed ones (Zhang, Mao, Gong et al., 2003).

5 The role of lifestyle and other factors mediating the link between unemployment and health

Similar to earlier research in other countries, Chinese research suggests that unemployment leads to a behavioural change towards a poorer lifestyle including heavier smoking and drinking as well as less physical exercise, which will result in the deterioration of health in the long run (Zhang, Guo & Yang, 1999).

Prolonged unemployment was also found to be associated with a higher risk of somatic and mental ill-health including the prevalence of lifestyle risk factors such as smoking and problem drinking.

Unemployment is a grave life stressor, but other factors such as additional stress due to some other negative life incidents such as the reduction of income, the changing of social support, etc. can still aggravate the negative effect of unemployment (Hagen, 1983; Hamilton, Merrigan et al., 1997). The research of Viinamäki et al. (1999) showed that not only does unemployment cause a decline in self-confidence, but interpersonal relationships may also change. Thus, the health consequences of unemployment may be confounded by a wide range of factors including e.g. gender, pre-health status, duration of unemployment, income change, education, social support, civil status and region of residence.

Pre-existing mental and physical health problems are likely to have an impact on health after a person is laid off. The health-based selection into unemployment will affect the health status of the laid-off workers. It is generally known that persons suffering from ill-health run a greater risk of unemployment, and that healthier people are more likely to be employed. So there may be a possibility of a causal link between the ill-health prior to unemployment and subsequent mental or physical health problems found during unemployment. A person may also become unemployed because of a psychological problem but he/she may develop some other health problems because of unemployment. There is a complex dynamic interaction between psychological distress and unemployment, which must be studied from a holistic viewpoint.

Economic status is strongly associated with the social class of the family of origin, education as well as job history and current occupation. These may constitute risks or buffers in the relationship of unemployment to various dimensions of health such as physical functional capacity, morbidity and mortality. A very problematic situation may be caused when a specific age group is under high economic pressure exerted by the family and cannot easily find a new job in the case of unemployment because of a low level of education. This may be confounded by other factors, such as gender (Zhang, Guo & Yang, 1999; Chen, Zhao & Wang, 1999). The unemployed people between the ages of 30-49, i.e. in their best working years, were more sensitive to mental distress than other age groups based on the SCL-90 questionnaire investigations (Zhang, Guo & Yang, 1999; Li, He, Lin et al., 2000; Chen, Zhao & Wang, 1999; Wang, Zhang & Ti, 2000; Chen & Guo, 2000). This is in agreement with the mass of international literature. Though the probability of unemployment is still higher among those aged 40+, youth unemployment is rising in China (Li, He, Lin et al., 2000; Wang, Zhang & Ti, 2000; Viinamäki, 1993). A comparison of male and female unemployed persons by the mean SCL-90 scores was conducted by Chen and his colleagues (2000). A significant difference between the genders was found in somatic symptoms, compulsive behaviour, interpersonal oversensitivity and anxiety after unemployment (Chen & Guo 2000).

It has been stated that the job loss causes a strong negative impact on health by Chen and Tan. With some time passing, mental health improved and the situation remained stable for 1.5 years. After this, if no re-employment took place, mental health would again start to deteriorate. The results showed that the score of SCL 90 of the group whose duration of unemployment was shorter or equal to six months went down from the initial shock level, but the opposite was true for those whose unemployment was longer than six months (Figure 1) (Chen & Guo, 2000; Tan, Liu, Li et al., 2000).

The results demonstrate the classic pattern indicated by Jahoda and others in Europe of the 1930s as well as Wacker and Kieselbach in the 1970s: There is a path of acute response - accepting the fact - chronic response.

Figure 1. **Relationship of SCL-90 score with unemployment time (Tan, Liu, Li et al., 2000)**

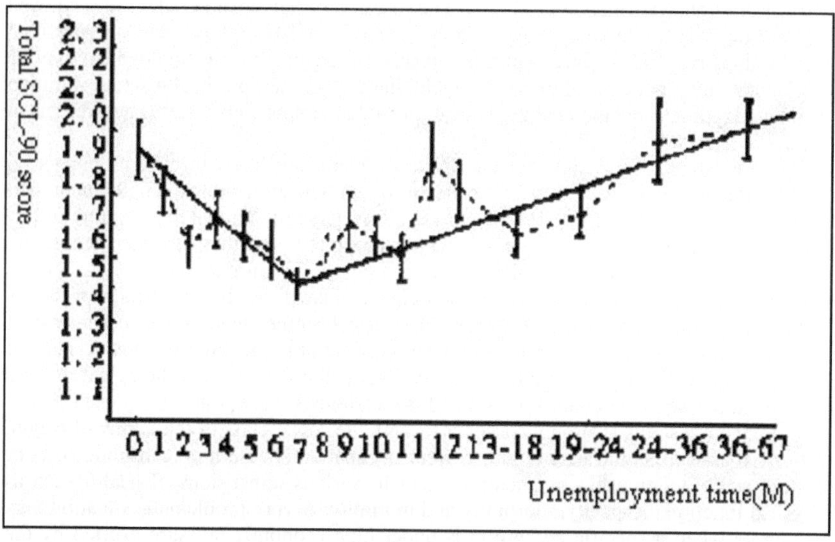

One of the important findings from the Chinese research is that the social support or family support is a very effective measure against a worsening of mental health status (Chen, Zhao & Wang, 1999; Li & Lu, 2001). The unemployed who received support had usually only a mild negative change of mental health status (Shao, 1999; Li & Lu, 2001; Xu, Xiao & Chen, 2002). This result is also in compliance with the body of international literature.

6 Conclusions

Unemployment is considered to be a public health concern in China, since deteriorations in the health of the unemployed persons is anticipated by international and Chinese research. The problems will become more prominent in China's economy for the years to come, compounding hindrances to sustainable economic development and causing social instability as well. The existence of large numbers of unemployed will exert a severe pressure on social security. Unemployment is unavoidable. Besides helping the unemployed persons to find new jobs by various policies and measures, we should also pay more attention to their mental and physical health status.

Great efforts have been made by the Chinese government to pave ways out of unemployment. Both the government and scientists of various orientations will proceed hand in hand in drawing further attention to the adverse health effects of unemployment. This approach is shown e.g. by an article by Xu Zhenghui from the Job Security Branch of the National Federation of Labour, entitled Strategic Thinking on How to Alleviate Labour Market Contradictions in China (Xu, 1995).

The massive lay-off of workers from unproductive branches of economy will be on the Chinese government's agenda, but at the same time, the government intends on helping unemployed persons get a new job. The function of labour market training and similar schemes is to educate laid-off people with respect to the culture of the new Chinese market economy. The Chinese government has been attempting to create more new jobs to relieve the social tension of unemployment. A goal has been set to help the laid-off get back to work as soon as possible and keep the unemployment rate below 4.7 percent.

There are still few reports regarding the impact of health care interventions on the link between unemployment and health. It is necessary to outline possible strategies in this respect; they would include the provision of accessible and adequate health care for unemployment people, new development of capacity in health care; cross-administrative co-operation; more research. There is a clear function for health care in reducing the negative health impact of unemployment and ensuring that poor health does not act as a barrier to the re-employment of unemployed people.

References

Chen, C.G., Zhao, C.M. & Wang, B. (1999). Study on the state of mental health and physiological training of 286 laid-off workers. *Journal of Xi'an Physical Education, 16*(3), 82-85. (in Chinese)

Chen, L., Zhang, Y.P. & Li, W.H. (2004). Study on the psychological condition of the unemployed and some suggestion against these problems. *Chinese Clinical Psychology, 12*(2), 183-184. (in Chinese)

Chen, Q.Z. & Guo, W.B. (2000). A study on the mental health of dismissed workers. *Health Psychology Journal, 8*(4), 465-468. (in Chinese)

Chen, Y.J. (1999). Analysis of mental health status of laid-off workers. *Anthology of Medicine, 18*(1), 96-97. (in Chinese)

Hagen, D.Q. (1983). The relationship between job loss and physical and mental illness. *Hospital and Community Psychiatry, 34*(5), 438-441.

Hamilton, V.H., Merrigan, P. et al. (1997). Estimating the relationship between mental health and unemployment. *Health Economy, 6*(4), 397-406.

Li, Q. & Lu, Y. (2001). Social support and health mental status of young laid-off workers. *Youth Study, 7*, 12-17. (In Chinese).

Li, Y.P., He, S.Z., Lin, Y.X. et al. (2000). Survey on the mental health condition of dismissed workers. *Health Psychology Journal, 8*(1), 80-82. (in Chinese)

Liu, G.Q. & Tan, D.L. (2000). Analysis of factors affecting the psychological health of unemployed workers. *Journal of North Sichuan Medical College 2000, 15*(3), 76-77. (in Chinese).

Shao, G.P. (1999). Survey on the mental health status of laid-off workers. *Chinese Mental Health, 13*(1), 35. (in Chinese)

Shi, J.G. & Tao, M.Z. (2000). An analysis of the laid-off workers's MMPI test results. *Chinese Journal of Behavioral Medical Science, 9*(1), 21-22. (in Chinese)

Tan, D.L., Liu, G.Q., Li, J. et al. (2000). Correlational analysis between the unemployment duration and psychological health. *Journal of North Sichun Medical College, 15*(3), 74-76. (in Chinese)

Tian, C.S. & Sun, F. (2003). An investigation of the laid-off worker in psychological health status, social support and ways of coping. *Occupation and Health, 19*(9), 3-5. (in Chinese)

Viinamaki, H., Koskela, K., Niskanen, L., Arnkill, R. & Tikkanen, J. (1999). Unemploymen and mental well being: A factory closure study in Finland. *Acta Psychiatrica Scandinavica, 88*, 429-433.

Wang, H.P. (2002). Stress and psychological status of laid-off workers in state-owned factories. *China Mental Health, 16*(6), 30-31. (in Chinese)

Wang, H.P., Zhang, J.J. & Ti, K.X. (2000). Mental health status of laid-off workers. *Youth Study 2000, 4*, 30-37. (in Chinese)

Xu, H.L., Xiao, S.Y. & Chen, J.P. (2001). Study of factors affecting the mental health level among the laid-off workers. *Chinese Clinical Psychology, 9*(3), 178-181. (in Chinese)

Xu, H.L., Xiao, S.Y. & Chen, J.P. (2002). Study on motivation and risk factors of laid-off workers. *Chinese Psychological Health Journal, 16*(2), 96-99. (in Chinese)

Xu, H.L., Xiao, S.Y. & Chen, J.P. (2001). Study on the mental health level of laid-off workers. *Chinese Clinical Psychology, 9*(4), 263-265. (in Chinese)

Xu, Z.H. (1995). Strategic thinking on how to alleviate labour market contradictions in China. *Reform in China, 10*, 38-39. (in Chinese)

Yan, F., Uhlemann, T., Weng, Z.H. et al. (2001).Qualitative study on health problems of vulnerable people and how they cope with their problems. *Chinese Health Resources, 4*(5), 208-210. (in Chinese)

Zhang, X.L., Guo, N.L. & Yang, Z.Y. (1999). Investigation on health behaviour and mental health status of laid-off workers. *Chinese Journal of Behavioural Medical Science, 8*(4), 310. (in Chinese)

Zhang, J.J., Mao, Z.Z., Gong, Z.P. et al. (2003). Comparison of health equity between laid-off workers and employed workers. *Chinese Health Management, 1*, 12-13. (in Chinese)

The Relationship between Downsizing, Psychosocial Stress at Work and Health[1]

Nico Dragano & Johannes Siegrist

Introduction and theoretical approach

The process of economic globalization in combination with progress in information technology continues to exert a profound impact on transnational division and organisation of work and on the development of an increasingly international labour market. Growing competition between corporations and companies and aggravated pressure towards maximising return on investment go along with these processes which in turn augment employees' work pressure and their risk of being laid off. Organisational downsizing is one of the typical consequences of economic globalization, given the cost saving effects of cuts in personnel. Employees who 'survive' a cut-down are often faced with major changes of their work environment, especially increased job demands, task reorganisation and re-composition of work teams (Allen, Freeman & Russel, 2001; Burke & Cooper, 2000; Kivimäki, Vahtera, Pentti & Ferrie, 2000). Moreover, negative health effects of downsizing on the remaining staff were demonstrated in several recent longitudinal studies. Health indicators used in these studies include sickness absence (Kivimäki et al., 2000; Vathera, Kivimäki & Pentti, 1997, Westerlund, Ferrie, Hagberg, Jeding, Oxenstierna & Theorell, 2004a), mortality risks (Vahtera, Kivimäki, Pentti, Linna, Virtanen & Ferrie, 2004), musculoskeletal complaints (Kivimäki, Vahtera, Ferrie, Hemingway & Pentti, 2001a), hospital admission (Westerlund et al., 2004a), disability pensions (Vahtera, Kivimäki, Forma, Wikström, Halmenmäki, Linna & Pentti, 2005), self-rated health (Kivimäki, Vahtera, Pentti, Thomson, Griffiths & Cox, 2001b) and depression (Grunberg, Moore & Greenberg, 2001; Moore, Grunberg & Greenberg, 2004).

In part, these negative effects on health are attributed to higher levels of exposure to stressful work associated with organisational downsizing (Moore et al., 2004; Herttin, Nilsson, Theorell & Larsson, 2004; Shannon, Woodward, Cunningham, McIntosh, Lendrum, Brown & Rosenbloom, 2001; Westerlund, Theorell & Alfredsson, 2004b), although few studies have tested this hypothesis in more detail. One such study was conducted in Finland in an attempt to analyse to what extent an increase in work-related

---

1 Parts of this contribution have been published as follows: Dragano, N., Verde, P. E. & Siegrist, J. (2005). Organisational downsizing and work stress: Testing synergistic health effects in employed men and women. *Journal of Epidemiology and Community Health, 59*, 694-699.

stress accounted for the link of downsizing with sickness absence (Kivimäki et al., 2000). The researchers found, that a higher work-related stress in employees who were faced with downsizing accounted for about 50% of the association between downsizing and sickness absence. Yet, separate and combined effects of organisational downsizing and work-related stress on health have rarely been analysed in a systematic way. Such an approach would be important for two reasons. Firstly, as downsizing is not always associated with an increase in work-related stress, it is interesting to estimate its net effect on health. Secondly, for preventive reasons, we would like to know whether employees who are simultaneously exposed to downsizing and work stress show health problems beyond those produced by each exposure separately. Knowledge concerning the synergy of exposure effects may provide a rationale for worksite health promotion measures in vulnerable groups of workers.

However, in order to test the hypothesis of separate and combined effects on health produced by downsizing and work-related stress, the latter notion needs to be defined both in conceptual and methodological terms. Today, few jobs are defined by high physical demand and exposure to a noxious work environment. Mental and emotional demands are increasingly important with the spread of information and communication technology and the growth of services. Work at high speed, monotonous work, and permanent and direct contact with clients are highly prevalent characteristics of today´s working life, at least in Europe (Paoli & Merllié, 2001). These characteristics are often combined with irregular work hours or shift work, with a de-standardization of job arrangements and increased job insecurity.

To understand those particular aspects of modern work that have a direct impact on health, theoretical concepts are needed. Such concepts aim at defining stressful work so that it can be identified at a generalized level to allow for its identification in a wide range of occupations. These concepts are then translated into operational measures with the help of social research methods (questionnaires, observation techniques, etc.). Several concepts of psychosocial work-related stress have been developed and tested in recent past (for an overview see e.g., Antoniou & Cooper, 2005; Dunham, 2001; Perrewé & Ganster, 2002). In our research the theoretical model of 'effort-reward imbalance' has been used. We describe its main features and argue why it is considered particularly appropriate in the context of the current analysis.

The effort-reward imbalance model is concerned with stressful work (Siegrist, 1996), and this model builds on the notion of social reciprocity, a fundamental principle rooted in an 'evolutionary old grammar' of interpersonal exchange. Social reciprocity lies at the core of the contractual interpretation of employment, which defines obligations or tasks to be performed in exchange with adequate rewards. These rewards include money, esteem and career opportunities, including job security. Contractual reciprocity operates through norms of return expectancy, where efforts spent by employees are reciprocated by equitable rewards from employers. The effort-reward imbalance model claims that lack of reciprocity occurs frequently under specific conditions and that failed reciprocity in terms of high cost and low gain elicits strong negative emotions with a special propensity to sustained autonomic and neuroendocrine activation and their adverse long-term consequences for health. According to the theory, contractual nonreciprocity is expected, if one or several of the following conditions of work are given: dependency, strategic choice, and over commitment.

Dependency refers to the structural constraints observed in certain types of employment, especially in unskilled or semi-skilled work or the working condition of elderly employees, employees with restricted mobility or limited work ability, and workers with short-term contracts. In all these instances, incentives of paying non-equitable rewards are high for employers, while the risk that employees would reject an unfair contractual transaction is low. High cost and low gain at work due to lack of alternative choices in the labour market is relatively frequent in modern economies that are characterized by a globalized labour market, mergers and organisational downsizing, rapid technological change, and a high level of job instability.

Strategic choice is a second condition of non-symmetrical exchange. Here, people accept high cost/low gain conditions of their employment for a certain time, often voluntarily, because they in this way want to improve their career chances and foresee some rewards at a later stage. This pattern is frequently observed in early stages of professional work and in jobs that are characterized by heavy competition. As anticipatory investments are made on the basis of insecure return expectancy, the risk of failure after long-term efforts is considerable.

Thirdly, there may be psychological reasons for a mismatch between efforts and rewards at work. People characterized by a motivational pattern of *over-commitment* to work may strive for a high achievement because of their underlying need for approval and esteem. Although an excessive effort often is not met with adequate rewards, overcommitted people tend to maintain their level of involvement. There is reason to believe that this motivational style affects how the work-related demands are appraised. Perceptual distortion prevents overcommitted people from accurately assessing cost-gain relations. As a consequence, they underestimate the demands, and overestimate their coping resources, not being aware of their own contribution to non-reciprocal exchange. Overcommitment to work may be elicited and reinforced by various job environments, and it is often experienced as self-rewarding over a period of years in occupational trajectories. However, in the long run, overcommitted people are susceptible to exhaustion and adaptive breakdown.

In sum, the model of effort-reward imbalance maintains that non-symmetric exchange at work is frequent under these structural and personal conditions, and that people experiencing dependency, strategic choice or over commitment, either separately or in combination, are at elevated risk of suffering from chronic stressful experiences and their long-term effects on health. This model seems particularly suitable to explore negative health effects in combination with organisational downsizing because it addresses stressful experience related to a lack of alternative choices in the labour market and continued frustration in combination with low job security.

Meanwhile, this model has been tested in a number of prospective and cross-sectional observational studies, in case-control studies with patients and healthy controls, and with experimental and natural design. Overall evidence indicates that an exposure to effort-reward imbalance at work significantly increases the risk of cardiovascular disease and depression (Siegrist, 2005). Moreover, several risk factors of cardiovascular and metabolic disorders, addictive behaviours and conditions of ill health are more prevalent among employees who suffer from this type of stress (for review see Tsutsumi & Kawakami, 2004; van Vegchel, de Jonge, Bosma & Schaufeli, 2005). In the next section, we describe results of a recent study on the relationship of downsizing and effort-reward imbalance at work to ill health in a large sample of employed men and women. The final section discusses the findings and their policy implications.

1 Studying separate and combined health effects of organisational downsizing and work-related stress

Methods

This study is based on the data from the regular survey by the Federal Institute of Vocational Training and the Institute for Employment Research (1998-99), comprising a 0.1% random sample of the German workforce (Dostal, Jansen & Parmentier, 2000). The response rate was 61%, and the research material gained consisted of 34.343 men and women aged 16 to 85 years. For the study described in this article, we focused on the age group 16 to 59 years to reduce the bias based on early retirement among older work force. We also focused on employees working with their employer for at least two years when studying the experience of downsizing. Finally, we excluded self employed people as well as civil servants, given their special legal status, from the research material. These restrictions resulted in a group of 22,559 respondents, 12,240 men and 10,319 women.

Data was collected by standardised questionnaires by trained interviewers face-to-face at the participants´ homes. Major topics of the questionnaire were job record and vocational training, but some aspects of the current work environment were also investigated. In addition, symptoms and complaints often experienced in association with work were assessed.

In this survey, the original measurement developed for the effort-reward imbalance was not available (Siegrist, Starke, Chandola, Godin, Marmot, Niedhammer & Peter, 2004). To measure the extrinsic components 'effort' and 'reward', proxy measures were developed from the available list of self-reported work characteristics. Eight items as proxies to the original items of the scale 'effort' were selected (sample questions: "How frequently do you work under time pressure?" "How frequently do you have to work additional hours?"). Similarly, seven items measuring reward at work were included, covering salary, prospects of promotion and collegial support (sample questions: "How satisfied are you with your opportunities for further training?" "How satisfied are you with your salary/income?"). The response scales for all questions were dichotomised, and two one-dimensional scales were constructed (internal consistencies of the scales 'effort' and 'reward' were α=0.70 and α=0.67). Factor loadings of the items based on exploratory principal component analysis ranged from 0.48 to 0.69. On the basis of the theory, a ratio of the two sum scales was generated to estimate the effect of high effort in combination with low reward (Siegrist et al., 2004). Scores in the upper quartile of the ratio were defined as an exposure to stressful work in terms of this model. Two items were used to measure downsizing, i.e. the question of whether during the past two years a reduction of staff or layoff took place in the respondent's company, and the question whether the respondent's own work was affected by this event. Exposure to downsizing was defined as simultaneous occurrence of both conditions.

Instead of introducing the two exposure variables 'downsizing' and 'effort-reward imbalance' separately into a regression model and studying their interaction by the product of the variables, a composite variable was constructed in order to test different combinations of exposures. This procedure was originally proposed by Rothman (1986) and has been applied in occupational epidemiology (Hallqvist, Diderichsen, Theorell, Reuterwall & Ahlbom, 1998). For this purpose, the two dichotomised exposure variables 'work-related stress' (upper quartile vs. remaining categories), and 'downsizing' (yes/no), were combined in the following way: (1) neither downsizing, nor work stress, (2) downsizing, but no work stress, (3) work stress, but no downsizing, (4) both downsizing and work stress present.

Work related symptoms were assessed using a checklist of 20 items. The list was compiled by a panel of occupational health experts, based on psychometrically validated questionnaires (Fahrenberg, 1995; Franke, 1995; von Zerssen & Koeller, 1976). Items included musculo-skeletal pain, sleep disorders, depressed mood, breathlessness, skin irritation and psychosomatic symptoms. Participants were asked to evaluate whether, and to what extent, all these symptoms were experienced during or after work. Rather than focusing on single complaints, a sum score of symptom load was calculated (Cronbach's α=0.76). A threshold of three or more symptoms was chosen to define a group of participants with self- reported ill-health.

The following covariates were included: the participants' socio-economic status was defined by the level of education and occupational category. Highest educational degree was categorized into three groups according to the International Standard Classification of Education (UNESCO, 1997): (1) "no degree or lower secondary", (2) "upper secondary", (3) "post-secondary or tertiary". Occupational category was divided into "blue collar" and "white collar" employment. Furthermore, two composite variables of physically and chemically risky work were defined, based on a series of Likert-scaled items, (1) physically demanding work (heavy lifting, vibrations, stressful posture; Cronbach's α=0.73), and (2) occupational hazards (noise, heat or cold, dust, smoke, gases, toxic substances; Cronbach's alpha = 0.83). Additional work characteristics were the number of weekly working hours and job insecurity (i.e. probability of losing the job). Finally, the economic-regional context of downsizing was included by distinguishing between Eastern or Western German respondents.

Multivariate logistic regression models were fitted, calculating odds ratios (OR) and 95% confidence intervals (95% CI) for different levels of covariate adjustment. Model I displays crude effects, whereas Model II is adjusted for socio-economic status and age. In Model III, physical and chemical exposures and working time are additionally included, whereas Model IV contains all covariates. As an increasing body of evidence suggests a gender-specific effect of stressful work on health, men and women were analysed separately (Östlin, 2002; Weidner & Cain, 2003).

Results

Altogether 14 per cent of men and 12 per cent of women reported having experienced downsizing during the past two years. Almost every second respondent who experienced downsizing scored high on effort-reward imbalance at work, compared to every fifth respondent among those without the experience of downsizing. More importantly, about 60 per cent of men and 68 per cent of women who were simultaneously exposed to downsizing and work stress experienced 3 and more symptoms, whereas frequencies were considerably lower among those with one exposure only and were lowest in the group of men and women who had experienced neither downsizing nor work stress (23 and 25 per cent respectively $p < 0.001$).

As these findings might be confounded by sociodemographic characteristics or additional working conditions, a multivariate logistic regression analysis was performed to estimate the odds ratios of symptom load according to exposure status, with stepwise control of covariates.

Results displayed in Table 1 indicate a consistent increase in risk according to the exposure status both in men and women. Adjusting for the confounder effects results in a minor decrease of the general effects found, but they all remain statistically significant. Concerning the separate effects of the two factors, the effort-reward imbalance is more strongly associated with the symptom load than downsizing. Yet, elevated risks are observed in the combined exposure groups, with odds ratios of 4.4 in men and 5.4 in women in the fully adjusted model.

Table 1. Combined exposure to downsizing and work stress in relation to work-related symptoms - the results of multivariate logistic regression, odds ratios and 95% confidence intervals.

Men	Model I		Model II		Model III		Model IV	
	OR	95% CI	OR	95% CI	OR	95% CI	OR	95% CI
no downsizing no work stress	1		1		1		1	
yes downsizing no work stress	1.73	1.49-2.01	1.67	1.43-1.95	1.58	1.35-1.85	1.53	1.30-1.79
no downsizing yes work stress	3.33	3.01-3.67	3.67	3.32-4.06	3.10	2.79-3.45	3.06	2.75-3.41
yes downsizing yes work stress	5.14	4.42-5.96	5.61	4.81-6.54	4.60	3.92-5.39	4.41	3.75-5.18
Women	**Model I**		**Model II**		**Model III**		**Model IV**	
	OR	95% CI	OR	95% CI	OR	95% CI	OR	95% CI
no downsizing no work stress	1		1		1		1	
yes downsizing no work stress	1.95	1.64-2.31	1.90	1.60-2.27	1.72	1.44-2.07	1.71	1.43-2.06
no downsizing yes work stress	3.62	3.25-4.03	3.89	3.49-4.34	3.27	2.92-3.66	3.26	2.91-3.65
yes downsizing yes work stress	6.61	5.53-7.89	7.04	5.88-8.43	5.41	4.49-6.52	5.37	4.45-6.47

Model I: unadjusted; Model II: adjusted for age, east/west residency, education, occupational status; Model III: adjusted for age, east/west residency, education, occupational status, physical demands and occupational hazards, weekly working hours; Model IV: adjusted for age, east/west residency, education, occupational status, physical demands and occupational hazards, weekly working hours, job insecurity

In additional analyses (results not shown in detail) it was tested if the strong relationship between symptoms and the combined exposure to work stress and downsizing is explained by the simple addition of the independent effect of work stress plus the effect of downsizing. That was not the case, as we found that the combined risk exceeds a simple additive effect, indicating a synergistic interaction between both exposures.

2 Discussion and policy implications

When discussing the significance of these results we should keep in mind that they are based on a cross-sectional study and that no objective health measure was available. Moreover, the theoretical model underlying this research was assessed by proxy measures rather than the original psychometrically validated effort-reward imbalance questionnaire. It must be pointed out that this survey was not originally designed to test our main hypothesis as it served descriptive purposes of the agency which collected the data. Moreover, we cannot rule out the possibility that reported statistical associations of work-related exposures and symptoms are subject to some reporting bias by respondents and may be influenced by specific personality characteristics, moods and contexts (Kahneman et al., 1999; Spector et al., 2000).

On the other hand, these limitations are balanced by several strengths of this study. Firstly, we carried out a systematic analysis of both separate and combined effects of two important conditions of modern working life, organisational downsizing and work-related stress, on health. Secondly, this study introduced a theory-based measure of chronic work stress and effort-reward imbalance. This model also successfully predicted ill-health in a number of prospective and cross-sectional studies in different occupational groups and countries and using various health indicators (Tsutsumi & Kawakami, 2004; van Vegchel et al., 2005). Thirdly, the current report is based on a large sample representing a broad spectrum of occupations that excludes only self-employed persons and civil servants. Whereas a majority of studies on downsizing and health are derived from Scandinavian countries, it is not known to what extent their findings can be generalized to other western European economies. This is one of few reports testing the hypothesis with a large data set on employed men and women in Germany. Finally, although our measure of work-related symptom load has not been validated by a medical examination it is known that the symptom load is a strong predictor of sickness absence and disability pension, two core health indicators of enterprises (Ferrie, Head, Shipley, Vahtera & Marmot, 2005; Karpansalo, Manninen, Kauhanen, Lakka & Salonen, 2004).

What are the policy implications of the results of this study? Firstly, it is important to notice that investment in health-promoting work environment can now be supported by theoretical models, such as the effort-reward imbalance model, that identify some 'toxic' components of work environment. Distinct measures of personnel and organisational development are available. They include e.g., investment in training and requalification, tailor-made promotion, improved in-cash (e.g., models of gain sharing) and in-kind rewards (e.g., strengthening the esteem and respect; leadership training), and measures of enhancing organisational justice and fairness. The effects of such measures on health and well being of employees need to be investigated in intervention studies. Preliminary results from some studies point to beneficial outcomes (Marmot, Siegrist & Theorell, 2005; see also Tsutsumi & Kawakami, 2004).

The second policy implication concerns the definition of primary target groups of interventions in enterprises and other organisations. Primary target groups are employees with an elevated health risk due to a high degree of exposure to stressful work. Employees who simultaneously experience organisational downsizing and a recurrently frustrating work situation of high 'cost' and low 'gain' should be a primary target group. It is expected that increased preventive interventions such as theory-guided health promotion at the work place and in these groups might reduce their health risks and increase their productive engagement. Such preventive interventions can also reduce sickness absence and various compensation claims.

We see that the enterprises - especially medium and large scale ones - should bear the primary responsibility of the worksite health promotion. Nevertheless, a third policy implication of this research points to the need of international and national regulations concerning downsizing, mergers and re-structuring of enterprise, at the expense of workers' rights and well-being. It is encouraging to see a growing interest among organisations such as the European Commission, the World Health Organisation, and the International Labour Organisation (ILO), to address these issues. For instance, the European Commission has recently proposed an agenda for Corporate Social Responsibility, and a number of other initiatives support this approach at the global level, such as ILO's Tripartite Declaration on Multinational Enterprises and Social Policy, and the OECD Guidelines for Multinational Enterprises. In all these initiatives, the protection and promotion of health of employees is considered an important task of employment policy. Increased efforts are needed at different levels of intervention to reduce the gap between scientific evidence on negative effects of work-related stress and policy measures that aim at promoting health and well being of working populations.

References

Allen, T.D., Freeman, D.M. & Russel, J.E. (2001). Survivor reactions to organisational downsizing: Does time ease the pain? *Journal of Occupational and Organisational Psychology, 74*, 145-164.

Antoniou, A.S. & Cooper, C.L. (Eds.). (2005). *Research companion to organisational health psychology.* Cheltenham: Edward Elgar.

Burke, R.L. & Cooper, C.L. (2000). *The organisation in crisis.* Oxford: Blackwell.

Dostal, W., Jansen, R. & Parmentier, K. (2000). *Wandel der Erwerbsarbeit: Arbeitssituation, Informatisierung, berufliche Mobilität und Weiterbildung* [The changing world of work: Work environment, information technology, mobility and qualification]. Nürnberg: Institut für Arbeitsmarkt- und Berufsforschung.

Dunham, J. (2001). *Stress in the workplace. Past, present and future.* London: Whurr.

Fahrenberg, J. (1995). Somatic complaints in the German population. *Journal of Psychosomatic Research, 39*, 809-817.

Ferrie, J.E., Head, J., Shipley, M.J., Vahtera, J. & Marmot, M.G. (2005). A comparison of self-reported sickness absence with absences recorded in employers' registers: Evidence from the Whitehall II study. *Journal of Occupational and Environmental Medicine, 62*, 74-79.

Franke, G. (1995). Symptom-Checkliste von L.R. Derogatis – Deutsche Version [The Derogatis Symptom-Checklist – German Version]. Weinheim: Beltz.

Grunberg, L., Moore, S. & Greenberg, E. (2006). Differences in psychological and physical health among layoff survivors: The effect of layoff contact. *Journal of Occupational and Organisational Psychology, 6*, 15-25.

Hallqvist, J., Diderichsen, F., Theorell, T., Reuterwall, C. & Ahlbom, A. (1998). Is the effect of job strain on myocardial infarction risk due to interaction between high psychological demands and low decision latitude? Results from Stockholm Heart Epidemiology Program (SHEEP). *Social Science and Medicine, 46*, 1405-1415.

Hertting, A., Nilsson, K., Theorell, T. & Larsson, U.S. (2004). Downsizing and reorganisation: Demands, challenges and ambiguity for registered nurses. *Journal of Advanced Nursing, 45*, 145-154.

Kahneman, D., Diener, E. & Schwarz, N. (1999). *Well-being: The foundations of hedonic psychology.* New York: Russel Sage Foundations.

Karpansalo, M., Manninen, P., Kauhanen, J., Lakka, T.A. & Salonen, J.T. (2004). Perceived health as a predictor of early retirement. *Scandinavian Journal of Work, Environment & Health, 30*, 287-292.

Kivimäki, M., Vahtera, J., Pentti, J. & Ferrie, J.E. (2000). Factors underlying the effect of organisational downsizing on health of employees: A longitudinal cohort study. *British Medical Journal, 320*, 971-975.

Kivimäki, M., Vahtera, J., Ferrie, J.E., Hemingway, H. & Pentti, J. (2001). Organisational downsizing and musculoskeletal problems in employees: A prospective study. *Journal of Occupational and Environmental Medicine, 58*, 811-817.

Kivimäki, M., Vahtera, J., Pentti, J., Thomson, L., Griffiths, A. & Cox, T. (2001). Downsizing, changes in work, and self-rated health of employees: A 7-year 3-wave panel study. *Anxiety, Stress & Coping, 14*, 59-73.

Marmot, M., Siegrist, J. & Theorell, T. (2005). Health and the psychosocial environment at work. In M. Marmot & R. G. Wilkinson (Eds.), *Social determinants of health*. Oxford: Oxford University Press.

Moore, S., Grunberg, L. & Greenberg, E. (2004). Repeated downsizing contact: The effects of similar and dissimilar layoff experiences on work and well-being outcomes. *Journal of Occupational Health Psychology, 9*, 247-257.

Paoli, P. & Merllié, D. (2001). *Third European survey of working conditions*. Dublin: European Foundation for the Improvement of Living and Working Conditions.

Perrewé, P.L. & Ganster, D.C. (Eds.). (2002). *Historical and current perspectives on stress and health*. Amsterdam: JAI Elsevier.

Rothman, K.J. (1986). *Modern Epidemiology*. Boston: Little, Brown & Co.

Shannon, H.S., Woodward, C.A., Cunningham, C.E., McIntosh, J., Lendrum, B., Brown, J. & Rosenbloom, D. (2001). Changes in general health and musculoskeletal outcomes in the workforce of a hospital undergoing rapid change: A longitudinal study. *Journal of Occupational Health Psychology, 6*, 3-14.

Siegrist, J. (1996). Adverse health effects of high effort – low reward conditions at work. *Journal of Occupational Health Psychology, 1*, 27-43.

Siegrist, J. (2005). Social reciprocity and health: New scientific evidence and policy implications. *Psychoneuroendocrinology, 30*, 1033-1038.

Siegrist, J., Starke, D., Chandola, T., Godin, I., Marmot, M., Niedhammer, I. & Peter, R. (2004). The measurement of effort-reward imbalance at work: European comparisons. *Social Science and Medicine, 58*, 1483-1499.

Spector, P.E., Zapf, D., Chen, P.Y. & Frese, M. (2000). Why negative affectivity should not be controlled in job stress research: Don't throw out the baby with the bath water. *Journal of Organisational Behaviour, 21*, 79-95.

Tsutsumi, A. & Kawakami, N. (2004). A review of empirical studies on the model of effort-reward imbalance at work: Reducing occupational stress by implementing a new theory. *Social Science and Medicine, 59*, 2335-2359.

UNESCO (1997). *International standard classification of education. ISCED 1997*. Paris: UNESCO.

Vahtera, J., Kivimäki, M. & Pentti, J. (1997). Effect of organisational downsizing on health of employees. *Lancet, 350*, 1124-1128.

Vahtera, J., Kivimäki, M., Pentti, J., Linna, A., Virtanen, M. & Ferrie, J.E. (2004). Organisational downsizing, sickness absence, and mortality: 10-town prospective cohort study. *British Medical Journal, 328*, 555.

Vahtera, J., Kivimäki, M., Forma, P., Wikström, J., Halmeenmäki, T., Linna, A. & Pentti, J. (2005). Organisational downsizing as a predictor of disability pension: The 10-town prospective cohort study. *Journal of Epidemiology and Community Health, 59*, 238-242.

Vegchel, N. van, Jonge, J. de, Bosma, H. & Schaufeli, W. (2005). Reviewing the effort-reward imbalance model: Drawing up the balance of 45 empirical studies. *Social Science and Medicine, 60*, 1117-1131.

Weidner, G. & Cain, V.S. (2003). The gender gap in heart disease: Lessons from Eastern Europe. *American Journal of Public Health, 93*, 768-770.

Westerlund, H., Ferrie, J., Hagberg, J., Jeding, K., Oxenstierna, G. & Theorell, T. (2004a). Workplace expansion, long-term sickness absence, and hospital admission. *Lancet, 363*, 1193-1197.
Westerlund, H., Theorell, T. & Alfredsson, L. (2004b). Organisational instability and cardiovascular risk factors in white-collar employees. *European Journal of Public Health, 14*, 37-42.
Östlin, P. (2002). Gender inequalities in health: The significance of work. In S.P. Wamala & J. Lynch (Eds.), *Gender and social inequities in health* (pp. 43-65). Lund: Studentlitteratur.
Zerssen, D. von & Koeller, D.M. (1976). *Die Beschwerden-Liste* [The health complaints list] Weinheim: Beltz.

Organisational Downsizing and Psychosocial Work Characteristics: A Longitudinal Study of Japanese White-Collar Workers

Yasumasa Otsuka[1], Yuko Yamate & Shotaro Kosugi

Introduction

Downsizing, defined as the reduction of workforce to improve organisational performance (Kozlowski, Chao, Smith, & Hedlund, 1993), has become a significant characteristic of working life in Japan. Figure 1 shows the changes in the Gross Domestic Product of Japan (GDP) from 1978 to 2004. It increased constantly until 1991 when it started to slow down. In 1998, it registered a slump for the first time since World War II (Cabinet Office, Government of Japan, 2005). The unemployed population in Japan increased sharply in the early 1990s (Figure 2; Japan Ministry of Health, Labour and Welfare, 2004). In 2002, the number of unemployed people was over 3.5 million, signifying that over 5% of the people in the labour force could not find any jobs.

In 2002, the Japan Institute of Labour reported the results of a questionnaire survey of 10,761 companies with over 300 employees. The report showed that 17.5% of the companies had carried out organisational downsizing over the past three years, 25.4% were in the process of downsizing and 9.0% were planning to downsize within the next few years (Japan Institute of Labour, 2002). Furthermore, 70% of the companies that had downsized or were in the process of downsizing gave financial difficulties as the main reason. About half of the Japanese companies had carried out organisational downsizing to cut expenses mainly through the reduction of employed personnel.

1 To whom correspondence should be addressed, Address: Hiroshima University Graduate School of Education,1-1-1 Kagamiyama, Higashi-hiroshima, Hiroshima 739-8524 Japan.
E-mail: yasumasa-otsuka@hiroshima-u.ac.jp

Figure 1. **Gross Domestic Product (GDP) of Japan from 1978 to 2004 (Cabinet Office, Government of Japan, 2005)**

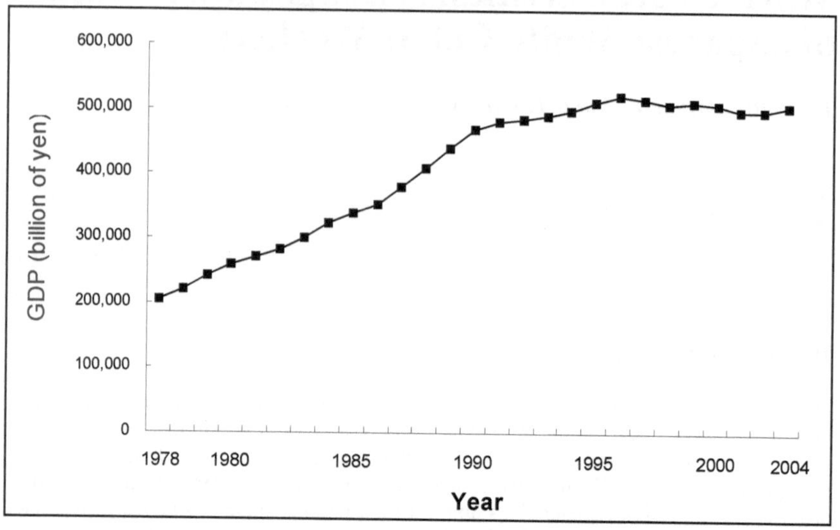

Figure 2. **Changes in the number of unemployed people in Japan from 1978 to 2004 (Japan Ministry of Health, Labour and Welfare, 2004)**

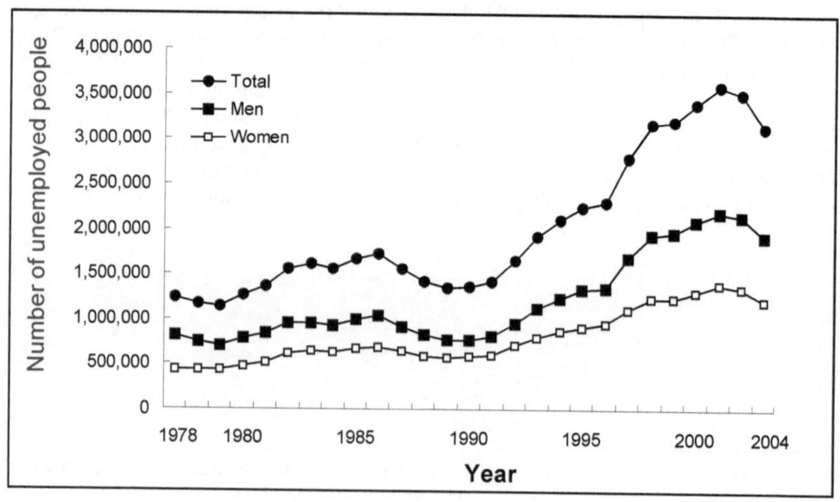

Previous studies conducted in several countries have revealed that organisational downsizing has deleterious consequences for the health of those employees who continue to be employed. For example, in the Whitehall II study, Ferrie, Shipley, Marmot, Stansfeld & Smith (1995) found that the self-reported health status of employees anticipating privatization tended to deteriorate as compared with that of the rest of the cohort. In a sample comprising 981 Finnish local-government employees, the rate of medically certified sickness absence (3 days or more) was found to be 2.3 times greater after major downsizing than after minor downsizing (Vahtera, Kivimäki & Pentti, 1997). For psychosocial work characteristics, major downsizing was associated with increased levels of physical work demands and job insecurity, and decreased levels of job control and skill discretion (Kivimäki, Vahtera, Pentti, Thomson, Griffiths & Cox, 2001; Kivimäki, Vahtera, Pentti & Ferrie, 2000).

A number of studies concerning the effects of organisational downsizing on psychosocial work characteristics have been conducted worldwide; however, to the best of our knowledge, only a few of those studies focused on Japanese working populations. In this study, we carried out a longitudinal survey to examine the relationship of organisational downsizing with psychosocial work characteristics among white-collar workers in a construction company in Japan. We hypothesized that organisational downsizing is associated with poor psychosocial work characteristics.

1 Methods

Participants

We initiated a longitudinal study in 1997 to investigate the impact of organisational downsizing on the psychosocial work characteristics in the head office of a construction company located in Tokyo.

Owing to a drop in their total annual revenue (Figure 3), the organisation had to resort to two types of downsizing during the five-year study period (Figure 4). The first phase of downsizing spanned the period between April 1998 and March 2000. The management approached senior employees with offers of early retirement in lieu of an increased retirement allowance. This was the first sign of downsizing. About 200 employees accepted early retirement offers. The next round of downsizing took place between September and October 2000. The management announced a bigger cut in the retirement pay after January 2001. Subsequently, about 600 senior employees accepted the management's early retirement offer.

In March 1997, a baseline questionnaire was administered to 1,817 employees at the head office of a construction company located in Tokyo. Of the respondents, 384 were working with the organisation even five years later at the time of the follow up survey (May 2001), and 353 of them (91.9%) responded to the second survey. After excluding 73 subjects who had missing responses in the questionnaire, the responses of the remaining 280 full-time workers were analyzed. At the time of the baseline survey, the mean age of the participants was 39.2 years (±SD 7.29) for men (N = 220) and 27.4 years (±SD 4.57) for women (N = 60).

Figure 3. **The company's total annual revenue from 1997 to 2001**

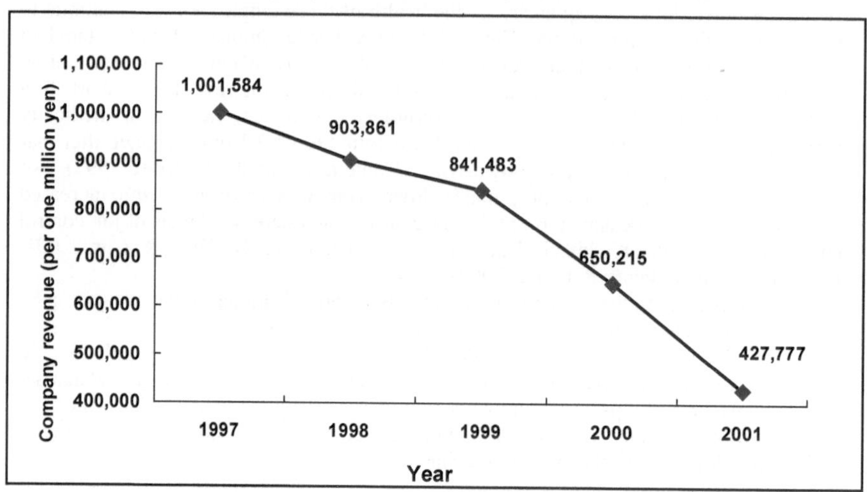

Figure 4. **Change in the number of employees from 1997 to 2001**

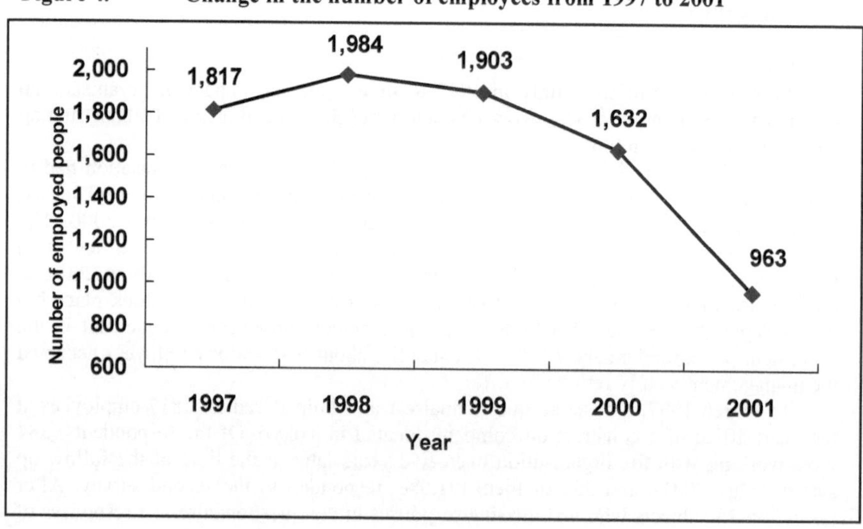

Measurements
The Job Stress Scale (Kosugi, 2000; Shimazu & Kosugi, 2003) was used to assess the levels of psychosocial work characteristics among participants. In this study, we used four job stressor scales (overload, cognitive demand, role ambiguity, and insufficient authority), six stress reaction scales (fatigue, irritability, anger, interpersonal sensitivity, cardiovascular symptoms, and depressive symptoms), five coping strategy scales (active coping, distancing, seeking social support, resignation, and restraint), and three social support scales (work, non-work, and family). The response categories ranged from (1) strongly disagree to (5) strongly agree. Cronbach's alpha coefficients for the participants were .679, .700, .680, and .852 for job stressor scales, .713, .635, .893, .815, .859, and .906 for stress reaction scales, .871, .622, .715, .753, and .572 for coping strategy scales, and .903, .930, and .904 for social support scales at the baseline, and .639, .734, .693, and .834 for job stressor scales, .728, .652, .870, .785, .862, and .897 for stress reaction scales, .859, .670, .714, .736, and .571 for coping strategy scales, and .914, .929, and .903 for social support scales at the follow up (in order of appearance).

Statistical analysis
A repeated-measures multivariate analysis of variance (MANOVA) was conducted on each of the psychosocial work characteristics to assess the possible main effect of time by sex. The SAS software package, version 8, was used for all analyses by GLM procedure.

2 Results

Demographic characteristics of the 280 participants who remained in employment are shown in Table 1. Among men, the number of managers in 2001 was twice of that in 1997, and the mean working hours in 2001 were shorter than that in 1997. Among women, however, the number of managers was almost zero in both study periods. Mean working hours in 2001 were greater as compared with 1997.

Table 2 presents the results of the MANOVA among men with their statistical significance. Organisational downsizing was associated with increased levels of anger ($p < .05$), depressive symptoms ($p < .05$), seeking social support ($p < .001$), resignation ($p < .001$), and restraint ($p < .001$), and decreased levels of role ambiguity ($p < .05$) and insufficient authority ($p < .001$). No significant main effects were found in social support.

Among women, the MANOVA showed significant effects of organisational downsizing for job stressors and coping strategies (Table 3). Organisational downsizing was associated with increased levels of cognitive demand ($p < .05$), seeking social support ($p < .001$), resignation ($p < .001$), and restraint ($p < .001$), and decreased levels of insufficient authority ($p < .01$).

Table 1. Demographic characteristics of the participants

	Men (N = 220)				Women (N = 60)			
Year	1997		2001		1997		2001	
Variable	N	(%)*	N	(%)*	N	(%)*	N	(%)*
Age, years								
20 – 29	16	(7.3)	2	(0.9)	44	(73.3)	17	(28.3)
30 – 39	100	(45.5)	62	(28.2)	13	(21.7)	40	(66.7)
40 – 49	92	(41.8)	98	(44.5)	3	(5.0)	3	(5.0)
50 –	12	(5.5)	58	(26.4)	0	(0.0)	0	(0.0)
Occupation								
Manager	98	(44.5)	183	(82.3)	0	(0.0)	1	(1.7)
Subordinate	122	(55.5)	23	(11.4)	60	(100.0)	59	(98.3)
Unknown	0	(0.0)	14	(6.4)	0	(0.0)	0	(0.0)
Mean working hours								
Shorter (≤ 9 h/day)	121	(55.0)	181	(82.3)	50	(83.3)	35	(58.3)
Longer (> 9 h/day)	99	(45.0)	25	(11.4)	10	(16.7)	24	(40.0)
Unknown	0	(0.0)	14	(6.4)	0	(0.0)	1	(1.7)
Marital status								
Married	187	(85.0)	183	(83.2)	14	(23.3)	19	(31.7)
Single	33	(15.0)	23	(10.5)	46	(76.7)	40	(66.7)
Unknown	0	(0.0)	14	(6.4)	0	(0.0)	1	(1.7)

* Figures do not always add up to 100% due to rounding or missing.

3 Discussion

The main purpose of this study was to examine the relationship between organisational downsizing and psychosocial work characteristics among Japanese white-collar workers who continue in the organisation after downsizing. We found that these relationships were different in the two sexes.

Among men, organisational downsizing was associated with increased levels of psychological stress reactions such as anger and depressive symptoms. These findings are almost identical to the results of previous studies. For example, Tsutsumi, Kayaba, Theorell & Siegrist (2001) found that employees who were threatened by job loss were more likely to display depressive symptoms than those who did not face the organisational downsizing. Kivimäki et al. (2001) revealed that organisational downsizing predicted long-lasting decline in self-rated health status in a cohort of 132 men and 418 women who worked in Finnish local government. As demonstrated by our study site, organisational downsizing is an inevitable way to ensure survival in the Japanese economy; however, it may also adversely affect the stress reactions of those people who continue working with the organisation.

However, among men, the levels of job stressors were stable or decreased after organisational downsizing. The decreased levels of qualitative job stressors (role ambiguity and insufficient authority) may have resulted from the experience of the previous five years and the increased responsibility from promotions, which led to more clarity about the employees' roles and gave them more decision latitude and authority. On the other hand, the levels of quantitative job stressors (overload and cognitive demand) did not change, probably because of the reduction in the total working hours after downsizing. The company encouraged its employees to finish and leave work within the scheduled time to save on overtime expenses. As a result, 82.3% of the male employees worked less than nine hours a day after downsizing. Although further study might be required, creating an appropriate personnel management system during downsizing may prevent the levels of qualitative and quantitative job stressors from increasing.

Table 2. Summary (F values) of the MANCOVA results for men

		Year				
		1997		2001		
Variables	Range	M	(SD)	M	(SD)	F
Job stressors						
Overload	7-35	22.4	(3.73)	22.1	(3.31)	1.33
Cognitive demand	6-30	19.7	(3.83)	19.8	(3.90)	0.25
Role ambiguity	6-30	15.5	(3.88)	14.9	(3.84)	5.51 *
Insufficient authority	9-45	21.3	(5.66)	19.9	(5.36)	13.53***
Stress reactions						
Fatigue	5-25	12.7	(3.19)	12.8	(3.15)	0.61
Irritability	4-20	13.0	(2.62)	13.0	(2.59)	0.02
Anger	6-30	14.7	(4.24)	15.2	(4.24)	4.36 *
Interpersonal sensitivity	7-35	16.5	(4.48)	16.3	(3.98)	0.74
Cardiovascular symptoms	5-25	9.7	(3.80)	9.9	(3.89)	0.70
Depressive symptoms	10-50	22.7	(6.57)	23.5	(6.51)	4.31 *
Coping strategies						
Active coping	9-36	22.9	(4.68)	23.0	(4.27)	0.07
Distancing	7-28	10.5	(2.71)	10.7	(2.69)	1.42
Seeking social support	5-20	7.9	(3.11)	9.8	(2.88)	65.78 ***
Resignation	5-20	7.3	(2.93)	8.4	(2.41)	23.40 ***
Restraint	5-20	10.5	(2.31)	11.2	(2.41)	15.95 ***
Social support						
Work	5-20	14.3	(2.58)	14.3	(2.57)	0.00
Non-work	5-20	14.5	(3.15)	14.3	(2.89)	0.75
Family	5-20	17.2	(2.81)	17.0	(2.82)	1.09

N = 220; * $p < .05$ *** $p < .001$

Organisational downsizing was associated with increased levels of cognitive demand in female workers. This finding seems to be in contrast to that of their male counterparts. One possible explanation for this is that most of the female employees were not the target of retrenchment because their mean age was significantly lower than that of the men. In the office that we studied, there were about 1,800 employees working in 1997, but only about 900 employees were left in 2001. This signifies that nearly 50% of the workers had accepted early retirement packages. In the Cooperative Wage Study (CWS) Russell (1995) pointed out that downsizing had entailed job expansion for the remaining workers. Consistent with this, reduction of personnel in our study site might have led to higher demands for female workers who mainly engaged in supportive tasks of male workers, their cognitive demand is likely to be increased.

Organisational downsizing was associated with increased levels of non-active or emotional coping strategies such as seeking social support, resignation, and restraint in both the sexes. In 'survivor' literature, coping strategies have been found to being related to the survivors' attitudes and behaviour (Noer, 1993; Havlovic, Bouhillette, & van der Wal, 1998; Armstrong-Stassen, 1994). Armstrong-Stassen (2005) revealed that survivors reported a significant decrease in the use of control-oriented coping strategies in the downsized period compared with the period prior to the downsizing. Uncontrollable events such as organisational downsizing may tend to provoke destructive coping strategies (Shimazu, Shimazu, & Odahara, 2004; Spector, 2002).

Table 3. Summary (F values) of the MANCOVA results for women

		Year				
		1997		2001		
Variables	Range	M	(SD)	M	(SD)	*F*
Job stressors						
Overload	7-35	17.6	(4.13)	18.8	(4.41)	3.23
Cognitive demand	6-30	13.3	(3.47)	14.5	(4.20)	4.78 *
Role ambiguity	6-30	16.8	(4.25)	16.3	(3.96)	0.90
Insufficient authority	9-45	25.9	(6.61)	23.7	(4.99)	8.24 **
Stress reactions						
Fatigue	5-25	13.6	(3.47)	13.6	(3.40)	0.01
Irritability	4-20	12.8	(2.68)	12.6	(2.89)	0.32
Anger	6-30	16.4	(5.72)	16.7	(4.69)	0.28
Interpersonal sensitivity	7-35	17.1	(4.18)	16.6	(4.35)	2.88
Cardiovascular symptoms	5-25	9.7	(4.24)	9.6	(4.15)	0.03
Depressive symptoms	10-50	23.1	(7.81)	23.8	(6.93)	0.72
Coping strategies						
Active coping	9-35	18.7	(5.90)	19.5	(5.19)	1.74
Distancing	7-28	12.2	(2.56)	11.9	(3.08)	0.85
Seeking social support	5-20	7.0	(3.46)	10.8	(3.43)	67.72 ***
Resignation	5-20	7.9	(3.14)	10.0	(2.95)	17.71 ***
Restraint	5-20	23.1	(3.15)	11.3	(2.56)	5.56 *
Social support						
Work	5-20	14.9	(3.83)	15.2	(3.38)	0.30
Non-work	5-20	17.5	(2.33)	17.4	(2.54)	0.12
Family	5-20	17.4	(2.56)	17.7	(2.31)	1.27

N = 60; * p < .05 *** p < .001

When the coping strategies are destructive or maladaptive, a critical negative work event can set in motion a chain of events capable of producing a downward spiral in performance and job satisfaction (Brown, Westbrook, & Challagalla, 2005). Although further study may be needed, psychological stress management interventions to promote active coping strategies among the survivors may have a positive effect on their performance.

Limitations

Several limitations of this study were noticed. First, this study coincided during the process of organisational downsizing. Although the severity of downsizing probably occurred after the conclusion of our study, we did not have any data for the period of our study. Second, the selection bias could not be controlled. The subjects of our study were the survivors, those who chose to remain with the company. We suspect that these survivors differed from those who left the company, in terms of ability, future vision, family and financial situation, etc. The results of this study only apply to the group who chose to stay on with the company. Third, our data was collected from only one worksite, and thus one cannot easily make general conclusions based on them. Fourth, other important psychosocial work characteristics such as job insecurity or future ambiguity were not assessed in the study.

Conclusions

Organisational downsizing appears to be a significant trend in the Japanese economy. This study indicates that organisational downsizing may lead to (1) increased levels of stress reactions (anger and depressive symptoms) and coping strategies (seeking social support, resignation, and restraint) and decreased levels of job stressors (role ambiguity and insufficient of authority) among men, and (2) increased levels of cognitive demand and coping strategies (seeking social support, resignation, and restraint), and decreased levels of insufficient authority among women. An appropriate personnel management system and sensitive treatment of the people who remain with the organisation are required when downsizing takes effect. Strategies to prevent negative effects of downsizing may include improved time management, according broad latitude to the employees, and promoting active coping strategies through the psychological stress management system.

Acknowledgements

This study was supported by Grant-in-Aid for Young Scientists (B), from the Japan Ministry of Education, Science, Sports, and Culture (grant number 17790401).

References

Armstrong-Stassen, M. (1994). Coping with transition: A study of layoff survivors. *Journal of Organisational Behavior, 15,* 597-621.

Armstrong-Stassen, M. (2005). Coping with downsizing: A comparison of executive-level and middle managers. *International Journal of Stress Management, 12,* 117-141.

Brown, S., Westbrook, R.A. & Challagalla, G. (2005). Good cope, bad cope: Adaptive and maladaptive coping strategies following a critical negative work event. *Journal of Applied Psychology, 90,* 792-798.

Cabinet Office, Government of Japan (2005). *Kokumin Keizai Keisan* [Business outlook survey]. Tokyo: Cabinet Office, Government of Japan.

Ferrie, J.E., Shipley, M.J., Marmot, M.G., Stansfeld, S. & Smith, G.D. (1995). Health effects of anticipation of job change and non-employment: Longitudinal data from the Whitehall II study. *British Medical Journal, 311,* 1264-1269.

Havlovic, S.J., Bouhillette, F. & van der Wal, R. (1998) Coping with downsizing and job loss: Lessons from the Shaughnessy hospital closure. *Canadian Journal of Administrative Sciences, 15,* 322-332.

Japan Institute of Labour (2002). *Kigyo no risutora to koyou: Jigyo saikouchiku to koyou ni kansuru tyousa, saishusyoku no jokyo ni kansuru tyousa* [Restructuring and employment: Survey on reorganisation, employment, and rework]. Tokyo: Japan Institute of Labour.

Japan Ministry of Health, Labour and Welfare (2004). *Shokugyo antei gyoumu toukei* [Survey on employment trends]. Tokyo: Japan Ministry of Health, Labour and Welfare.

Kivimäki, M., Vahtera, J., Pentti, J. & Ferrie, J.E. (2000). Factors underlying the effect of organisational downsizing on health of employees: Longitudinal cohort study. *British Medical Journal, 320,* 971-975.

Kivimäki, M., Vahtera, J., Pentti, J., Thomson, L., Griffiths, A. & Cox, T. (2001). Downsizing, changes in work, and self-rated health of employees: A 7-year 3-wave panel study. *Anxiety, Stress, and Coping, 14,* 59-73.

Kosugi, S. (2000). Sutoresu sukeiruno issei jissi niyoru shokuba mentaru herusu katsudou no jissai: Shinrigakuteki apurouchi niyoru shokuba mentaru herusu katsudou [Mental health activities in the workplace by a general checkup using the Job Stress Scale]. *Sangyo Sutoresu Kenkyu, 7,* 141-150.

Kozlowski, A.W., Chao, G.T., Smith, E.M. & Hedlund, J. (1993). Organisational downsizing: Strategies, interventions, and research implications. *International Review of Industrial and Organisational Psychology, 8,* 263-332.

Noer, D.M. (Ed.). (1993). *Healing the wounds*. San Francisco: Jossey-Bass.

Russell, B. (1995). The subtle labor process and the great skill debate: Evidence from a potash mine-mill operation. *Canadian Journal of Sociology, 20,* 359-385.

Shimazu, A. & Kosugi, S. (2003). Job stressors, coping, and psychological distress among Japanese employees: Interplay between active and non-active coping. *Work and Stress, 17,* 38-51.

Shimazu, A., Shimazu, M. & Odahara, T. (2004). Job control and social support as coping resources in job satisfaction. *Psychological Reports, 94,* 449-456.

Spector, P.E. (2002). Employee control and occupational stress. *Current Directions in Psychological Science, 11,* 133-136.

Tsutsumi, A., Kayaba, K., Theorell, T. & Siegrist, J. (2001). Association between job stress and depressive symptoms among Japanese employees threatened by job loss in a comparison between two complementary job-stress models. *Scandinavian Journal of Work, Environment, and Health, 27,* 146-153.

Vahtera, J., Kivimäki, M. & Pentti, J. (1997). Effect of organisational downsizing on health of employees. *Lancet, 350,* 1124-1128.

A Follow-Up Study on Predictors for Long-Term Unemployment and Lifestyle Changes of Unemployed Persons Following Bankruptcy in Japan

Tatsuya Ishitake, Tsunetaka Matoba, Akira Shigemoto & Kaori Nagatomi

Introduction

Employment status in Japan has become unstable as a result of prolonged economic recession and company restructuring. The unemployment rate has increased correspondingly, reaching 5% according to a government report in 2002 (Japanese Ministry of Public Management, Home Affairs, Post and Telecommunications, 2003). This is about 2.5 times greater than in the early 1990s. The continuing recession implies that unemployment will increase further. Also, the number of persons unemployed for at least one year has increased. In 2001, long-term unemployment exceeding one year reached 25% of total unemployed persons. Several reports have stated that the long-term unemployed are more vulnerable to negative health effects because of economic and psychological distress (Geyer & Peter, 2003; Ytterdahl et al., 1999). According to national statistics, suicide in Japan was running at more than 30,000 persons per year in these five years (Health and Welfare Statistics Association, 2002). It is plausible that some suicides are directly related to unemployment. In particular, depression due to unemployment is a main cause of suicide (Blakely et al., 2003; Dooley et al. 2000; Kposowa, 2001). It is therefore important to clarify the effects of unemployment on mental health in Japan.

Individual unemployment is positively correlated with subsequent ill-health. The effect of unemployment on physical and psychological health has been extensively studied in Western European countries in the last 20 years. Unemployment leads to an increase in morbidity and an unhealthy lifestyle (Claussen, 1993; Leino-Arjas et al., 1999; Weber, 1997). Several researchers have looked at the effect of such long-term unemployment. Leino-Arjas et al. (1999) reported that several variables could predict long-term unemployment and the consequence of long-term unemployment on health. There is little evidence about this matter in Japan. The present study aims to investigate some predictors of long-term unemployment and examine the effects of long-term unemployment on changes in lifestyle and mental health condition, based on a 4-year follow-up study.

1 Subjects and methods

1.1 Study sample

A Japanese shoemaking company went bankrupt in April, 1998. In July 1998, after company resuscitation was considered, the company was required under the Corporate Reorganisation Law to reduce its workforce by 760 persons. A voluntary redundancy scheme was offered and accepted by 760 persons, who became unemployed. Of this group, 473 persons who belonged to a voluntary association for the aim of exchanging information about re-employment and promoting mutual friendship, participated in this study.

1.2 Methods

A questionnaire about health status and lifestyle was sent to all participants in November, 1998 (baseline data), November, 1999 (1 year later), and November 2002 (4 years later) by mail. The questionnaire asked about re-employment situation, subjective symptoms, lifestyle including eating, smoking, drinking, exercise habits, and was described in our previous study (Matoba et al., 2003). 252 people (126 men, 126 women) who responded to all surveys in full were selected for this analysis.

Mental status was assessed according to a Self-rated questionnaire for depression (SRQ-D), consisting of 18 items with 4 possible answers: “no”, “sometimes”, “frequently”, and “always”. These answers were respectively scored as 0, 1, 2, and 3. Six out of 18 questions were used as dummy questions (Abe et al., 1972; Rockliff, 1969). According to the criteria by Rockliff and Abe, a score of 9 points or less was within normal range, between 10 and 15 was borderline, and 16 points or more were regarded as light depression or masked depression.

Statistical analysis was by a McNemar test, to look for statistical significance in changes of lifestyle habits between the two different surveys. An X^2-test was performed for changes of lifestyle habits between both genders. A multiple logistic regression model was used for detecting some predictors of long-term unemployment and poor mental health. Differences were taken to be statistically significant when p was less than 0.05. These statistical analyses were conducted with the JMP computer program (SAS Institute, USA; version 5).

2. Results

2.1 Re-employment status during 4 years follow-up by gender (see Figure 1).

Fifty-five percent of men had found a full-time job four years later. But, 20% were still unemployed. About one-quarter were part-time workers. Of the women, only 6% had found a full-time job during the follow-up period. About 50% were still unemployed. There was a significant difference in re-employment status according to gender.

3 Predictors of long-term unemployment

Table 1 shows that several factors contributed to the re-employment one year later. In our multiple logistic regression model, long-term unemployment (one year or more) was predicted by age (5 years, odds ratio 2.1, 95% CI 1.3-3.4), by gender (women vs. men, odds ratio 3.4, 95% CI 1.9-6.1). However, subjective symptoms, past history and sleep quality were not significant predictors of long-term unemployment. Other variables related to lifestyle habits such as eating, drinking, smoking, and exercise were not associated with long-term unemployment.

Figure 1. Re-employment status by gender

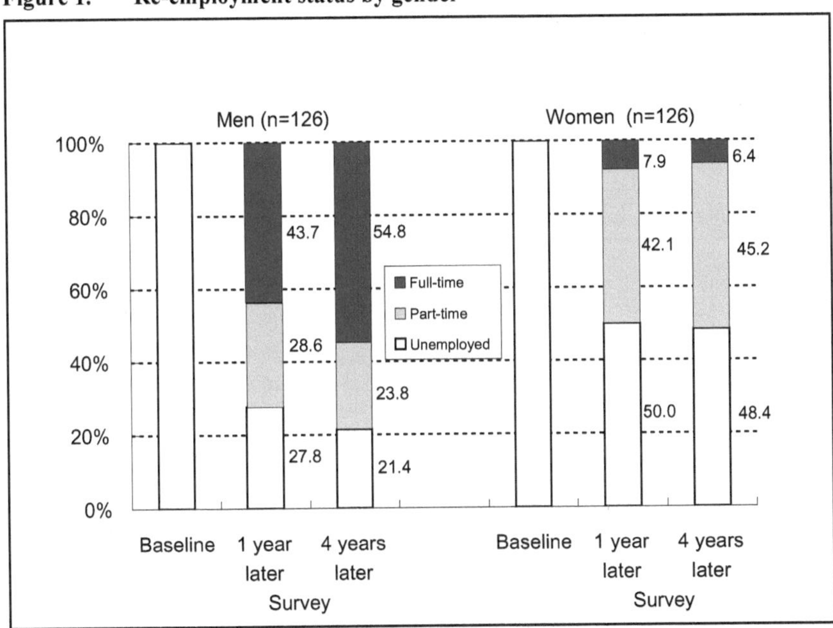

Table 2 shows which variables can predict long-term unemployment status, four years later. For predicting long-term unemployment (4 years), the factors of age and gender were significant: age (5 years; odds ratio 3.7, 95% CI 1.9-7.0), gender (women; odds ratio 7.9, 95% CI 3.7-17.2). We did not find any other significant relationship.

Table 1. Predictors of unemployment at the survey 1 year later from baseline data by multiple logistic regression

Variable (Reference category)	Category	Odds ratio (95% CI)	P-value
Age	5 years	2.08 (1.29 – 3.36)	0.0027
Gender (men)	women	3.42 (1.92 – 6.09)	<0.0001
Past history (no)	yes	0.98 (0.65 – 1.48)	0.914
Subjective symptom (no)	yes	0.85 (0.57 – 1.28)	0.447
Sleep (good)	poor	1.14 (0.75 – 1.73)	0.543
Eating (good)	poor	1.08 (0.71 – 1.65)	0.708
Drinking (no)	yes	1.12 (0.70 – 1.81)	0.632
Smoking (no)	yes	1.00 (0.63 – 1.59)	0.987
Exercise (yes)	no	0.90 (0.58 – 1.39)	0.639

Table 2. Predictors of unemployment at the survey 4 years later from baseline data by multiple logistic regression.

Variable(Reference category)	Category	Odds ratio (95% CI)	P-value
Age	5 years	3.65 (1.90 – 7.03)	<0.0001
Gender (men)	women	7.98 (3.70 – 17.20)	<0.0001
Past history (no)	yes	0.77 (0.49 – 1.21)	0.254
Subjective symptom (no)	yes	0.72 (0.46 – 1.12)	0.144
Sleep (good)	poor	0.75 (0.47 – 1.19)	0.216
Eating (good)	poor	1.26 (0.79 – 2.02)	0.327
Drinking (no)	yes	0.85 (0.44 – 1.29)	0.303
Smoking (no)	yes	0.93 (0.55 – 1.56)	0.785
Exercise (yes)	no	1.26 (0.78 – 2.04)	0.354

4 Predictors of poor mental health

We examined the effect of long-term unemployment on mental health (Table 3). As usual, long-term unemployment was defined as lasting more than one year. The SRQ-D score, according to the criteria by Rockliff and Abe, was applied to this analysis. In a multiple logistic regression model, poor mental health was predicted significantly by subjective symptoms (yes, odds ratio 2.97, 95% CI 1.2-7.0), and by drinking habit (yes, odds ratio 3.2, 95% CI 1.2-8.5). However, long-term unemployment itself was not a significant predictor of poor mental health in our data.

Table 3. Predictors of poor mental health status at the survey 4 years later from baseline data by multiple logistic regression. Mental health status was evaluated by SRQ-D scale

Variable (Reference category)	Category	Odds ratio (95% CI)	P-value
Age	5 years	1.37 (0.50 – 3.78)	0.543
Gender (men)	women	1.26 (0.50 – 3.26)	0.633
Past history (no)	yes	1.28 (0.61 – 2.56)	0.546
Subjective symptom (no)	yes	2.97 (1.26 – 6.99)	0.013
Drinking (no)	yes	3.24 (1.23 – 8.51)	0.017
Smoking (no)	yes	0.62 (0.25 – 1.54)	0.305
Exercise (yes)	no	1.88 (0.86 – 4.11)	0.115
Unemployment period (short)	long	0.80 (0.36 – 1.76)	0.576

5 Healthy lifestyle habits

To examine the relation between lifestyle changes and re-employment status, we selected subsamples of our subjects. We examined the relation between re-employment and lifestyle in three distinct groups based on re-employment status in the survey taken four years later: 1) unemployed (n=57), 2) part-time workers (n=49), 3) full-time workers (n=50). Figure 2a/b shows the proportion having healthy lifestyle traits, based on 7 factors: no drinking, no smoking, regular exercise, regular diet, nutritional balance, not eating between meals, and salt intake, for the baseline and at the survey four years later, by gender. There was no significant difference in the proportion having healthy traits across the three groups.

In the continuing unemployment group (Figure 2-A), the proportion having good health traits in the data from four years later had not deteriorated, at least in comparison with the baseline data. Nutritional balance and diet regularity were improved, but this change was not significant. We suggest that long-term unemployment does not influence individual lifestyle in middle-aged men and women. Lifestyle was maintained reasonably accurately for four years. In the other two groups of part-time workers and full-time workers (figure 2-B, C), similar tendencies were observed in the changes of lifestyle of both men and women. There was no deterioration in lifestyle health traits during the four year follow-up.

Figure 2. Change of healthy lifestyle about 7 factors between baseline data and 4 years later data by gender

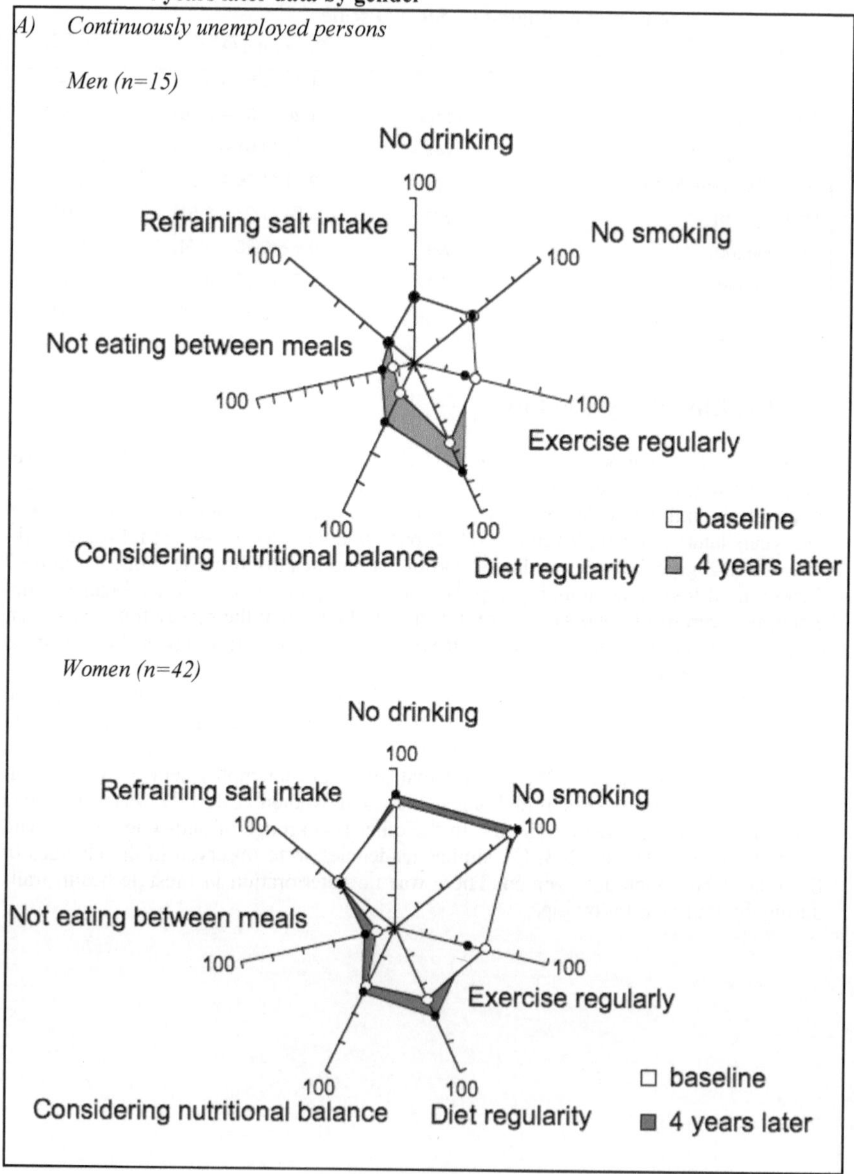

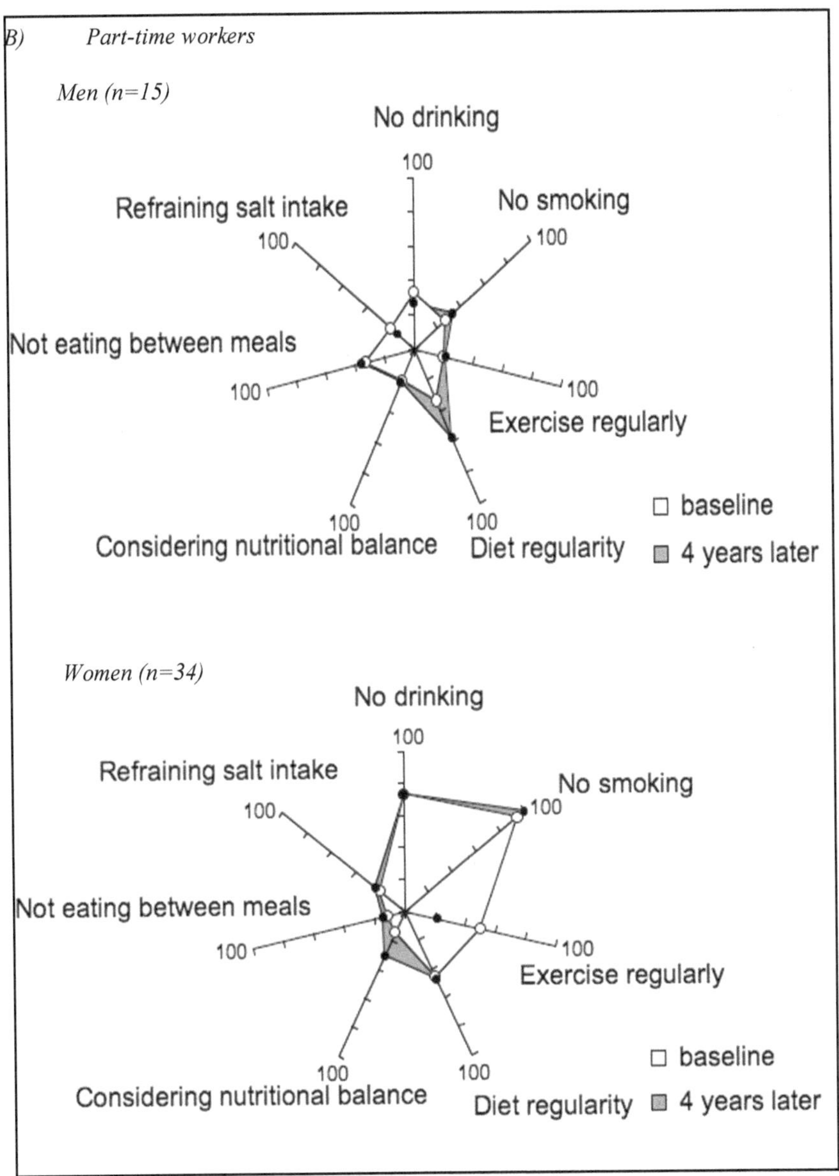
B) Part-time workers
Men (n=15)
No drinking
No smoking
Refraining salt intake
Not eating between meals
Exercise regularly
Considering nutritional balance
Diet regularity
100
baseline
4 years later
Women (n=34)
No drinking
No smoking
Refraining salt intake
Not eating between meals
Exercise regularly
Considering nutritional balance
Diet regularity
100
baseline
4 years later

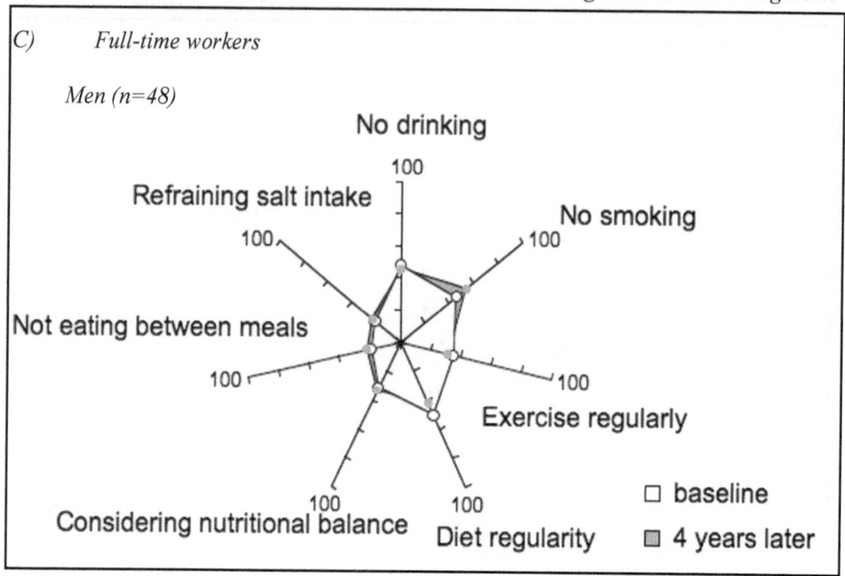

6 Discussion

We have performed a prospective cohort study to investigate predictors and consequences of long-term unemployment with regard to lifestyle and mental health in persons unemployed due to company closure in Japan. We found that age and gender are significant predictors of long-term unemployment, and poor mental health was predicted by subjective symptoms and a drinking habit. Favourable lifestyle changes were observed among the unemployed workers relative to the other groups comprising part-time workers and full-time workers.

Several researchers have conducted cohort studies for predictors and consequences of unemployment (Ferrie et al., 2001; Hammarström & Janlert, 2003; Leino-Arjas et al., 1999, Reine et al., 2004; Wooley et al., 2002). These are mostly studies of young adults. Poor mental health and bad lifestyle traits, such as smoking and heavy alcohol drinking, were both predictors and consequences of long-term unemployment. However, we did not find that poor mental health was a significant factor in relation to long-term unemployment. What might explain this discrepancy? Adverse effects of long-term unemployment on health may depend on age, being less in middle-aged persons (Osler et al., 2003). Another explanation could be the existence of the special support association for the unemployed to exchange information about re-employment and for mutual friendship. This association may be important in helping people to cope with stress-related unemployment. However, the long-term unemployed suffer more from poor health because of economic and psychological distress (Geyer & Peter, 2003; Ytterdahl et al., 1999), so that a special health care system should be maintained for these persons (Matoba & Ishitake, 1998). Maintenance of financial security, provision of proactive health care and retraining for re-employment can all reduce the impact of unemployment on health (Dooly, Prause & Ham Rowbottom, 2000).

In general, unemployment leads to an increase of unhealthy behaviour, such as increased alcohol and tobacco consumption, worse dietary habits, and reduced rates of physical exercise (Claussen, 1993; Janlert, 1997; Lahelma et al., 1997; Wilson & Walker, 1993). However, there was no definite deterioration on lifestyle in our follow-up study. Leino-Arjas et al. (1999) have found a significant reduction in alcohol consumption and greater exercise in workers with prolonged unemployment. According to our 2-year follow-up survey on the same populations, healthy lifestyle and good health were maintained in long-term unemployed persons (Matoba et al., 2003). Greater health promotion among the long-term unemployed could be important in preserving healthy lifestyle habits. Standard factors such as age, gender, educational and cultural background must also be taken into consideration. In conclusion, age and gender were proved to be significant predictors of long-term unemployment. Subjective symptom and alcohol drinking habits could relate to poor mental health. However, long-term unemployment had no significant effect on lifestyle or mental health. We recommend that some supporting and counselling systems for both mental and somatic conditions should be available to unemployed persons for an interval following company restructuring or downsizing.

References

Abe, T., Tsutsui, S., Nanba, K., Nishida, K., Nozawa, A., Kato, Y. & Saito, T. (1972). Masked depression no screening test to si te no sitsumonhyo (SRQ-D) ni tsuite. *Seishin Shintai Igaku, 12,* 243-247. (in Japanese)

Blakely, T.A., Collings, S.C. & Atkinson, J. (2003). Unemployment and suicide. Evidence for a causal association? *Journal of Epidemiology and Community Health, 57,* 594-600.

Claussen, B. (1993). A clinical follow up of unemployed I: Lifestyle, diagnosis, treatment and re-employment. *Scandinavian Journal of Primary Health Care, 11* , 211-218.

Dooley, D., Prause, J. & Ham Rowbottom, K.A. (2000). Underemployment and depression: Longitudinal relationships. *Journal of Health and Social Behavior, 41,* 421-436.

Ferrie, J.E., Martikainen, P., Shipley, M.J., Marmot, M.G., Stansfeld, S.A. & Smith, G.D. (2001). Employment status and health after privatisation in white collar civil servants: Prospective cohort study. *British Medical Journal* 322, 647-651.

Geyer, S. & Peter, R. (2003). Hospital admissions after transition into unemployment. *Sozial- und Präventivmedizin, 48,* 105-114.

Janlert, U. (1997). Unemployment as a disease and diseases of the unemployed. *Scandinavian Journal of Work, Environment and Health, 23* (Suppl 3), 79-83.

Hammarström, A. & Janlert, U. (2003). Unemployment - an important predictor for future smoking: A 14-year follow-up study of school leavers. *Scandinavian Journal of Public Health, 31,* 229-232.

Health and Welfare Statistics Association (2002). Suicide statistics. *Journal of Health and Welfare Statistics, 49,* 53-55. (in Japanese)

Kposowa, A.J. (2001). Unemployment and suicide: A cohort analysis of social factors predicting suicide in the US National Longitudinal Mortality Study. *Psychological Medicine, 31,* 127-138.

Lahelma, E., Rahkonen, O., Berg, M.A., Helakorp, S., Prättälä, R., Puska, P. & Uutela, A. (1997). Changes in health status and health behavior among Finnish adults 1978-1993. *Scandinavian Journal of Work, Environment and Health, 23* (Suppl 3), 85-90.

Leino-Arjas, P., Liira, J., Mutanen, P., Malmivaara, A. & Matikainen, E. (1999). Predictors and consequences of unemployment among construction workers: Prospective cohort study. *British Medical Journal, 319,* 600-605.

Matoba, T., & Ishitake, T. (1998). A strategy for health promotion among unemployed people in Japan. *International Archives of Occupational and Environmental Health, 72* (suppl), 31-33.

Matoba, T., Ishitake, T. & Noguchi, R. (2003). A two-year follow-up survey of health condition and life style follow-up in Japanese unemployed persons. *International Archives of Occupational and Environmental Health. 76*, 302-308.

Ministry of Public Management, Home Affairs, Post and Telecommunications, Statistics Bureau & Statistics Center (2003). *Labour Force Survey 2003.*

Osler, M., Christensen, U., Lund, R., Gamborg, M., Godtfredsen, N. & Prescott, E. (2003). High local unemployment and increased mortality in Danish adults: Results from a prospective multilevel study. *Occupational and Environmental Medicine, 60*(11), e16.

Reine, I., Novo, M. & Hammarström, A. (2004). Does the association between ill health and unemployment differ between young people and adults? Results from a 14-year follow-up study with a focus on psychological health and smoking. *Public Health, 118*, 337-345.

Rockliff, B.W. (1969). A brief self-rating questionnaire for depression (SRQ-D). *Psychosomatics, 10*, 236-243.

Weber, A. & Lehnert, G. (1997). Unemployment and cardiovascular diseases: A causal relationship? *International Archives of Occupational and Environmental Health. 70*, 153-160.

Whooley, M., Kiefe, C.I., Chesney, M.A., Markovitz, J.H., Matthews, K. & Hulley, S.B. (2002). Depressive symptoms, unemployment, and loss of income. *Archives of Internal Medicine, 162*, 2614-2620.

Wilson, S.H. & Walker, G.M. (1993). Unemployment and health: A review. *Public Health, 107*, 153-162.

Ytterdahl, T. (1999). Routine health check-ups of long-term unemployment in Norway. *International Archives of Occupational and Environmental Health, 72* (Suppl), 38-39.

Differences in the Patterns of Primary Health Care Use Between the Employed and the Unemployed[1]

Juha Nyman, Unto Häkkinen, Pirkko Alha & Ilmo Keskimäki

Introduction

The Finnish health service system is based on the principles of the Nordic welfare model: the coverage for all Finnish residents, the principle of universality, the strong public sector in the provision of services, tax funding, and equal treatment. The municipalities are responsible for organising primary health care for their inhabitants, and the majority of services are provided by the municipal health centres. The physician visits in primary health care are distributed as follows: visits to municipal health centres cover 77% of outpatient visits to a doctor, private-sector occupational health care cover 13% of the visits, and the visits to private sector general practitioners cover 10% of all visits (Statistical Yearbook on Social Welfare and Health Care, 2003, p. 117). For the residents of the municipality the municipal health centre offers health education and counselling, medical examinations, screenings, and treatment given or supervised by a physician. In 2002 the number of physician visits at health centres was about 1.8 per inhabitant, and visits to other health care professionals 2.9 per inhabitant.

In 1993 the Finnish health care policy was reformed: the central government regulation was reduced, and the municipalities were given a wider freedom to choose how to organise the health services for their inhabitants. In many municipalities the visits to a doctor became chargeable, and visits to health centres are now charged a certain fee per visit or an annual sum for treatment by a doctor. Visits to a public health nurse or preventive health care remained free of charge. There is a fixed maximum payment in a calendar year; after this limit the visits at the health centre become free of charge.

Employers are responsible for arranging preventive healthcare for their employees. The employers may arrange for medical check-ups, treatment, and rehabilitation. The municipal health centres provide occupational health services to the employers who request these services but the employers have a right to organise the occupational health services themselves or buy them from a private supplier. In Finland, occupational health care discontinues in case of redundancy and the redundant workers have to use regular municipal health centre services. Although Finns are relatively satisfied with their

1 The study was financially supported by the Juho Vainio Foundation

health care, the functioning of health centres have been criticised for long waiting times and for a poor continuity of care.

From 1990 to 1994 Finland experienced a major recession where a short- and long-term unemployment became common. The general unemployment rate remained high and in some sectors unemployment still continued to grow during the entire decade. The proportion of the long-term unemployed increased: By the year 1996 one in three of those unemployed had not been employed for over one year. In 2000, 253,000 persons were registered as unemployed in Finland, corresponding to the general unemployment of rate of 9.8% (Statistical Yearbook of Finland, 2001, p. 372). In this study, our aim is to describe the status of health among unemployed people in Finland and their need for health services, and to investigate how these needs are being met.

1 Unemployment and ill-health

Long-term unemployment often leads to economic marginalisation and social exclusion of the individuals and families involved and the status of health is worse among unemployed people than among the employed (Grzywacz & Dooley, 2003). While unemployment undermines health and promotes unhealthy behaviour, the workers who feel themselves ill or lead an unhealthy lifestyle also face an increased risk of involuntary dismissal if downsizing takes place, and decreased chances of reemployment (Nyman, 2002). Redundant male workers experience more psychical and physical hardships than women following job loss (Levenstein et al., 2001).

Unemployment and ill-health have been found to be related according to a broad range of health measures: physiological responses, unhealthy behaviour, self-reported ill-health, deficient functional ability, and finally mortality (Arnetz et al., 1987; Arnetz et al., 1991; Claussen, 1999; Shortt, 1996; Morris et al., 1994; Montgomery et al., 1998; Martikainen, 1996). More than forty years ago Kasl et al. (1968), Kasl and Cobb (1970) and later Mattiasson (1990) showed in their prospective studies that among men who were threatened by unemployment the levels of serum cholesterol and blood pressure rose, increasing the overall risk of cardiovascular disease. Those who succeeded in finding a job had decreased cholesterol and blood pressure levels. Correspondingly the most prevailing causes of death among the unemployed are cardiovascular disease and cancer (Morris et al., 1994).

Unemployment and its consequences: lack of money, low self-esteem, depression, pessimism, and stress increase, among others, increases the likelihood of unhealthy behaviour. The longer the duration of unemployment, the higher the increase of unhealthy living habits, such as problem drinking – particularly among men – regular smoking, and overweight. These all predispose to physical frailty and functional deterioration (Hammarström, 1988; Power & Estaugh, 1990; Dooley et al., 1992; Janlert & Hammaström, 1992; Ferrie et al., 2001). Unemployment remains a significant risk factor for hypertension among men even after adjustment of health risk behaviour such as overweight and a sedentary lifestyle (Levenstein et al., 2001).

Earlier research indicates that increasing social resources and support decreases demand on mental health services (Sherbourne, 1988). The individuals' experiences are affected by the duration of their unemployment, the negative consequences of being unemployed, the person's financial situation, and supportive social networks. Informal support buffers the health effects of unemployment whereas those not receiving support have more symptoms of illness. An important source of social support is marriage, partnership or family (Hintikka et al., 1999).

2 Changed pattern of health service use

The previous investigations in different countries have reported 1.6-3.0-fold increases in aggregate health service use among the unemployed (Jin et al., 1986; Studnicka et al., 1991; Kraut et al., 2000). Yuen and Balarajan (1989) analysed the effects of unemployment on consultation rates and reported that, adjusted for age, occupation, area of living, and having a longstanding illness, unemployment was associated with the probability of an increased number of consultations with general practitioners. The consultation pattern was dependent on the length of unemployment. Studnicka (1991) reported that the overuse was demonstrated among older unemployed men but not necessarily among unemployed women or younger unemployed persons.

However, there is also some evidence that at least some groups of unemployed discontinued their use of services as a result of unemployment. Ferrie et al. (2001) found that even though the unemployed had worse than average physical and psychical health, or more reported symptoms, they did not necessarily have more longstanding illnesses diagnosed by a physician. Furthermore, Cook et al. (1982) found that the unemployed were more likely than the employed to suffer from diseases not diagnosed previously until the medical examinations of the study. Some researchers, especially in Finland, have reported that the unemployed reduce their demand on health services (Virtanen, 1993). Contrary to his expectations, Virtanen found that those among his clients who found a job made more primary health care visits than those who were permanently unemployed. The number of visits increased following re-employment and declined after a new job loss.

Workers in different occupations are exposed to different physical, chemical, biological, ergonomic, or psychosocial effects. Consequently, many workers having a high level of exposure to noxious substances at work may suffer from work-related illnesses. As a worker then loses his or her job, such exposure decreases, which might be a reason why many of those who once worked in demanding workplace environments need less health services after their lay-off. Iversen et al. (1989) found that among displaced male shipyard workers the risk of admission to hospital due to cardiovascular disease increased, but that the risk due to accidents decreased after the displacement. Stress at the workplace is associated with an increase of sick leave and thus more visits to the doctor. The unemployed do not require an official doctor's certificate for absence from work but they are encouraged to get certificates if they have to stay away from jon training because of their own or their child's illness.

At present in Finland the public sector covers 80% of the health services, private for-profit corporations 17%, and the non-profit voluntary sector 3%. In 2001, the private sector accounted for 25% of all medical services and for 35% of specialised doctor services (Statistical Yearbook on Social Welfare and Health Care, 2001, p. 103). If a patient chooses to use private sector services, the public health insurance compensates an average of only 30% of expenses. However, not all health services are compensated by the public health insurance. If the patient has a voluntary private health insurance he can pay most of the expenses through this insurance. If the patient is not insured she or he has to bear the expenses.

During the 1990s, the medical expenses of households increased in Finland (Klavus, 2000, pp. 70-71). The lowest income decile spent 8% of their incomes on health care whereas the highest decile spent only 3% of their disposable incomes. Irrespective of the fact that the Finnish health care system has universal coverage based on residence, one consequence of job loss is reduced availability of health services. Employers organise preventive and secondary occupational health services for their staff. For the unemployed, easily available, employer-provided, occupational health services (health checks, support, treatments) are no longer available. The continuity of care (the presence of a personal doctor) is found to be an important determinant in the rate of using physician services (Häkkinen et al., 1995).

3 Research problem

In Finland the public sector is almost free of charge, but unfortunately access to public sector health centres is not always satisfactory. At the same time, the private sector is increasing its share in the health service market. However, the high charges of private sector services make them largely inaccessible to the unemployed. Furthermore, it is more difficult to get a referral to specialised care at the health centre. For these reasons it is unclear whether the health needs of the unemployed coincide with the available health services. Previous studies have usually described aggregate health service use and not differentiated the reasons for visits to the doctor or public health nurse. This study aims at describing the patterns exhibited by the unemployed in their health care use, access to medical care, and the participation in prevention practices. Häkkinen (2002) investigated the visits to a physician among the labour force and reported that unemployment decreased the probability of such use. It is reasonable to assume that the unemployed might use less health services than the employed although they have more health needs.

It might be that consultations in the Finnish health care are structurally linked to being employed: for example, employees have to submit a sickness certificate for a sick leave, or poor work motivation is likely to increase absenteeism and also consultation rates among the employed. The specific question in this study is therefore: Is there more unmet need of health services among the unemployed than among the employed?

4 Data and methods

Based on the nation-wide Health 2000 survey data, this paper describes and analyses the relation between employment status and the use of health care. The sampling frame was adults 30 years old and over, living in Finland. The stratified sample was collected from five university hospital regions such that in each of them 16 health centre districts were sampled as clusters. Health centre districts were selected in the sample by probability-based proportional size-sampling in each stratum except for the 15 largest health centre districts, which were selected in the sample with probability of 1. The health centre districts, which represented 160 municipalities, were the ultimate sampling units and the participants were taken by systematic sampling from these health centre districts. This was done in order to minimise costs: distances remained shorter than in the case of probability sampling. A detailed description of the study sample and recruitment methods has been published by Aromaa and Koskinen (2004).

Data on health status were obtained in health interviews and health examinations. The study began with a health interview at the participant's home. All interviewees were asked to come for the health examination at the appointed time. The structured survey included questions on occupational and employment history, health and illness, and health care use. The health examination consisted of clinical physical measurements. The data was gathered between the years 1999 and 2001.

In the Health 2000 survey the whole sample comprised 8,028 persons. Of these, 87% were interviewed and 79% participated in the health examination. Only employed or unemployed participants between the ages of 30 and 64 were surveyed. A total of 4,377 persons were selected for the analyses of this study. We divided employment status into three categories: employed at a full-time job, unemployed for less than 12 months and unemployed for 12 months or more. For the most part, the variables used in the analysis are categorical. The demographic variables in the statistical models were categories of gender, age, level of education, region i.e. university hospital district, marital status, number of children, having a specific personal doctor, distance to a doctor, having a primary public health nurse available for contact, and family income adjusted for family size using the OECD equivalent scale (OECD 1982).

Education was a composite variable with three categories: basic, secondary, and higher education. The categories were defined as follows. Those respondents, who had no vocational training beyond a vocational course or on the job training and who had not taken the matriculation examination, were categorised as having only basic education. Completion of vocational school was defined as secondary education. Those respondents who had a matriculation examination but no vocational training beyond a vocational course were also categorised as having a secondary education. Higher vocational education at a polytechnic or university put one into the higher education group.

The following health-related controlling variables were obtained directly from the interview: self-rated health status, chronic disease (do you have an illness which impairs your ability to work or your functional health?), mental illness (alcohol abuse, anxiety, depression, psychosis). Data on disease categories concentrated on the disease categories most important in public health terms: cardiovascular, respiratory, and musculoskeletal diseases. The definition of regular smoker follows the World Health Organisation's recommendation: daily smoking is smoking on a regular basis during which the person has smoked at least 100 cigarettes for one year or more and has smoked during the day of the interview or on the day before that.

Health examinations were carried out at health centres. The measurements made in this study were weight, blood pressure (twice from the upper right arm, examinee sitting), serum cholesterol, serum triglyceride, and serum blood sugar.

Dependent variables were the following: self-assessed met or unmet need of health services, consultation with a physician during the past 12 months, consultation with a physician because of mental problems, number of visits to a physician, number of visits to a physician due to mental problems, and health examinations (blood pressure, serum cholesterol, and sum variable of health check-ups) made during the past 5 years.

Using a dichotomous response variable (visit or no visit to a doctor) we fit models as a logistical regression. With a dependent variable, which measured the number of visits to a doctor during the previous year, we fit models as a non-binomial regression. The variable selection was based on univariate analyses (a 0.25 level was used as a screening criterion). Estimates were generated from the data set using weighted results, which permit the estimation of models representative to the whole population by taking into account the sampling method and non-respondents.

5 Results

The study data comprised 4,377 respondents between the ages 30 and 65 years old who were in the labour force during the interview. In the data, 3,846 respondents (87.9%) were employed, 279 (6.4%) were short-term unemployed (less than 12 months unemployment), and 252 (5.7%) were long-term unemployed (at least 12 months unemployment). The proportion of men was 48.4% and women 51.6%.The age categories were 55 to 64 years old (n=575), 45 to 54 years old (n=1,558), and 30 to 44 years old (n=2,076).

The mean age of the long-term unemployed was higher than that of the other groups. At the time of the interviews 37.6% of the long-term unemployed were 55 years old or over (13.9% of the employed, 15.9% of the short-term unemployed). The long-term unemployed were twice as likely to be divorced, separated or widowed (50.0%) than the employed (22.6%). About 82% of the long-term unemployed, 60% of the short-term unemployed, and 52.6% of the employed had no underage children. Of the long-term unemployed 52.3% had no more than basic education, whereas only 22.9% of the employed where in this category. The mean family income adjusted for the family size according to the OECD equivalent scale was 10,268 Finnish marks (1,711 euro) in the employed group, 5,097 marks (849 euro) in the short-term unemployed group, and 3,952 marks (658 euro) in the long-term unemployed group.

Of the employed, 54% (83% of the both unemployed groups) did not have a primary nurse or a health visitor to contact if needed. There was no such difference when questioned if the municipal health centre had appointed a personal/family doctor for the interviewee. This applied to every other person in each group.

Among the long-term unemployed, about 50% of the respondents felt that their overall health was only average or poor whereas only one fourth of the employed felt their health to be average or poor (Table 1). Of all respondents, 38% had a disabling chronic disease. The most common diseases were cardiovascular and musculoskeletal diseases, and diseases of the nervous system and sensory organs. While the unemployed reported more often than the employed a chronic illness diagnosed by a doctor, they reported nearly three times as often as the employed to have mental ill-health (i.e. alcohol abuse, anxiety, depression, or psychosis). Among the most common symptoms was depression.

Table 1. Self-reported overall health, chronic illness and mental illness by employment status

	Employed	Unemployed < 12 months	Unemployed ≥ 12 months	Total	χ^2
Health average, rather poor, or poor	24.0% n=876	39.9% n=103	50.7% n=115	26.5% n=1099	p < 0.001
At least one chronic illness	35.8% n=1320	47.4% n=131	56.4% n=129	37.7% n=1580	p < 0.001
Self reported mental illness (depression, anxiety, alcohol abuse, or psychosis)	8.7% n=308	24.2% n=67	24.6% n=62	10.8% n=437	p < 0.001

According to the results, all common risk factors for cardiovascular diseases were more common among the unemployed than the employed (Table 2). Of the unemployed, 22% were obese whereas less than a fifth of the employed were obese. This difference did not reach statistical significance between the different groups (χ^2 = 3.18, p = 0.246). Hypertension was observed among one-third of the long-term unemployed and only in 19% of the employed. The mean serum total cholesterol concentration was 5.9 mmol/l among the employed, 5.9 mmo/l in the short-term unemployed, and 6.2 mmol/l in the long-term unemployed. The recommended level of serum cholesterol was then 5.0 mmol/l or lower. Nearly half of the long-term unemployed, but only one third of the employed, had an increased serum cholesterol concentration of 6.5 mmol/l or greater. The recommended level of serum triglyceride is less than 2.0 mmol/l. The mean serum triglyceride was 1.5 mmol/l in the employed but about 1.7 in both unemployed groups. The difference between the groups was statistically significant (χ^2 = 16.00, p < 0.001). The mean serum glucose was the same among the employed, and for the short-term and long-term unemployed.

Table 2. Prevalence of daily smoking, obesity, hypertension, serum cholesterol and serum triglyceride, by employment statuses

	Employed	Unemployed < 12 months	Unemployed ≥ 12 months	Total	χ^2
Daily Smoking	26.7% n=983	40.9% n=112	45.5% n=104	28.7% n=1199	p < 0.001
BMI ≥ 30 kg/m²	18.9% n=689	22.3% n=61	22.3% n=48	19.3% n=798	p = 0.246
Blood pressure Syst ≥ 160 or Diast ≥ 95 mmhg	18.9% n=532	24.6% n=66	30.6% n=69	19.9% n=817	p < 0.001
Increased serum cholesterol concentration ≥ 6.5 mmol/l	29.6% n=1077	32.3% n=88	44.6% n=102	30.6% n=1267	p < 0.001
Increased triglyceride concentration ≥ 2.0 mmol/l	23.6% n=853	30,5% n=82	33.1% n=75	24.6% n=1010	p < 0.001

After the questions on chronic illnesses, the respondents were asked whether they receive sufficient care for the illnesses they have. The need for care as a whole, as well as satisfaction in care, was assessed. The need for care was classified as being inadequately met if the interviewee responded that he/she needed care more than he/she had got. The unemployed reported more often than the employed not to have received sufficient treatments for illnesses. Of the employed, 11% reported having an illness for which they did not get sufficient care, compared to 16% of the short-term unemployed and 21% of the long-term unemployed.

The health interview enquired about the frequency of health examinations, and visits to a physician. These questions covered the previous year. The proportion of those who had not used the health services at all during the past year were calculated as well the number of visits in the two different groups. Of the long-term and short-term unemployed, about 38% had not visited a physician during the past year (31% of the employed). Of those who had visited the health services, the average of outpatient visits to doctor was 3.9 per person in the employed group, 4.1 among the short-term unemployed, and 4.7 among the long-term unemployed. Women used all kinds of ambulatory health services more than men.

Of the employed, 79% had been in some sort of health examination during the past 5 years (65% of the short-term unemployed and 59% of the long-term unemployed). The people in the employed group had been 1.6 times in a health examination whereas the long-term unemployed had been 1.2 times in a health examination during the past five years. Of the employed, 5% had used health services because of mental problems (12% of the short-term unemployed and 16% of the long-term unemployed). Among those who had used health services because of mental problems, the mean number of visits per year was about 9 in the employed group but 13 among the long-term unemployed group.

Among those who had visits in health centres during the past year, the number of visits ranged from 1 to 40. The mean number of outpatient visits in health centres was 2.3 per person in the employed group, 2.9 in the group of short-term unemployed, and 3.5 in the group of long-term unemployed. On the other hand, the employed had on average 2.6 visits to occupational health care, whereas the short-term unemployed had only 1.9 and the long-term unemployed 1.7 visits to occupational health care. In addition, the employed had more outpatient visits to private sector health services (mean 2.1) than either of the unemployed groups (1.9). The long-term unemployed made more visits to outpatient department (3.1) than either of the other groups (the employed 2.2).

5.1 Multivariate analyses

The unmet need was associated with unemployment. Adjusted for state of health and region, the long-term unemployed reported to have an illness for which they felt they did not get sufficient care more often than other groups (OR 1.5 compared to the employed) (Table 3).

Table 3. Results of logistic regression on probability of having an illness for which has not got sufficient care

	OR	95% CI
Unemployed < 12 months	1.1	0.88-1.71
Unemployed ≥ 12 months	1.5*	1.05-2.22
University hospital region, Helsinki	1.0	
University hospital region Turku	1.1	0.81-1.58
University hospital region Tampere	1.3	0-97-1.70
University hospital region Kuopio	1.3	0.93-1.76
University hospital region Oulu	1.5**	1.11-1.97
General health state average or poor	2.2***	1.75-2.73
Chronic disease	1.9***	1.59-2.38

Note. *$p < .05$. **$p < .01$. ***$p < .001$

In addition, those suffering from ill health and a chronic disease felt they did not get sufficient care. There was also the Oulu region where the people felt that their care was defective.

After control for gender, family income, general health, and chronic disease, the likelihood of the short-term unemployed not having visited a doctor during the past 12 months was 50% higher than for employed people (Table 4). In addition, those who suffered from poor health and/or a chronic disease were more likely to visit a physician. The increase of 1000 Finnish marks in the household's annual income increased the probability of visiting a doctor by 2%.

Table 4. Results of logistic regression on the probability of no visits to a doctor during the past 12 months

	OR	95% CI
Unemployed < 12 months	1.5*	1.07-1.96
Unemployed ≥ 12 months	1.4	0.99-2.00
Male	2.2 ***	1.91-2.62
Family incomes	0.98**	0.96-0.99
Self-rated health status average or poor	0.6 ***	0.47-0.67
Chronic disease	0.4***	0.36-0.52

Note. *$p < .05$. **$p < .01$. ***$p < .001$
Note. As the income per family member increase by 1000 Fin mark (about 167 Euro) the likelihood of visiting a doctor increases 2%.

The number of visits to a physician per year due to own illness or a pregnancy was 3.9 in the group of employed, 4.1 in the short-term unemployed and 4.7 in the group of long-term unemployed. However, there was no association between the employment status and the number of visits when the demographic and health variables were controlled for about 16% of the long-term unemployed had visited the health services because of mental problems (12% of the short-term unemployed and 5% of the employed). After controlling for the demographic, health, and health behaviour variables, the short-term unemployed were twice as likely to use the health services because of mental problems, and the long-term unemployed were three times more likely, both compared to the employed (Table 5).

Table 5. Results of logistic regression on probability of having used health services during the past 12 months because of mental problems

	OR	95% CI
Unemployed < 12 months	2.1***	1.37-3.16
Unemployed ≥ 12 months	3.2***	2.01-5.05
Female	2.1***	1.58-2.73
Age from 35 to 44 years	1.0	
Age from 45 to 54 years	1.0	0.69-1.29
Age from 55 to 64 years	0.6*	0.34-0.91
Marital status single	2.0***	1.48-2.57
Higher education	1.0	
Secondary education	0.9	0.64-1.13
Basic education	0.5***	0.35-0.75
Obesity	1.5*	1.13-2.11
Average or poor general health state	2.0***	1.51-2.70
Chronic disease	1.8***	1.36-2.35

Note. $*p < .05$. $**p < .01$. $***p < .001$

Among those who had visited a physician the mean number of visits to a health services due to mental problems was 0.4 among the employed (1.0 in the short-term unemployed and 1.8 in the long-term unemployed).The number of visits to health services during the past 12 months due to mental problems was almost six times higher among the long-term unemployed relative to the employed (Table 6).

Table 6. Results of non-binomial regression on number of visits to health services during the past 12 months because of mental problems

	IRR	95% CI
Unemployed < 12 months	1.8	0.94-3.31
Unemployed ≥ 12 months	5.8***	2.52-13.31
Female	3.8***	2.20-6.46
Marital status single	2.4***	1.40-4.05
Average or poor general health state	2.8***	1.76-4.48

Note. $*p < .05$. $**p < .01$. $***p < .001$

The health-check variable was a composite variable of health examinations related to a driving licence, new job, age category, pregnancy, or a health examination focused to the unemployed people. Of the employed, 21% had not have any health-check up during the past 5 years (35.8% of the short-term unemployed and 41% of the long-term unemployed). Additionally, the mean number of health check-ups was higher (1.6) in the employed group than in the long-term unemployed group (1.2).

Adjusted for the demographic variables and health variables, the odds ratio for the short-term unemployed not to have any health examination during the five years was 1.9, and for the long-term unemployed the corresponding odds ratio was 2.6 relative to the employed (Table 7 on the next page).

Table 7. Results of logistic regression on probability of having no health-check-ups during the past five years

	OR	95% CI
Unemployed < 12 months	1.9***	1.39-2.50
Unemployed ≥ 12 months	2.6***	1.93-3.44
Female	0.5 ***	0.41-0.56
Age 35 to 44 years	1.0	
Age 45 to 54 years	0.7***	0.54-0.78
Age 55 to 64 years	0.5***	0.40-0.62
Family incomes	0.98**	0.96-0.99
No primary nurse or health visitor	1.6***	1.35-1.86

Note. *$p < .05$. **$p < .01$. ***$p < .001$

For the follow-up of blood pressure, the short-term unemployed had a higher probability (OR 1.4) of not having their blood pressure measured during the past five years than the employed. Correspondingly, both the short- and long-term unemployed had undergone tests for serum cholesterol less often than the employed (OR 1.5, Table 8).

In the case of serum glucose measurements, the differences between the groups were in the same direction but not statistically significant in the full model.

Table 8. Results of logistic regression on measurement of blood pressure and serum cholesterol during the past 5 years

	Blood pressure not measured during the past 5 years		Serum cholesterol not measured during the past 5 years	
	OR	95% CI	OR	95% CI
Unemployed < 12 months	1.4*	1.01-1.84	1.5**	1.10-1.91
Unemployed ≥ 12 months	1.1	0.76-1.47	1.5**	1.11-2.15
Female			1.2**	1.05-1.34
Age 35 to 44 years	1.0		1.0	
Age 45 to 54 years	0.9	0.77-1.07	0.6***	0.51-0.70
Age 55 to 64 years	0.6***	0.48-0.79	0.4***	0.35-0.54
Family incomes	0.98*	0.96-0.99	0.97***	0.96-0.98
Obesity	0.6***	0.53-0.76	0.8***	0.67-0.92
Chronic disease	0.7***	0.59-0.87	0.8***	0.68-0.89
No primary nurse or health visitor	2.1***	1.73-2.44	2.1***	1.84-2.40

Note. *$p < .05$. **$p < .01$. ***$p < .001$

6 Conclusions

The present study is based on an extensive nationally representative Finnish health survey, Health 2000, comprising both interview and health examination data. The survey had in general a relatively low non-response rate and the research data included a large number of demographic, health and health care variables. However, there were some differences in responding according to employment. 9.5% of the long-term unemployed but only 4% of the employed failed to answer the question on the self-reported health state. Moreover, 14% of the long-term unemployed did not respond to the question about sufficiency of health care they received, which is more than the non response in the other groups. The disparate responding rates for these questions may have had some effect on our results.

Among the unemployed there were more unmet physical health needs than among the employed. Our results indicate that the unemployed had more health risks than the employed. The unemployed led an unhealthier life; they smoked more regularly, had more often elevated blood-pressure and serum cholesterol levels, and were more often obese. But the probability of having undergone health examinations or tests for these risk factors was lower for them than for the employed. Moreover, the unemployed received less preventive health services than the employed.

The average number of visits to a physician was higher among the unemployed than among the employed, but the mean age of the unemployed was higher than that of the employed. Adjusted for background variables, the short-term unemployed were more likely not to visit a doctor. However, the probability of the long-term unemployed to have used health services because of anxiety, depression, alcohol abuse or psychosis was three times higher than the employed. Furthermore, the estimation of the increased number of visits to a doctor among the long-term unemployed was five times higher than the corresponding estimation by the employed. This over-utilisation of health services by the long-term unemployed in Finland will impose a substantial burden on services.

Considering the persons in poor health, their incapacity prevents them from entering the labour force and from participating in the labour markets (Malo & García-Serrano, 2001). While physical ill-health after a layoff may be an important explanation for the failure to regain work among manual workers particularly, for the unemployed, poor health may similarly have some repressive repercussions in seeking a job as well (Dahl, 1993). The degree of functional capacity also determines whether one will have to give up working (Vahtera & Pentti, 1997). If a downsizing takes place, a decreased functional capacity will pose an increased risk of losing the job. Due to these various factors, it is not surprising that the level of health is poorer among the unemployed than the employed. Our study suggests that the unemployed may also have recognised their higher physical health needs. According to the results, the unemployed had more self-evaluated unmet needs from health services than the employed. A potential solution to this situation in Finland might be to intensify personal health monitoring in municipal health services, for instance, by appointing a personal or family nurse or health visitor for all municipal residents. This proposal is supported by our finding that if a person did not have a primary nurse or a health visitor to contact, this increased his or her risk of missing health check-ups and suffering from increased blood-pressure and serum cholesterol levels.

The results of our study are in accordance with the Finnish Sociobarometer, which has criticised the low impact of public social and health services in supporting disadvantaged groups such as the long-term unemployed (e.g. Sosiaalibarometri, 2003). The study suggests that health services respond to the needs of the unemployed people only to a limited extent and the reduced supply of and demand on health services for the unemployed have a considerable impact on public health.

References

Arinen, S., Häkkinen, U., Klaukka, T., Klavus, J., Lehtonen, R. & Aro, S. (1998). *Health and the use of health services in Finland.* Helsinki: National Research and Development Centre for Welfare and Health.

Arnetz, B.B., Brenner, S.-O., Levi, L., Hjelm, R., Petterson, I.-L., Wasserman, J. et al. (1991). Neuro-endocrine and immunologic effects of unemployment and job insecurity. *Psychotherapy and Psychosomatics, 55,* 76-80.

Arnetz, B.B., Wasserman, J., Petrini, B., Brenner, S.-O., Levi, L., Eneroth, P. et al. (1987). Immune function in unemployed women. *Psychosomatic Medicine, 49*(1), 3-12.

Aromaa, A. & Koskinen, S. (Eds.). (2004). *Health and functional capacity in Finland. Baseline results of the health 2000 health examination survey.* Publications of the National Public Health Institute B12. Helsinki: Hakapaino Oy.

Claussen, B. (1999). Alcohol disorders and re-employment in a 5-year follow-up of long-term unemployed. *Addiction, 94*(1), 133-138.

Cook, D., Bartley, M., Cummings, R. & Shaper, A. (1982). Health of unemployed middle-aged men in Great-Britain. *Lancet, 5,* 1290-1294.

Dahl, E. (1993). Social inequality in health – the role of the healthy worker effect. *Social Science and Medicine, 36*(8), 1077-1086.

Dooley, D., Catalano, R. & Hough, R. (1992). Unemployment and alcohol disorder in 1910 and 1990: Drift versus social causation. *Journal of Occupational and Organisational Psychology, 65,* 277-290.

Ferrie, J., Martikainen, P., Shipley, M., Marmot, M., Stansfeld, A. & Smith, G. (2001). Employment status and health after privatisation in white collar civil servants: Prospective cohort study. *British Medical Journal, 322,* 647-651.

Grzywacz, J.G. & Dooley, D. (2003). "Good jobs" to "bad jobs": Replicated evidence of an employment continuum from two large surveys. *Social Science and Medicine, 56,* 1749-1760.

Hammarström, A., Janlert, U. & Theorell, T. (1988). Youth unemployment and ill health: Results from a 2-year follow-up study. *Social Science and Medicine, 26*(10), 1025-1033.

Hintikka J., Koskela, T., Kontula, O., Koskela, K. & Viinamäki, H. (1999). Men, women, and marriages: Are there differences in relation to mental health? *Family Therapy, 26*(3), 213-218.

Häkkinen, U., Rosenqvist, G. & Aro, S. (1995). *Economic depression and the use of physician services in Finland. Themes from Finland.* Helsinki: National Research and Development Centre for Welfare and Health.

Häkkinen, U. (2002). Change in determinants of use of physician services in Finland between 1987 and 1996. *Social Science and Medicine, 55,* 1523-1537.

Iversen, L., Sabroe, S. & Damsgaard, M. (1989). Hospital admissions before and after shipyard closure. *British Medical Journal, 299,* 1073-1076.

Janlert, U. & Hammarström A. (1992). Alcohol consumption among unemployed youths: Results from a prospective study. *British Journal of Addiction, 87,* 703-714.

Jin, R., Shah, C. & Svoboda, T. (1995). The impact of unemployment on health: A review of the evidence. *Canadian Medical Association, 153*(5), 529-540.

Kasl, S.V., Cobb, S. & Brooks, G. (1968). Changes in serum uric acid and cholesterol levels in men undergoing job loss. *Journal Of the American Medical Association, 206*(7), 1500-1507.

Kasl, S.V. & Cobb, S. (1970). Blood pressure changes in men undergoing job loss: A preliminary report. *Psychosomatic Medicine, 32*(1), 19-38.

Klavus, J. (2000). *Empirical studies on the measurement of distribution in health care.* Stakes research report 108. Saarijärvi: Gummerus.

Kraut, A., Mustard, C., Walld, R. & Tate, R. (2000). Unemployment and health care utilization. *Scandinavian Journal of Work Environment and Health, 26*(2), 169-177.

Levenstein, S., Smith, M. & Kaplan, G. (2001). Psychosocial predictors of hypertension in men and women. *Archives of Internal Medicine, 161*(10), 1341-1346.

Malo, M.A. & García-Serrano, C. (2001). *An analysis of the employment status of the disabled persons using the ECHP data.* Retrieved from www.employment-disability.net.

Martikainen, P.T. & Valkonen, T. (1996). Excess mortality of unemployed men and women during a period of rapidly increasing unemployment. *Lancet, 348*(5), 909-912.

Mattiasson, I., Lindgärde, F., Nilsson, J.Å. & Theorell, T. (1990). Threat of unemployment and cardiovascular risk factors: Longitudinal study of quality of sleep and serum cholesterol concentrations in men threatened with redundancy. *British Medical Journal, 301*, 461-466.

Montgomery, S.M., Cook, D.G., Bartley, M.J. & Wadsworth, M.E. (1998). Unemployment, cigarette smoking, alcohol consumption and body weight in young British men. *European Journal of Public Health, 8*, 21-27.

Morris, J.K., Cook, D.G. & Shaper, A.G. (1994). Loss of employment and mortality. *British Medical Journal, 308*, 1135-1139.

Nyman, J. (2002). *Does unemployment contribute to ill-being: Results from a panel study among adult Finns, 1989/90 and 1997*. Publications of the National Public Health Institute, A4. Helsinki: Hakapaino Oy.

OECD (1982). *The OECD list of social indicators*. Paris: OECD.

Power, C. & Estaugh, V. (1990). Employment and drinking in early adulthood: A longitudinal perspective. *British Journal of Addiction, 85*, 487-494.

Sherbourne, C.D. (1988). The role of social support and life stress events in use of mental health services. *Social Science Medicine, 27* (12), 1393-1400.

Shortt, S. (1996). Is unemployment pathogenic? A review of current concepts with lessons for policy planners. *International Journal of Health Services, 26*(3), 569-589.

Sosiaalibarometri (2003). [Sociobarometer]. Helsinki: The Finnish Federation for Social Security and Health.

Statistical Yearbook of Finland (2001). *Statistics Finland*. Keuruu: Otavan Kirjapaino Oy.

Statistical Yearbook on Social Welfare and Health Care (2003). *National research and development centre for welfare and health*. Saarijärvi: Gummerus Kirjapaino Oy.

Studnicka, M., Studnicka-Benke, A., Wögerbauer, G., Rastetter, D., Wenda, P., Gathmann, P. & Ringel, E. (1991). Psychological health, self-reported physical health and health service use. *Social Psychiatry and Psychiatric Epidemiology, 26*, 86-91.

Vahtera, J. & Pentti J. (1997). Uhkia vai mahdollisuuksia? [Threats or possibilities?]. Helsinki: Finnish Institute of Occupational Health.

Virtanen, P. (1993) Unemployment, re-employment and the use of primary health care services. *Scandinavian Journal of Primary Health Care, 11*, 228-233.

Virtanen, P. (1994). An epidemic of good health at the workplace. *Sociology of Health and Illness, 16*(3), 394-401.

Yuen, P. & Balarajan, R. (1989). Unemployment and patterns of consultation with the general practitioner. British Medical Journal, 298, 1212-1214

Job Insecurity, Centrality of Work and Relations to Work-Related Attitudes

Kathleen Otto[1] *& Claudia Dalbert*[2]

Background

In the past, occupational biographies were typically characterised by full-time employment within a single organisation; people qualified for an occupation early in life and pursued that occupation until their retirement (Arthur, 1994). These "bounded" careers, and the feelings of security and stability associated with them (Goffee & Scase, 1992), have become increasingly rare over the past decades (Arthur & Rousseau, 1996). Information technology and increasing global competition (e.g., Blossfeld & Mayer, 1988; Burchell, 1992) have brought radical changes in both organisational structures and individual employees' careers (e.g., Briscoe et al., 2006; Brown, 1995). It has been suggested that 21st century careers will be "protean" (Briscoe et al., 2006), meaning that they will be driven by individuals themselves, rather than by a single organisation. Many occupational biographies now involve unemployment spells, temporary layoffs, involuntary part-time (under-) employment, temporary employment contracts and casual labour. Thus, careers have become less predictable, less structured, and consequently less secure (Arnold, 2001).

In Germany, the positive economic developments of the past few years have not yet led to a consistent drop in unemployment (e.g., Institut für Wirtschaftsforschung Halle, 2006). On the contrary, unemployment rates remain high, at approximately 10 per cent in the west and 19 per cent in the east of the country (Microcensus Germany, 2007). There are substantial differences in the unemployment rates of the individual federal states, ranging from a low of 7 per cent to a high of more than 20 per cent (Federal Statistical Office and the statistical Offices of the Länder, 2007).

Despite these changes in occupational structures, gainful employment remains highly valued (e.g., Jahoda et al., 1994). It serves vital functions such as providing a livelihood, structuring the daily routine, strengthening self-confidence, and offering social rewards (i.e., co-operation, social contact, appreciation). Being accepted by superiors, subordinates or colleagues, and feeling self-confident and proud of one's performance can help to satisfy individuals' basic needs within an employment context. Furthermore, unemployment is commonly associated with negative stigmas (Kulik, 2001)

1 University of Leipzig, Germany
2 Martin Luther University of Halle-Wittenberg, Germany

such as laziness or apathy. It may even be seen as a 'contagious disease' (Letkemann, 2002). In fact, numerous studies have identified unemployment to be a major stressor that may impair mental health (e.g., Axelsson & Ejlertsson, 2002; Bergmann, 1994; Donatella & Maass, 2000; Dzuka & Dalbert, 2002; Frese & Mohr, 1978; Jackson, 1999; Payne & Hartley, 1987; Wang, 2004; for a review, see Häfner, 1990). Such negative consequences seem to become more severe as a function of the length of the unemployment spell (Warr & Jackson, 1984, 1987). Note, however, that only gainful employment exerts a positive influence on mental health, while purely voluntary work, i.e. work on a non-paid basis as housework (Richter & Nitsche, 2002; for a review, see Mohr & Otto, 2005) does not.

1 Job insecurity: Summarising the literature

Labour market changes have also put more employed individuals at risk of losing their jobs (Sverke & Hellgren, 2002). It is now difficult to open a newspaper or watch television without encountering reports of mass layoffs. In particular, organisational downsizing in the automobile industry, the telecommunications sector and the manufacturing sector is a frequent topic of media discussion. In fact, downsizing "appears to be the standard solution in organisational attempts at improving organisational effectiveness and reducing labour costs" (Sverke & Hellgren, 2002, p. 25f). Unsurprisingly, downsizing increases job insecurity among employees (e.g., Kivimäki et al., 2001). Thus, job insecurity is associated with processes of organisational change (Klandermans & van Vuuren, 1999).

In general, job insecurity is defined as the felt powerlessness of an individual to sustain continuity in a threatening work situation (Greenhalgh & Rosenblatt, 1984). This definition highlights the subjective character of job insecurity (Sverke et al., 2006). The focus is not on the actual risk of losing one's job, but on the subjectively perceived level of job insecurity, which may be even higher in those who later "survive" downsizing than in those who are actually made redundant (e.g., Kivimäki et al., 2000). Although job insecurity has not been investigated as extensively as unemployment itself, it seems to have very similar negative consequences for the individual (Otto & Dalbert, 2005). Specifically, research has shown that the threat of job loss can be seen as a psychological stressor (e.g., Sverke et al., 2002; Roskies et al., 1993) that may be associated with a decrease in self-reported mental health (for a review, see de Witte, 1999), with anxiety, psychosomatic complaints, fatigue symptoms, depressed mood and even depressive symptoms (e.g., Hellgren et al., 1999; Kivimäki et al., 2001; Mohr, 2000; Pelfrene et al., 2003; Price et al., 2002), and with decreased self-esteem (Kinnunen et al., 2003).

The negative consequences of job insecurity for work-related attitudes and behaviour have also been investigated. Studies indicate that job insecurity is negatively associated with work-related attitudes such as job involvement, job satisfaction, and organisational commitment (for a review, see Sverke et al., 2002). Conversely, high job security is associated with higher organisational commitment and lower intentions to leave the organisation (Finegold et al., 2002). In the same vein, the threat of job loss has been found to be positively correlated with turnover intentions (Hellgren et al., 1999). Given that recent research indicates that withdrawal behaviour - including turnover intentions -

costs approximately 16.5 per cent of a company's pre-tax income each year (Sagie & Birati, 2002), these findings should be of great interest to organisations.

Some employers think that they can achieve higher performance by threatening their employees with job loss - the threat is assumed to spur employees to work harder in the hope of securing their jobs. In line with this argumentation, an experiment by Probst (2002) indicated that a group threatened by layoffs was more productive than a control group. Importantly, however, the threatened group also exerted less work effort and the quality of their work was lower. In addition, it is unclear whether this result can be generalised to the world of work. A meta-analysis summarising several correlational studies failed to show that job insecurity is related to job performance in any way (Sverke et al., 2002).

There has been an extended debate in the scientific literature about the changing value of work (for a review, see Neuberger, 1994). Whereas hard work, a conservative orientation, and loyalty towards the organisation were the key values in the 1940s and 1950s, aspects such as flexibility, independence, relaxation and leisure activities (work-life balance) are now more central (e.g., Inglehart, 1989; Simon et al., 1992). Gebert & von Rosenstiel (2002) argue that, on a global level, there has been an increase in employees' 'post-materialistic' values over the last decades, and a decrease in materialistic values.

Nevertheless, as noted above, gainful employment serves a number of adaptive functions (e.g., Jahoda et al., 1994). A paid job remains central to most people's lives. Not all individuals evaluate employment in the same way, however. Those with a high career orientation may evaluate their work as more important than their private life, whereas those with a stronger family orientation may put work in second place (Moya et al., 2000). The construct used to measure the importance of work is known as work involvement or centrality of work (Kanungo, 1982) and is defined as an emotional dimension reflecting work as the key task in life (Lodahl & Kejner, 1965). We expected centrality of work as a general attitude towards work to be more or less independent of respondents' perceptions of their current job security and the actual labour market situation (Moser & Schuler, 1993), and hypothesised that the maladaptive effects of job insecurity would be particularly strong in individuals who value work as an important part of their life.

In sum, although several studies have confirmed the negative effects of job insecurity on both mental health (De Witte, 1999) and work-related attitudes (Sverke et al., 2002), we sought to broaden the scope of this research by investigating some work-related attitudes that have not yet been a focus of research in this context. Specifically, we investigated how job insecurity relates to occupational commitment (Study 1), career satisfaction and willingness to make concessions in working life (Study 2). In addition, we explored the interplay between job insecurity and centrality of work in explaining these three work-related attitudes.

2 Study 1: Research design and data

In our first study, we explored the relationship between job insecurity, centrality of work and occupational commitment. The data were gathered in 2002 and 2003 by questionnaires distributed in various institutions and through private contacts. Our sample comprised 227 employees (59 per cent female) in 13 federal states. The mean age was approximately 34 years; 145 respondents were born and lived in eastern Germany and 80 in western Germany.

We measured *job insecurity* using the item "The risk of losing my job is high" from the Job Descriptive Questionnaire (Neuberger & Allerbeck, 1978), which was developed on the basis of the Job Descriptive Index (Smith et al., 1969). Responses were given on a scale ranging from 1 (= strongly disagree) to 6 (= strongly agree). The average job insecurity rating was slightly below the theoretical scale mean ($M = 2.5$). East Germans perceived a significantly higher risk of losing their jobs ($M = 2.7$) than West Germans ($M = 2.2$; $p < .01$). This difference reflects the particularly difficult labour market situation in eastern Germany. It is in line with previous research indicating that eastern Germans have stronger preferences for secure jobs than western Germans (Maier et al., 1994). Note that there were no substantial differences between the two groups in terms of the length of the current period of employment (tenure) or the position in organisational hierarchy (i.e., whether or not they held leadership positions).

Our respondents worked in different occupational fields: (a) as white-collar or blue-collar workers in the automobile industry, (b) as skilled craftsmen (e.g., roofers, bricklayers or locksmiths), (c) in banking or the information and communications sector (e.g., in call centres), and (d) in various professions (e.g., engineers, teachers, lawyers, physicians). Casual inspection showed the occupational profiles of the eastern and western subgroups to be equivalent. Furthermore, the proportion of respondents in each subgroup who had completed professional training was equal, as was the frequency of unemployment.

There was only one significant difference between the two subsamples: the eastern German employees had somewhat higher educational qualifications than their western German counterparts ($p < .01$). On average, the eastern Germans had a high school diploma based on 12-13 years of education, whereas the western Germans had a similar diploma based on 11 years of education. Research has shown that the actual risk of unemployment decreases with the level of educational qualification (Christensen, 2001). Nevertheless, although the eastern Germans had somewhat better educational qualifications than the western Germans, they reported higher levels of perceived job insecurity. In fact, our results showed that job insecurity and educational level were independent of each other.

3 Job insecurity and occupational commitment

Occupational commitment (see, e.g., Blau, 2003; Meyer et al., 1993) is defined as the "psychological link between a person and his or her occupation that is based on an affective reaction to that occupation" (Lee et al., 2000, p. 800). Developing identification with a specific occupation is one of the most important developmental tasks in adolescence (Erikson, 1976; Havighurst, 1972), and commitment to this occupation later serves as a point of reference for vocational decisions (Heinz, 2002). In a flourishing economy, strong occupational commitment may provide companies with loyal employees, but precisely this attribute may be of limited use to employees when the labour market situation is poor as the case is in eastern Germany. Research has shown that strong occupational commitment decreases the likelihood of changing occupations (Blau, 2000). It can thus be concluded that individuals with strong occupational commitment may avoid making the adaptations necessary to remain competitive in the labour market.

Although occupational commitment is a work-related attitude that may affect employability (Watts & Sultana, 2004), not one of the 50 studies on job insecurity covered in a recent meta-analysis (Sverke et al., 2002) used it as an outcome variable. For this reason, our first study investigated the relationship between job insecurity and occupational commitment.

We assessed *occupational commitment* by averaging 3 items: "I really have an interesting occupation", "I would change my occupation if I could" (reversed coded), and "My job is absolutely my dream job" (α = .86). The scale was developed ad hoc and mainly on the basis of a questionnaire by Weyer et al. (1980) measuring subjective strain and dissatisfaction at work. Again, responses were given on a scale ranging from 1 (= strongly disagree) to 6 (= strongly agree). Respondents' occupational commitment was well above the theoretical scale mean (M = 4.6).

Comparison of the two subsamples revealed that the eastern Germans showed higher occupational commitment (M = 4.8) than the western Germans (M = 4.4; p = .02). The finding that the subsample of eastern German employees reported both higher perceived job insecurity and stronger occupational commitment might seem surprising. However, it is in line with findings from a recent study showing that, despite the more difficult labour market situation, eastern Germans report more organisational commitment than their counterparts in the west (Rigotti et al., 2007). As expected, perceived job insecurity was negatively associated with occupational commitment ($r = -.18$; $p < .05$).

4 Job insecurity, centrality of work and occupational commitment

We used the four-item work involvement subscale of the German involvement scale by Moser & Schuler (1993; α = .67) to assess *centrality of work* as defined by Lodahl & Kejner (1965). The subscale included items such as "The most important things I experience have to do with my work" and "Most things in life are more important than work" (reversed coded). Again, responses were given on a scale ranging from 1 (= strongly disagree) to 6 (= strongly agree). The average centrality of work rating was below the theoretical mean and nearly similar in both subsamples (West: M = 2.9; East: M = 3.0). As expected, job insecurity - as a measure reflecting the perceived job situation - and centrality of work - as a general work-related attitude - were uncorrelated in the two subsamples.

We further explored the negative relationship between job insecurity and occupational commitment by considering the centrality of work as a moderator variable. We argued that the negative influence of job insecurity would be stronger for employees with high centrality of work than for those with low centrality of work. In other words, we expected the negative relationship between job insecurity and occupational commitment to be stronger in respondents with high centrality of work. To test this hypothesis, we conducted a moderated multiple regression analysis (see, Aiken & West, 1991), regressing occupational commitment on job insecurity, centrality of work and the interaction term of the two. Because we found a significant correlation between regional background (eastern vs. western Germans) and occupational commitment, we controlled for a possible linear effect of regional background. The continuous variables were centred before the interaction term was calculated (see Aiken & West, 1991). The results of the moderated regression analysis are presented in Table 1.

Table 1. Hierarchical regression analysis for occupational commitment on regional background, job insecurity and centrality of work (N = 227)

Variable	*B*	*SE B*	β
Step 1			
Regional background (1 = west)	-0.39	0.18	-.15*
Constant	-4.76		
Step 2			
Regional background (1 = west)	-0.48	0.18	-.19**
Job insecurity	-0.19	0.06	-.21**
Constant	-4.79		
Step 3			
Regional background (1 = west)	-0.45	0.17	-.18*
Job insecurity	-0.18	0.06	-.20**
Centrality of work	-0.31	0.09	.24***
Constant	-4.78		
Step 4			
Regional background (1 = west)	-0.38	0.17	
Job insecurity	-0.19	0.06	
Centrality of work	-0.31	0.08	
Job insecurity x Centrality of work	-0.22	0.06	
Constant	-4.75		

Note. $R^2 = .02$ for Step 1 ($p < .05$); $\Delta R^2 = .04$ for Step 2 ($p < .01$); $\Delta R^2 = .06$ for Step 3 ($p < .001$); $\Delta R^2 = .06$ for Step 4 ($p < .001$).
* $p < .05$; ** $p < .01$; *** $p < .001$.

Altogether 18 per cent of the variance in occupational commitment was explained by the regression equation. After controlling for the significant effect of regional background revealing eastern Germans to be more committed to their occupation than western Germans, findings showed that job insecurity, centrality of work and their interaction significantly predicted occupational commitment. The implications of this interaction are illustrated in Figure 1, which shows regression lines for employees with high

(M + SD = 3.9) and low (M – SD = 2.0) centrality of work when regional background was controlled. In line with our hypothesis, job insecurity was negatively associated with occupational commitment for employees with high centrality of work, but the same effect was not observed for those with low centrality of work. In fact, job insecurity did not explain the occupational commitment of this subgroup. In other words, only respondents with high centrality of work and no or low perceived job insecurity reported high occupational commitment.

We used a single item to measure job insecurity in this study. Previous studies implementing similar measures have shown them to have validity. For example, researchers have asked respondents to gauge the estimated probability of losing their job within the next year on a scale from 0 to 100 (Roskies et al., 1993) or to evaluate the probability of losing the job in the near future on a 5-point scale ranging from highly improbable to highly probable (Mohr, 2000). Unidimensional and multidimensional scales have also been developed to investigate job insecurity (for a review, see Reisel & Banai, 2002). Future studies should apply one of these more reliable measures to assess the relationship between job insecurity and work-related attitudes such as occupational commitment.

5 Study 2: Research design and data

Our second study served three major aims. First, we sought to replicate our findings on the meaning of job insecurity and centrality of work for work-related attitudes. Second, we broadened the scope of our research by addressing two other work-related attitudes, namely career satisfaction and willingness to make concessions in working life. Finally, we used a more reliable measure than the single-item measure administered in Study 1 to assess job insecurity.

The second questionnaire study was conducted between July and October 2006. A total of 245 participants (59 per cent women) completed the questionnaire. Of these participants, 12 per cent were born and lived in western Germany, 79 per cent were in paid employment and 21 per cent were freelancers. The respondents' ages ranged from 20 to 63 years (M = 39). The employed respondents reported that they had been working in their current organisation for an average of about 10 years. The regional unemployment rates in the administrative districts covered by the sample ranged from 3 to 22 per cent (M = 16). Respondents had been unemployed for an average of 3.5 months over the last 10 years, with most respondents (65 per cent) not having experienced unemployment at all.

Job insecurity was assessed by a four-item scale (de Witte, 2000; $\alpha = .87$). It includes, for instance, the following items: "I am concerned about being made redundant", "I feel uncertain about the future of my job". The response scale ranged from 1 (= strongly disagree) to 5 (= strongly agree). As in the first study, the average job insecurity rating was slightly below the theoretical scale mean (M = 2.4). In this sample, the perceived risk of job loss was not significantly associated with either the length of unemployment in the last decade or the current unemployment rate in the administrative district. Given that job insecurity reflects the specific situation of the employing organisation (e.g., downsizing) and one's own experiences as well as the regional unemployment rate, this result is understandable.

Figure 1. Change in occupational commitment as a function of centrality of work and job insecurity

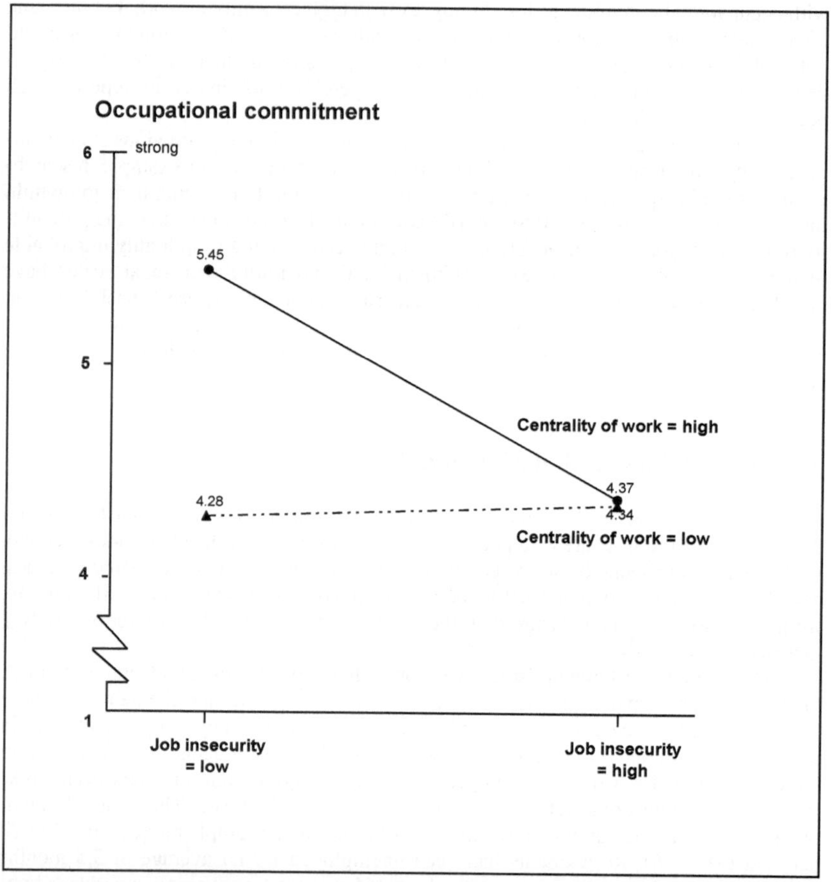

Centrality of work was again assessed by the four-item work involvement subscale of the German involvement scale by Moser & Schuler (1993; $\alpha = .65$). This time the response scale ranged from 1 (= strongly disagree) to 7 (= strongly agree). On average, reported centrality of work was somewhat below the theoretical mean (M = 3.6). Centrality of work was positively associated with the regional unemployment rate ($p < .05$) and negatively associated with perceived job insecurity ($p < .05$). Thus, respondents who felt at risk of losing their jobs also reported that work was less central to their life.

6 Job insecurity, centrality of work and career satisfaction

Career satisfaction can be seen as an important psychological outcome associated with job achievement and accomplishment (Holland, 1997). A perceived threat of job loss may diminish satisfaction with one's career. However, a search of the PsycINFO database failed to identify a single published study relating job insecurity to career satisfaction. Thus, we sought to highlight the relationship between job insecurity and career satisfaction in our second study.

Career satisfaction was operationalised by a well-known scale developed by Greenhaus et al. (1990; α = .84). The 5-item measure included items such as "I am satisfied with the success I have achieved in my career", and "I am satisfied with the progress I have made towards achieving my overall career goals"; responses were given on a 5-point scale ranging from 1 (= strongly disagree) to 5 (= strongly agree). The average career satisfaction rating was somewhat above the theoretical mean (M = 3.5).

In line with our hypotheses, job insecurity and career satisfaction were negatively associated ($p < .05$). In contrast, the regional unemployment rate was not associated with career satisfaction. Thus, respondents who felt at higher risk of losing their jobs reported a less satisfying career trajectory to date. Moreover, as could be expected, centrality of work and career satisfaction were positively correlated ($p < .01$).

Table 2. Hierarchical regression analysis for career satisfaction on job insecurity and centrality of work (N = 245)

Variable	*B*	*SE B*	β
Step 1			
Job insecurity	-0.10	0.04	-.16*
Constant	-3.48		
Step 2			
Job insecurity	-0.08	0.04	-.12
Centrality of work	-0.14	0.04	.23***
Constant	-3.48		
Step 3			
Job insecurity	-0.08	0.04	
Centrality of work	-0.14	0.04	
Job insecurity x Centrality of work	-0.004	0.04	
Constant	-3.48		

Note. R^2 = .02 for Step 1 ($p < .05$); ΔR^2 = .05 for Step 2 ($p < .001$); $\Delta R^2 < .001$ for Step 3 ($p > .05$).
* p < .05; ** p < .01; *** p < .001.

As in Study 1, we again performed a moderated regression analysis with job insecurity, centrality of work and the interaction of the two as potential predictors of career satisfaction. The results of the regression analysis are presented in Table 2. Altogether 7 per cent of the variance in career satisfaction was explained by the regression equation. Job insecurity had a negative impact on career satisfaction, but the relationship was no longer significant when centrality of work was introduced as additional predictor. The interaction between the two was not significant either. Thus, only centrality of work substantially explained career satisfaction. The more respondents valued work a central part of their lives, the more satisfied they were with their career. Contrary to our expectations, however, career satisfaction was not explained by job insecurity.

7 Job insecurity, centrality of work and willingness to make concessions

Finally, one way of adapting to changing occupational demands is to enhance one's employability (Watts & Sultana, 2004) by being willing to make certain concessions, such as working in jobs for which one is overqualified (Brixy & Christensen, 2002). Perceived job insecurity can be assumed to increase people's willingness to accept worse occupational conditions. Again, however, a search of the PsycINFO database failed to identify any study examining the relationship between job insecurity and willingness to make concessions at work. To close this gap in literature, we examined this work-related attitude in our second study.

Table 3. Hierarchical regression analysis for willingness to make concessions on job insecurity and centrality of work (N = 245)

Variable	*B*	*SE B*	β
Step 1			
Job insecurity	-1.04	0.30	-.22**
Constant	-8.50		
Step 2			
Job insecurity	-1.01	0.30	-.22**
Centrality of work	-0.14	0.28	-.03
Constant	-8.50		
Step 3			
Job insecurity	-1.01	0.30	
Centrality of work	-0.14	0.28	
Job insecurity x Centrality of work	-0.45	0.26	
Constant	-8.57		

Note. $R^2 = .05$ for Step 1 ($p = .001$); $\Delta R^2 = .001$ for Step 2 ($p > .05$); $\Delta R^2 = .01$ for Step 3 ($p < .10$).
* $p < .05$; ** $p < .01$; *** $p < .001$.

We assessed the *willingness to make concessions* by presenting respondents with a list of 22 constraints, and asking which they would accept in order to secure their jobs. The items covered concessions of various types, such as "lower salary", "going without a pay rise", "poorer opportunities for career advancement", "longer working hours", "poorer working conditions (e.g., work environment, social climate, noise)" and "working below one's qualification level". Respondents were allowed to select any number of items from the 22 listed. The results of a factor analysis for dichotomous data (see Kubinger, 2003) indicated that all the concessions represented one dimension. Thus, all 22 items were summed. Respondents chose 8.5 items on average, meaning that they would accept more than one third of the constraints listed. The most frequently endorsed item was "going without a pay rise" (72.2 per cent); the least frequently endorsed item was "poorer working conditions" (9 per cent).

Figure 2. **Change in willingness to make concessions as a function of centrality of work and job insecurity**

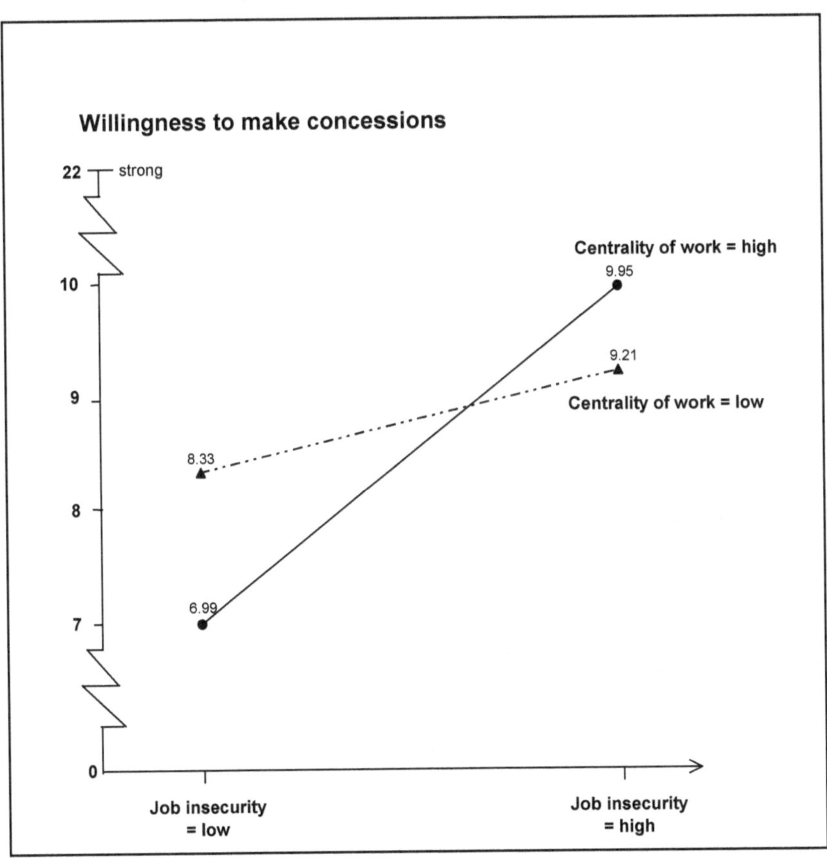

As expected, job insecurity and willingness to make concessions were positively associated ($p = .001$). However, willingness to make concessions was not significantly associated with either the duration of unemployment during the past decade or with the unemployment rate in the administrative district.

Finally, we further explored the positive relationship between job insecurity and willingness to make concessions by considering centrality of work as a moderating variable. We hypothesized that the positive relationship between the two variables would be even stronger in respondents with high centrality of work. To test this hypothesis, we again conducted a moderated multiple regression analysis. The results of this analysis are presented in Table 3.

Overall, 8 per cent of the variance in willingness to make concessions was explained by the regression equation. Job insecurity, as well as the interaction of job insecurity with centrality of work, significantly ($p < .10$) predicted this work-related attitude. The implications of this interaction are illustrated in Figure 2. The higher the respondents' ratings of their job insecurity, the more willing they were to make concessions, but this relation was weaker for individuals with low centrality of work scores.

8 Discussion

We conducted two studies to provide empirical evidence for the effects of job insecurity and its interaction with centrality of work on three work-related attitudes: occupational commitment, career satisfaction and willingness to make concessions. We found that the higher the perceived threat of job loss, the weaker the reported occupational commitment. This relationship was moderated by the centrality of work: only respondents who felt their work to be central to their life reduced their occupational commitment when confronted with job insecurity. The occupational commitment of respondents with weak centrality of work was lower, and this relationship held independently of their perceptions of job insecurity. Why do individuals who strongly endorse the centrality of work decrease their occupational commitment in response to insecurity? We would like to speculate that it may be a problem-oriented coping reaction. Individuals who value their work strongly and perceive a high risk of losing it may be motivated to defend the centrality of work by trying to do something else. A recent study shows that the lower the occupational commitment, the higher the probability of changing occupations (Blau, 2000). Thus, decreased occupational commitment in combination with high centrality of work might foster occupational mobility, the aim being to stay employed.

Our second study revealed a similar pattern of results for the willingness to make concessions. The higher the perceived threat of job loss, the higher the respondents' willingness to make and accept concessions in working life. This relationship was moderated by the centrality of work: Respondents who felt their work to be central to life showed increased willingness to accept worse occupational conditions when confronted with job insecurity. However, although we suggested that willingness to make concessions may help to increase employability (Watts & Sultana, 2004), it is important to bear in mind that not every job is a good one. In Germany, for example, it takes an average of some 13 years to overcome a mismatch between beginning to work below one's qualification level and finding a position that matches one's qualifications (Scherer, 2004). Thus, willingness to make concessions may be one way of staying employed, but it could potentially lead to lower career attainment.

Job insecurity did not explain career satisfaction. In fact, the evaluation of work as being more or less central to one's life seems to be more important than perceived job insecurity. One explanation for this finding is that careers develop over an extended period, during which those who value work highly might work harder to advance their career - resulting in higher satisfaction with their career trajectory.

Nevertheless, our studies have several shortcomings. Firstly, both used a cross-sectional design, meaning that no causal conclusions can be drawn. A longitudinal study could test whether an increase in perceived job insecurity prompts a decrease in occupational commitment or an increase in willingness to make concessions - or whether both occur simultaneously, triggered by the same cause (e.g., demotion). Secondly, all of our data were gathered by means of self-reports, which may have caused carry-over effects.

Our studies suggest that the interplay between subjective job insecurity and centrality of work is important in explaining work-related attitudes. The results are in line with the idea that employees cope with perceived job insecurity by increasing their availability to the job market, namely by reducing their occupational commitment and increasing their willingness to make concessions. To our knowledge, these were the first studies to focus on three neglected work-related attitudes: occupational commitment, career satisfaction, and willingness to make concessions. The results are in line with the notion that these work-related attitudes are very sensitive to changes in the working situation and should be taken into account in future studies investigating the consequences of job insecurity.

References

Aiken, L.S. & West, S.G. (1991). *Multiple regression: Testing and interpreting interactions.* Sage, London.

Arnold, J. (2001). Careers and career management. In N. Anderson, D.S. Ones, H.K. Sinangil & C. Viswesvaran (Eds.), *Handbook of industrial, work, and organisational psychology* (Vol. 2, pp. 115-132). London: Sage Publications.

Arthur, M.B. (1994). The boundaryless career: A new perspective for organisational inquiry. *Journal of Organisational Behavior, 15,* 295-306.

Arthur, M.B. & Rousseau, D.M. (1996). A career lexicon for the 21st century. *Academy of Management Executive, 10,* 28-39.

Axelsson, L. & Ejlertsson, G. (2002). Self-reported health, self-esteem and social support among young unemployed people: A population-based study. *International Journal of Social Welfare, 11,* 111-119.

Bergmann, B. (1994). Erleben und Bewältigen von Arbeitsunsicherheit in Sachsen [Experiencing and coping with job uncertainty in Saxony]. In L. Montada (Ed.), *Arbeitslosigkeit und soziale Gerechtigkeit* (pp. 214-232) . Frankfurt/Main: Campus.

Blau, G. (2000). Job, organisational, and professional context antecedents as predictors of intent for interrole work transitions. *Journal of Vocational Behavior, 56,* 330-345.

Blau, G. (2003). Testing for a four-dimensional structure of occupational commitment. *Journal of Occupational and Organisational Psychology, 76,* 469-488.

Blossfeld, H.-P. & Mayer, K. U. (1988). Arbeitsmarktsegmentation in der Bundesrepublik Deutschland [Labour market segmentation in the Federal Republic of Germany]. *Kölner Zeitschrift für Soziologie und Sozialpsychologie, 40,* 262-283.

Briscoe, J.P., Hall, D.T. & Frautschy DeMuth, R.L. (2006). Protean and boundaryless careers: An empirical exploration. *Journal of Vocational Behavior, 69,* 30-47.

Brixy, U. & Christensen, B. (2002). *Wie viel würden Arbeitslose für einen Arbeitsplatz in Kauf nehmen?* [How much would unemployed individuals put up with to get a job?] (IAB-Kurzbericht Nr. 25). Nürnberg: Bundesanstalt für Arbeit.

Brown, P. (1995). Cultural capital and social exclusion: Some observations on recent trends in education, employment, and the labour market. *Work, Employment, and Society, 9,* 29-51.

Burchell, B. (1992). Towards a social psychology of the labour market: Or why we need to understand the labour market before we can understand unemployment. *Journal of Occupational and Organisational Psychology, 65,* 345-354.

Christensen, B. (2001). Mismatch – Arbeitslosigkeit unter Geringqualifizierten [Mismatch – unemployment among low-qualified workers]. *Mitteilungen aus der Arbeitsmarkt- und Berufsforschung, 34,* 506-514.

De Witte, H. (1999). Job insecurity and psychological well-being: Review of the literature and exploration of some unresolved issues. *European Journal of Work and Organisational Psychology, 8,* 155-177.

De Witte, H. (2000). Arbeidsethos en Jobonzekerheid: Meeting en Gevolgen voor Welzijn, Tevredenheid en Inzet op het Werk. In R. Bouwen, K. De Witte, H. De Witte, & T. Tailleu (Eds.), Van Groep Naar Gemeenschap. Liber Amicorum Prof. Dr. Leo Lagrou (pp. 325-350). Leuven: Garant.

Donatella, M. & Maass, A. (2000). Unemployment and life satisfaction: The moderating role of time structure and collectivism. *Journal of Applied Social Psychology, 30,* 1095-1108.

Dzuka, J. & Dalbert, C. (2002). Mental health and personality of Slovak unemployed adolescents: The impact of belief in a just world. *Journal of Applied Social Psychology, 32,* 732-757.

Erikson, E.H. (1976). *Identität und Lebenszyklus* [Identity and life cycle]. Frankfurt: Suhrkamp.

Federal Statistical Office and the Statistical Offices of the Länder (2007). *Unemployment rate.* [Online]. Retrieved May 15, 2007, from http://www.statistik-portal.de/Statistik-Portal/en/en_jb02_jahrtab 13.asp

Finegold, D., Mohrman, S. & Spreitzer, G.M. (2002). Age effects on the predictors of technical workers' commitment and willingness to turnover. *Journal of Organisational Behavior, 23,* 655-674.

Frese, M. & Mohr, G. (1978). Die psychopathologischen Folgen des Entzugs von Arbeit: Der Fall Arbeitslosigkeit [Psychological effects of losing a job: The case of unemployment]. In M. Frese, S. Greif & N. Semmer (Eds.), *Industrielle Psychopathologie (Schriften zur Arbeitspsychologie, 23).* Bern: Huber.

Gebert, D. & von Rosenstiel, L. (2002). *Organisationspsychologie. Person und Organisation* [Organisational psychology: Person and organisation]. Stuttgart: Kohlhammer.

Goffee, R. & Scase, R. (1992). Organisational change and the corporate career: The restructure of managers' job aspirations. *Human Relations, 45,* 363-385.

Greenhalgh, L. & Rosenblatt, Z. (1984). Job insecurity: Toward conceptual clarity. *Academy of Management Review, 3,* 438-448.

Greenhaus, J.H., Parasuraman, S. & Wormley, W. (1990). Effects of race on organisational experiences, job performance evaluations, and career outcomes. *Academy of Management Journal, 33,* 64–86.

Häfner, H. (1990). Arbeitslosigkeit – Ursache von Krankheit und Sterberisiken [Unemployment – cause of disease and mortality risks]. *Zeitschrift für Klinische Psychologie, 1,* 1-17.

Havighurst, R.J. (1972). *Developmental task and education.* New York: McKay.

Heinz, W.R. (2002). Transition discontinuities and biographical shaping of early work careers. *Journal of Vocational Behavior, 60,* 220-240.

Hellgren, J., Sverke, M. & Isaksson, K. (1999). A two-dimensional approach to job insecurity: Consequences for employee attitudes and well-being. *European Journal of Work and Organisational Psychology, 8,* 179-195.

Holland, J.L. (1997). *Making vocational choices: A theory of careers.* Odessa, FL: Psychological Assessment Resources, Inc.

Inglehart, R. (1989). *Kultureller Umbruch* [Cultural upheaval]. Fankfurt/Main: Campus Verlag.

Institut für Wirtschaftsforschung Halle (2006). *Die Lage der Weltwirtschaft und der deutschen Wirtschaft im Frühjahr 2006* [The situation of the world economy and the German economy in spring 2006]. [Online]. Retrieved October 22, 2006, from http://www.iwh-halle.de/

Jackson, T. (1999). Differences in psychosocial experiences of employed, unemployed, and student samples of young adults. *The Journal of Psychology, 133,* 49-60.

Jahoda, M., Lazarsfeld, P. & Zeisel, H. (1994). *Die Arbeitslosen von Marienthal: ein soziographischer Versuch über die Wirkungen langandauernder Arbeitslosigkeit* [The unemployed of Marienthal: A sociographic essay on the effects of long-term unemployment]. Frankfurt: Suhrkamp.

Kanungo, R.N. (1982). Measurement of job and work involvement. *Journal of Applied Psychology, 67,* 341-349.

Kinnunen, U. Feldt, T. & Mauno, S. (2003). Job insecurity and self-esteem: Evidence from cross-lagged relations in a 1-year longitudinal sample. *Personality and Individual Differences, 35,* 617-632.

Kivimäki, M., Vahtera, J., Pentti, J. & Ferrie, J. E. (2000). Factors underlying the effect of organisational downsizing on health of employees: Longitudinal cohort study. *British Medical Journal, 320,* 971-975.

Kivimäki, M., Vahtera, J., Pentti, J., Thomson, L., Griffiths, A. & Cox, T. (2001). Downsizing, changes at work and self-rated health of employees: A 7-year 3-wave panel study. *Anxiety, Stress, and Coping, 14,* 59-73.

Klandermans, B. & van Vuuren, T. (1999). Job insecurity: Introduction. *European Journal of Work and Organisational Psychology, 8,* 145-153.

Kubinger, K.D. (2003). On artificial results due to using factor analysis for dichotomous variables. *Psychology Science, 45,* 106-110. Available from: http://www.univie.ac.at/Psychologie/diagnostik/Service/software/stat.htm#jmp05

Kulik, L. (2001). Assessing job search intensity and unemployment-related attitudes among young adults: Intergender differences. *Journal of Career Assessment, 9,* 153-167.

Lee, K., Carswell, J.J. & Allen, N.J. (2000). A meta-analytic review of occupational commitment: Relations with person- and work-related variables. *Journal of Applied Psychology, 85,* 799-811.

Letkemann, P. (2002). Unemployed professionals, stigma management, and derivative stigmata. *Work, Employment and Society, 16,* 511-522.

Lodahl, T., & Kejner, M. (1965). The definition and measurement of job involvement. *Journal of Applied Psychology, 49,* 24-33.

Maier, G.W., Rappensperger, G., Rosenstiel, L. v. & Zwarg, I. (1994). Berufliche Ziele und Werthaltungen des Führungsnachwuchses in den alten und neuen Bundesländern [Professional goals and values of future managers in east and west Germany]. *Zeitschrift für Arbeits- und Organisationspsychologie, 38,* 4-12.

Meyer, J.P., Allen, N.J. & Smith, C.A. (1993). Commitment to organisations: Extension and test of a three-component conceptualization. *Journal of Applied Psychology, 78,* 538-551.

Microcensus (2007). *Erwerbstätigkeit* [Gainful employment]. [Online]. Retrieved May 15, 2007, from http://www.destatis.de/cgi-bin/printview.pl

Mohr, G. (2000). The changing significance of different stressors after the announcement of bankruptcy: A longitudinal investigation with special emphasis on job insecurity. *Journal of Organisational Behavior, 21,* 337-359.

Mohr, G. & Otto, K. (2005). Langzeiterwerbslosigkeit: Welche Interventionen machen aus psychologischer Sicht Sinn? [Long-term unemployment: Which interventions make sense from a psychological perspective?] *Zeitschrift für Psychotraumatologie und Psychologische Medizin, 3,* 45-63.

Moser, K. & Schuler, H. (1993). Validität einer deutschsprachigen Involvement-Skala [Validity of a German involvement scale]. *Zeitschrift für Differentielle und Diagnostische Psychologie, 14,* 27-36.

Moya, M., Expósito, F. & Ruiz, J. (2000). Close relationships, gender and career salience. *Sex Roles, 42,* 825-846.

Neuberger, O. (1994). *Personalentwicklung* [Personnel development]. Stuttgart: Enke.

Neuberger, O. & Allerbeck, M. (1978). *Messung und Analyse von Arbeitszufriedenheit* [Measuring and analyzing job satisfaction]. Bern: Huber.

Otto, K. & Dalbert, C. (2005). *Unemployment and job insecurity as threats to mental health and occupational confidence.* Manuscript submitted for publication.

Payne, R. & Hartley, J. (1987). A test of a model for explaining the affective experience of unemployed men. *Journal of Occupational Psychology, 60,* 31-47.

Pelfrene, E., Vlerick, P., Moreau, M., Mak, R. P., Kornitzer, M. & De Backer, G. (2003). Perceptions of job insecurity and the impact of world market competition as health risks: Results from Belstress. *Journal of Occupational and Organisational Psychology, 76*, 411-425.

Price, R.H., Choi, J.N. & Vinokur, A.D. (2002). Links in the chain of adversity following job loss: How financial strain and loss of personal control lead to depression, impaired functioning, and poor health. *Journal of Occupational Health Psychology, 7,* 302-312.

Probst, T.M. (2002). Layoffs and tradeoffs: Production, quality, and safety demands under the threat of job loss. *Journal of Occupational Health Psychology, 7,* 211-220.

Reisel, W.D. & Banai, M. (2002). Comparison of multidimensional and global measure of job insecurity: Predicting job attitudes and work behaviors. *Psychological Reports, 90,* 913-922.

Richter, P. & Nitsche, I. (2002). Langzeiterwerbslosigkeit und Gesundheit – Stabilisierende Effekte durch Tätigkeiten außerhalb der Erwerbsarbeit [Long-term unemployment and health – Stabilizing effects of activities out of gainful employment]. *Zentralblatt für Arbeitsmedizin, 52,* 194-199.

Rigotti, T., Otto, K. & Mohr, G. (2007). East-West differences in employment relations, organisational justice and trust: Possible reasons and consequences. *Economic and Industrial Democracy, 28,* 212-238.

Roskies, E., Louis-Guerin, C. & Fournier, C. (1993). Coping with job insecurity: How does personality make a difference? *Journal of Organisational Behavior, 14,* 617-630.

Sagie, A. & Birati, A. (2002). Assessing the costs of behavioral and psychological withdrawal: A new model and an empirical illustration. *Applied Psychology: An International Review, 51,* 67-90.

Scherer, S. (2004). Stepping-stones or traps? The consequences of labour market entry positions on future careers in West Germany, Great Britain and Italy. *Work, Employment and Society, 18,* 369-394.

Simon, H., Wiltinger, K., Sebastian, K.-H. & Tacke, G. (1995). *Effektives Personalmarketing: Strategien, Instrumente, Fallstudien* [Effective personnel marketing: Strategies, instruments, case studies]. Wiesbaden: Gabler.

Smith, P.C., Kendall, L.M. & Hulin, C.L. (1969). *The measurement of satisfaction in work and retirement.* Chicago, Ill.: Rand McNally.

Sverke, M. & Hellgren, J. (2002). The nature of job insecurity: Understanding employment uncertainty on the brink of a new millenium. *Applied Psychology: An International Review, 51,* 23-42.

Sverke, M., Hellgren, J., & Näswall, K. (2002). No security: A meta-analysis and review of job insecurity and its consequences. *Journal of Occupational Health Psychology, 7*, 242-264.

Sverke, M., Hellgren, J., & Näswall, K. (2006). Arbeitsplatzunsicherheit: Überblick über den Forschungsstand [Job insecurity: Review of the state of research]. In B. Bandura, H. Schellschmidt, & C. Vetter (Eds.), *Fehlzeiten-Report 2005. Arbeitsplatzunsicherheit und Gesundheit* (pp. 59-92). Berlin: Springer Verlag.

van Ruysseveldt, J., Manshoven, J., De Witte, H., van den Bergh, O., Bundervoet, J. & van Hootegem, G. (2003). Stress en welzijn in de banksector. *Over.Werk. Tijdschrift van het Steunpunt WAV, 13,* 174-178.

Wang, J.L. (2004). Perceived work stress and major depressive episodes in a population of employed Canadians over 18 years old. *The Journal of Nervous and Mental Disease, 192,* 160-163.

Warr, P. & Jackson, P. (1984). Men without jobs: Some correlates of age and length of unemployment. *Journal of Occupational Psychology, 57,* 77-85.

Warr, P. & Jackson, P. (1987). Adapting to the unemployed role: A longitudinal investigation. *Social Science and Medicine, 25,* 1219-1224.

Watts, A. G. & Sultana, R. G. (2004). Career guidance policies in 37 countries: Contrasts and common themes. *International Journal for Educational and Vocational Guidance, 4,* 2004.

Weyer, C., Hodapp, V. & Neuhäuser, S. (1980). Weiterentwicklung von Fragebogenskalen zur Erfassung der subjektiven Belastung und Unzufriedenheit im beruflichen Bereich [Development of questionnaire scales to measure subjective strain and dissatisfaction in the occupational context] *Psychologische Beiträge, 22,* 335-355.

Mortality of Unemployed Men and Women in Relation to their Former Occupation in Finland in 1996–2000

Tiina Pensola & Veijo Notkola

Introduction

Research from different countries and different periods has shown that mortality is higher in the unemployed population than in people at work (Moser, Fox et al., 1984; Iversen, Andersen et al., 1987; Stefansson, 1991; Martikainen & Valkonen, 1996; Pensola, Ahonen et al., 2004). It has been shown that this association between unemployment and elevated mortality is related to the experience of unemployment and related factors that have a direct effect on the risk of death (Moser, Fox et al., 1984; Morris, Cook et al., 1994; Mathers & Schofield, 1998). Some of the factors at play are related both to the risk of unemployment and the risk of death; one example is poor health. Furthermore, these factors may be related not only to job loss, but also to the duration of unemployment (Stewart, 2001). Prolonged unemployment has been found to increase mortality even years after the unemployment episode (Jin, Shah et al., 1995; Martikainen & Valkonen, 1996; Pensola, 2003).

The excess mortality found among the unemployed is likely attributable not only to personal characteristics such as age, sex, education and social class, but to macro and meso level factors. For instance, the relationship between unemployment and mortality is bound to be affected by macro-economic factors, such as the national unemployment rate and economic growth, and at the meso level by the occupational unemployment rate (Iversen, Andersen et al., 1987; Martikainen & Valkonen, 1996; Julkunen, 2001).

The occupational unemployment rate may be related to mortality through both selective and causative mechanisms. The likelihood of unemployment and re-employment varies from one occupation to another. If the joblessness rate is high in an occupation where unemployment usually is low, it is possible that the unemployed constitute a more heterogeneous group and less selected group than at times of low unemployment. If there has been only minor selection into unemployment according to health-related factors, it is likely that mortality in this group of the unemployed is not elevated to the same extent as in the occupation where selection effects are substantial. However, if the unemployment rate remains high and opportunities for re-employment therefore low, mortality may increase with a growing duration of unemployment. That may be attributable to selection into long-term unemployment: These people are likely to be older and less educated and to have a lower capacity for work and more chronic diseases. However, unemployment may also have direct and indirect effects on health through its impact on incomes, health-related behaviour, health care, self- esteem and ability to plan for the future. For both reasons, it is likely that the number of those unemployed on a long-term basis in any given occupation will be related to the level of mortality in that group.

The unemployment rate in the overall labour force may affect the occupational mortality of both the employed and unemployed through complex structural and mediating social and biological factors (Bartley & Owen, 1996). Prolonged economic contraction may increase the likelihood of unemployed people with deteriorated health being selected out of the labour force. This may have a positive impact on the health of unemployed persons.

Recession in Finland

Finland experienced severe economic contraction in 1991–95. During this recession the unemployment rate among men increased from 3.6% in 1990 to 8.0% in 1991 and further to 18.1% in 1994; for women the corresponding figures were 2.7, 5.1, and 14.8%. After 1994 the unemployment rate among men began to fall off, but among women the figures continued to rise to 15.1% in 1995. The most significant change was the growth of long-term unemployment. Whereas in 1990 no more than 3% of the jobless population had been out of work for 12 months (which translates to less than one in a thousand of the whole labour force), in 1995 the majority of all those out of work had been unemployed for 12 months or more.

Different occupations were affected by the recession very differently. In some occupations, large numbers lost their jobs due to factory closures and downsizing, while in others selection had greater effect. For instance, the number of police officers, religious professionals, teachers and medical doctors who lost their jobs remained quite small. In some occupations (e.g., among psychiatric nurses and technicians) unemployment was moderate but below the national level, in other occupations the jobless rate was very high: examples include unskilled manual workers, typists and accounting assistants, welders and many occupations in construction.

What is more, the recession did not hit all sectors of the economy at the same time, but it progressed in waves. The first to be affected was the export sector, followed by the private service and public sector. Many banks did not begin downsizing until after 1995, when many industries had already pulled through the worst recession and national GDP was back on a growth track. The occupations that were first affected by unemployment may have a larger proportion of long-term unemployed and therefore a higher mortality rate.

Aims of the study

Because of the high overall level of unemployment, the long duration of the recession and the large number of long-term unemployed, it is important to study the mortality of unemployed persons in Finland. The aims of this study are to:

- analyse occupational mortality among the unemployed;
- find out how the length of unemployment is related to mortality in different causes of death;

- and identify the occupations where mortality among the unemployed is the highest.

1 Data and methods

Compiled by Statistics Finland, the dataset for this analysis consists of longitudinal data from censuses and different registries. It covers all Finnish persons who at the end of 1995 were aged 25–64 and who were in the labour force. At the end of the follow-up the study cohort was aged 30–69. On the whole, the cohort consists of 214,053 unemployed men and 178,649 unemployed women. Among the unemployed men, 31% were aged 25-34, 29% were aged 35-44, 26% were aged 45-54 and 14% were aged 55-64 in 1995. The corresponding figures for women were 29%, 28%, 25% and 18%. Death records by cause of death for 1996-2000 were linked to the data by means of personal identification numbers. Information on occupation, industry and labour force status is based on the situation at the end of 1995.

Occupational information on the unemployed persons is either based on self-report data from Ministry of Labour job applicants' registers or obtained from tax returns. Occupations are classified on the basis of Statistics Finland's occupational classification as used in the longitudinal census data file. This classification is hierarchically organised so that occupations are classified at the 1-, 2-, and 3-digit levels, and some occupations at the 4-digit level. In this study, the 1-digit level is referred to as the industry (main occupational branch) and the 2-digit level as the occupational group; reference to specific occupations at the 3- or 4-digit levels are made only occasionally. The main occupational branches (1-digit level) and person-years for women and men are shown in Table 1. No former occupation was found for 7% of men and 10% of women. Occupation is not known for approximately 1% of all employed persons. Because of differences in the risk of unemployment between different occupations, the distribution of industries among the unemployed differs from that among the employed (see the percentages of person-years for the unemployed in different industries in Table 1). The risk of unemployment also varies by gender partly due to segregation in the labour market. Among both men and women, the numbers out of work were lower than average in technical, health care, pedagogical, managerial and clerical occupations, and higher than average in agriculture and manufacturing. Among men the numbers out of work were also lower than average in sales and service work, whereas among women the corresponding figures were higher than average.

For persons who were out of work in 1995, the total duration of unemployment is based on the number of months they were unemployed in 1993, 1994 and 1995. Those who were out of work for 1 to 12 months during this three-year period are classified as short-term unemployed. The long-term unemployed were divided into two groups: those who were unemployed from 13 to 24 months and those unemployed from 25 to 36 months. The unemployment of persons out of work in 1993 but employed in 1995, is not recorded here. Furthermore, we do not know if an unemployed person was reemployed during the mortality follow-up, that is, during 1996–2000, or on the other hand if someone who was employed at the beginning of the study lost his or her job during the follow-up.

Table 1. Person-years, deaths, unemployment rate[a], and age-standardized death rates (per 100 000 person-years) by main occupational branches for men and women in 1996–2000

Main occupational branches	Person-years in 1000s		Unem-ployment	Deaths	
	Number	%	%[a]	Number	Death rate
MEN					
Technical, health care & pedagogical	123	12	10.4	594	482.1
Managerial & clerical	43	4	10.9	203	433.9
Sales	53	5	17.8	366	651.6
Agriculture & forestry	63	6	41.7	555	804.2
Mining & quarrying	4	0,4	28.6	48	944.9
Transport & communications	83	8	19.1	670	800.4
Manufacturing	550	52	29.8	4932	892.8
Service	51	5	17.5	356	821.8
Unknown	77	7	64.2	645	1003.5
All	1049	100	22.0	8372	801.6
WOMEN					
Technical, health care & pedagogical	160	18	10.1	245	189.9
Managerial & clerical	170	19	15.0	397	240.3
Sales	85	10	21.3	212	207.2
Agriculture & forestry	32	4	49.2	76	226.8
Mining & quarrying	0	---	----	0	----
Transport & communications	16	2	15.0	61	360.0
Manufacturing	112	13	25.7	342	281.6
Service	220	25	24.0	675	302.3
Unknown	92	10	70.2	250	299.7
All	887	100	18.6	2258	256.3

[a] Unemployment rate is based on the number of person-years for the unemployed

During the follow-up 8,372 men and 2,258 women died (see Table 1). Their causes of death are classified according to the Tenth Revision of the International Classification of Diseases (STAKES 1999).

- All causes (A00-Y89)
- All diseases
- Neoplasms (c00-d48)

- Cardiovascular diseases (i00-i425, i427-i52)
- Other diseases
- External causes
- Suicides (x60-x84, y87.0)
- Other accidents and violence

In order to control for the effects of differences in the age structures of the occupational groups, directly age-standardized death rates (DR) were calculated for different occupations. In addition, we have used comparative mortality indexes (MI) (and their 95% confidence intervals), which were obtained by dividing the standardized death rates for the industry or occupational group by the death rate for the reference population (Armitage, 1980; Rothman, 1986; Checkowaj, Pearce et al., 1989). An MI of less than 100 indicates under-mortality in that particular industry or occupational group. Accordingly, an MI of over 100 is indicative of excess mortality in that group. For instance, MI 250 reflects a 150% excess mortality, i.e. mortality is 2.5 times higher than in the reference population.

We have used different reference population for females and males: for females a female labour force and for males a male labour force. This is because the mortality of males is higher than that of females. In the employed population the mortality of males is twice as high as the mortality of females, while in the unemployed population male mortality is three times as high as in females. When the mortality of the unemployed is compared to that of the employed within an industry or occupational group, relative rates ($RR = DR^{unemployed} / DR^{employed}$) are used.

2 Results

On average, the mortality of unemployed males was three times as high as that of employed males, and the mortality of unemployed females twice as high as that of employed females. The higher mortality of the unemployed is also seen in life expectancy figures at age 25: the life expectancy for an employed man at age 25 was 51.4 years and for an employed woman 57.1 years, but for an unemployed man 45.9 years and for an unemployed woman 55.5 years. In other words, unemployment reduced the life expectancy of men by 5.6 years and that of women by 1.7 years.

The mortality of unemployed men and women was higher than average in all branches of industry, but there were also marked industry differences (Figure 1). In the industry with the highest excess mortality (manufacturing in men and service and transport in women), the death rate was approximately twice as high as in the branch with the lowest excess mortality (administration in men and technical, health care and pedagogical work in women). In men, the death rate correlated with the proportion of person-years of unemployed persons in the branch of industry, but among women this association was very weak (see Table 1).

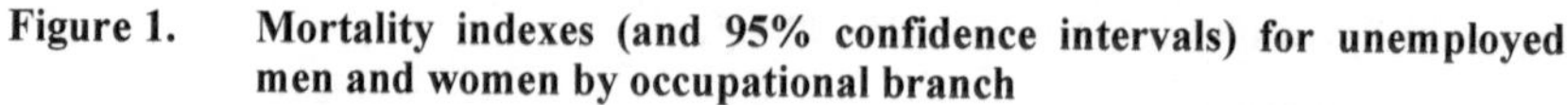

Figure 1. Mortality indexes (and 95% confidence intervals) for unemployed men and women by occupational branch

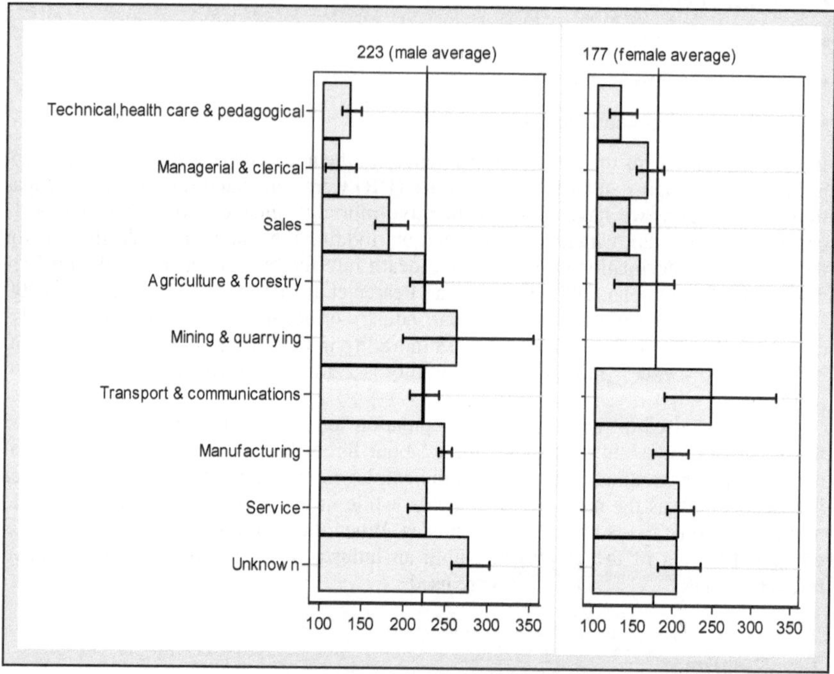

Table 2 and 3 provide mortality information for the occupational groups with the highest excess mortality rates in the whole labour force. The highest mortality rates among unemployed men were found in manual work with lower skills requirements. In these and some other manual or service occupations, the mortality of employed men was also elevated when compared to all employed men (see column 12 in Table 2). Such occupations included paper and cardboard mill workers, electricians, welders, engine mechanics and sheet metal workers, various occupations in construction, postmen and sorters, packers, warehouse workers and waiters. However, although mortality in these occupations was higher than the average for the employed, the mortality of the unemployed was considerably higher (see columns 2 and 12): for instance, the life expectancy for an unemployed unskilled manual worker at age 25 was 10 years shorter than for an employed unskilled manual worker.

Mortality was also elevated among unemployed skilled manual workers and lower and upper non-manual workers. There were only a few occupations that showed no excess mortality in comparison with the whole labour force. These were business managers, other health and nursing professionals such as physiotherapists, computer

Table 2. Mortality indexes (MI = 100, mortality in the whole male labour force) for unemployed and employed men, and proportion of the excess mortality attributable to different causes of death and relative mortality rate (RR) for unemployed men

	Unemployed men											Employed
	Deaths			Proportion (%) of the excess mortality attributable to different causes of death[a]								Deaths
	n	MI	95% CI	All	Diseases	Neo-plasms	Cardio-vascular	Other	External	Suicide	Other	MI
	1	2	3	4	5	6	7	8	9	10	11	12
Deep drilling	11	675	367, 1242	100	106	-4	70	40	-6	-2	-4	103
Transport service work	17	578	332, 1006	100	57	10	23	25	43	5	38	144
Other unskilled manual work	861	514	479, 552	100	60	6	22	32	40	7	33	280
Agricultural and horticultural work, animal husbandry	134	351	296, 418	100	59	12	26	21	41	8	34	123
Deck and engine-room crew	42	351	259, 477	100	73	28	19	26	27	12	15	141
Postal services and couriers	63	306	237, 395	100	62	9	15	38	38	6	32	114
Mining and quarrying work etc.	20	295	188, 462	100	79	4	54	21	21	-6	27	125
Chemical processing, paper and cardboard mill work	58	284	205, 395	100	59	26	22	11	41	9	31	125
Packing and wrapping work	27	283	187, 429	100	55	6	8	42	45	16	29	166
Occupation not specified	645	279	258, 303	100	55	6	18	31	45	10	34	129
Home and institutional housekeeping	70	276	195, 393	100	40	6	18	16	60	22	38	237
Dock and warehouse work	237	272	239, 311	100	56	11	14	31	44	6	37	135
Iron and metal work	1102	263	247, 280	100	55	5	23	27	45	12	33	114
Graphic work	82	255	202, 321	100	65	19	20	26	35	8	27	98
Shoe and leather work	15	254	149, 432	100	33	-13	15	32	67	14	53	74
Building maintenance and cleaning	161	249	212, 291	100	58	12	22	24	42	11	31	113
Other construction work	839	245	228, 263	100	58	6	23	28	42	12	31	147

Painting and lacquering work	210	239	208,274	100	52	10	13	28	48	13	35	110
Smelting, metal and foundry work	30	238	161,352	100	61	19	20	21	39	3	37	113
Cutting, sewing and upholstering occupations	19	236	149,372	100	42	-6	16	32	58	37	21	96
Watchmen and security guards	44	235	172,322	100	77	20	31	26	23	3	20	96
Forestry workers and log floaters	280	221	194,250	100	66	17	32	17	34	15	19	120
Road transport work	509	212	194,232	100	63	13	25	25	37	15	23	125
Sales representatives	101	210	172,257	100	88	9	45	34	12	6	6	95
Stationary engine and machine work	292	209	186,236	100	71	14	30	28	29	9	20	109
Restaurant service work	43	207	145,296	100	14	-10	1	23	86	38	48	146
Electrical work	286	199	176,224	100	71	9	29	33	29	7	21	118
Traffic supervising work	15	198	113,348	100	26	-21	12	35	74	14	60	92
Food and beverage work	59	194	148,254	100	41	13	23	5	59	43	17	104
Rubber, plastic or cement products machine operators	93	191	154,236	100	64	25	19	21	36	11	25	106
Other animal husbandry and fishing	17	189	117,308	100	50	14	6	29	50	11	39	101
Precision mechanics	19	189	119,300	100	48	10	-9	47	52	31	22	94
Ship's officers	15	187	105,335	100	52	19	-3	36	48	0	48	128
Shop and market salespersons and demonstrators	205	181	157,209	100	74	17	15	43	26	11	15	83
Managerial work in agriculture, forestry and horticulture	125	172	143,208	100	67	11	41	15	33	13	19	74
Wood work	685	172	159,186	100	52	6	22	23	48	18	30	108
Travel guides, undertakers, sports persons etc.	32	167	118,237	100	48	0	5	43	52	28	24	114
Other mining and quarrying work	16	165	100,273	100	24	32	20	-28	76	29	48	148
Artistic, literary and entertainment work	67	160	125,205	100	51	13	-3	41	49	11	37	90
Chemistry, physics and biology work	34	159	111,227	100	85	-3	40	48	15	-5	21	81
Accounting, book keeping, statistical & finance clerks	94	140	113,173	100	56	-7	5	59	44	6	38	91
Supervision and executive work in the technical field	267	138	121,158	100	93	30	27	36	7	-7	14	82
Accountants, social science, archivists, librarians etc.	61	130	101,168	100	47	-27	20	53	53	37	17	77
Technical planning, administrative and research work	107	127	104,154	100	64	-7	11	60	36	-15	51	65
ALL	8372	223	218,229	100	60	8	23	29	40	11	30	71

[a,b,c] see Table 3

operators and education methods specialists. However, mortality among the unemployed in these occupations was higher than among employed men. For instance, mortality among unemployed business managers (MI 85 (51,140) 21 deaths) was 57% higher than among employed business managers (MI 54 (44,66) 116 deaths). Furthermore, although the unemployed in the health sector did not have a higher than average mortality rate, the mortality of unemployed psychiatric nurses was very high (MI 1208 (217-6710), 6 deaths).

The excess mortality of unemployed men was attributable to different causes in different occupations. On average, 60% of the excess mortality of unemployed men could be attributed to their higher mortality from diseases. The contribution of neoplasms was minor, approximately 8%. In a few occupations, however, men showed a higher mortality rate from neoplasms: these included deck and engine room crew and machine operators working with rubber, plastic and cement products. Mortality from cardiovascular diseases was high among unemployed men, with the exception of men who had been engaged in planning, administrative and research work in technical fields and in accounting, book keeping, statistical and finance work. Almost one-third or 29% of total excess mortality was attributable to "other diseases", which mainly consist of alcohol-related diseases. Approximately 40% of total excess mortality in this group can be attributed to external causes. There were only some upper non-manual occupations where unemployed men did not have an elevated suicide risk. Compared to the whole male labour force, mortality from suicides was 2.1 times higher and mortality from other external causes 2.9 times higher among unemployed men.

In women, most occupations with the highest mortality rates were less skilled manual occupations in manufacturing and the service sector. The mortality of unskilled manual workers was also elevated among employed women, but among unemployed women it was high even when compared to other unemployed women. When compared to employed unskilled manual workers, the mortality of unemployed women in this occupational group was seven times higher, and their life expectancy at age 25 was 5.5 years shorter. In many other occupations the mortality of employed women did not deviate from the average for all employed women (column 12), but was much lower than for unemployed women in the same occupational group (column 2 and 12). The one exception to the rule was restaurant service work – the mortality of both employed and unemployed waitresses was at the same level.

There were only very few occupations where unemployed women did not show excess mortality in comparison with the whole female labour force. These occupations were mainly in chemistry, physics and biology work, in pedagogic work (teachers) and in health care work. However, even in these occupations the mortality of the unemployed clearly exceeded the mortality rates for employed women.

On average, 67% of the excess mortality of unemployed women can be attributed to diseases. Half of this is attributable to other diseases than neoplasms and cardiovascular diseases, the two major causes of death among employed women. As observed for men, 'other diseases' in women consist mainly of alcohol-related causes. It was not possible in this dataset to distinguish alcohol-related diseases from accidental alcohol poisonings, which are included among external causes. Of all the deaths among unemployed women, 18% were related to alcohol. The contribution of 'other diseases' varied between occupational groups. For instance, it was greater among women who had been in hygiene and beauty work. Among these 26 women, 22 had been hairdressers. The mortality of employed hairdressers was almost as high as that of unemployed hairdressers (MI for employed hairdressers 254 and for unemployed hairdressers 301). However, their excess mortality was mainly attributable to neoplasms, whereas the contribution of other diseases to the total excess was greater among unemployed hairdressers.

Table 3. Mortality indexes (CMI = 100, mortality in the whole female labour force) for unemployed and employed women, and proportion of the excess mortality attributable to different causes of death and relative mortality rate (RR) for unemployed women

Occupational group	Unemployed women											Employed
	Deaths			Proportion (%) of the excess mortality attributable to different causes of death[a]								Deaths
	No	MI	95% ci	All	Diseases	Neoplasms	Cardio-vascular	Other	External	Suicide	Other	MI
	1	2	3	4	5	6	7	8	9	10	11	12
Other unskilled manual work	51	688	520, 911	100	76	22	15	39	24	3	21	240
Painting and lacquering work	8	440	211, 917	100	57	18	11	28	43	30	13	101
Postal services and couriers	24	317	205, 490	100	54	14	22	18	46	3	43	117
Laundering, dry cleaning and pressing work	14	313	178, 551	100	88	30	23	35	12	-4	15	58
Building maintenance and cleaning	371	293	262, 327	100	77	19	24	34	23	4	19	100
Chemical processing and related work	8	277	121, 637	100	72	31	29	11	28	15	13	77
Computer work	16	253	154, 415	100	71	-1	20	53	29	-5	34	133
Hygiene and beauty treatment	26	250	168, 373	100	74	0	27	48	26	1	24	254
Woodwork	18	243	148, 400	100	76	60	-3	18	24	4	20	97
Dock and warehouse work	23	242	151, 388	100	54	32	17	5	46	32	14	106
Textile work	17	228	132, 397	100	73	46	18	10	27	5	22	97
Food and beverage work	43	228	161, 322	100	82	24	24	34	18	-2	20	98
Stationary engine and machine work	8	228	102, 508	100	71	7	53	11	29	-6	35	82

Postal and telecommunications work	25	211	130, 343	100	87	63	15	10	13	8	5	83
Packing and wrapping work	26	200	126, 318	100	81	28	44	9	19	0	19	114
Rubber and plastic-products machine operators	16	186	110, 315	100	55	-30	74	10	45	18	27	82
Real-estate, business services and securities	10	179	96, 335	100	56	-16	23	50	44	36	8	89
Secretarial and typing work	62	172	134, 222	100	38	2	-9	45	62	28	34	91
Agricultural and horticultural work	45	169	124, 230	100	79	11	13	55	21	0	21	90
Accounting, book keeping, statistical & finance clerks	269	166	147, 188	100	58	16	14	29	42	14	28	87
Restaurant service work	71	166	131, 211	100	69	11	27	31	31	13	18	123
Accountants, social science, archivists, librarians etc.	79	155	124, 195	100	46	-10	18	38	54	18	36	80
Managerial work in agriculture, forestry, horticulture	30	150	103, 219	100	49	33	15	1	51	15	36	85
Clerical work	32	147	102, 214	100	102	14	2	86	-2	4	-6	87
Home and institutional housekeeping	181	143	123, 167	100	49	-18	31	36	51	21	30	89
Shop and market salespersons and demonstrators	183	140	118, 166	100	89	56	17	15	11	-7	18	84
Medical and nursing work	76	129	101, 166	100	-21	-45	-2	26	121	44	76	74
ALL	2258	177	169, 186	100	67	14	20	33	33	9	24	86

Note. [a] the proportion of the excess mortality that is attributable to a specific cause of death is calculated with a following formula: 100*(d^i - 100*D/CMI * p) / (D – 100*D/CMI)
where, d^i = the observed number of deaths in the cause i, and D is the total number of deaths, MI = the mortality index, and p = the proportion of deaths in the cause I in the whole female labour force.

When compared to the whole female labour force, mortality from suicides among unemployed women was 2.2 times higher and mortality from other external causes 2.8 times higher. On average, 9% of the total excess mortality could be attributed to suicides and 24% to other external causes. For instance, unemployed nurses and social workers had an elevated suicide risk. However, in many occupational groups there were no deaths due to suicides. In general, due to the small numbers of deaths in many occupational groups, mortality patterns varied very widely indeed.

Among the unemployed, mortality was related to the total duration of unemployment spells. Among those men and women who had been out of work for less than 13 months, mortality was approximately 40% higher than in the whole labour force (MI for men 140 (130,150) and for women 133 (120,148)). Mortality among those who had been unemployed from 13 to 24 months was almost at the same level (MI for men 153 (146,161) and for women 133 (122,144). However, in cases where unemployment had lasted from 25 to 36 months, mortality was almost three times higher among men (295 (286,304)) and 2.5 times higher among women (258 (242,275)) than in the sex-specific labour force on average.

The association between the duration of unemployment and mortality was found for various causes of death among both men and women (see Figure 2). There was no mortality difference between those who had been out of work for less than a year and those who had been unemployed from one to two years. However, for all causes of death mortality was significantly higher among those who had been unemployed for at least two years. In other diseases and external causes, the excess mortality of the long-term unemployed is particularly high, more than six times higher as compared to others.

Figure 2. Relative mortality of unemployed men and women compared to the mortality of the employed by length of unemployment in main causes of death

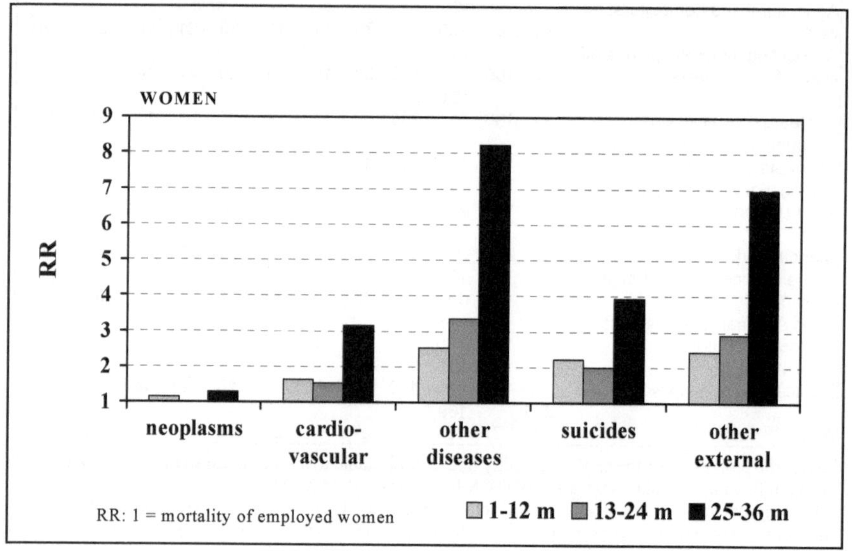

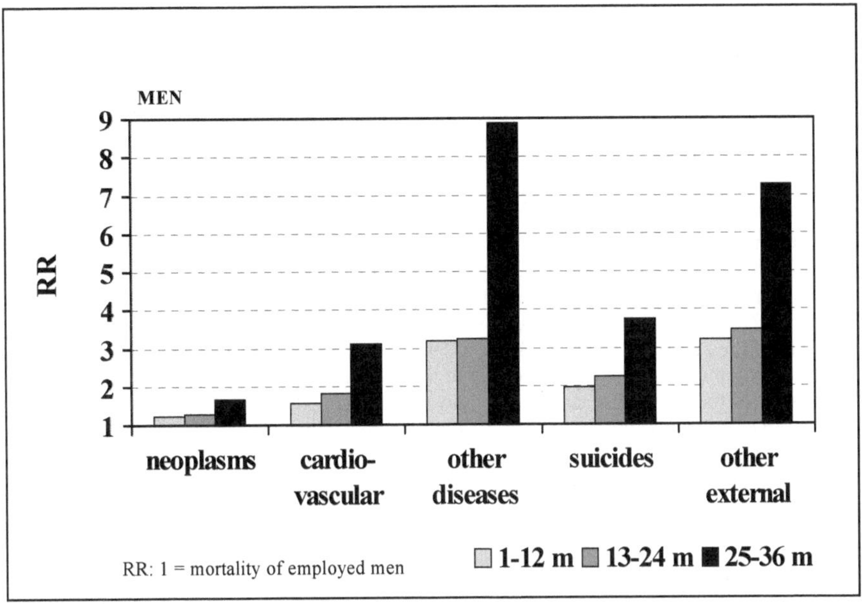

3 Discussion

According to these results the mortality of unemployed women and men was higher than that of employed women and men from all causes of death and in all branches of industry. Among the unemployed, however, there were occupational mortality differences. In men, the highest mortality rate among the unemployed was found in the same occupations as among the employed. In women, the occupations with the highest mortality differed between the employed and unemployed. On average, about 60-70% of the total excess mortality was attributable to diseases and the remaining 30-40% to suicides and accidents and violent causes. However, the contribution of different causes of death to the total excess varied between occupations and sexes.

3.1 Unemployment and mortality

There has been long-standing debate over the possible associations of changes in the unemployment rate with the population mortality level (Brenner & Mooney, 1983; Charlton, Bauer et al., 1987; Valkonen and Martikainen, 1995; Brenner, 2005; Catalano & Bellows, 2005; McKee & Suhrcke, 2005; Tapia Granados, 2005). However, on the basis of observed associations in ecological data, it is not possible to draw conclusions on the effect of unemployment on health and death risk of the jobless. Nevertheless, comprehending the association between unemployment and mortality as such and in relation to mortality of the employed will become more profound if economic cycles are taken into account. Furthermore, coverage of social and health policies, and medical innovations, availability of other resources and changes in standard of living in different population groups are likely to mediate the (short- and long-term) effects of economic changes on health (Catalano & Bellows, 2005; Edwards, 2005; McKee & Suhrcke, 2005). In our cohort the unemployment rate was 18% compared to just 3% in the earlier cohort in 1990. This six-fold increase in unemployment was not associated with mortality at the population level: both the unemployed and employed persons showed a lower mortality and thus a higher life expectancy in 1996-2000 than in 1991-95 (Pensola, Ahonen et al., 2004). However, the change was somewhat smaller among the unemployed, and their excess mortality in comparison to the employed persons increased from 190% to 214%.

Unemployment may be an indicator of a chain of other factors and events with negative health effects, possibly starting from childhood and youth (Pless, Cripps et al., 1989; Spruit, 1989; Fergusson & Horwood, 1998; Pensola, 2003). In this chain, the "experience of unemployment itself is not necessarily the only or the most harmful aspect" of the cumulative effect (Bartley, 1988, p. 62) that is reflected in elevated mortality among the unemployed. The effect of these confounding factors on unemployment risk is likely to be stronger, and thus relative mortality is likely to be higher among the unemployed at times when the overall jobless rate is lower (Iversen, Andersen et al., 1987; Martikainen & Valkonen, 1996; Julkunen, 2001). Conversely, in the manual class the likelihood of unemployment was greater among men with a limiting long-standing illness during recession than at times of economic stability (Bartley & Owen, 1996). This kind of selection into unemployment may have a negative effect on mortality in manual occupations during economic decline. In this study, the occupations showing the highest mortality rates were the same among unemployed and employed men. It is likely that the high mortality of unemployed persons with a certain occupational background partly reflects the hazards directly related to work, but to a greater extent the effects of health-related behaviours and lifestyles common in these occupations. The effect of these risks may be intensified when a person has lost his or her job.

On the other hand, the experience and consequences of unemployment may differ somewhat according to the person's occupational position. Persons in unskilled and semi-skilled jobs in the manufacturing, construction and service sectors are more likely to have insecure and low-paid jobs, poor working conditions, and experiences of unemployment during stable economic situations. During economic contraction, people in the middle social classes will also encounter undesirable job and financial events that are further related to their health (Catalano & Dooley, 1983).

One hypothesis explaining the relationship between unemployment and mortality refers to the psychosocial effects of unemployment on health. It is possible that the psychological effects of unemployment are less severe at times of mass unemployment when losing one's job is not as easily considered a personal and stigmatising failure (Hammarström, 1994). However, even if unemployment is less stigmatising, its possible impacts on psychosocial stress do not necessarily differ from those seen during low unemployment. In addition, a high unemployment level may affect the unemployed person's circumstances in other respects, too, including his or her prospects of re-employment, which in some unemployed groups may be connected to the mortality level as well.

All things considered, it is likely that the changes, which swept the economy in the 1990s, involved various factors with contradicting effects on mortality. At first glance our results are not consistent with the observations, which indicate a lower level of mortality with higher unemployment rates. However, if the unemployed in 1995 are divided into groups according to the duration of their unemployment spells, it turns out that the excess mortality of short-term unemployed men was 97% compared to 315% among men who had been unemployed for over 24 months. In the previous follow-up the contribution of the long-term unemployed to the 190% excess mortality of unemployed men must have been minor due to their very small number. This suggests that the excess mortality of the short-term unemployed might in fact have dwindled in the 1990s. However, we were unable to establish to what extent the lower mortality among the short-term unemployed in this compared to the previous follow-up was attributable to the higher prevalence of the unemployed in 1995 and to what extent to different selection in these two study cohorts.

Although no relationship was seen between the unemployment rate and the population mortality rate, it is probable that macro-economic factors such as structural unemployment, the composition of the labour force and households, and income differences and changes in these factors, have affected health and mortality among the long-term unemployed in our data (Brenner & Mooney, 1983; Wadsworth, 1997).

Long-term unemployment

In these datasets mortality increased in all occupations, and from all causes of death, when unemployment had lasted for two years. The only exception was seen in the category of male finance and administration managers (37 deaths), who did not show excess mortality in comparison to the whole male labour force after 24 months out of work. Their mortality did, however, exceed the mortality of employed managers. There is also earlier evidence on the significance of length of unemployment in terms of mortality (Jin, Shah et al., 1995; Martikainen & Valkonen, 1996; Pensola, 2003).

The high excess mortality of the long-term unemployed may be attributable to various factors. Even if the risk of unemployment is more evenly spread out between different occupations than before, it is unlikely that spells of unemployment have lengthened to the same extent. Even though it has been shown that underlying illness has only a minor effect on selection to unemployment, there is evidence that in the complex relationship between unemployment and illness, unemployment has an effect on health (Jin, Shah et al., 1995; Montgomery, Cook et al., 1999). One possible factor affecting re-employment and thus the length of unemployment is the individual's health, and this relationship varies by social class (Bartley & Owen, 1996).

Regardless of the original reasons lying behind long-term unemployment, the severity of the problems related to unemployment tends to increase with the length of unemployment (Beckett, 1988). In Finland the amount of unemployment benefits decreases substantially after 500 days out of work. In the 1990s changes to unemployment security, social assistance and housing allowances very much undermined the economic security of the unemployed. No inflation adjustments were made to housing allowances in spite of soaring costs of living. As a consequence the economic situation of the long-term unemployed has deteriorated during the follow-up (Kangas & Ritakallio, 2003). On the whole, income differentials have increased in Finland since 1995 (Kautto & Uusitalo, 2004). In another Finnish study concerning the same period as ours, it was found that economic difficulties accounted for the relationship discovered between unemployment and reduced self-esteem and well-being (Ervasti, 2003).

In addition, differences in the health care services available for the employed and unemployed are more pronounced with the prolongation of unemployment. Most people in gainful employment are covered by occupational health services, while the unemployed mainly rely on the public health care system – which is also part of the occupational health service. In addition, unemployment may obstruct persons from detecting their symptoms and seeking care in the early phases of diseases such as breast cancer (Catalano, Satariano et al., 2003). In our study, women who had been unemployed for 24–36 months showed a 30% higher mortality from neoplasms than employed women.

The category of long-term unemployed includes persons from all age groups. However, persons aged 45 and over are overrepresented, and in the age group 55–64 a substantial proportion of the long-term unemployed are women (Martelin, Karvonen et al., 2004). Among these long-term older unemployed are persons who had worked for years or even decades without earlier spells of unemployment. Despite their ample experience, they have a lower level of education and may lack the skills they would need to secure a new job. Furthermore, they may be victims of age discrimination, which affects men and women differently. (Beckett, 1988; Bolinder, 2000; Rouvinen, 2003). The poorer prospects of re-employment in a job equivalent to the one from which one has been made redundant, may have negative effects on health directly or via health-related behaviours. Among persons aged 45 and over, it has been found that even short-term unemployment is associated with reduced working capacity. On the other hand, women aged 30-44 who had experienced a short spell of unemployment, had a higher working capacity than women who had not experienced job loss (Pensola, Järvikoski et al., 2006). It is likely that limited working capacity affects the risk of unemployment and the length of unemployment (Lahelma & Mannila, 1984). However, reversed causation is also possible: unemployment is likely to have negative effects and re-employment positive effects on mental well-being and working capacity (Lahelma, 1989; Pensola, Järvikoski et al., 2006).

Things to be considered in studies on unemployment and mortality

The findings described above may indicate that the relationship between unemployment and health depends on the timing of unemployment in the individual's work history. For instance, for some highly educated young women short-term unemployment may actually provide an opportunity to look for a better job, while for an aged woman unemployment hardly is a voluntarily move and a short-term spell of unemployment may easily become long-term. Mel Bartley pointed out in her paper in 1988 that there is no undifferentiated experience of unemployment which is invariably health damaging in itself; unemployment should always be studied in the context of economic issues, taking into the account individual work histories (Bartley, 1988). In this study we did not have access to these histories. Therefore part of the association between unemployment and mortality may actually reflect the effects of a person' being in a secondary sector characterized by job insecurity, lower wages and poorer working conditions rather than the effects of unemployment per se. Job loss during a period of recession may result in a person having to move to a secondary sector: an unemployment experience increases the probability of subsequent unemployment (Bartley & Plewis, 2002). With respect to these problems, we were unable in this study to make a distinction between those who had been out of work before the recession and those who had lost their job for the first time during the period of high unemployment.

In addition, we followed unemployed persons in 1995 and were unable to detect changes in their labour force status during the follow-up. Therefore employees who were made redundant in 1996–2000 are included in this study as employed persons. This concerns, for instance, many finance associate professionals, tellers and other counter clerks who were made redundant in the latter part of the 1990s. Correspondingly, persons who were unemployed during the recession in 1991–94 but who were employed or out of the labour force (for instance, the long-term unemployed aged 60 or over can apply for unemployment pension) in the census week in 1995 were not followed in this study. Therefore we were unable to study whether persons experiencing long-term unemployment in 1993–95 differed in terms of mortality according to their labour force status in 1995. Findings from other studies have shown that unemployment even several years earlier is associated with an increased risk of limiting long-term illness, poor health and mortality (Grundy & Holt, 2000; Bartley & Plewis, 2002; Pensola, 2003).

In 1995 the official number for the long-term unemployed in Finland stood at almost 140.000 (Martelin, Karvonen et al., 2004). Here the number was over 160.000 because those who were unemployed in the census week and who had been temporarily employed or in labour market training for a few months in 1995 were also regarded as long-term unemployed if they had been out of work for over 12 months during the previous three years. These persons showed elevated mortality in this study: it seems that the category of long-term unemployed should be expanded to include not only persons who have been unemployed for more than a certain number of consecutive months, but also those with a fragmentary working career, i.e. with repeated spells of unemployment (perhaps due to interruptions in labour market training), or who are employed in jobs not commensurate with their skills (Keskitalo & Mannila, 2004).

4 Concluding remarks

The health and mortality consequences of unemployment on a population and individual level are likely to vary according to individual resources and macro-economic factors (Bolinder, 2000; Kurvinen, 2002). The most fecund data will include information not only on age and occupation, but also on the individual's whole employment history. Analyses of unemployment and mortality should always be carried out in a wider macro-social and -economic context. It is likely that in a different economic situation, we would find occupational mortality differences among the unemployed, but it is possible that the occupational groups showing the highest mortality would be partly different from those found here. Predominantly, the numbers who are unemployed on a long-term basis in any given occupation will be related to the level of mortality in that group. In Finland in 1996–2000, long-term unemployed women and men showed particularly high mortality, from both diseases and external causes. This is something that warrants greater public attention in Finland.

Acknowledgements

We are grateful to Statistics Finland for the data used in this paper.

References

Armitage, P. (1980). *Statistical methods in medical research.* Oxford: Blackwell Scientific Publications.

Bartley, M. (1988). Unemployment and health: Selection or causation - a false antithesis? *Sociology of Health and Illness, 10*(1), 41-67.

Bartley, M. & Owen, C. (1996). Relation between socioeconomic status, employment, and health during economic change, 1973-93. *British Medical Journal, 313*, 445-449.

Bartley, M. & Plewis, I. (2002). Accumulated labour market disadvantage and limiting long-term illness: Data from the 1971-1991 Office for National Statistics' Longitudinal Study. *International Journal of Epidemiology, 31*, 336-341.

Beckett, J.O. (1988). Plant closings: How older workers are affected. *Social Work, 33*(1), 29-33.

Bolinder, M. (2000). Arbetslösas anspråk och anspråkens betydelse för chansen att få ett bra jobb. *Sociologisk forskning, 2*, 5-34.

Brenner, M.H. (2005). Commentary: Economic growth is the basis of mortality rate decline in the 20th century-experience of the United States 1901-2000. *International Journal of Epidemiology, 34*(6), 1214-1221.

Brenner, M.H. & Mooney, A. (1983). Unemployment and health in the context of economic change. *Social Science and Medicine, 17*(16), 1125-1138.

Catalano, R. & Bellows, B. (2005). Commentary: If economic expansion threatens public health, should epidemiologists recommend recession? *International Journal of Epidemiology, 34*(6), 1212-1213.

Catalano, R. & Dooley, D. (1983). Health effects of economic instability: A test of economic stress hypothesis. *Journal of Health and Social Behavior, 24*, 46-60.

Catalano, R., Satariano, W. et al. (2003). Unemployment and the detection of early stage breast tumors among African Americans and non-Hispanic whites. *Annals of Epidemiology, 13*(1), 8-15.

Charlton, J.R.H., Bauer, R. et al. (1987). Unemployment and mortality: A small area analysis. *Journal of Epidemiology and Community Health, 41*, 107-113.

Checkowaj, H. & Pearce, N. et al. (1989). *Research methods in occupational epidemiology*. New York: Oxford University Press.

Edwards, R.D. (2005). Commentary: Work, well-being, and a new calling for countercyclical policy. *International Journal of Epidemiology, 34*(6), 1222-1225.

Ervasti, H. (2003). Työttömyys elämäntilanteena. In O. Kangas (Ed.), *From bust to boom. Finnish society in the 1990s* (pp. 119-151). Helsinki: The Social Insurance Institution.

Fergusson, D. & Horwood, L. (1998). Early conduct problems and later life opportunities. *Journal of Child Psychology and Psychiatric and Allied Disciplines, 39*(8), 1097-1108.

Grundy, E. & Holt, G. (2000). Adult life experiences and health in early old age in Great Britain. *Social Science and Medicine, 51*, 1061-1074.

Hammarström, A. (1994). Health consequences of youth unemployment: Review from a gender perspective. *Social Science and Medicine, 38*(5), 699-709.

Iversen, L., Andersen, O. et al. (1987). Unemployment and mortality in Denmark, 1970-80. *British Medical Journal, 295*(6603), 879-884.

Jin, R., Shah, C. et al. (1995). The impact of unemployment on health: A review of the evidence. *Canadian Medical Association Journal, 153*(5), 529-540.

Julkunen, I. (2001). Coping and mental well-being among unemployed youth - a Nothern European perspective. *Journal of Youth Studies, 3*, 261-278.

Kangas, O. & Ritakallio, V.-M. (2003). Moniulotteisen köyhyyden trendit 1990-luvulla. In O. Kangas (Ed.), *From bust to boom. Finnish society in the 1990s* (pp. 49-91). Helsinki: The Social Insurance Institution.

Kautto, M. & Uusitalo, H. (2004). Welfare policy and income distribution: The Finnish experience in the 1990s. In M. Heikkilä & M. Kautto (Eds.), *Welfare in Finland* (pp. 83-102). Helsinki: STAKES - National Research and Development Centre for Welfare and Health.

Keskitalo, E. & Mannila, S. (2004). Activation policy - an answer to the problem of long-term unemployment and exclusion? In M. Heikkilä & M. Kautto (Eds.), *Welfare in Finland* (pp. 103-122). Helsinki: STAKES - National Research and Development Centre for Development and Health.

Kurvinen, A. (2002). Resources and labour market orientation of the unemployed. In P. Koistinen & W. Sengenberger (Eds.), *Labour flexibility - a factor of economic and social performance of Finland in the 1990s*. Tampere: Tampere University Press.

Lahelma, E. (1989). *Unemployment, re-employment and mental well-being*. Helsinki: Rehabilitation Foundation.

Lahelma, E. & Mannila, S. (1984). Pitkäaikaistyöttömien elämäntilanteen eriytyminen [Differentation of the life situation of the long-term unemployed]. *Sosiaalilääketieteellinen aikakauslehti - Journal of Social Medicine, 21*, 285-292.

Martelin, T., Karvonen, S. et al. (2004). Welfare of the working-age population. In M. Heikkilä & M. Kautto (Eds.), *Welfare in Finland* (pp. 55-79). Helsinki: STAKES - National Research and Development Centre for Welfare and Health.

Martikainen, P. & Valkonen, T. (1996). Excess mortality of unemployed men and women during a period of rapidly increasing unemployment. *Lancet, 348*, 909-912.

Mathers, C. & Schofield, D. (1998). The health consequences of unemployment: The evidence. *The Medical Journal of Australia, 168*, 178-182.

McKee, M. & Suhrcke, M. (2005). Commentary: Health and economic transition. *International Journal of Epidemiology, 34*(6), 1203-1206.

Montgomery, S.M., Cook, D. et al. (1999). Unemployment pre-dates symptoms of depression and anxiety resulting in medical consultation in young men. *International Journal of Epidemiology, 28*, 95-100.

Morris, J., Cook, D. et al. (1994). Loss of employment and mortality. *British Medical Journal, 308* (6937), 1135-1139.

Moser, K. & Fox, A. et al. (1984). Unemployment and mortality in the OPCS longitudinal study. *Lancet*, 1324-1329.

Pensola, T. (2003). *From past to present*: Effect of lifecourse on mortality, and social class differences in mortality in middle adulthood. Helsinki: The Population Research Institute.

Pensola, T., Ahonen, H. et al. (2004). *Ammatit ja kuolleisuus - työllisten ja työttömien kuolleisuus ammatin mukaan 1996-2000* [Occupational mortality among the employed and unemployed in 1996-2000]. Helsinki: Statistics Finland.

Pensola, T., Järvikoski, A. & Järvisalo, J. (2006). [The association of unemployment and other risks for social marginalisation with work ability]. In R. Gould, J. Ilmarinen, J. Järvisalo & S. Koskinen (Eds.), *[Dimensions of Work Ability]* (pp. 223-240). Helsinki: Finnish Centre for Pensions, National Pension Institute, National Public Health Institute, Finnish Institute for Occupational Health.

Pless, I.B., Cripps, H.A. et al. (1989). Chronic physical illness in childhood: Psychological and social effects in adolescence and adult life. *Developmental Medicine and Child Neurology, 31*, 746-755.

Rothman, K. (1986). *Modern Epidemiology*. Boston: Little, Brown and Company.

Rouvinen, M. (2003). *Elämää pankkityön jälkeen* [Life after a career in banking]. Tampere: Tampere University Press Oy Juvenes Print.

Spruit, I.P. (1989). Vulnerability and unemployment: A process to ill health and constraints on intervention strategies in the Netherlands. In B. Starrin, P.G. Svensson & H. Wintersberger (Eds.), *Unemployment, poverty and quality of working life*. Berlin: Edition Sigma

STAKES (1999). *Tautiluokitus ICD-10* [Classification of diseases, 10th revision]. Turenki: Turengin tekstipalvelu.

Stefansson, C.G. (1991). Long-term unemployment and mortality in Sweden, 1980-86. *Social Science and Medicine, 32*(4), 419-423.

Stewart, J. (2001). The impact of health status on the duration of unemployment spells and the implications for studies of the impact of unemployment on health status. *Journal of Health Economics, 20*(5), 781-796.

Tapia Granados, J.A. (2005). Increasing mortality during the expansions of the US economy, 1900-1996. *International Journal of Epidemiology, 34*, 1194-1202.

Wadsworth, M.E.J. (1997). Changing social factors and their long-term implications for health. *British Medical Bulletin, 53*(1), 198-209.

Valkonen, T. & Martikainen, P. (1995). *The association between unemployment and mortality: Causation or selection? Adult mortality in developed countries: From description to explanation* (pp. 201-222). Oxford: Clarendon Press.

Unemployment, Therapy and Relapses of Alcohol-Addicted Inpatients: Results from a Prospective Longitudinal Study[1]

Dieter Henkel, Uwe Zemlin & Peer Dornbusch

1 Background

Representative statistical data[2] on patients treated for addiction[3] show that the unemployment rate of the alcohol addicted has increased from 9% in the year 1975 to 37% in 2003, by far the greatest increase taking place in the first half of the 1980s (Figure 1). Moreover, it is striking that the unemployment rate of the addicted patients has increased much more than the general unemployment rate in Germany. This can be traced back to different causes:

1. Even though no empirical facts exist, it may be assumed that in times of high unemployment rates accompanied by tighter personnel selection in private business and public administration, alcoholics become unemployed disproportionately more frequently due to longer times of absence from work associated with addiction and illness, decreases in work performance, disciplinary conflicts etc., so that the rate of unemployed alcoholics has certainly increased more than the overall rate of the unemployed.

2. Furthermore, the unemployment rates in those groups have increased above the average since the beginning of the 1980s (Federal Agency of Work, Statistics on Unemployment 1980-2004), to which the unemployed treated for alcohol addiction frequently belong: older than 50 years, low vocational qualification and limitations of functional capacity (see 2.5.1, Table 1).

---

1 Acknowledgement: This research project was financially supported by the Ministry of Sciences, Assia, Wiesbaden

2 The data refer to the pension insurances financing about 80% of all addiction treatments („medical rehabilitation“) in Germany (Zobel et al.2004).

3 The term „addiction therapy" refers to „medical rehabilitation“ in the entire text as well as in all figures and tables.

Figure 1. Unemployment rate (%) of alcohol addicted patients (ICD-10 F10.2) at the beginning of treatment and general unemployment rate in Germany 1975-2003.
Data source Association of the German Pension Insurances, Statistic of Rehabilitation 1975-2003 and Federal Ministry of Labour and Economy 2004, pp. 26

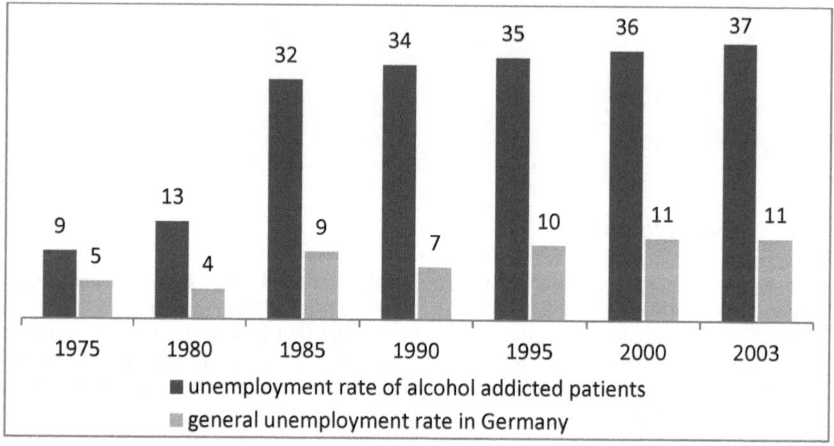

3. Accumulation effects due to treatment repetitions also play a considerable role. For, as will be shown in more detail later compared to the employed, the unemployed alcohol addicted more frequently take up repeated addiction treatments due to higher rates of relapse after therapy. The unemployment rate solely among those receiving treatment for the first time is considerably lower, in recent years probably by more than one third (see 2.5.1, Table 1).

4. Long-term unemployment (lasting longer than one year) has increased disproportionately since the beginning of the 1980s (Federal Agency of Work, Statistic of Unemployment 1980-2004) and as a result of this probably also the rate of unemployed alcoholics being in need of treatment. For longitudinal studies show that pronounced alcohol problems already present before the start of unemployment in the course of lasting unemployment intensify significantly more frequently than under conditions of continuous employment. This has been proven for men at least (research review Henkel, 1998).

5. It is possible that the increased unemployment rate in addiction treatment facilities may be traced back to intensified interventions of job centres as well. For if an addiction problem exists, unemployment benefits may be withdrawn if no treatment is commenced. In the year 2001, in about 27.500 (9%) of the overall 310,000 job centre medical experts' reports the first diagnosis comprised "psychological and behaviour disorders due to psychoactive substances" (ICD-10 F1) (Hollederer, 2002). However, no comparable data from earlier years exist.

Disproportionately high rates of unemployment can be found for both men and women alike (Figure 2). However, the rates vary. 40% of the male alcohol addicted in treatment were unemployed, whereas only 25% of the female ones. Moreover, the extent of unemployment is considerably high not only in the alcohol addicted group but in all groups with substance use disorders. In view of the absolute number of cases, though, the unemployed alcoholics comprise the largest group by far (Figure 3). Those addicted to psychoactive medical drugs are not depicted in Figure 3 due to too a small number of cases: 425 patients, of these 111 (26%) unemployed had been in addiction therapy (medical rehabilitation) in the year 2003.

Figure 2. Unemployment rates (%) at the beginning of addiction therapy according to gender and type of substance addiction in the year 2003. Men (*n* 40.341), women (*n* 10.782), data source Association of the German Pension Insurances, Statistic of Rehabilitation 2003

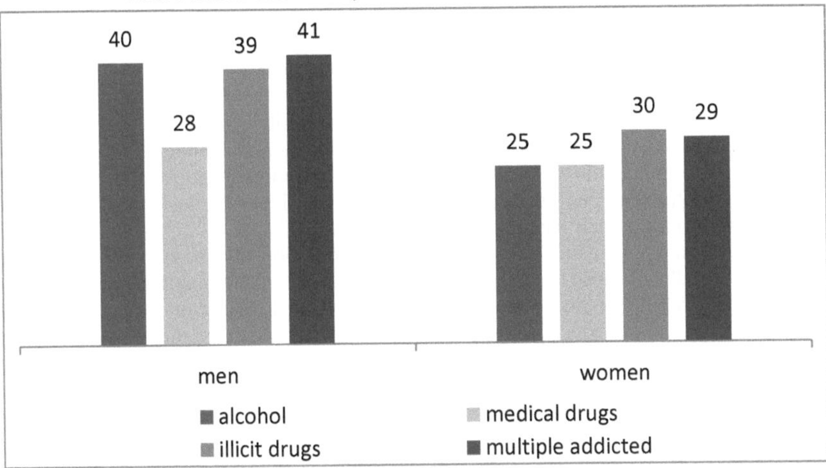

1.1 Vocational promotion and integration

As a reaction to growing unemployment rates amongst clients of addiction therapy since the beginning of the 1990s, various measures were developed for the promotion of vocational integration in Germany, which by now belong to the standard service catalogue of treatment for substance addiction and are already being implemented during therapy: counselling the patient for the return to work, accompanying to work, testing his or her work abilities, work training, counselling for questions of vocational rehabilitation and job search training (review Zemlin, 2005).

Figure 3. Absolute numbers of the unemployed persons in addiction therapy according to the type of substance addiction in 2003. (Association of the German Pension Insurances, Statistics of Rehabilitation 2003).

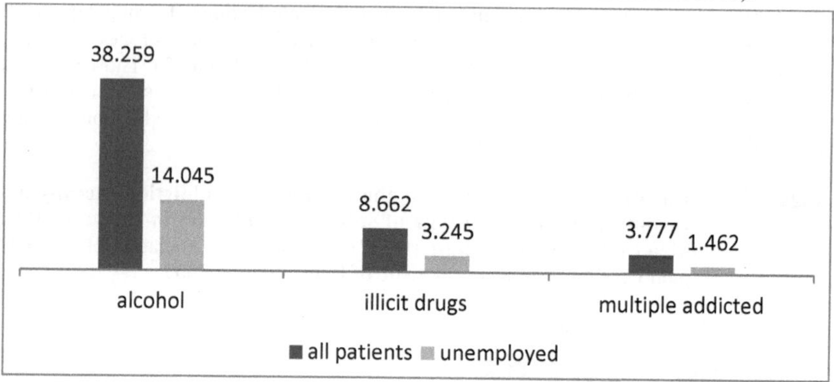

But the effectiveness of these measures become very limited due to the restrictive state of the labour market in general and frequent labour market handicaps of the unemployed treated for addiction, e.g., high age and little vocational qualifications. On top of that come the frequent and grave relapses of the unemployed addicts (see 2.5.3, Table 3), further decreasing their chances for reintegration.

Representative statistical data on in-patient addiction treatment show that in the 1990s about 92% of those unemployed at admission were also unemployed when discharged, only 5% had a job again or underwent vocational promotion measures and 3% were unfit for work (Sedos, 1995-2000). The data clearly proves the poor chances of effecting a positive change in the situation of the unemployed during treatment.

In an analysis of the rehabilitation statistics of workers' pension insurance, carried out by Bütefisch (2004), the employment status of patients (alcohol addiction ICD-10 F10.2) was examined by means of insurance contributions made after completing addiction treatment in the year 2000 until 2002. The patients had the status "mandatory insured". During this period only 66% of the men and 61% of the women paid their contributions without any gaps. 26% and 31% respectively paid occasionally or not at all, because they were temporarily or continuously unemployed or in marginal employment and therefore exempt from paying insurance (Figure 4).

Hümmelink and Grünbeck (2002), evaluating data from the salaried employees' mandatory insurance, arrived at similar results (ICD-10 F10.2, n 5.776). Furthermore, they found disproportionately low employment rates in those patient groups that exhibited the following features: aged over 50 years, low income (as an indicator of low vocational qualification), long periods of being unfit for work before the beginning of therapy and dropping out of treatment.

Figure 4. **Vocational integration (%) in the 2 years following addiction therapy 2000-2002, alcohol addiction (ICD-10 F10.2). Compulsory pension insurance of workers, men (*n* 14.541), women (*n* 1.425) (Bütefisch 2004)**

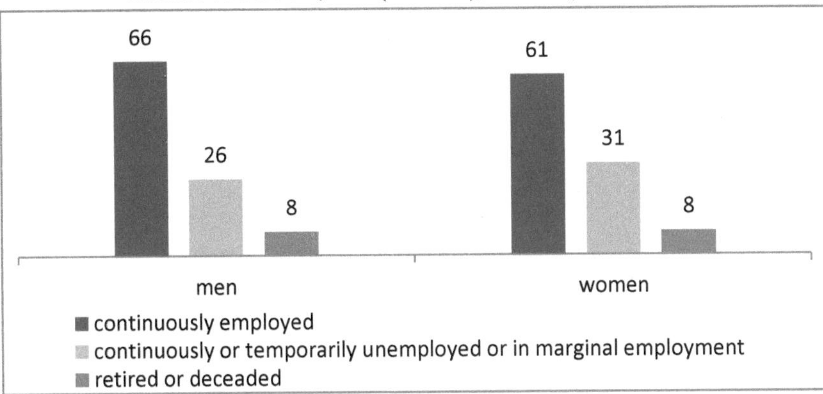

The studies of Bütefisch (2004) and Hümmelink and Grünbeck (2002), however, do not show the net result of vocational integration because data on the unemployment and employment rate at the beginning of therapy are lacking. This, however, is made possible by the data of Zobel et al. (2004) from 12 clinics for alcohol addiction therapy (n 7.824) in West and East Germany in the year 2001. According to them, the unemployment rate stood at 32% at the beginning of therapy and at 29% one year after completion of therapy. During the follow-up time, 36% of the follow-up respondents were unemployed at least once.

1.2 Risk of relapsing according to employment status

A number of studies exist on the risk of relapsing for the unemployed in comparison to the employed (Missel et al., 1998; Zemlin et al., 1999; Kluger et al., 2003). They refer to alcohol addiction (ICD-10 F10.2) und unanimously indicate a higher rate of relapsed patients among the unemployed at the time of follow-up (Figure 5). On the other hand, the employment status at the beginning of therapy had no significant influence. But in view of the follow-up time the results cannot be interpreted unequivocally. For while the relapse was surveyed during the one-year follow-up period, the employment status was surveyed at the follow-up time point. Hence, it remains unclear which employment status existed at the time of relapse. And thus it remains uncertain if unemployment and relapse really coincided, i.e. if the relapses occurred under conditions of unemployment.

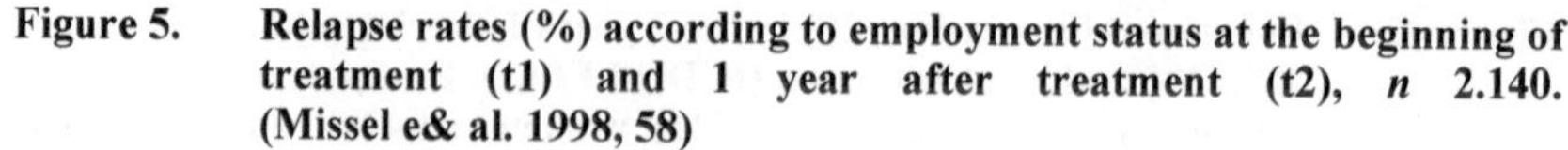

Figure 5. Relapse rates (%) according to employment status at the beginning of treatment (t1) and 1 year after treatment (t2), *n* 2.140. (Missel e& al. 1998, 58)

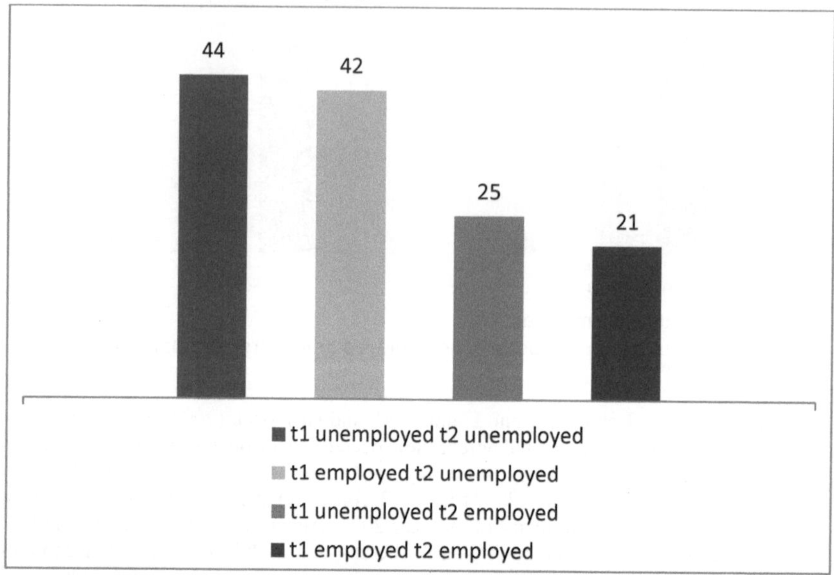

2 The ARA-Study

In addition to the present practice of promoting vocational reintegration it is necessary to develop a relapse prevention program for the unemployed. It should have as its goal the strengthening of psychosocial resources of the unemployed so as to maintain abstinence. This should already begin during addiction therapy. The ARA project was directed towards the development of such measures. This, however, requires a systematic and differentiated analysis of the connection between unemployment and relapse.

2.1 Aims

The study investigated: 1) What chances do the unemployed have for to achieve both main goals of addiction therapy: abstinence and reintegration into work? 2) What conditions influence the risk of the unemployed relapsing? 3) Can these conditions be modified by a preventive interaction, already starting during therapy as much as possible? 4) What practical consequences are derivable from the results for more effective treatment?

2.2 Design and subjects

The study was carried out in the clinic of Wilhelmsheim (near Stuttgart, South Germany) specializing in the treatment of substance use disorders. The clinic offers an ongoing scientifically evaluated treatment program comprising psychotherapeutic, medical und sociotherapeutic measures as well as systematic approaches to vocational promotion. The study had a prospective longitudinal design: t1 beginning of treatment, t2 end of treatment, t3 follow-up 6 months and t4 12 months after treatment. The duration of treatment when therapy is completed normally averaged 12.5 weeks. At t1 the study sample included all patients (n = 929) who were discharged in the year 2002, among whom at the beginning of treatment 397 were unemployed and 435 employed.

2.3 Measures

In addition to the measurement of sociodemographic as well specific characteristics of addiction and unemployment, standardized questionnaires were used for surveying the following traits: life satisfaction (AHG-Wissenschaftsrat, 1997), financial strain (Brinkmann, 1984), self-efficacy (Schwarzer & Jerusalem, 1989), coping (Jerusalem 1993), self-esteem (Badura, Kaufold, Lehmann et al., 1987), symptoms of mental disorders (SCL90-R Franke, Jäger & Stäcker, 1995), social support (Puls, Ulbrich & Wienold, 2000), social participation (Zemlin, Dornbusch & Henkel, 2001), break down of time structures (Puls, Ulbrich & Wienold, 2000), subjective meaning of work (Henkel, Zemlin & Dornbusch, 2001), expectations concerning the effects of abstinence after treatment (Ackermann & Munckes, 1998), alcohol abstinence self-efficacy (DiClemente, Carbonari, Montgomery & Hughes, 1994). In addition to that, a survey was made on the arrangement of leisure time, critical life events (e.g., divorce, illness) and contact with self-help groups after treatment. Furthermore comorbidity i.e. alcohol addiction (ICD-10 F10.2) and mental disorders (ICD-10 F2-F9) were investigated. The follow-ups were conducted by post for 90% of the follow-up respondents and for 10% by telephone.

The different indicators of relapse are cited in Table 3. It also considered those forms of alcohol consumption situated between the two extreme poles of abstinence and addiction, since it is not possible to infer a priori that a relapse is always a return to earlier addictive behaviour. Hence a distinction was made between the following forms of alcohol consumption: 1) addictive consumption (ICD-10 F10.2), 2) harmful consumption: impairments to health, to efficiency, to the ability to judge or to relationships with other people, but no alcohol addiction, 3) risky consumption: men >30g/day, women >20g/day (QF-Index[4]) (British Medical Association 1995), but no harmful consequences of alcohol consumption and no alcohol addiction, (4) harmless consumption: men ≤30g/day, women ≤20g/day and no alcohol addiction. The assignment to these categories was not carried out by the patients themselves but was derived in each case as calculated from the information given in the follow-up questionnaire.

4 The QF-Index (quantity-frequency-index) was calculated in the following manner: quantity of pure alcohol at a typical drinking day multiplied by the number of drinking days per week divided by 7 days. The conversion in pure alcohol was calculated according to the standards of the DGSS: 1 litre beer = 40g, 1 litre wine/sparkling wine/fruit wine = 86g and 1 litre spirit = 250g.

2.4 Statistical procedures

Statistical analysis was made with the SPSS Version 12.0. The Pearson χ^2-test or Wilcoxon test was used with categorical variables, and the t-test with the numerical and normally distributed ones. All tests were carried out two-tailed. For the analysis of relapse, multiple logistic regression analyses were conducted. The odds ratios (*OR*) indicate the factor by which the risk of relapse is increased or decreased for the respective group examined compared to the reference group. Moreover the 95% confidence intervals (*CI*) were calculated.

2.5 Results

2.5.1 Patient characteristics at t1

Table 1 shows that the unemployed had a more chronic addiction problem. Compared to the employed they were alcohol addicted for a longer period of time and had already been in addiction therapy considerably more frequently. In addition, the prevalence rate of comorbidity was significantly higher among the unemployed. They had lower educational and vocational qualifications as well as a lower per capita income and more frequently lived without partnership. The length of unemployment until the start of treatment (Figure 6), together with high age and low degrees of qualification, clearly indicates the limited chances of the unemployed finding work in the current state of the labour market.

Table 1. Characteristics of the unemployed (U) and employed (E) at t1

	U	E	value	p
n	397	435	χ^2, t	
Men/women %	78.3 / 21.7	73.6 / 26.4	2.6	n.s.
Alcohol addiction (ICD-10 F10.2) %	98.7	97.6	2.1	n.s.
Duration of alcohol addiction *(M / SD)*, years	12.9 / 8.2	11.3 / 7.4	2.8	0.005
Previously treated for alcohol addiction %	41.9	21.8	42.9	0.000
Comorbidity: alcohol addiction (ICD-10 F10.2) and mental disorders (ICD-10 F2-F9) %	25.9	17.9	7.8	0.006
Age *(M / SD)*	43.6 / 7.3	44.9 / 7.5	2.7	0.007
>50 years old %	18.4	25.3	-	-
At the most lower secondary school %	70.3	58.2	26.9	0.000
No vocational training %	23.4	13.5	23.8	0.001
Household income per capita %				
<1.000 €	69.7	36.8	71.8	0.000
>1.000 €	30.3	63.2		
Living in partnership %	47.6	64.1	24.6	0.000

Figure 6. **Length of unemployment (%) until t1, t1-unemployed (n 397)**

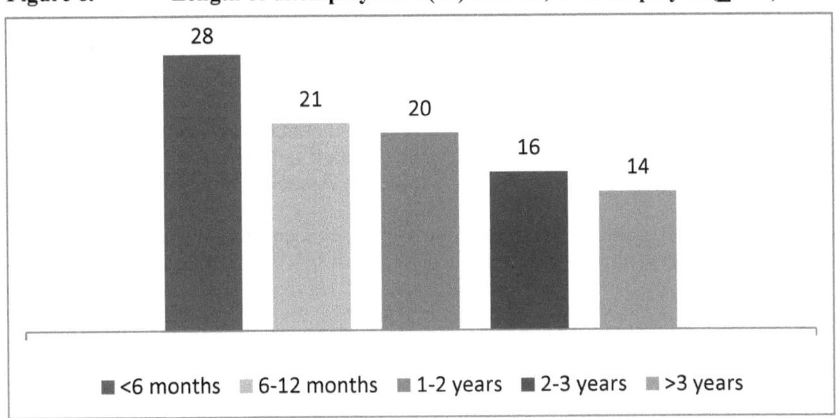

2.5.2 Samples at t1-t4 and drop-out analysis

Table 2 shows the response rates at different times of measurement. The sample size of the unemployed and employed from t3 onwards refer to those persons who were continuously unemployed or employed during the follow-up periods. They were chosen for the analysis of relapse and abstinence respectively, since only in the case of constant employment status does it become unequivocal whether relapse or abstinence are actually connected with unemployment or employment respectively. For these groups however no drop-out analyses were possible since it remains unknown which of the follow-up respondents were continuously unemployed or employed.

Table 2. **Samples at t1-t4, unemployed (U), employed (E)**

	Total		U		E	
	n	%	n	%	n	%
t1	929	100.0	397	100.0	435	100.0
Dropping out of treatment	156	16.8	70	17.6	36	8.3
t2	773	83.2	327	82.4	399	91.7
			continuous unemployed		continuous employed	
t3	682	73.4	212	31.1% of 682	334	49.0% of 682
t4	614	66.1	175	28.5% of 614	323	52.6% of 614
t3+t4	557	60.0	128	23.0% of 557	257	46.1% of 557

Drop-outs in the follow-up periods were investigated by means of the t1-samples. Compared to the t1-employed, the response rates of the t1-unemployed were always lower by about 10 to 15 percentage points. All diagnostic measures were used in order to examine the possible drop-out rate. It turned out that for both the unemployed and employed alike all follow-up samples reflected the t1-samples in a representative manner (for details see Henkel, Zemlin & Dornbusch, 2003, 2004a, 2004b).

2.5.3 Relapse rates in t4

In the following, results of the t4-follow up of the continuous by unemployed and employed will be described, since it is scientific standard to measure the effectiveness of addiction treatment one year after therapy (German Association of Addiction Research and Therapy).

Looking first at the results for all follow-up respondents as a reference (Table 3), it turns out the relapse rate for the unemployed is about twice as high as that of the unemployed. Moreover, significantly more unemployed regularly took (i.e. at least once a week) psychoactive medicine (sedatives, hypnotics, stimulants) and illicit drugs. Consumption of alcohol *and* medicine as well as alcohol *and* illicit drugs occurred more frequently amongst the unemployed, too. In the case of harmless and of risky or harmful consumption (both categories were merged because of the small number of cases) results of the unemployed and employed lie close together, whereas the prevalence rates of addictive alcohol consumption differ significantly. The unemployed follow-up participants met three times more frequently the criteria of the ICD-10 F10.2 alcohol addiction syndrome (39% to 13%). Even in the case of smoking and nicotine addiction the differences are to the disadvantage of the unemployed, this however not significantly.

Taking those who relapsed as a reference, Table 3 shows that the unemployed consumed alcohol distinctly more seldom in a non-addictive, i.e. harmless or risky or harmful manner. Amongst the unemployed the addictive consumption dominated unequivocally with a prevalence rate of 76%. In this way, the almost threefold higher rate of those drinking alcohol each day (26% to 10%) can be explained.

Since continuously unemployed and employed were compared, for the first time without doubt the ARA-study proves that relapses occur much more frequently and also in considerably more serious forms under conditions of unemployment than under conditions of employment. It must be emphasized that the presented results refer to follow-up respondents. As already mentioned the non-response rates of the t4-continuous unemployed and employed are unknown. But they probably are unequally high. This is certainly what the t4-response rates of the t1-unemployed and the t1-employed suggest. Hence in reality even greater differences in the relapse rates do exist to the disadvantage of the unemployed since we can assume that the relapse rates of the non-respondents are disproportionately high.

2.5.4 Changes in health impairments, alcohol consumption and alcohol addiction from t1 to t4

Indeed, the treatment results measured by relapsing were considerably worse for the unemployed than for the employed. But this does not mean that no positive changes had taken place. Compared to the beginning of therapy at t1, highly significant positive changes at t4 were found among the unemployed as well as the employed relapsing respondents (Table 4). In both groups prevalence rates of health impairments due to alcohol consumption and of alcohol addiction had significantly decreased. But the results for the unemployed were less positive. The alcohol addiction rate decreased and the average amount of alcohol consumed per day (QF-Index) only significantly less for the employed respondents.

Table 3. Indicators of relapsing, t4-continuous unemployed (U) and emloyed (E), t4-respondents

	U	E	value	p
n	175	323	χ^2	
Reference t4-respondents				
Relapse to alcohol, medical drugs or illicit drugs %	54.9	26.6	39.0	0.000
Relapse to alcohol %	50.9	24.1	36.3	0.000
Types of alcohol consumption, reference t4-respondents				
Harmless consumption %	8.0	5.9	13.3	0.011
Risky or harmful consumption %	4.0	5.0		
Addictive consumption (ICD-10 F10.2) %	38.9	13.3		
Consumption of alcohol, medical drugs, illicit drugs, reference t4-respondents				
Medicine at least once per week %	15.4	3.4	23.2	0.000
Illicit drugs at least once per week %	8.0	2.8	7.0	0.009
Alcohol and medical drugs at least once per week %	12.6	2.5	20.4	0.000
Alcohol and illicit drugs at least once per week %	6.9	1.9	8.1	0.006
Types of alcohol consumption, reference t4-relapsed respondents				
Harmless consumption %	15.7	24.4	10.4	0.016
Risky or harmful consumption %	7.9	20.5		
Addictive consumption (ICD-10 F10.2) %	76.4	55.1		
Frequency of alcohol consumption, reference t4-relapsed respondents				
Daily consumption %	25.8	9.6	12.9	0.012
Tobacco smoking, reference t4-respondents, nicotin addiction only smokers				
Smokers %	81.8	70.5	3.2	*n.s.*
Middle or high nicotine addiction (Fagerström) %	53.5	46.3	0.9	*n.s.*

2.5.5 Stability and instability of abstinence and relapse from t3 to t4

During the entire 12-month follow-up period significant negative and positive changes, i.e. giving up abstinence and overcoming alcohol relapses, had taken place with a relevant part of the t3+t4-continuous unemployed and employed (Figure 7). However, whereas among the unemployed positive and negative changes occurred in an approximately equal frequency, among the employed the positive ones predominated. These results suggest that under conditions of long-term unemployment the risk of relapses is significantly higher than under conditions of continuous employment.

Table 4. Health impairments due to alcohol (1), addictive consumption of alcohol (ICD-10 F10.2) (2) and alcohol consumption g/day (QF-Index) (3) at t1 and t4. (U) Unemployed at t1 and continuously unemployed in t4 and (E) employed at t1 and continuously employed in t4, relapsed t4-respondents.

	U				E			
n	139				254			
	t1	t4	value	p	t1	t4	value	p
(1)	80.4%	57.1%	2.8*	0.005	86.7%	37.8%	4.7*	0.000
(2)	98.4%	81.0%	3.2*	0.002	99.2%	53.1%	4.8*	0.000
(3) *M / SD*	117 / 79	86 / 99	1.7**	*n.s.*	110 / 85	35 / 21	4.5**	0.000

Note. * Wilcoxon-test, ** t-test

Figure 7. Stability and instability of abstinence and relapse from t3 to t4, t3+t4-continuous unemployed (n = 128) and employed n = 257), t3+t4-respondents

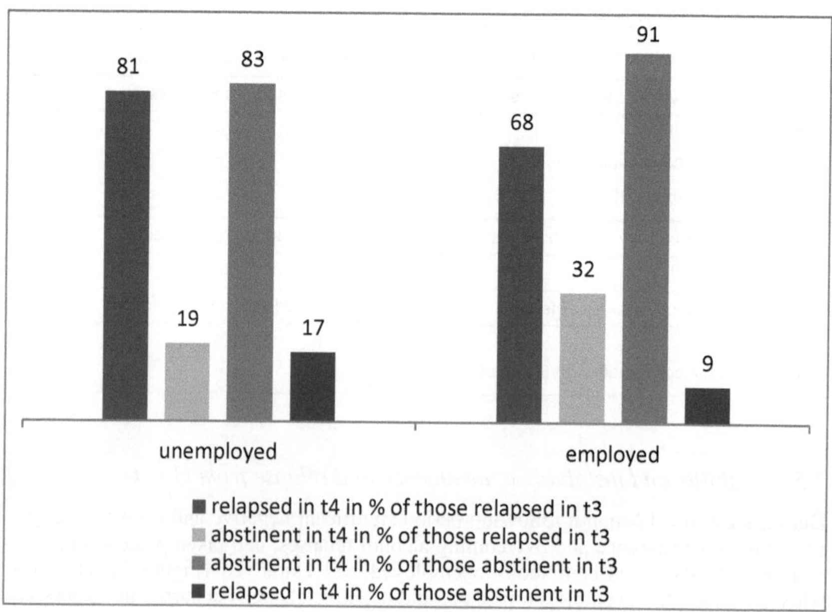

2.5.6 Vocational integration

In addition to the data already reported in Table 2, the net result of vocational integration can be demonstrated by the following results: Of the 327 t1-unemployed completing treatment only 44 (14%) had a job again when treatment ended, and of the 399 employed 24 (6%) lost their job. Of the t3+t4-follow-up respondents who were continuously unemployed in the first six months after therapy, 80% remained without work in the following six months as well. Taking into consideration that out of these almost every second person was continuously unemployed for one year already before intake (about 27% even longer than 2 und 12% longer than 3 years), then for some of the unemployed processes of labour market exclusion had taken place which are difficult to reverse.

Comparing the unemployment rate at the beginning and one year after treatment, Figure 8 shows that the unemployment rate has decreased by only 7 percentage points and the employment rate has only increased by 6 percentage points. The decline of the unemployment rate is probably overrated due to the relative low t4-response rate of the t1-unemployed.

Figure 8. Unemployment and employment rate (%) at t1 and at the end of t4, unemployed (n = 175), employed (n = 323), t4-respondents.

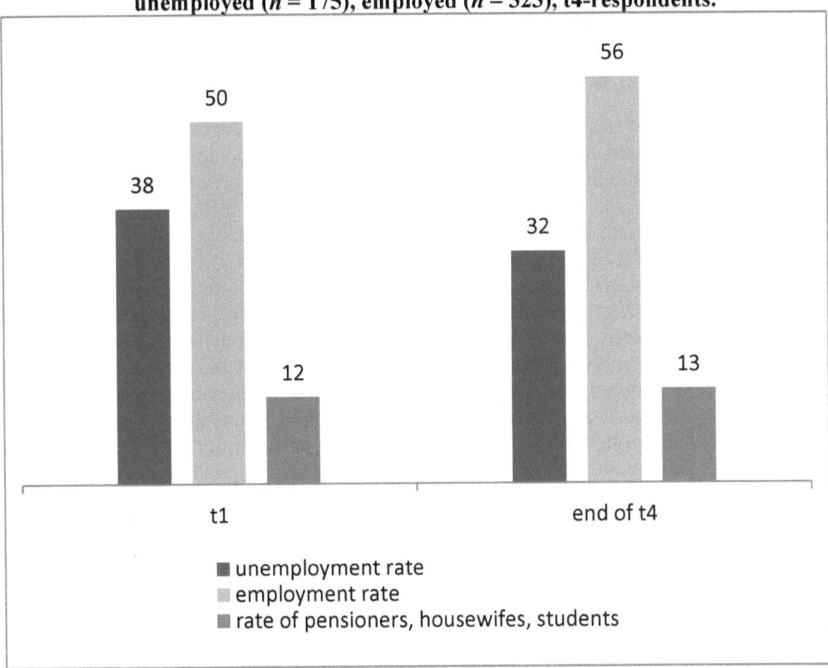

In view of these results it might be critically objected that a part of the unemployed actually are not interested in regaining a job any more, e.g., because of demoralisation effects, since in many cases unemployment had lasted for a long time already. But only about 3% of the unemployed had given up looking for a job in the time after treatment. Furthermore the unemployed and employed did not differ significantly with regard to subjective meaning of work (e.g., "work gives me a meaning to life", "work mediates social contacts I would otherwise not find"), so that an erosion of subjective meaning of work did not take place in spite of long-term unemployment.

After treatment 41 (19%) of the continuously unemployed at t3 and 32 (18%) of the continuously unemployed at t4 had taken part in vocational rehabilitation. Their relapse rates were distinctly lower compared to those who did not participate in such measures (t3-unemployed 42% vs. 60% t4-unemployed: 44% vs. 59%). These included various types of measures, from a short PC course to a long-term retraining. The results prove that effects protecting against relapse are linked with such measures, since they support social integration, improve self-esteem, provide time structures and vocational perspectives.

In the end it should be mentioned that at least among those employed who were continuously employed during the first 6 months after treatment, vocational integration had a positive outcome almost without exception. Of these 93% remained employed in the t4-follow-up period. In addition, this also means that a relapse does not easily result in job loss, since the relapse rate nevertheless was to about 25% even in this group of the t3+t4-continuous employed.

2.5.7 Differences between relapsed and abstinent unemployed

Until now the risk of relapse has been the focus in the present report. On the other hand, it must be emphasized that about 50% of the unemployed completing treatment normally remained abstinent. For the development of a relapse prevention program it is of great importance to examine how the relapsed unemployed differ from the abstinent unemployed at different survey times. To do this the t3-sample of the continuous unemployed was chosen for the following reasons:

Addiction research has proven that in almost all cases the first relapse already occurs within the first six months after treatment (e.g., Zobel et al., 2004). The data of the ARA study additionally show that the unemployed relapse much earlier than the employed (Figure 9). For 36% of the unemployed and for 21% of the employed, the initial relapse to alcohol already happened within the first month after treatment. Hence, relapse prevention measures in general, and especially for the unemployed must refer to early stages of the rehabilitation process and, as far as possible, must be carried out already during treatment. Therefore it would be best if factors protecting against relapse could be identified already at t1 and t2, however at least at t3.

Figure 9. First relapse (%), by the month after treatment (M), relapsed t3-continuously unemployed (n = 118) vs. employed (n = 90) at t3

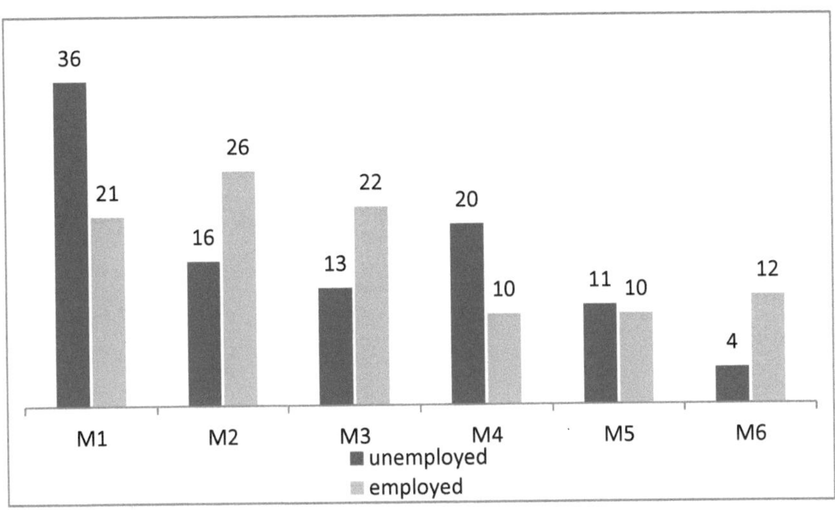

The sample comprises 181 who were continuously unemployed at t3 (without patients who dropped out of treatment since from them no t2-data were available). Of these, 91 (50%) relapsed and 90 (50%) remained abstinent. In order to compare both groups all diagnostic measures of the t1-, t2- and t3-surveys were used. Factors with no statistically significant differences at any time of measurement are not mentioned here, e.g., age, gender, income, duration of alcohol addiction, subjective meaning of work and comorbidity i.e. alcohol addiction combined with mental disorders, except depression (Table 5). The main results were as follows (for details see Henkel et al. 2004b).

Table 5. Psychosocial situation at t1, t2 and t3 of the relapsed (R) (n = 91) and abstinent subgroups of those with unsteady unemployment at t3 (A) (n = 90)

	t1			t2			t3		
Questionnaires/ scales	*M* R/A	Value *t*	*p*	*M* R/A	Value *t*	*p*	*M* R/A	Value *t*	*p*
Life satisfaction* in general	3.7 3.3	1.9	*n.s.*	3.0 2.5	2.8	0.005	3.6 2.5	6.0	0.000
Satisfied with… health status	3.5 2.9	2.8	0.006	2.5 2.3	1.2	*n.s.*	3.3 2.2	5.4	0.000
working situation	5.1 5.3	1.1	*n.s..*	4.7 4.6	0.6	*n.s.*	4.9 4.7	0.9	*n.s.*
financial situation	4.2 3.9	1.2	*n.s.*	4.2 3.6	2.2	0.027	4.4 3.9	2.3	0.022
social activities	3.7 3.4	1.8	*n.s.*	3.3 2.8	2.5	0.012	3.7 3.0	4.0	0.000
familiy situation	3.2	1.6	*n.s.*	3.1	3.0	0.003	3.3	4.8	0.000

	2.8			2.4			2.3		
social contacts with friends	3.0 2.7	1.1	*n.s.*	2.8 2.4	2.3	0.025	3.1 2.5	3.2	0.002
arranging leisure time	3.5 3.1	2.1	0.034	2.9 2.4	2.9	0.004	3.4 2.5	4.9	0.000
Self-efficacy	2.6 2.8	2.5	0.015	3.0 3.2	2.1	0.040	2.7 3.1	3 6	0.000
Self-esteem**	2.3 2.2	1.9	*n.s.*	1.9 1.7	2.2	0.027	2.2 1.9	2.8	0.007
Active coping	3.4 3.5	0.5	*n.s.*	3.5 3.7	2.6	0.012	3.1 3.4	3.5	0.001
Somatic complaints SCL90-R	8.1 6.2	2.1	0.039	5.3 4.7	0.8	*n.s.*	8.8 6.5	2.2	0.027
Depression SCL90-R	10.3 10.5	0.1	*n.s.*	6.8 4.8	2.2	0.026	13.9 9.0	3.5	0.001
Break down of time structures	2.6 2.5	1.0	*n.s.*	-	-	-	2.4 2.0	2.7	0.007
Financial strain	1.3 1.4	0.4	*n.s.*	-	-	-	1.3 1.2	1.5	*n.s.*
Social support	2.6 2.6	0.1	*n.s.*	-	-	-	2.7 2.8	1.1	*n.s.*
Negative abstinence Expectations	3.9 3.1	3.8	0.000	3.3 2.7	3.2	0.002	-	-	-
Alcohol abstinence self-efficacy	3.6 3.8	1.6	*n.s.*	4.0 4.4	3.3	0.001	-	-	-

Note. * high value = low satisfaction, ** high value = low self-esteem

In comparison to the abstinent the relapsed were unemployed for longer periods before the treatment (*M* 17.8 as compared to 13.7 months, *t* 2.8, *p* 0.021), relapsed more often during treatment (23% to 3%, χ^2 17.1, *p* 0.000) and had more frequently already received addiction therapy at least once before (50% to 31%, χ^2 12.1, *p* 0.017).

Table 5 shows that in view of the psychosocial situation at t1 relatively few differences existed to the disadvantage of those relapsing: satisfaction with health status, with arranging leisure time, self-efficacy, somatic complaints and negative expectations concerning abstinence after treatment. In both groups values from t1 to t2 improved in almost all aspects in a highly significant manner (not included in Table 5, for details see Henkel et al. 2004b). But as Table 5 shows, those relapsing often scored lower at t2 than those who were abstinent (exception: satisfaction with health status and with job situation and somatic complaints). The number of significant differences increased from t1 to t2. This means that those relapsing profited from treatment in a substantial manner, but distinctly less than the abstinent, so that at the beginning of the follow-up time they were in a more unfavourable psychosocial situation.

Even therapists assessed treatment effects for both groups differently, without knowing who remained abstinent or relapsed after treatment, and also without being aware of the employment status in the follow-up time. Out of the unemployed relapsing only 36% were categorized as "considerably improved", of the abstinent group the corresponding share was 56.4% ($\chi^2$16.0, *p* 0.001).

Also at t3 (Table 5) the psychosocial situation of those relapsing was considerably worse (except for: satisfaction with job situation, financial strain and social support). Further analysis of the data prove that in the group of those relapsing the sustainable effects of treatment were significantly less at t2 and t3 with regard to life satisfaction in general, health status and leisure time, self-efficacy, coping and depression (not included in Table 5, for details see Henkel et al. 2004b).

Those relapsing were socially less integrated: they lived more seldom in partnership (42% compared to 61% of the abstinent unemployed, χ^2 9.4, *p* 0.025), were more often divorced or separated (17% to 7%, χ^2 4.3, *p* 0.039) and had less regular contact with self-help or support groups (23% to 40%, χ^2 15.6, *p* 0.000). In addition to that, they more often spent their time watching TV (75% to 51%, $\chi^2$10.8, *p* 0.005).

The results might explain why some unemployed relapsed while others remained abstinent, therapy research proves that those showing a higher risk of relapse are more discontent with their life situation, have higher levels of depression, show less self-efficacy, have fewer coping resources at their disposal, do not live in a partnership, are socially less integrated, have problems with spending their leisure time actively and meaningfully and do not take part in any self-help or support groups (research review Körkel, 1996, 1999).

2.5.8 High risk groups

Two high risk groups have so far been identified in our study: unemployed patients dropping out of treatment and unemployed persons previously treated for addiction.

2.5.8.1 Patients dropping out of treatment

As could already be seen in Table 2, many more unemployed than employed persons dropped out of treatment (18% compared to 8%). Whereas in the case of the unemployed persons completing treatment regularly the average length of treatment amounted to 93.2 days (employed 83.3 days), dropping out of treatment occurred after 47.2 days on average (employed 51.9 days). These differences are not significant.

Several studies record that there is an increased risk of relapse for those dropping out of therapy (e.g., Kluger et al., 2003). The ARA study, however, shows this is only valid for the unemployed persons (Figure 10). Here, the difference in relapse rates proved to be highly significant (χ^2 35.2, *p* 0.001).

Figure 10. Relapse to alcohol consumption (%) according to the way treatment ended, t1-unemployed (*n* = 257) and those employed at t1 (*n* = 351), t3-respondents

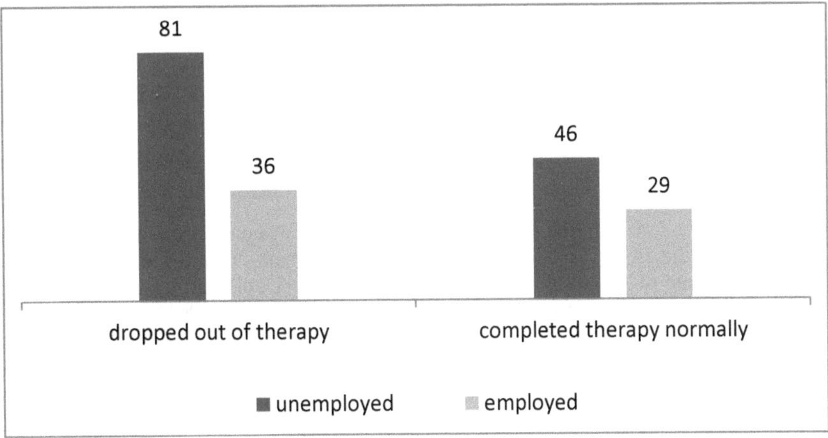

The reasons why far more unemployed than employed dropped out of treatment are not completely known. About two thirds of the unemployed patients gave up therapy by themselves. One third were discharged by the clinic due to insufficient compliance, e.g., because of repeated consumption of psychoactive substances during treatment. Also of possible importance is that the unemployed persons, especially the long-term unemployed persons, because of their unstructured everyday life and lower degree of activity, find it more difficult to conform to the structured time management of a clinic and to produce the often high degree of personal activity demanded within the therapeutic process. What might also play a role is that the unemployed, compared to the employed persons, anticipate a poorer cost-benefit ratio. In this way they may be less motivated to undergo treatment. While the employed can keep their job with a successful treatment, the unemployed are simply left with vague hopes of improving their chances of vocational reintegration with addiction therapy.

These hypotheses, however, cannot be tested with the ARA data. At least it could be proved that, as compared to the unemployed completing treatment regularly, the unemployed dropping out of treatment had a significantly fewer alcohol abstinence, a weaker coping behaviour and considerably higher values for the entire spectrum of mental disorders (e.g., depression, anxiety and psychotic symptoms). Exactly these differences at t1, too, were found as compared to the employed persons who dropped out of therapy (for details see Henkel et al., 2004a).

2.5.8.2 Patients previously treated for addiction

The question of whether a connection exists between the risk of relapsing and the number of previous treatments for addiction was also analyzed with the t3-sample. Table 6 shows that 40% of the unemployed persons had previously been treated, as compared to just 20% of the employed persons.

Table 6. Number of previous treatments for addiction, for these unemployed and employed at t3

	Unemployed		Employed	
	n	%	n	%
Number of previous treatments for addiction				
0	128	60.4	270	80.8
1	56	26.4	46	13.8
2 or more	28	13.2	18	5.4
Total	212	100	334	100

Addiction therapy research has already documented that as compared to those receiving treatment for the first time, those with repeated treatments have a disproportionately high relapse rate (e.g., Zemlin et al., 1999; Kluger et al., 2003). But as the following results show, this is correct only for the unemployed (Figure 11). Whereas for the employed, relapse rates even decreased, they significantly increased for the unemployed from 49% to 75% (χ^2 7.0, p 0.031).

Figure 11. **Relapse rates (%) according to the number of previous treatments for addiction, the unemployed (*n* = 212) and the employed (*n* = 334)**

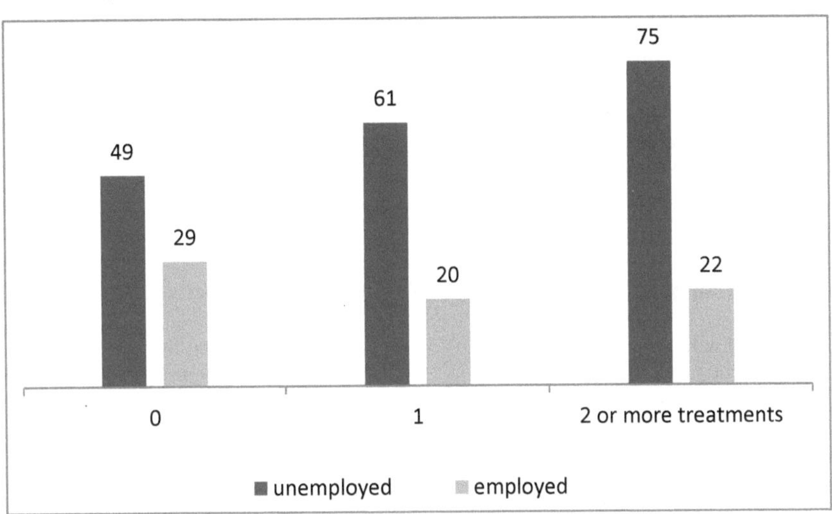

2.5.8.3 Multiple regression analysis

Multiple regression analysis was carried out in order to assess whether dropped out of therapy and repeating treatment represent independent risk factors for relapsing. Table 7 shows that this is the case. In comparison to being unemployed and completing therapy regularly the probability for relapsing was significantly higher for dropped out of therapy (*OR* 7.39, *CI* 2.46-22.22). And having received already 2 or more treatments increased the risk as well (*OR* 3.56, *CI* 1.39-9.13). The results for the employed are not significant.

The reasons for why those with multiple treatment periods represented a high risk group have not yet been examined. Possibly their life situation successively worsened between treatments with interdependent effects between relapse and unemployment reinforcing each other, with the result that the need for taking up treatment reemerged, but may have been the chances for a successful treatment were increasingly reduced.

Table 7. Multiple logistic regression, odds ratios (*OR*) and 95% confidence intervals (*CI*) for relapsing, for the unemployed and the employed at t3

	OR	*95% CI*	*p*
Unemployed (n 212)			
Dropping out of treatment	7.39	2.46-22.22	0.000
Number of previous treatments			
1	1.70	0.87-3.29	0.119
2 or more	3.56	1.39-9.13	0.008
Employed (n 334)			
Dropping out of treatment	0.76	0.24-2.40	0.639
Number of previous treatments			
1	0.48	0.20-1.13	0.092
2 or more	0.71	0.22-2.21	0.548

Note: Reference: completing therapy normally = *OR* 1.00, number of previous treatments 0 = *0R* 1.00

6 Conclusions for the practice of addiction therapy

The results emphasize the necessity of increasing and intensifying both programs for employment promotion as well as concepts for relapse prevention by strengthening psychosocial resources of the unemployed.

The results clearly show the poor chances for vocational reintegration under the current restrictive conditions of the labour market, even then when measures for employment promotion are implemented, as is the case in the treatment program of the Clinic of Wilhelmsheim. Nevertheless, measures for employment promotion must and can be further systematized and intensified, for resources deriving from a closer link of psychosocial with employment rehabilitation are not yet exhausted (Zemlin, 2005).

In addition to that, concepts and programs must be developed in order to prevent the unemployed from dropping out of treatment, to increase social support for the unemployed, especially by ensuring a smooth inclusion into self-help and support groups as well as by activating other social support systems and to intensify treatment for the unemployed in certain areas of mental health (e.g., depression, coping, self-efficacy and self-esteem). And finally, intensive supportive and promotional measures must be created for the unemployed repeating treatment, since for them there is a high risk of being caught in a process of relapsing and employment exclusion that is difficult to reverse.

References

Ackermann, K. & Munckes, P. (1998). *Abstinenz-Erwartungs-Fragebogen* [Abstinence Expectancy Questionaire]. Tübingen.

AHG-Wissenschaftsrat. (1997*). Skala Lebenszufriedenheit* [Life Satisfaction Scale]. Hilden.

Badura, B., Kaufold, G., Lehmann, H., Pfaff, H., Schott, T. & Waltz, M. (1987). *Leben mit dem Herzinfarkt. Eine sozialepidemiologische Studie* [Living after a heart attack. A socioepidemiological study]. Berlin: Springer.

Brinkmann, C. (1984). Die individuellen Folgen langfristiger Arbeitslosigkeit: Ergebnisse einer repräsentativen Längsschnittuntersuchung [Individual impacts of long-term unemployment: Results of a representative longitudinal study]. *Mitteilungen aus der Arbeitsmarkt- und Berufsforschung, 4,* 454-473.

British Medical Association (1995). *Alcohol: Guidelines on sensible drinking*. London: British Medical Association.

Bütefisch, Th. (2004). *Sozialmedizinische 2-Jahresprognose (Alkoholabhängigkeit) in der Arbeiterrentenversicherung 2000-2002* [2-years follow-up after treatment of alcohol addicted patients 2000-2002, compulsury pension insurance for workers]. Unpublished manuscript, Verband Deutscher Rentenversicherungen, Frankfurt a. M.

DiClemente, C., Carbonari, J., Montgomery, R. & Hughes, S. (1994). The Alcohol Abstinence Self-Efficacy Scale. *Journal of Studies on Alcohol, 55,* 141-148.

Federal Agency of Work (2004). *Arbeitslosenstatistik 1975-2004* [Statistic of unemployment 1975-2004]. Nürnberg.

Federal Ministry of Labour and Economy. (2004). *Statistisches Handbuch 2004: Arbeits- und Sozialstatistik* [Statistical handbook 2004: Labour and social affairs]. Berlin.

Franke, G., Jäger, H. & Stäcker, K. (1995). Die Symptom-Check-Liste SCL90-R im Einsatz bei HIV-infizierten Patienten [Symptom Check List SCL90-R]. *Zeitschrift für Differentielle und Diagnostische Psychologie, 16,* 195-208.

Henkel, D. (1998). Arbeitslosigkeit, Alkoholkonsum und Alkoholabhängigkeit: Forschungsergebnisse, Defizite, Hypothesen [Unemployment, alcohol consumption and alcohol addiction. A review of research]. *Abhängigkeiten, 4*(3), 9-29.

Henkel, D., Zemlin, U. & Dornbusch, P. (2001). *Skala Subjektive Bedeutung der Erwerbsarbeit* [Subjective meaning of work scale]. Wilhelmsheim.

Henkel, D., Zemlin, U. & Dornbusch, P. (2003). Analyse rückfallbeeinflussender Bedingungen bei arbeitslosen Alkoholabhängigen (ARA-Projekt) – Teil I: Einführung in die Thematik, Projektziele, Untersuchungsanlage und Ergebnisse zu Beginn der Suchttherapie [Analysis of risk factors for relapse among unemployed alcohol addicted patients – Part I: Introduction, study design and results at the beginning of treatment]. *Sucht aktuell, 10*(2), 5-14.

Henkel, D., Zemlin, U. & Dornbusch, P. (2004a). Analyse rückfallbeeinflussender Bedingungen bei arbeitslosen Alkoholabhängigen (ARA-Projekt) - Teil II: Ergebnisse des Therapieverlaufs und der 6-Monatskatamnese [Analysis of risk factors for relapse among unemployed alcohol addicted patients – Part II: Treatment effects and results of the 6-month follow-up]. *Sucht aktuell, 11*(1), 21-32.

Henkel, D., Zemlin, U. & Dornbusch, P. (2004b). Analyse rückfallbeeinflussender Bedingungen bei arbeitslosen Alkoholabhängigen (ARA-Projekt) – Teil III: Abstinenz und Rückfall in der 12-Monatskatamnese, Veränderungen im Katamneseverlauf, Unterschiede zwischen Rückfälligen und Abstinenten in der Aufnahme-, Entlass- und Katamnesediagnostik sowie zwischen Erst- und Wiederbehandelten [Analysis of risk factors for relapsing among unemployed alcohol addicted patients – Part III: Results of the 12-month follow-up, differences between relapsed and abstinent patients at the beginning and at the end of treatment and at follow-up time, differences between patients with and without previous treatments of addiction]. Sucht aktuell, 10(2), 11-22.

Hollederer, A. (2002). Arbeitslosigkeit und Gesundheit. Ein Überblick über empirische Befunde und die Arbeitslosen- und Krankenkassenstatistik [Unemployment and health. A review of empirical findings and data from the unemployment and health insurance statistic]. *Mitteilungen aus der Arbeitsmarkt- und Berufsforschung, 35*(3), 411-428.

Hümmelink, R. & Grünbeck, P. (2002). Sozialmedizinische Prognose nach stationärer Sucht-Rehabilitation. Aktuelle Auswertungen von Routinedaten der BfA [2-years follow-up after treatment of alcohol addicted inpatients. An Analysis of statistical data from the compulsury pension insurance for employees]. *Sucht aktuell, 10*(2), 26-29.

Jerusalem, M. (1993). Coping. In G. Westhoff (Ed.), *Handbuch psychosozialer Messinstrumente* (pp. 187-188). Göttingen: Hogrefe.

Kluger, H., Funke, W., Bachmeier, R., Brünger, M., Herder, F., Medenwaldt, J., Missel, P., Weissinger, V. & Wüst, G. (2003). Effektivität der stationären Suchtrehabilitation – FVS-Katamnese des Entlassjahrgangs 2000 von Fachkliniken für Alkohol- und Medikamentenabhängige [Effectiveness of inpatient treatment of addicted – FVS follow up from clinics for alcohol and medical drug addicted 2000]. *Sucht aktuell, 10*(1), 14-23.

Körkel, J. (1996). Neuere Ergebnisse der Katamneseforschung. Folgerungen für die Rückfallprävention [New results of follow-up research. Conclusions for relapse prevention]. *Abhängigkeiten, 2*(3), 39-60.

Körkel, J. (1999). Rückfälle Drogenabhängiger: Eine Übersicht [Relapses of addicted patients. A review]. *Abhängigkeiten, 5*(1), 24-43.

Missel, P., Braukmann, W., Buschmann, H., Dehmlow, A., Herder, F., Jarreis, R., Ott, E.S., Quinten, C., Schneider, B. & Zemlin, U. (1998). Effektivität und Kosten in der Rehabilitation Abhängigkeitskranker [Effectiveness and costs of rehabilitation of addicted patients]. In Fachverband Sucht (Ed.), *Suchttherapie unter Kostendruck* (S. 41-66), Geesthacht: Neuland.

Puls, W., Ulbrich, T. & Wienold, H. (2000). *Skalen zur Arbeit, Arbeitslosigkeit und Gesundheit* [Scales concerning work, unemployment and health]. Münster: Institut für Soziologie der Universität Münster.

Schwarzer, R. & Jerusalem, M. (1989). Anxiety and Self-Concept, Antecedents of Stress and Coping. A longitudinal Study with German and Turkish Adolescents. *Personality and Individual Differences, 10*, 785-792.

Schwarzer, R. & Leppin, A. (1989). *Sozialer Rückhalt und Gesundheit* [Social support and health]. Göttingen: Hogrefe.

Sedos. (1994-2000). *Jahresstatistik der stationären Suchtkrankenhilfe in der Bundesrepublik Deutschland* [Statistic of inpatient treatment of addiction in Germany]. Hamm: Deutsche Hauptstelle gegen die Suchtgefahren.

Verband Deutscher Rentenversicherungsträger [Association of the German Pension Insurances]. (1975-2003). *Statistik Rehabilitation 1975-2003* [Statistic of Rehabilitation]. Frankfurt a.M.

Zemlin, U., Dornbusch, P. & Henkel, D. (2001). *Skala Soziale Partizipation* [Social participation scale]. Wilhelmsheim.

Zemlin, U., Herder, F. & Dornbusch, P. (1999). Wie wirkt sich die durch die Spargesetze bedingte Verkürzung der Behandlungsdauer in der stationären Rehabilitation Alkohol- und Medikamentenabhängiger auf den Behandlungserfolg von stationär Erstbehandelten und stationären Therapiewiederholern aus? Ergebnisse einer prospektiven Katamneseuntersuchung. *Sucht aktuell, 2*, 1-18.

Zemlin, U., Schneider, B., Braukmann, W., Buschmann, H., Dehmlow, A., Herder, F., Jarreis, R., Missel, P., Ott, E., Quinten, C. & Roeb, R. (1999). Effektivität in der Rehabilitation Abhängigkeitskranker: Ergebnisse einer klinikübergreifenden 1-Jahreskatamnese in fünf Fachkliniken [Effectiveness of the rehabilitation of addicted patients. Results of 1 year follow-up in five clinics specialized for addiction treatment]. *Praxis Klinische Verhaltensmedizin und Rehabilitation, 47*, 60-73.

Zemlin, U. (2005). *Teilhabe am Arbeitsleben und der Gesellschaft fördern: Empfehlungen und Forderungen aus Sicht der Behandler* [Promotion of vocational and social participation for addicted patients]. Vortrag auf dem Kongress des Fachverbands Sucht, Mai 2005 .

Zobel, M., Missel, P., Bachmeier, R., Brünger, M., Funke, M., Herder, F., Kluger, H., Medenwaldt, J., Weissinger, V. & Wüst, G. (2004). Effektivität der stationären Suchtrehabilitation – FVS-Katamnese des Entlassjahrgangs 2001 von Fachkliniken für Alkohol- und Medikamentenabhängige [Effectiveness of inpatient treatment of addicted – FVS follow up from clinics for alcohol and medical drug addicted 2001]. *Sucht aktuell, 11*(1), 11-20

Unemployment and the Effects of Activation Policy on Quality of Life and Self-Performance

Mika Ala-Kauhaluoma & Antti Parpo

Introduction

This article addresses Finnish active social policy measures and legislation as the term goes, the reform of rehabilitative work. The reform aims at co-ordinating active labour market policy measures and basic income policy for long-term unemployed and unemployed youth as a joint effort of employment and social authorities. It imposes new obligations both on these authorities and on the target groups of the policy reform.

The purpose of the Act on Rehabilitative Work Experience is to improve the prospects of the long-term unemployed receiving labour market support benefit or social assistance in finding work and to promote their participation in training and benefiting from other measures that promote employment. The goal of the legislative reform - promotion of employment and prevention of exclusion - are to be reached through a closer cooperation between labour and social welfare authorities, and measures designed to activate the unemployed. In the activation plan, compared for every unemployed work seeker, the focus is on the life situation of the unemployed and the purpose is helping them to get a job.

The aim of the project, carried out by the Rehabilitation Foundation and the National Research and Development Centre for Welfare and Health (STAKES), was to evaluate the effects of this new policy. In this article we examine the effects of the activation plan and the activation measures involved in the plan from the perspective of quality of life and self-performance.

1 Research context: Activation policy

Activation programs and the increased responsibility of social benefits recipients for their own well-being have been a common trend in social and employment policy in many western European countries during the past decades. Originally, the activation policy or so-called workfare policy was launched in the USA in the 1970s (Dwyer, 2000), while measures to activate the unemployed have been adopted in some European countries from the beginning of the 1990s. The main objective of the preferred activation policy in Europe is to create opportunities for social and labour market integration through participation in training, education, rehabilitation, activation projects etc. In contrast, workfare primarily means an obligation imposed on target groups such as the unemployed to take up an offer of work or training with the threat of negative sanctions (Kosonen, 1998).

Countries such as Denmark and the Netherlands have led the way in this new activation policy in Europe. Activation policy with an emphasis on active labour market policy (ALMP) does have a long history in Scandinavia, especially in Sweden. However, the tradition and aims of active labour market policy are not exactly the same as those of the new activation policies in Europe, with active labour market policy being used more for promoting structural changes in economic and labour markets. Aims towards full employment and the better movement of the labour force and economic growth have been an important part of classic active labour market policy (Rehn, 1988). In the new activation policy paradigm, emphasis has been put on resolving social issues due to unemployment.

At present, the broadly expanded activation policy trend in western European countries is a remarkable social innovation. A number of theoretical and empirical analyses of this new activation policy have been published, and they include an extensive and heterogenic group of analyses and typologies of the policy measures. As a result of these variations, approaches to activation policy in general require a relatively high level of abstraction. From a general point of view, most researchers think that activation policy is a policy for decreasing the level of unemployment. Activation policy is also seen as a new kind of welfare policy where the relationship between work and social security has been reconstructed (Deacon, 2002). In the new activation policy paradigm, well-being is more related to the work or activities of individuals. There has been a desire to cut down the number of passive social benefit recipients - the so-called 'free riders'.

The basic assumption behind the activation policy is that it is better for individuals to be active than to receive social benefits and spend the time in passive leisure. According to the thesis behind the new policy paradigm, work gives the best social security and it is also the best way to prevent social exclusion. Activation policies can include:

- Legal interventions with regard to service providers (an obligation to offer activation programmes) or assistance recipients (right of access to programme participation or obligation for recipients to accept work or training);
- Financial incentives for employers or recipients (e.g. subsidised work, in-work benefits or reduction of benefits);
- Work integration services (e.g. counselling, job placement, work orientation, education and training, public employment, subsidised work) or
- Services to promote social integration (participation in socially useful work, voluntary work in NGOs etc.) (Hanesch et al., 2001, pp. 123-124; cf. Gilbert & van Voorhis, 2001.)

The long-term aims of activation policy are to increase the employment rate, the well-being of the unemployed, and to decrease the dependency on social security (Heikkilä, 1999). Activation policy can also be seen through the emphasis placed on the values of the work ethic and through the changed principle of the source of the welfare of individuals. Typically, it is argued that activation policy not only benefits individuals, but also the broader society. From the perspective of the unemployed, an activation policy can increase the resources that are important in the labour market and prevent the negative effects that unemployment has on well-being. From the perspective of the wider society, an activation policy is expected to increase the supply to the labour force, strengthening integration within society and ultimately decreasing the costs of unemployment (Hvinden, 1999).

Experiences of the activation measures adopted in many European countries are not unambiguous. Despite this, it is possible to find evidence that activation policy has had an important role in those European countries where high levels of unemployment have been reduced. Countries such as the Netherlands, Denmark and the UK are such examples of countries where a lot of effort has been invested in activation with a resulting decrease in unemployment. In addition reduce unemployment, in the Netherlands and Denmark it has also been possible to sustain high level of social security (Clasen & van Oorshot, 2002, pp. 236-238).

1.1 Development and justifications of Finnish activation policy in the 1990s [1]

In Finland, activation policy can be traced back to two different types of policy making: ALMPs and policy interventions towards social assistance recipients. Historically, ALMPs - of which the Swedish tradition dating back to the 1940s is the most famous example - have two main aims, namely to promote long-term economic growth and to support vulnerable groups in the labour market (e.g., Sihto, 2001). However, Finnish public discussion has mostly concentrated on the latter aim, nowadays often referred to as special employment policy aiming to guarantee and increase the employment of vulnerable groups. In practice, this means that a wide range of active labour market measures, such as vocational rehabilitation, training and special employment schemes based on e.g. incentives or direct job creation, have been available for this purpose (Mannila, 2001). During the 1990s, before the Act on Rehabilitative Work Experience, the Finnish employment administration introduced a number of new policy measures consistent with the activation policy approach: including, for instance, a benefit called labour market support combined with training. However, during the 1990s, unemployment was increasingly linked with social assistance recipiency in Finland. Today, unemployment is the reason for providing social assistance for approximately 50% of the cases, while the corresponding figure was 20% at the beginning of the 1990s. Long-term social assistance dependency has also made the social administration look for solutions that are socially integrative and provide employment for the clients. At the local level a number of initiatives of this kind were created during the latter part of the 1990s.

In Finland, the public and scientific discussion addressing incentives to work gained strength in the mid-1990s, a few years later than in many other European and Nordic countries. In Finnish social policy the early 1990s was a period of 'savings policy' accelerated by economic recession and the alleged public expenditure crisis, while the cuts in social expenditure made in the late 1990s had a more ideological basis

---

1 Chapter 1.1 and 1.2 are mainly based on the following paper: Keskitalo E., Ala-Kauhaluoma M. & Mannila S. (2002) An evaluation of a proactive social policy reform and local activation practices in Finland. (http://www.europeanevaluation.org/conferences/past/seville_2002/employment_pap.html)

(Heikkilä & Uusitalo, 1997; Kosunen, 1997). 'Making work pay' became an important guiding principle in the reform of social protection in the years to come. Consequently, in the reforms taken, access to earnings-related unemployment benefits was made more stringent, means-tested labour market support was made more conditional for unemployed youth, and sanctions were imposed on social assistance recipients who refused an offer of work or training.

In the late 1990s, long-term unemployment and risk of social exclusion among unemployed became a more serious political concern in Finland, and combating poverty and social exclusion was for the first time included in the Government programme during Prime Minister Paavo Lipponen's II Government (1999-). Traditional ALMPs were regarded as insufficient in alleviating the structural long-term unemployment. In 1999, a ministerial working group on proactive social policy was set up by the Government to prepare new tools for activating long-term unemployed and other minimum income recipients. The Act on Rehabilitative Work Experience was based on the proposal of the working group.

The proposal of the working group on proactive social policy was not accepted by a general consensus. Trade unions were against the new obligations to be imposed on the unemployed. Instead, they maintained that what were needed to alleviate structural unemployment was investment in ALMPs and more individualised services for the unemployed. The Association of Finnish Local and Regional Authorities also objected the proposal because of the new tasks – provision of the rehabilitative work experience - assigned to municipalities. Such new tasks without sufficient resources were seen to cause problems at the local level. The proposal was debated in the Finnish media to an extraordinary degree. The proposed Act on Rehabilitative Work Experience was modified slightly to take into account the opinion of the social partners, and the Act was finally approved by the Finnish Parliament in December 2000 to take effect in September 2001.

It was mainly individual-level justifications that were offered in public for the introduction of activation policy in Finland in the mid-1990s, with the emphasis on combating dependency on social welfare and on work incentives. In contrast, new activation measures aim, at least officially, more at creating new tools for solving structural problems of unemployment. Interestingly, the new activation measures introduced by the Act on Rehabilitative Work Experience are targeted at individuals, seeking to activate the unemployed. In other words, these new measures are mostly supply-side measures. Hence, structural unemployment is being tackled with individual-level tools. As Lødemel et al. (2001) point out, different interpretations of the causes of unemployment and labour market exclusion in principle relate to different policy responses. In political rhetoric, however, different justifications are often used to reinforce one another. In Finland, activation policy is motivated by a mix of concepts, and different justifications reinforce one another, which cannot be regarded as untypical of activation policies in general.

1.2 New activation instruments (introduced by the act on rehabilitative work experience)

The Act on Rehabilitative Work Experience continues and strengthens the activation measures that were increasingly introduced in the latter part of the 1990s concurrently in employment policies and in social welfare. In many countries, for administrative reasons, activation policies in social welfare and employment services constitute two different 'worlds of activation'. In Finland, the Act on Rehabilitative Work Experience covers both employment and social welfare services and seeks added value through joint efforts of the two branches of administration. This is a complicated problem: Employment authorities are centrally administered, while social welfare services are administered at the local level, within the municipal autonomy. The new Act defines the target groups by means of benefit and age criteria, and the activation of unemployed persons receiving labour market support or social-assistance follows the same procedure. Moreover, employment authorities and municipal social workers are obliged to co-ordinate the activation measures by means of a common activation plan to be negotiated in a tri-partite process (client, employment counsellor, social worker).

The Act on Rehabilitative Work Experience extends the scope of activation measures beyond employment policies, as it makes activation a municipal responsibility as well as increases the workfare element of last-resort social assistance. The Act provides a new basis for the rights and obligations between the state/municipalities and the welfare client. The client's rights and obligations are negotiated during an action plan process. An action (activation) plan constitutes a new kind of co-operation forum for employment and social welfare authorities. The plan is a signed contract where pathways towards the labour market are negotiated and agreed on. Vocational rehabilitation in Finland has traditionally included similar teamwork-based rehabilitation plans. However, these rehabilitation plans are merely recommendations and like the new activation plan, they do not define the rights and obligations of the public sector.

The activation plan is meant to provide a step-by-step path towards employment and to mainstream ALMPs. The Act introduces a new municipal activation measure 'rehabilitative work experience' as the last resort if the ALMPs are not applicable or have not been effective. Rehabilitative work experience is seen as a social service as defined in the Finnish Social Welfare Act and not as employment. It can only be organised by bodies governed by public law (municipalities, associations, foundations, religious communities). The contents of rehabilitative work experience are defined locally and agreed on with the client on an individual basis. The duration of rehabilitative work experience is flexible and varies from 3 to 24 months, 1 to 5 days a week. In addition to the previous minimum income maintenance (labour market support, social assistance), the participants receive a small supplementary benefit and compensation for travel expenses. While drafting the activation plan the client's need for social and health care services and rehabilitation will also be assessed and appropriate services of these kinds will be included in the plan. Unemployed persons do not have any subjective right to demand either social and health care services or ALMP measures. The services are provided 'within the framework of public resources'.

The Act defines the target groups of its provisions and their rights and obligations. Clients that are to participate in the activation process come by two routes: from employment offices and from social welfare offices. Previously, a job search plan was drafted for unemployed persons who were entitled to labour market support. Now, this plan is replaced with the activation plan defined in more detail in the legislation. The activation process is triggered for those under 25 years of age at an earlier stage than for those over 25. Those under 25 must participate in the process of activation if they have received labour market support for a period of 180 days during the past 12 calendar months, basic unemployment allowance or earnings-related unemployment allowance for the maximum period (500 days), or social assistance on the grounds of unemployment as the main source of income during the past 4 months. Those over 25 must participate in the activation process if they have received labour market support for at least 500 days, labour market support for at least 180 days after the maximum period (500 days) of daily unemployment allowance, or social assistance on the grounds of unemployment as a main source of income during the past 12 months.

Participation in drawing up the activation plan is obligatory for persons meeting the criteria defined in the Act. In addition, participation in rehabilitative work experience is obligatory for those under 25 years of age, if rehabilitative work experience is included in the activation plan. This means that a the person without good reason refuses to participate in drawing up the activation plan or in rehabilitative work experience (if this has been agreed on in the activation plan), he or she will lose the labour market support for a two months' period and the amount of social assistance will be reduced by 20% and after repeated refusals by 40%. The sanctions here are the same as those applied in the case of refusal of a job or training offer with regard to labour market support and social assistance recipients. The Act on Rehabilitative Work Experience also specifies the grounds for an acceptable refusal to participate in drawing up the activation plan and in rehabilitative work experience.

The Act on Rehabilitative Work has much in common with the activation and workfare measures introduced in many European countries in the 1990s. The Finnish Act on Rehabilitative Work could be seen to represent a case of centralised European programmes if the frame of analysis used by Lødemel and Trickey (2001, pp. 279-281) in the seven-country comparison of workfare is applied. In this study, the Danish, Dutch and British programmes represented centralised European programmes, while the German, Norwegian and French programmes were grouped under decentralised programmes. The key indicator of the centralised programmes was the universality of the scheme and the size of the target group. In Finland, the relevant Act is universal, targeting all unemployed persons having received either labour market support or social assistance for a specified period of time. In 2001, the size of the target group was estimated to be around 70 000, a later estimate being around 100 000.

Another common characteristic of the centralised programmes was the wide range of placement options. Basically, the Finnish Act on Rehabilitative Work Experience allows a wide range of causes for unemployment and labour market exclusion to be recognised and a variety of tailored strategies to be applied.

The sanctions of the activation programmes are applicable to a varying degree, depending on the type of the programme. Centralised programmes define sanctions accurately, leaving less room for local discretion. This also applies to the Finnish Act, where the process of activation and the rights and obligations are standardised and codified at the national level. The basis of the rights and obligations and related sanctions is specified in the Act. This is different from the Nordic social welfare tradition, where a relatively large scope has existed for local discretion within the national universal framework (Eardley et al., 1996). From the client's perspective, a stronger codification and standardisation may be beneficial, clarifying his/her rights of participation. However, it is questionable whether such uniform sanction policy can be attained: traditionally, there has been plenty of variation in the sanction policies of both social welfare administration and employment authorities due to differences in local administrative cultures.

As in Finland, other European activation programmes are also generally targeted to young and long-term unemployed persons. Young people are usually targeted first and more rigorously than older groups. Workfare is easily accepted in the case of young persons without any work history. In the context of the Finnish proactive social policy reform, it was easier to introduce workfare for young uninsured unemployed. Older unemployed were protected more efficiently by trade unions. Besides, the Act on Rehabilitative Work Experience only concerned the means-tested part of the income support for the unemployed, i.e. labour market support and last-resort social assistance: Unemployed jobseekers whose income maintenance is based on unemployment insurance are not included. In Finland, the target group of the new activation legislation is large and relatively heterogeneous due to the high number of unemployed that participate in the activation scheme. The large numbers of participants make the implementation process resource-intensive and also qualitatively demanding. Furthermore, the high number of clients leave room for selection at the local level. This may bring up the question of 'creaming off' instead of focusing on the most vulnerable cases.

2 Quality of life and self-performance

According to prevailing knowledge based on various studies, opportunities in the labour market can also be expected to mean opportunities for developing personal time, social relationships and psychological well-being (Jahoda, 1982; Gallie et al.., 1994). Typically, it was assumed that employed persons have a higher level of well-being than the unemployed. This simple interpretation is, however, inadequate to explain all the variations that individuals may experience in different labour market situations. It might be that unemployment is desirable in the sense that it enables individuals to withdraw from a stressful and busy working life. It is also possible that being in work can be a source of depression and a low level of psychological well-being. Not all results of research support the assumption that the unemployed automatically have a lower level of well-being. For example, Goul Andersen (2001) has empirically demonstrated that some individuals have a higher level of well-being in long-term unemployment than during work.

In such cases, stereotypical assumptions on what effect different labour market positions have on well-being are reversed. Hence, it can surely be expected that variations within the level of well-being between the unemployed and the employed can be considerable at the individual level, but in general, it is also possible to argue that unemployment is more often associated with a low level of well-being, with employment usually connected to a higher level of well-being. Therefore, it can be said that efforts for increasing levels of employment and decreasing levels of unemployment can have an overall positive impact on the social and psychological well-being for those individuals who find a job. It can be expected that getting a job improves, in most cases, the quality of life of those who were previously unemployed.

This hypothetical connection between labour market status and social and psychological well-being is the focus of this article. Our aim is to assess the connection between the reform measures of rehabilitative work and the social and psychological well-being of individuals. We will attempt to discover how the participation in activation measures based on rehabilitative work reform affects the experiences of the activated unemployed. More precisely, we will examine the impact of activation measures on the self-performance and quality of life of the unemployed.

One way to approach the concept of self-performance is to relate it to individual expectations, in other words, the capability of coping and interacting with one's environment and how an individual manages to carry out the various tasks and activities of one's life. Self-performance is closely related to the concept of self-respect, where individuals with a low level of self-performance - as with low levels of self-respect - can be seen to act only with partial effectiveness in daily life, whereas individuals with a high level of self-performance can be seen to have more self respect. It can also be argued that individuals who have a high level of self-performance mostly feel themselves able to interact effectively with their environment (Smith & Wallston, 1991).

Self-performance is examined in this article by using Wallston's 'four items scale' (see Smith & Wallston 1991). We focus on how capable individuals feel themselves to be and what kind of expectations they have concerning their ability to interact effectively with their environment. The scale of self-performance can vary from a lowest score of 4 to a highest score of 24, where low values indicate a low level of self-performance and high values a high level of self-performance.[2]

As with the concept of self-performance, the concept of the quality of life is not unambiguous. From a wider perspective, the quality of life can be connected to the concept of well-being. A good quality of life is paralleled by a fulfilment of personal needs and the experiences of an adequate level of personal resources. The quality of life becomes concrete through the subjective feelings and experiences of individuals. Generally, quality of life is something that makes our life feel like it is worth living, though on an individual level, interpretations on which elements are more strongly related to the quality of life can vary greatly. A high quality of life may be due to good social relationships whereas, in other cases, it may be due more to the extent that an individual has succeeded in the labour market or achieved professional status.

Important factors that have been associated with quality of life include: good health, social relationships, economic well-being, and the capabilities of the individual in general (Chubon, 1999; Allardt, 1976). Any measurement of the quality of life must be related to individual values and requires both an objective and subjective defining of the essential elements of the quality of life. An objective way to define quality of life is based on the valuations and sophisticated assumptions of the researchers about the important factors of the quality of life. The subjective approach to defining the quality of life is based on the factors that individuals have determined to be the important elements in their life and its quality. In Nordic countries, quality of life is usually examined objectively, as it is in this article, with the subjective approach evident but less in use (Kainulainen, 1998).

Quality of life is examined in this article by using a scale for the quality of life formulated by Chubon (1999). The quality-of-life scale includes 20 items in the form of questions. The questions concern the health, self-confidence, and autonomy of individuals. The range of the scale is 20–140, where scores over 100 indicate a good quality of life, while scores below 80 indicate a poor quality of life.[3]

---

2 The Cronbach Alpha value for the sum variable of self-performance was in the first survey period (T1) 0.71 and in the second survey period (T2) 0.77. Presented values were almost equal with the values presented in study carried out by Smith *et al.* (1991). Smith et al. applied also the four items self perfomance scale.

3 Cronbach Alpha value for the sum variable of the quality of life was in the first survey period (T1) 0.91 and in the second survey period (T2) 0.92. Presented values are in line with the earlier studies carried out by using the method which is applied also in this article (see Chubon 1999).

3 Research data and method

The study aimed at an overall assessment of the implementation of the Act and the effectiveness of the measures involved, using different theoretical and methodological approaches. The extensive part of the research examined how the activation measures in the Act affect the welfare resources of the unemployed.

The study was based on a quasi-experimental research frame. A longitudinal design was applied to follow the impact of the measures on an individual level, and a comparison group was composed for the research group on the basis of the waiting lists for activation measures kept by the civil service. The impact of the activation process on clients' employment status and life in general was analysed quasi-experimentally at two different stages. The longitudinal design was as follows: The first survey (T1) was carried out on the sample group immediately after the activation plan, the follow-up (T2) took place 6 to 10 months after the plan had been drawn up. The research group consisted of unemployed persons who took part in the activation plan programme. The control group was formed by the unemployed entitled to the activation plan but who had not yet been invited to participate in this activation measure.

The data were collected by questionnaires and a quantitative statistical analysis was done. The selection process of the research material can also be studied on an aggregate basis by means of the client register of the employment authorities. At stage T1, the study focused on both the unemployed for whom an activation plan had been prepared and on the civil servants engaged in the process. At stage T2 questionnaires were sent to the target group part that constituted the research group of unemployed persons at T1.

The research was based on a representative sample of Finnish municipalities: covering 57 municipalities and corresponding employment offices. The municipalities were chosen so as to cover small, medium-size and large municipalities and both high and low unemployment rates. The total number of the forms we got back was 3,158 in the first survey and 2,013 in the second survey. The relative response rate was 64 %.

4 Results

Results concerning the activation plan showed that it seems to have no clear impact on self-performance. The level of self-performance was more related to labour market status, the previous level of self-performance and experienced working ability (Table 1).

Table 1. Self-performance by explanatory variables[4]

		Regression coefficient	Std. reg. coefficient
Standard: 6,347			
Labour market status			
	Work	0,97**	0,07**
	Work-based activation	0,90***	0,08***
	Training and rehabilitation	0,31	0,02
Activation position			
	Activation plan done	0,15	0,02
Marital status			
	Single	-0,59*	-0,06*
Age		0,00	0,00
Gender			
	Woman	-0,07	-0,01
Vocational training			
	Vocational school	0,40	0,05
	College (at least)	0,00	0,00
Working capacity			
	Moderate	1,42***	0,17***
	Good	2,57***	0,30***
Unemployment duration while 24 month			
	7-12 month	0,15	0,01
	13-24 month	0,28	0,03
Community type	Densely populated community		
	Town	-0,45	-0,05
		-0,74*	-0,08*
Previous self-performance (T1)		0,56***	0,52***

R-square: 45,6 % *$p < 0.05$, **$p < 0.01$, ***$p < 0.001$.*

Further, marital status and the type of living area or municipality evidently had an impact on the level of self-performance. Self-performance was higher among those unemployed who had got a job or who had participated in work-based activation measures than in those who remained unemployed or took part in education-based activation measures. Activation measures such as education and rehabilitation did not have a clear positive or negative impact on self-performance. If we look at the impact of marital

4 Compared groups: Labour marked status = Unemployed; Activation position = Activation plan not done; Marital status = Married; Age = continuous variable; Vocational training= No vocational training; Working capacity = Bad; Unemployment duration in the last 24 months= under 6 months; Community type = Scattered settlement; Previous self- performance (T1) = Bad.

status on self-performance, we find that unmarried unemployed persons have a lower level of self-performance than those who were married. In addition, individuals with good working ability have a higher level of self-performance than individuals who had a lower level of self-perceived working ability. Finally, those individuals who had a high level of self-performance during the first survey period also had a higher level of self-performance in the time of second survey period.

Next, we turn to the effects of activation measures on the quality of life. In order to prove the effectiveness of the activation measures, it is essential to address the relationship between the labour market status and quality of life. Our results show that those people who participated in the activation measures that were work-based had a higher level of quality of life than those who - in the second survey period - were still unemployed, or who participated in activation measures that were based on education or rehabilitation, or who had got a job (Table 2).

By considering those factors that impact on the quality of life, it is hoped to obtain a more accurate and extensive understanding of the concept of quality of life. In Table 2 it can be seen how factors other than labour market status influence the quality of life. We can see that the activation plan as such is not a measure that has an impact on the quality of life. Moreover, professional education, age, duration of unemployment, gender, marital status, and type of living area are also not factors that predict the level of quality of life. However, factors such as the level of previous quality of life and working ability can be seen to have an effect. The higher the level of self-perceived working ability during the second survey period the higher is the quality of life.

The connection between the level of the quality of life and labour market status is not unambiguous in light of these results. Unemployed persons who participated in activation measures similar to work have a higher probability of reaching a higher quality of life than those who are unemployed or who have not participated in work-based activation measures or who got a job in the labour market. The reason for the fact the quality of life is not significantly better amongst the employed than amongst the unemployed is not easy to discern. The results run counter to the common paradigm on the general positive impacts of employment. Perhaps the jobs where respondents work are not in compliance with the hopes of respondents or perhaps the jobs need more or less working skills than the respondent possesses.

To be able to answer this unpredictable result, we would need more information on the work places of those who became employed and the jobs they do. With the data set we had for this study, sophisticated explanations or assumptions for the unexpected connections between labour market status and the quality of life are difficult to arrive at.

Table 2. The quality of life by explanatory variables[5]

		Odds Ratio	*Asymptotic 95 % CI*
Labour market status			
	Work	1,64	0,94 - 2,84
	Work-based activation	2,07***	1,35 - 3,18
	Training and rehabilitation	1,30	0,75 - 2,26
Activation position			
	Activation plan done	0,93	0,66 -1,33
Marital status			
	Single	0,77	0,55 - 1,08
Age		0,99	0,97 - 1,00
Gender			
	Woman	1,07	0,77 - 1,46
Vocational training			
	Vocational school	1,27	0,90 - 1,79
	College (at least)	1,17	0,72- 1,92
Working capacity			
	Moderate	1,23	0,78 - 1,97
	Good	2,26**	1,35 - 3,79
Unemployment duration while 24 month			
	7-12 month	1,78	0,89 - 3,59
	13-24 month	1,58	0,86 - 2,93
Community type			
	Densely populated community	1,22	0,79 - 1,88
	Town	0,87	0,53 - 1,44
Previous quality of life (T1)			
	Moderate	8,00***	5,77 - 11,11
	Good	39,57***	22,65 - 67,15

R-square: 79.3 %
*p < 0,05, **p < 0,01, ***p < 0,001.

5 Conclusions

The results concerning the activation plan as such can be interpreted in two different ways. Firstly and according to the critical point of view, the activation plan can be seen

5 Compared groups: Labour marked status = Unemployed; Activation position = Activation plan not done; Marital status = Married; Age = continuous variable; Vocational training = No vocational training; Working capacity = Bad; Unemployment duration in the last 24 months = under 6 months; *Community type = Scattered settlement;* Previous quality of life (T1) = Bad. Hosmer-Lemeshow: p = 0,168. *N = 1170.*

as a measure that has no effect at all on the self-performance of the unemployed. Secondly, and according to a more positive interpretation, the activation plan can be seen as a measure that increases opportunities for the unemployed to participate in different activation measures, and in this way, indirectly increases the self-performance of individuals. Without an activation plan, it would be more likely that at least some clients would have lacked an opportunity to take part in concrete activation measures such as supported employment and rehabilitative work, which proved to be factors improving the social and psychological well-being of the unemployed.

In particular, the benefits of the activation measures were best achieved through work-based measures. The unemployed who participated in work-based activation measures had a higher level of self-performance and quality of life than those who had stayed unemployed or who took part in education-based activation measures. Other factors that seemed to have impacted on the welfare of activated and non-activated persons were marital status, vocational training, and health. Married and unmarried couples, better educated persons, and people with good health on the whole had a better quality of life and self-performance than singles, less-educated persons and people with poor health. Activation measures such as training and social and health services had no major impacts on welfare among those activated. In this sense, the activation plan as a new activation method is akin to a tool of inclusion. It increases the possibility of getting work-based activation services and hence indirectly improves the social well-being of the unemployed.

Despite the indirect benefits of the activation plan and the direct benefits of the activation measures on social and psychological well-being, the impact of activation on employment in general was not evident. The employment rate in the open labour market was not higher among those who had participated in the activation plan compared to those who remained unemployed or who had no experience of activation. After 6–8 months of the implementation of the activation plan, only 8 per cent of those activated had got a market-based job, while 9 per cent of those who had not participated in the activation plan had managed to find a job (Ala-Kauhaluoma et al., 2004.) The act of rehabilitative work was therefore, at least in the short term, effective only in a social sense.

References

Ala-Kauhaluoma, M., Keskitalo, E., Lindqvist, T. & Parpo A. (2004). Työttömien aktivointi. Kuntouttava työtoiminta -lain sisältö ja vaikuttavuus [Activating the unemployed. Rehabilitating work experience – the content and effectiveness of the Act]. *Tutkimuksia* 141. Helsinki: STAKES.

Allardt, E. (1976). *Hyvinvoinnin ulottuvuuksia [Dimensions of welfare]*. Porvoo: Wsoy.

Andersen, G. (2001). *Coping with long-term unemployment: Economic security, labour market integration and well-being.* Paper prepared for ESF/EURESCO Conference Labour Market Change, Unemployment and Citizenship in Europe, Helsinki, 20–25. April 2001.

Chubon, R.A. (1999). *Manual for the Life Situation Survey*. University of South Carolina, School of Medicine. Rehabilitation Counseling Program.

Clasen, J. & van Oorschot, W. (2002). Work, welfare and citizenship: Diversity and variation within European (un)employment policy. In J.G. Andersen, J. Clasen, W. van Oorschot & K. Halvorsen (Eds.), *Europe' s new state of welfare* (pp. 233–245). Bristol: The Policy Press.

Deacon, A. (2002). *Perspectives on welfare. Ideas, ideologies and policy debates.* Buckingham: Open University Press.

Dwyer, P. (2000). *Welfare rights and responsibilities*. Bristol: Polity Press.

Eardley, T., Bradshaw, J., Ditch, J., Gough, I. & Whiteford, P. (1996). *Social assistance schemes in OECD countries*. Volume 1: Synthesis Report No 46. DSS Research Report: London HMSO.

Gallie, D., Gershuny, J. & Vogler, C. (1994). Unemployment, the house social networks. In D. Gallie, C. Marsh & C. Vogler (Eds.), *Social change and the experience of unemployment* (pp. 231–263). New York: Oxford University Press.

Gilbert, N. & van Voorhis, R.A. (Eds.). (2001). *Activating the unemployed. A comparative appraisal of work-oriented policies*. ISSA Series 3. New Brunswick: Transaction.

Hanesch, W., Stelzer-Orthofer, C. & Balzter, N. (2001). Activation policies in minimum income schemes. In M. Heikkilä & E. Keskitalo (Eds.), *Social assistance in Europe. A comparative study on minimum income in seven European countries. Synthesis Report* (pp. 122–151). Helsinki: STAKES.

Heikkilä, M. (1999). *A brief introduction to the topic. In linking welfare and work. european foundation for the improvement of living and working conditions*. Luxembourg: Office for the Official Publications of the European Communities.

Heikkilä, M. & Uusitalo, H. (Eds.) (1997). *The cost of cuts. Studies on cutbacks in social security and their effects in the Finland of the 1990s*. Helsinki: STAKES.

Hvinden, B. (1999). Activation: a Nordic Perspective. In Linking Welfare to Work.

European Foundation for the Improvement of Living and Working Conditions. Luxembourg: Office of Official Publications of the European Communities.

Hvinden, B. Heikkilä, M. & Kankare, I. (2001). Towards activation? The changing relationship between social protection and employment in Western Europe. In M. Kautto, J. Ftitzell, B. Hvinden, J. Kvist & H. Uusitalo (Eds.), Nordic Welfare States in the European Context. London: Routledge.

Jahoda, M. (1982). Employment and unemployment – A social-psychological analysis. New York: Cambridge University Press.

Kainulainen, S. (1998). *Elämäntapahtumat ja elämään tyytyväisyys eri sosiaaliluokissa* [The events of life and life satisfaction in different social classes]. Kuopion yliopiston julkaisuja E. Yhteiskuntatieteet 62. (University of Kuopio Publications E. Social Sciences 62.) Kuopio: Kuopion yliopisto/University of Kuopio.

Kosonen, P. (1998). Activation, incentives and workfare in four nordic countries. Paper presented at the Conference: Comparing Social Welfare Systems in Nordic Countries and France, Gilleleje, Denmark, 4-6 September 1998.

Kosunen, V. (1997). The recession and changes in social security in the 1990s. In M. Heikkilä & H. Uusitalo (Eds.), *The cost of cuts. Studies on cutbacks in social security and their effects in the Finland of the 1990s* (pp. 41–68). Helsinki: STAKES.

Lødemel, I. & Trickey, H. (Eds.) (2001). *An offer you can't refuse. Workfare in international perspective*. Bristol: The Policy Press.

Mannila, S. (2001). Unemployment and activation policy. Finnish experience. In T. Kieselbach, A.H. Winefield, C. Boyd & S. Anderson (Eds.), *Unemployment and helath, international and interdisciplinary perspectives* (pp. 285-302). Bowen Hills: Australian Academix Press.

Rehn, G. (1988). Ekonomisk politik vid full sysselsättning. [Economic policy and full employment]. In E. Wadensjö, Å. Dahlberg & B. Holmlund (Eds.), *Full Sysselsättning utan inflation [Full employment without inflation]* (pp. 53–64). Stockholm: Skrifter i urvall.

Sihto, M. (2001). The strategy of an active labour market policy. An analysis of its development in a changing labour market. *International Journal of Manpower, 22*, 683-706.

Smith, C. & Walsston, K. (1991). The mediational role of perceived competence in psychological adjustment to rheumatoid arthritis. *Journal of Applied Social Psychology, 21*, 1218–1247.

3. INTERVENTIONS TO LIMIT THE ADVERSE HEALTH IMPACT

Health Promotion for the Unemployed: Needs, Strategies and Evidence on Effectiveness and Efficiency

Thomas Elkeles & Wolf Kirschner

Introduction

Our paper will focus on five topics concerning the role health insurance companies could play with regard to health promotion activities aimed at the unemployed or persons threatened by unemployment. We will:

1. briefly describe the development of unemployment in Germany,
2. summarise research which provides support for the association between unemployment and health,
3. clarify legal and factual possibilities of health promotion activities for the unemployed in Germany,
4. identify and describe national and international practical health interventions in this group, summarise the outcome of these interventions with respect to their effectiveness and efficiency, and determine existing problems in interventions carried out,
5. and draw up final conclusions and recommendations for further projects as well as give guidance for future research.

Our paper is based on the results of a respected expert report for the Federal Association of Company Health Insurance Funds (Bundesverband der Betriebskrankenkassen) (Elkeles & Kirschner, 2004), which was prepared in October 2003 and a 2002 evaluation for the Fund for a Healthy Austria (Fonds Gesundes Österreich) (Kirschner et al., 2003).

1 Development of the unemployment rate in Germany

In Germany (see Figure 1), unemployment has been steadily increasing in the last forty years and characterized by ever higher levels of unemployment on a regular basis, indicating that cyclical economic developments were and still are affected by structural problems in the German economy and in the labour markets. Leading German economists do not expect a significant reduction of unemployment in light of the moderate future growth being predicted for the economy.

Figure 1. Unemployment rates, Federal Republic of Germany, 1965 - 2003

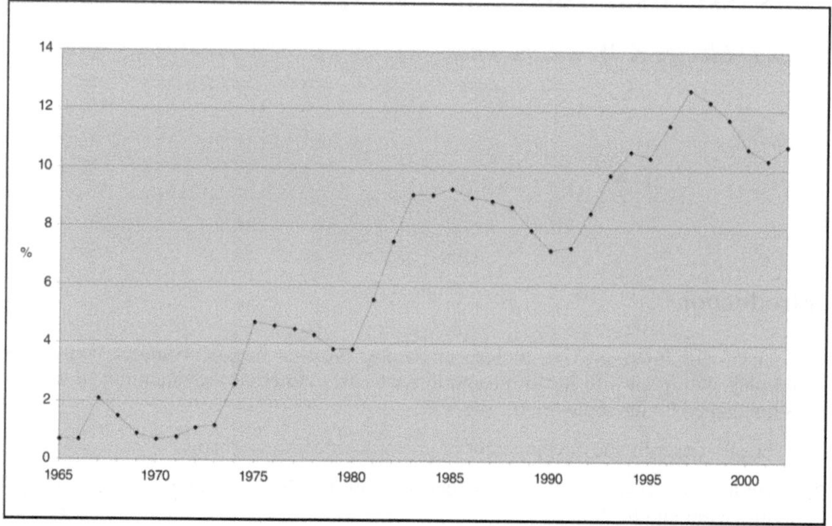

As in many other countries, unemployment affects particularly those persons who lack school education or vocational training - regardless of age. In Germany, unemployment rates differ drastically between eastern and western federal states with rates twice as high in the east. But there are also severe differences in the western federal states with rates decreasing from northern to southern states. The proportion of persons, who have been unemployed for more than 12 months, is 33% (see Figure 2); in 2003 this was 40% for those in Eastern Germany (see Figure 3). Realistically, this means that about 10% of the population are officially out of work. However, we have to bear in mind that the complete number and proportion of the unemployed in society at large is much higher due to legal regulations that apply to unemployment statistics. Additionally, with respect to the dynamics of entering and leaving the unemployment situation, large segments of the population have already experienced unemployment themselves. Furthermore, everybody knows people who are or were unemployed or who are threatened by unemployment in their current job. To summarise, a majority of Germans regard unemployment as being the main problem facing German society today.

Figure 2. Relative duration of unemployment in Germany, 2002 (in %)

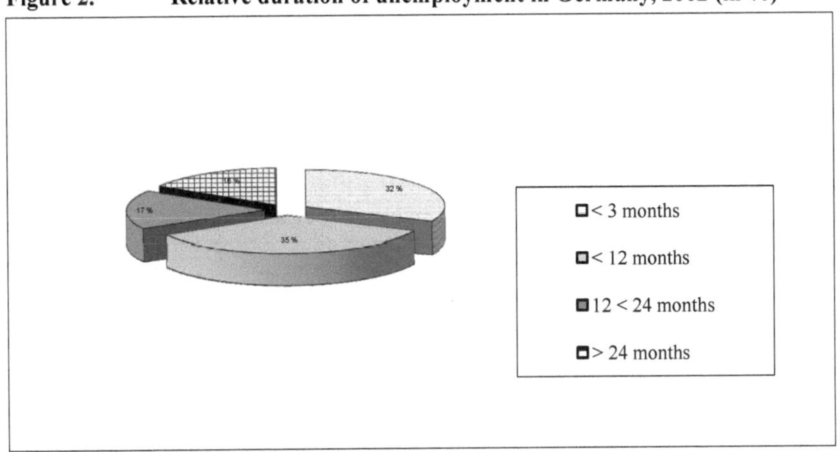

Figure 3. Proportion of Long-term-Unemployed in West and East Germany, 1992-2003 (%)

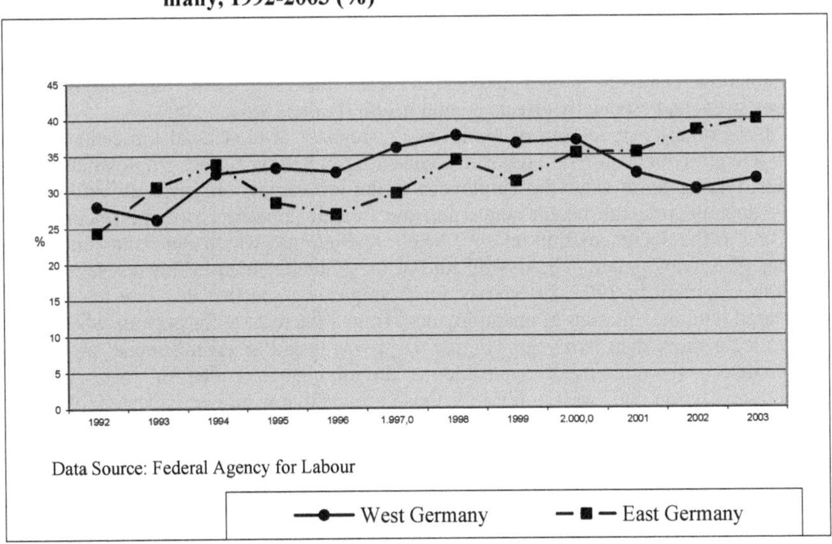

2 Health effects of unemployment – Social epidemiological evidence

Internationally and nationally there is overwhelming evidence that unemployment has adverse effects on health in terms of increasing risk behaviour, decreasing resources which in turn increase the risk of incident morbidity and/or further lead to a progression of already prevalent diseases. Even increased mortality rates have been demonstrated. With respect to the duration of unemployment a response relationship can be shown. Though there is strong descriptive epidemiological evidence for these associations, there are several intervening factors aggravating or diluting negative health effects. As many respective results have already been presented and discussed at this conference, we will focus on three examples: data with respect to self reported health status in survey data with health indicators from the job centers resp. the Federal Agency for Labour hospital discharge data on days of hospitalisation by underlying diagnoses.

As welfare recipients and long-term unemployed will be integrated legally in the new group of recipients of unemployment support II (Empfänger von Arbeitslosenhilfe II) beginning January 2005, we will finally show data on the health status of welfare recipients in Berlin. Table 1 shows several health effects of (long term) unemployment: While only 11% of employed persons rate their overall health status as not good or bad, this proportion rises to 16% for the short-term unemployed and to 33% for the long-term unemployed.

There are similar effects on handicaps in daily living, diseases and hospitalisation - though not altogether significant - and on the average rating of health and the overall life situation. Data not shown here demonstrate that (long term) unemployment increases stress and especially affects mental health (Paul & Moser, 2001).

In Germany persons out of work have to register at their local job centers to receive unemployment support and/or to counselling regarding further employment possibilities. During these visits the employees of the job centers are officially and unofficially documenting the health status and the overall appearance of the unemployed, focusing in the documentation on any “health restrictions” which can reduce the possibilities of reemployment. The overall rate of these documented health restrictions was 23% in Germany in 2003. In Saxony for example, we see that this rate has steadily increased with the duration of unemployment from 14% to 25% for persons who are out of work for more than two years (Figure 4). It was found in several model projects in the context of the ongoing reorganisation of the job centers (called *MOZART)* that this “health restriction rate” rose to 60% for those groups that were less educated, older and unemployed for a long time. This indicates that long term and older unemployed persons suffer from several complaints to a high degree, but they also tend to be severely and often chronically ill. This clearly indicates that many persons who are out of work for a long time need to be included in health promoting activities but they may also benefit from effective health management strategies combining targeted therapy, rehabilitation and health promotion measures.

Table 1. Self reported health status of employed and unemployed person in Germany

	Unemployed		Employed	
	>12 months	<12 months		
	%			
n=	122	131	35392	
Reported health status				
very good / good	33,2	43,5	50,5	
satisfactorily	32,8	40,5	38,9	
not good /bad	**32,8**	**16,0**	**10,7**	***/*
Handicaps by health status in dealing with daily activities				
not at all	50,0	54,2	68,1	
moderate	22,1	32,1	25,9	
severe	**27,9**	**13,7**	**6,0**	
At least one day in the last 4 weeks bed-ridden	12,4	15,3	7,8	***/n.s.
Hospital stay last 12 months	11,5	10,8	8,3	***/n.s
Average satisfaction with:				
health	**4,5**	**4 ,9**	**5,1**	***/*
life situation	**4,8**	**5,2**	**5,6**	***/*
Average age	44,5	40,4	42,2	

Note. *p < 0,05; ** p < 0,01; *** p < 0,001 (Chi-SquareTest)
a: column 1 + 2 vs. column 3
b: column 1 vs. column 2
c: Average of a 7-point scale
Data source: Cumulative data of the National Health Survey (West) 1984 to 1991, German Cardiovascular Prevention Study (GCP) (N=55.308). Source: Elkeles (1999).

Up to now we have dealt with rather subjective data. The scope of the medical problems in unemployed can be demonstrated by hospital discharge data from a health insurance company (Table 2). This data suggest that addictive behaviour increases with greater duration of unemployment, resulting in severe addiction and psychotic diseases. Other widespread chronic diseases such as diabetes and heart disease are also somewhat higher amongst the unemployed. These data may, however, reveal only the tip of the iceberg as members of the health insurance company GEK (Gmünder Ersatzkasse) are biased towards better education and middle and upper social classes.

Welfare recipients – most of them are male and long term unemployed - are one population group, which has one of the worst health status in the German society. A survey conducted in the city of Berlin in 1990 revealed that 22% of welfare recipients between the ages of 25 to 34 years rate their health status as neither good nor bad. This frequency of health status is common in the overall population in Berlin between the ages of 50 to 59 years (Table 3). 55%, that is the majority of the older welfare recipients, rated their health status as not good or bad.

Figure 4. Health restrictions of unemployed by length of unemployment (in %)

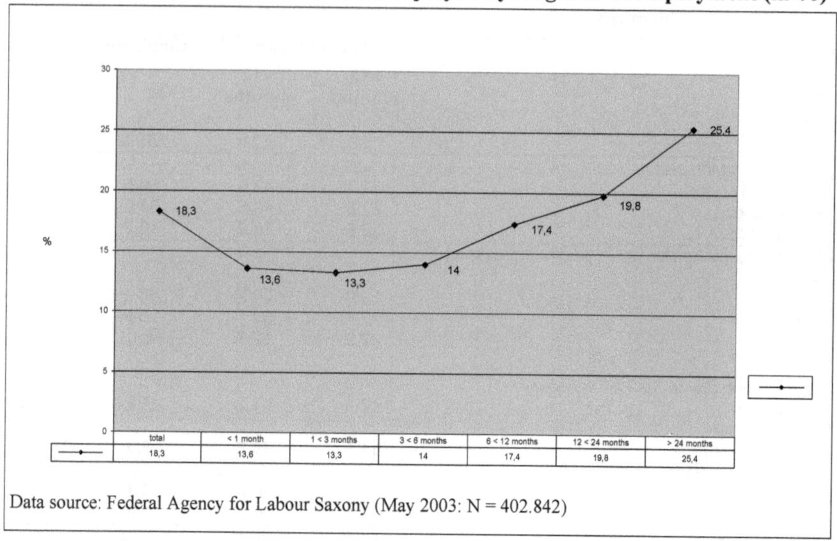

Data source: Federal Agency for Labour Saxony (May 2003: N = 402.842)

To summarize, unemployment and long-term unemployment have a considerably negative effect on health. Despite the descriptive epidemiological evidence, we do neither theoretically nor empirically understand the complete picture. Nor do we understand which mechanisms are involved. This means that we have only limited analytical knowledge concerning the risk or preventive factors in this process due to insufficient analytic research nationally and internationally (e.g. one would need to control some twenty moderating or confounding factors in the case of large scale and long-time cohort studies according to Kieselbach, 1988).

Due to this lack of knowledge, interventions in the field of health promotion for the unemployed cannot be based on confirmed analytic epidemiological data in terms of relative risks. The majority of interventions - if not all of them - in this field are – if one may say so, logically based on the belief that effective intervention strategies are expected to stop this process of worsening health. So in terms of intervention theories, interventions can only have the character of open experiments.

Table 2. Hospital days by underlining diagnosis of employed and unemployed

ICD 9	Days in hospitals	Unemployed	Employed	Ratio
		Days in 1000	Days in 1000	
303	Alcohol addiction	229	25	9,2
295	Psychoses	183	18	10,2
300	Neuroses	105	23	4,6
304	Medical drug addiction	66	3	22,0
291	Alcohol psychoses	43	4	10,8
414	Ischemic Heart Disesase	41	27	1,5
296	Affective Psychoses	36	12	3,0
722	Intervertebral disc degeneration	34	31	1,1
309	Psychogenic reaction	31	7	4,4
250	Diabetes	31	12	2,6
571	Chronic liver disease	30	3	10,0
301	Psychopathy	28	3	9,3

Data: Gmünder Ersatzkasse 2000

Table 3. Health status of welfare recipients and overall population in Berlin (1990)

Health status not good/bad				
	18-24 J.	25-34 J.	35-44 J.	45-55 J.
Welfare recipients n=497*	20%	**22%**	**50%**	**55%**
	to 29 J.	30-39J.	40-49 J.	50-59 J.
Population Berlin n=2732**	5%	8%	11%	**23%**

* Study on the health status of welfare recipients in Berlin 1990 (Kirschner, W. et al.).
** Survey on health and social situation in Berlin, 1991 (Kirschner,W. et al.).
[a] Data source: Kirschner et al. (1990).
[b] Data source: Kirschner et al. (1991).

3 Legal and factual possibilities of health interventions for the unemployed and welfare recipients

In Germany “Unemployment and Health” is still a rather scientific topic, which up to now has played only a marginal role in political debates. Health promotion in the unemployed was in past and is presently no “field of action” for the job centers or the Federal Agency for Labour. For years costly interventions in qualification processes were carried out by the job centers, but they were downsized dramatically since 2002 with the objective of lowering expenses. This decision was facilitated by the fact that these interventions were not sufficiently – if at all – evaluated, thus failing to demonstrate any evidence of their effectiveness and efficiency. Both make it very unlikely that health promotion for the unemployed will be a relevant intervention strategy in the job centers in the near future.

In the ongoing process of reorganising its job centers, Germany has adopted counselling- and profiling strategies from other countries with the overall objective to better match skills and qualifications of the unemployed to the needs of the enterprises. In these profiling processes health status and health restrictions have not yet been completely and thoroughly addressed. In reviewing the MOZART projects already mentioned, we found in only 3 of over 50 projects that health was part of the profiling process and only in one case were any regulations and recommendations made concerning what should be done if certain health problems or health restrictions should be detected. In addition, this exception concentrated mainly on the problem of alcohol addiction. In no case did we find anything resembling targeted health promotion and prevention. Therefore we can conclude that up to now health has not been regarded as an intervention tool in the counselling and profiling system. Additionally, some data privacy problems arise in a broader "screening" of health questions, as the job offices - as employers - are only allowed to ask health-related questions relevant to a specific workplace. In general job counselling - e.g. in contrast to employed pilots or bus-drivers - there is however a lack of concrete job requirements where the health status could or must be addressed.

In contrast to the Federal Agency for Labour the German welfare offices are familiar with addressing health questions and problems with their clientele. In the Code of Social Law (Sozialgesetzbuch) there is the possibility of offering additional health care services. But here too this perspective is strongly considered and applied mainly in the field of (apparent) addiction. There are, however, no statistics available on the kind and use of these services.

To summarise: Neither in the job centers and the Federal Agency for Labour nor in the welfare offices are there any concepts and regulations for prevention and health promotion for the unemployed. Health is regarded only in terms of limitations, not in terms of health that can be improved. Health is a necessary condition for further employment and against the background of 6 million persons out of work special interventions for the 23% with restricted health – in absolute terms: 1.2 million persons - such strategies make no sense since the other 4.8 million persons out of work without obvious health problems are waiting for jobs. Thus unemployed persons with health restrictions are classified as incapable of employment and without any perspective of re-entering the labour market. In the coming years we do expect the situation to change. Only over a greater course of time, say 15 to 20 years, could demographic changes alter this situation.

In Germany, primary prevention and health promotion is a service of the health insurance agencies for their members as well as the complete population. §20 of the Code of Social Law reads as follows: "The health insurance company shall provide services of primary prevention in their articles of association which are in accordance with the requirements stated in clauses 2 and 3. Services of primary prevention should improve health status and should especially contribute to narrowing social inequalities in health opportunities. The national associations of the health insurance companies in co-operation with independent analytic expertise mutually consult on prior operational fields and criterions for services on a regular basis according to clause 1, particularly with respect to needs, target populations, accessibility, contents and methods." (Kirschner et al., 1995)

For a better understanding of the practice of prevention and health promotion in Germany we would like to provide additional information and interpretations of this regulation.

Prevention and health promotion in Germany as services of the health insurance companies were initially established in 1989. Until 1996, when the relevant regulation was cancelled by the conservative-liberal coalition, the health insurance companies -

often in co-operation with external partners - had continuously developed and extended individual preventive services, mainly in form of courses addressing nutrition, drugs, physical activity and lower back pain. Only in the field of health promotion at the workplace was the individual approach exceeded. An independent scientific expert report (Kirschner et al., 1995) confirmed an altogether positive development with some critical remarks concerning aberrations with respect to ostensible marketing activities, insufficient assessment of needs and in particular very low standards in its evaluation and even documentation.

Box 1. § 20 of the code of social law

"The health insurance company should provide services of primary prevention in their articles of association, which are in concordance with the requirements stated in sentences 2 and 3. Services of primary prevention should improve the health status and should especially contribute to narrowing social inequalities in health chances. The national associations of the health insurance companies with involvement of independent analytic expertise resolve mutually and consistently upon prior operational fields and criterions for services according to sentence 1, especially with respect to needs, target populations, accessibility, contents and methods."

In 2000 the new §20 (see above) regulation was adopted with the political intention to:

1. strengthen quality of services, documentation and evaluation, and
2. extend services to population-based approaches, thereby reducing social inequalities in health, in contrast to merely providing individual services for members.

To realise these objectives, the national associations of the health insurance companies formulated precise guidelines as to which services could be offered in the frame of this regulation. As an important prerequisite for all interventions, the guidelines specify documented evidence on effectiveness and efficiency, which were to be substantiated by an expert report prior to the interventions instead of evaluating the interventions themselves. This "ex-ante proof" of effectiveness and efficiency has considerably hindered the further development of prevention and health promotion in Germany in the field of individual prevention as well as in the field of population-based prevention. This is because the experience in interventions and the evidence of evaluative results in health promotion is internationally too low to keep up with these high standards of evidence, especially in the field of interventions aiming at reducing socio-economic health differences (Gepkens & Gunning-Schepers, 1996).

Moreover, this position does not consider that evaluative research on social or health interventions means testing hypotheses, which includes a definite chance of error (alpha or beta).

In addition, evaluative research is highly complex with a high probability of deficiencies. To speak of evidence in the sense of effectiveness, we must have many interventions carefully evaluated by meta-analytical methods with a majority of consistent and positive results. These numerous interventions and evaluations do not exist worldwide for the unemployed.

There are, however, two other important weak points in the new §20 regulation, which obstruct interventions oriented toward population and environment. First of all, the federal states were not incorporated and the problems in involving competing health insurance companies in concerted regional and population-based actions and interventions were underestimated.

These are the most important reasons that the new §20 regulation, especially with its scope on non-individual interventions, was put into practice reluctantly. Since 2000, a few projects have been launched, mainly in schools and kindergartens and, as one result, new and additional structures for health promotion and prevention have actually been planned by the government. Less than 50 programmes addressing persons out of work are estimated for all of Germany. Here the Federal Association of Company Health Insurance Funds (Bundesverband der Betriebskrankenkassen) has led the way.

4 Evaluation of practical interventions with the unemployed

Part of our six-month expertise for the Federal Association of Company Health Insurance Funds (Bundesverband der Betriebskrankenkassen) involved a search of published projects on health promotion and primary prevention addressing persons out of work especially in Germany. Such a search can never be complete but it should nevertheless yield reliable evidence on the given structure of interventions. In our search we found totally 51 interventions. Only 14 of these were even marginally documented. Of the remaining 36 projects, two were still in the conceptual phase. The remaining 34 projects shall now be described according to:

a. intervention objectives (health and/or employment)
b. target populations
c. objectives and methods
d. institutional and financial framework
e. documentation and evaluation
f. number of participants / selection / access and accessibility
g. effectiveness and efficiency

The 36 projects are listed in Table 5 on the next page.

4.1 Intervention objectives (health and/or employment)

Health promotion projects for the unemployed can have uni- or bivariate objectives in promoting health or promoting re-employment or both (Figure 5). When we classify the projects according to the objectives described we find that with 18% only a minority of projects are focussing exclusively on health, while with 47% nearly half of the projects is oriented on exclusively re-employment while the rest aims at both objectives. So the majority of projects (82%) intend to improve re-employment chances.

Table 4. List of projects evaluated

1	Development
2	NAG Projekt
3	Arbeitslosigkeit und Gesundheit
4	Gesundheitsförderung bei Arbeitslosen
5	Job-Plan
6	Sozialagenturen
7	Kurssystem contra Langzeitarbeitslosigkeit
8	Massarbeit
9	Arbeit und Sozialhilfe (Spremberg)
10	Arbeit und Gesundheit (Forst)
11	Fit in den Tag (Wolfsburg)
12	Frauengesundheit (Oschersleben/Börde)
13	Gesundheitstisch (Berlin)
14	Neue Perspektiven – ein gesundheitsförd. Projekt für arbeitslose Menschen
15	Qigong-Kurs (Stuttgart)
16	Servicestelle Arbeit und Gesundheit (Spremberg)
17	Bündnis für Arbeit der Stadt Köln
18	Entwicklungs- und Vermittlungsassistenz für Dauerarbeitslose (EVA)
19	Berufliche Eingliederung von Sozialhilfeempfängern
20	S.A.V.E. Freising (Sozialamt Arbeitsamt Verbinden Entwickeln)
21	Projekt LOS
22	Fit für den Arbeitsmarkt
23	A walk on the wild site
24	ZALT Gesundheitswochen
25	Qu' est ce qui cloche chez elle?
26	Gesundheitsorientierte Selbstmanagement-Beratung bei Arbeitslosigkeit (GESA)
27	Selbstmanagement-Beratung und Gesundheitsförderung für Instabil-Beschäftigte und Arbeitslose (SEGEFIA)
28	Gesundheitsförderung und Kompetenzoptimierung in der Erwerbslosigkeit (Ge+Ko)
29	Berufliche Eingliederung und Arbeitsmaßnahme (BEAM)
30	Gesundheitlich orientierte Outplacementberatung
31	Werkstatt 90
32	Selbstverantwortung fördern durch motivierende Gesundheitsgespräche
33	Michigan Prevention Research Center (MPRC)
34	Työhön Job Search Program
35	Proudfoot, J. et al.
36	Muller

4.2 Target populations

Thirty per cent of the projects deal with long term unemployed and 12% are directed explicitly to persons out of work with existing health problems (Table 6). The projects cover all age groups, men and women, and a few projects are also orientated to short-term unemployed with a defined objective of improved re-employment chances.

Figure 5. Objectives of n=34 projects evaluated

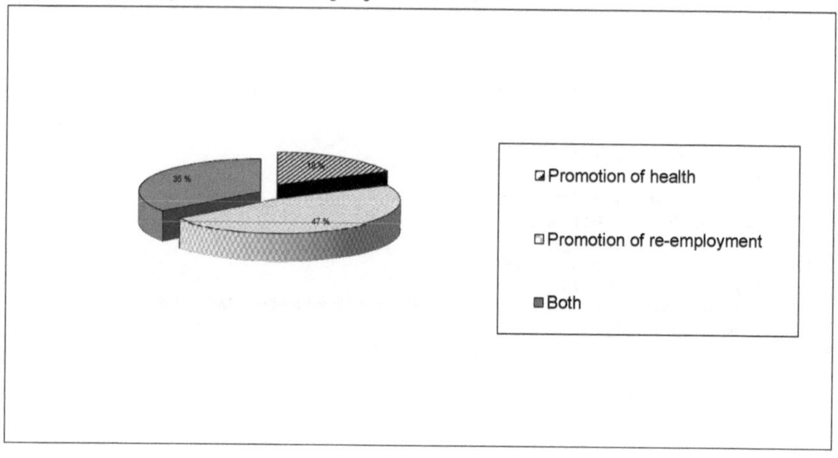

Table 5. Target population

	N	%
Long time unemployed	7	**21%**
Unemployed	5	15%
Social welfare recipients	5	15%
Long time unemployed with health restrictions	3	**9%**
Unemployed women	2	6%
Persons threatened by unemployment	2	6%
Short term unemployed	2	6%
Unemployed adolescents	1	3%
Unemployed of older age	1	3%
Social welfare recipients with health problems	1	3%
Members of health insurance companies unemployed	1	3%
Unemployed men	1	3%
Not clearly stated	3	9%
Total	34	100%

Source: Elkeles & Kirschner (2004), 211.

4.3 Objectives and methods

When looking at the stated objectives (Table 7) we find that, at 41%, improvements in the psycho-social situation is the most frequent objective, while 26% address health improvements as a primary objective and 33% primarily wish to improve their chances in the labour market.

When looking at the methods (Table 8) applied, we find that 79% focus on the traditional concepts and fields of individual health promotion, while the rest employ additional psychotherapeutic techniques. Altogether we do find a broad range of objectives and methods with a strong emphasis on stress and the psychosocial situation supplemented by traditional forms of health promotion activities.

Table 6. **Objectives**

Objectives	N	%
Improving employment chances		
Improving re-employment chances	12	22%
Improving job seeking skills	2	4%
Improving competences	4	7%
Total		**33%**
Psychosocial improvements		
Reducing stress and psychosocial complaints	8	15%
Psychosocial stabilisation	8	15%
Networking	3	6%
Motivation for self-helping activities	2	4%
Coping strategies	1	2%
Total		**41%**
Health		
Health promotion (undefined)	7	13%
Improving health	4	7%
Additional health services	1	2%
Improving health consciousness	1	2%
Health counselling	1	2%
Total		**26%**
Total (multiple nominations)	54	100%

Source: Elkeles & Kirschner (2004), 212.

4.4 Institutional and financial framework

Both have an extremely heterogeneous range with respect to institutions from university departments to private companies, health promotion institutes, self-help groups, welfare offices, job centers and health insurance companies. The same applies to the financial framework with budgets ranging from modest amounts up to millions of euro.

4.5 Documentation and evaluation

With respect to the heterogeneous institutional and financial framework, the standards of both important tools will also vary widely. Based on 22 projects either completed or in their final stage, we found (Table 9) just two reports on evaluation and furthermore 4 indications that an evaluation has been planned or is in the works. Naturally these results should not be overestimated but on the basis of our experience in gathering the respective information we have to conclude that the standards of documentation and evaluation are (very) low.

Table 7. Methods

Stress reduction	12	34%
Physical activity	8	23%
Reduce consumption of (any) drugs	4	11%
Nutritional improvements	3	9%
Traditional health promotion objectives		**77%**
Constructive thinking	1	3%
To enjoy	1	3%
Self- and time-management	1	3%
Personality development	1	3%
Cognitive behavioural therapy	1	3%
Social pedagogy measures	1	3%
Not clearly stated	2	6%
Therapeutic measures		**23%**
Total	35	100%

Source: Elkeles & Kirschner (2004), 212.

Table 8. Standards of documentation and evaluation

No documentation available*	8
Evaluation report	2
External evaluation	1
Internal evaluation	3
Preliminary analyses	1
Publications available	7
Total	**22**

*in the framework of the expertise (6 months)
Source: Elkeles & Kirschner (2004), 214.

4.6 Number of participants / selection / access and accessibility

The number of participants is – with some exceptions – low or very low. This raises questions with respect to the acceptance of these interventions by the unemployed. With respect to the overall determined proof of evidence of health promotion interventions by sufficient evaluation methods, it is astonishing that when participants of the projects are described, no thought is given to whether and to what extent the participants are biased with respect to social, motivational or health variables. And only in a few studies did we find information about the number of unemployed persons who had to be contacted to create the final study group, which is at least a simple possibility for indicating probable bias.

Getting access to persons out of work is one of the most practical problems in these intervention strategies compared to institutionalised populations at schools, hospitals or companies. At job centres and social-welfare offices this population may be easily contactable with the disadvantage, however, that interventions in these settings are not very popular in the opinion of the clientele. There is also the danger that taking part in these interventions could become obligatory, with refusal to take part resulting in a reduction in subsistence benefits, a well known instrument to coerce the unemployed in Germany. In one German intervention study in the context of job centers, the participants reported that they had been pushed to participate by the employees of the job centers. In fact the intervention turned out to be unsuccessful, which might indicate that health promotion under pressure will eventually fail, an obviously plausible conclusion.

4.7 Effectiveness and efficiency

Evidence on both outcome measures is low. When analysing six mental health projects, we find that, at least in terms of effectiveness, 50% reporting positive results, while the other half failed to reach the given objectives (Table 10). In addition, evaluative designs are often poor, which may indicate that the evaluative budgets were completely too small. Furthermore, we find in a majority of cases that the practice of self-evaluation often suggested insufficient experience in evaluation methods.

Table 9. Results on effectiveness of projects aiming at mental health improvements

	Results on effectiveness	
	Positive	Negative
Aktiva Dresden		X
EVA Siegen (Trube/Luschei)		X
MPRC Michigan (Caplan,Vinikur et al.)	X	
Työhön (Vuori et al.)		X
Proudfoot	X	
Muller	X	

Source: Elkeles & Kirschner (2004), p. 215.

4.8 Summary

Eight years after a comparable study by Gepkens and Gunning-Schepers (1996) - though it focused completely on interventions for the social disadvantaged and not specifically on the unemployed - we must - unfortunately - confirm some of their results and conclude that:

- experience from interventions in terms of the number of respective projects is still very limited
- interventions are focused on improvement in health as well as re-employment, with focus on the latter
- objectives and methods mirror the common practices in health promotion interventions supplemented by some additional psycho-therapeutic measures
- institutional and financial frameworks are highly heterogeneous

- standards of evaluation and even documentation are poor at best
- numbers of participants are – with some exceptions – small
- the breakdown of participants with respect to health and social status or participatory motivation is often unknown
- evidence with respect to effectiveness is limited, while efficiency was rarely under observation at all
- the only few positive effects detected in mental health interventions often suffer from methodological problems, e.g., selection bias in the intervention group, a point addressed only as an exception.

Altogether we have to conclude that the practice of health promotion interventions in the unemployed suffers from three important problems:

- inadequate theoretical foundations of interventions, e.g., with respect to the question as to why a certain intervention should yield a particular effect
- problems in program implementation plus
- problems in program evaluation.

Consequently, the hard-headed insistence on ex-ante proofs of effectiveness and efficiency before granting intervention projects is a chimera. On the contrary, we must emphasize that most of these interventions are experiments with unknown results. Funding of individual proposals should therefore be based on a prudent estimation and consideration of possible positive effects, and the respective hypotheses have to be thoroughly tested by an adequate evaluation of a sufficiently financed intervention.

5 Conclusions and recommendations for further projects and accompanying research

Reviewing once again the projects just described and taking into account that we have always been speaking of *interventions*, we detect an additional problem. In fact, in terms of intervention theory most of the projects described are not interventions at all but merely offers or services. The main difference between these two strategies is that offers and services lack any assessment of needs because they will be demanded by self-selective usage. On the contrary, real interventions are based on a thorough assessment of needs because participants without needs may already affect effectiveness and efficiency. Interventions must try to optimize and maximize the participation rates of persons with defined needs in order to realize a program coverage, which ensures efficiency. If not, the program can be effective but will not be efficient. All of these questions are of no interest in offers or services. When planning and developing real interventions in the field of health management or health promotion, we must first of all study the social-epidemiological needs of our target group based on adequate data. When putting the intervention into practice we have to provide need assessment strategies.

But even if we modestly confine ourselves to services and offers, we have to clarify the structure, effects and consequences of self-selective usage. Without this clarification, any so-called "evaluation" is useless because it can at best show positive results from test groups, which have not been sufficiently characterized. In Germany – and most likely in other countries as well – we know little about the usage of health promotion services (Table 11) but very little about the motivation to do so. There are, however, some indications that a majority of those using individual services are only those who are at least in need of them.

Additionally, we have to find out more about the acceptance of respective interventions or services by our target population since low acceptance and even lower potential usage are accompanied by inefficiency.

Finally, we have to remember that many persons out of work – especially those of older age and long-term unemployment – are often severely ill. For such persons, mere health promotion is a drop in the ocean. They are in need of integrative health management strategies combining therapeutic, rehabilitative and preventive measures, which have to be developed.

Table 10. Usage of health promotion services in Germany (in %)

	Total	Men	Women
Health Insurance compulsory	12,7%	7,3%	16,7%
High social class	12,6%	8,2%	18,0%
Private Health Insurance	11,9%	10,3%	14,4%
West Germany	11,5%	7,7%	15,1%
Middle social class	11,2%	7,1%	15,1%
Total	**10,5%**	**7,0%**	**13,8%**
Health Insurance compulsory (AOK)	7,2%	5,3%	8,8%
Low social class	7,1%	5,5%	8,3%
East Germany	6,4%	4,0%	8,6%
Unemployed	?	?	?

Data Source: National Health Survey 1998

For the future, respective programs should be developed more sophistically, implementation has to be better controlled and all phases of evaluation have to be improved, thus requiring additional funding. In addition, documentation and publication should be improved.

Anyone thinking that these proposals will meet no contradiction will be mistaken, at least with respect to the adequate data necessary or available for these analyses. In Germany there is a long tradition of social scientists and epidemiologists engaging in controversial debate - especially on the role of surveys in this context. Opponents of surveys argue that enough other data are already available for these purposes. We disagree, because the question is not whether you have data but whether you have the right data.

In our point of view, many of the questions regarding practice and research to be solved could be realized by carrying out a representative survey with the unemployed, thus clarifying:

- needs
- acceptance and potential usage of services, offers or interventions,
- and building up a data base for further development of strategies.

Last but not least, as an additional advantage, this database can be used as a control sample for all study or project groups, which would resolve the crucial problem of widely unknown participants and provide an opportunity for controlling bias.

References

Arbeitsamt Pforzheim (2003). *Arbeitsmarktprogramm Arbeitsamt Pforzheim.* Retrieved from http://www.arbeitsamt.de/pforzheim/aktuelles/amp_2003.pdf.

Arbeitsgemeinschaft der Spitzenverbände der Krankenkassen (2000). *Gemeinsame und einheitliche Handlungsfelder und Kriterien der Spitzenverbände der Krankenkassen zur Umsetzung von § 20 Abs.1 und 2.* Bergisch Gladbach: IKK-Bundesverband.

Arbeitsgemeinschaft der Spitzenverbände der Krankenkassen (2001). *Gemeinsame und einheitliche Handlungsfelder und Kriterien der Spitzenverbände der Krankenkassen zur Umsetzung von §20 Abs.1 und 2.* Bergisch Gladbach: IKK-Bundesverband.

Arbeitsgemeinschaft der Spitzenverbände der Krankenkassen und Medizinischer Dienst der Spitzenverbände (2003). *Dokumentation 2001. Leistungen der Primärprävention und der betrieblichen Gesundheitsförderung gemäß § 20 Absatz 1 und Absatz 2 SGB V.* Essen: MDS.

Bundesministerium für Wirtschaft und Arbeit (2001). *Einzelübersicht über die MoZart Projekte. Modellvorhaben zur Verbesserung der Zusammenarbeit von Arbeitsämtern und Trägern der Sozialhilfe* [Model projects to improve the co-operation between job centers and social welfare offices]. Retrieved from http://www.bma-mozart.de/index_ie.html.

Bormann, C. & Kneip, H. (2002). Arbeitslosigkeit und Gesundheit bei Frauen im Vergleich der alten und neuen Länder der Bundesrepublik Deutschland [Unemployment and health of women from old and new states of the Federal Republic of Germany – A comparison]. Problemaufriss und Präventionsmöglichkeiten. In A. Trojan & H. Döhner (Eds.), *Gesellschaft, Gesundheit, Medizin – Erkundungen, Analysen und Ergebnisse* (pp. 115-124). Frankfurt: Mabuse.

Braunmühl, C. von (1999). Modellvorhaben des Landes Brandenburg zu Arbeitslosigkeit und Gesundheit [Model project of Brandenburg toward Unemployment and Health]. *Sozialer Fortschritt, 48*(5), 123-130.

Brinkmann, C. & Potthoff, P. (1983). Gesundheitliche Probleme in der Eingangsphase der Arbeitslosigkeit [Health problems in the initial phase of unemployment]. *Mitteilungen aus der Arbeitsmarkt- und Berufsforschung, 16*(4), 378-389.

Brinkmann, C. (1984). Die individuellen Folgen langfristiger Arbeitslosigkeit [Individual consequences of long-term unemployment]. Ergebnisse einer repräsentativen Längsschnittuntersuchung. *Mitteilungen aus der Arbeitsmarkt- und Berufsforschung, 17*(4), 454-473.

Büchtemann, C.F. (1983). Infratest Sozialforschung. Die Bewältigung von Arbeitslosigkeit im zeitlichen Verlauf [Coping with unemployment]. Repräsentative Längsschnittuntersuchung bei Arbeitslosen, Abgängen aus Arbeitslosigkeit und beschäftigten Arbeitnehmern 1978-1982. *Forschungsberichte des Bundesministers für Arbeit und Sozialordnung, 85.* Bonn.

Büssing, A. (1993). Arbeitslosigkeit. Differentielle Folgen aus psychologischer Sicht [Unemployment. Differential consequences from a psychological view]. *Arbeit, 1*(2), 5-19.

Caplan, R.D., Vinokur, A.D., Price, R.H. & Ryn, M. van (1989). Job seeking, reemployment, and mental health. A randomized field experiment in coping with job loss. *Journal of Applied Psychology, 74*(5), 759-769.

Elkeles, T. (1999). Arbeitslosigkeit, Langzeitarbeitslosigkeit und Gesundheit [Unemployment, longterm unemployment and health]. *Sozialer Fortschritt, 48*(6), 150-155.

Elkeles, T. (2001). Arbeitslosigkeit und Gesundheitszustand [Unemployment and state of health]. In A. Mielck & K. Bloomfield (Eds.), *Sozialepidemiologie. Eine Einführung in die Grundlagen. Ergebnisse und Umsetzungsmöglichkeiten* (pp. 71-82). Weinheim/München: Juventa.

Elkeles, T. (2003). Arbeitende und Arbeitslose [Working and unemployed people]. In F.W. Schwartz, B. Badura, R. Busse, R. Leidl, H. Raspe, J. Siegrist & U. Walter (Eds.), *Das Public Health-Buch. Gesundheit und Gesundheitswesen* (2nd Ed., pp. 653-659). München/Jena: Urban and Fischer.

Elkeles, T. & Bormann, C. (1999). Arbeitslose [The unemployed]. In Bundesvereinigung Gesundheit e.V. (Ed.), *Gesundheit. Strukturen und Handlungsfelder* (pp. 1-20). Neuwied: Luchterhand.

Elkeles, T. & Bormann, C. (2002). Arbeitslose. [The unemployed]. In H.G. Homfeldt, U. Laaser, U. Prümel-Philippsen & B. Robertz-Großmann, (Eds.), *Gesundheit: Soziale Differenz - Strategien - Wissenschaftliche Disziplinen* (pp. 11-28). Neuwied: Luchterhand, Kriftel.

Elkeles, T. & Kirschner, W. (2004). Arbeitslosigkeit und Gesundheit. Intervention durch Gesundheitsförderung und Gesundheitsmanagement. Befunde und Strategien [Intervention by health promotion and health management. Results and strategies]. *BKK Bericht GuS, 3*. Bremerhaven: Wirtschaftsverlag NW.

Elkeles, T., Niehoff, J.U., Schneider, F. & Rosenbrock, R. (Eds.). (1991). *Prävention und Prophylaxe* [Prevention and prophylaxis]. Theorie und Praxis eines gesundheitspolitischen Grundmotivs in zwei deutschen Staaten. 1949-1990. Berlin: Edition Sigma.

Elkeles, T. & Seifert, W. (1993). Unemployment and health-impairments: Longitudinal analyses from the Federal Republic of Germany. *European Journal of Public Health, 3*(1), 28-37.

Elkeles, T. & Seifert, W. (1996). Immigrants and health. Unemployment and health – risks of labour migrants in the Federal Republic of Germany. 1984 – 1992. *Social Science and Medicine, 43*(7), 1035-1047.

Federal Agency of Labour (2004). *Statistics of the Federal Agency of Labour*. Retrieved from http://www.arbeitsamt.de.

Figgen, P. (2003). Das Kölner Job Center. MoZArt-Erfolgsmodell mit Zukunft. *MoZArt Newsletter, 3*(1), 3-5.

Gallo, W., Kasl, T. & Stanislav, V. (2001). The effect of job displacement on subsequent health. Proceedings of the 2000 Fourth International Conference of German Socio-Economic Panel Study Users (GSOEP). *Vierteljahrshefte zur Wirtschaftsforschung, 70*(1), 159-165.

Gepkens, A. & Gunning-Schepers, L.J. (1996). Interventions to reduce socioeconomic health differences. *European Journal of Public Health, 6*(3), 218-226.

Geyer, S. & Peter, R. (2003). Hospital admissions after transition into unemployment. *Sozial- und Präventivmedizin, 48*(2), 105–114.

Grobe, T., Dörning, H. & Schwartz, F.W. (1999). *GEK-Gesundheitsreport 1999 [GEK-Health-Report]. Auswertungen der GEK-Gesundheitsberichterstattung. Schwerpunkt Arbeitslosigkeit und Gesundheit, 12*. Sankt Augustin: Asgard.

Grobe, T. & Schwartz, F.W. (2003). Arbeitslosigkeit und Gesundheit [Unemployment and health]. Robert-Koch-Institut (Ed.), *Gesundheitsberichterstattung des Bundes, 13*. Berlin.

Haecker, G., Kirschner, W. & Meinlschmidt, G. (1990). Zur Lebenssituation von Sozialhilfeempfängern in Berlin (West) [The living situation of welfare recipients in Berlin]. Senatsverwaltung für Gesundheit, *Diskussionsbeiträge zur Gesundheitsforschung, 14*.

Henkel, D. (1992). *Arbeitslosigkeit und Alkoholismus* [Unemployment and alcoholism]. Epidemiologische, ätiologische und diagnostische Zusammenhänge. Weinheim: Deutscher Studien Verlag.

Hollederer, A. (2003a). *Der Gesundheitszustand von Arbeitslosen in der deutschen Arbeitslosenstatistik*. Unpublished manuscript.

Hollederer, A. (2003b). Arbeitslosigkeit und Gesundheit – Ein Überblick über empirische Befunde und die Arbeitslosen- und Krankenkassenstatistik [Unemployment and health]. *Mitteilungen aus der Arbeitsmarkt- und Berufsforschung, 36*(3), 411-428.

John, J., Schwefel, D. & Zöllner, H. (Eds.). (1983). *Influence of economic instability on health*. Berlin: Springer.

Kahl, H., Hölling, H. & Kamtsiuris, P. (1999). Inanspruchnahme von Früherkennungsuntersuchungen und Maßnahmen zur Gesundheitsförderung [The use of early diagnostic measures and measures of health promotion]. *Gesundheitswesen, 61*(Special issue 2), 163-168.

Kieselbach, T. (1988). Arbeitslosigkeit [Unemployment]. In R. Asanger & G. Wenninger (Eds.), *Handwörterbuch der Psychologie* (pp. 42-51). München: Psychologie Verlags Union.

Kieselbach, T., Klink, F., Scharf, G. & Schulz, S. (1998). *„Ich wäre ja sonst niemals an Arbeit rangekommen!" Evaluation einer Maßnahme für Langzeitarbeitslose.* [„I would have never had got a job again!" Evaluation of a reintegration intervention scheme for long-term unemployed] (Psychologie sozialer Ungleichheit, vol. 7). Weinheim: Deutscher Studien Verlag.

Kieselbach, T. (2001). Sozialer Konvoi und nachhaltige Beschäftigungsfähigkeit [Social convoy and sustainable employability]. Perspektiven eines zukünftigen Umgangs mit beruflichen Transitionen. In J. Zempel, J. Bacher & K. Moser (Eds.), *Erwerbslosigkeit. Ursachen, Auswirkungen und Interventionen* (pp. 381-396). Opladen: Leske and Budrich.

Kirschner, W. & Radoschewski, M. (1993). Gesundheits- und Sozialsurvey Berlin [Health and social survey]. *Diskussionsbeiträge zur Gesundheitsforschung, 17.*

Kirschner, W., Radoschewski, M. & Kirschner, R. (1995). *§20 SGB V Gesundheitsförderung, Krankheitsverhütung. Untersuchung zur Umsetzung durch die Krankenkassen* [Health promotion, prevention of disease. Study on the implementation by the health insurance companies]. Sankt Augustin: Asgard.

Kirschner, R., Elkeles, T. & Kirschner, W. (2003). *Evaluation of the activities of the Fonds Gesundes Österreich* [Fund for a healthy Austria] *1998-2001. Report on results*. Retrieved from http://www.fgoe.org/ Evaluationsberichtenglisch.pdf.

Klink, F. (1993). Hilfesuchverhalten erwerbsloser Frauen [Help-seeking behaviour of unemploed women]. In G. Mohr (Ed.), *Ausgezählt. Theoretische und empirische Beiträge zur Psychologie der Frauenerwerbslosigkeit* (pp. 205-251). Weinheim: Deutscher Studien Verlag.

Knab, H. (2003). MoZArt -Vorhaben in Rottweil zieht erste Bilanz. *MoZArt Newsletter, 3*, 5-7.

Kuhnert, P. & Kastner, M. (2002). Neue Wege in Beschäftigung. Gesundheitsförderung bei Arbeitslosigkeit [New ways in employment. Health promotion and unemployment]. In R. Geene, C. Gold & C. Hans (Eds.), Armut und Gesundheit. Gesundheitsziele gegen Armut. Netzwerk für Menschen in schwierigen Lebenslagen. *Materialien für Gesundheitsförderung, 10*, 336-364.

Lembcke, C. (2003). *Gesundheitsförderung für langzeitarbeitslose Frauen* [Health promotion for longterm unemployed women]. Unpublished dissertation, Technical University of Dresden.

Metz, A.M. & Kalytta, T. (2001). Modellprojekt Arbeits- und Gesundheitsförderung in der Prignitz. *Evaluation der Servicestelle Arbeit und Gesundheit, 3* [Model projects of health promotion and employment promotion in the region of Prignitz]. Unpublished research report, Institut für Psychologie der Universität Potsdam.

Morris, J.K., Cook, D.G. & Shaper, A.G. (1994). Loss of employment and mortality. *British Medical Journal, 308,* 1135-1139.

Muller, J. (1992). The effects of personal development training on the psychological state of long-term unemployed women. *Australian Psychologist, 27*(3), 176-180.

Niemeyer, G., Feige, L. & Kuhlmey, A. (2000). Bauhof-Zentrum für aktive Gesundheitsförderung. Evaluation. Wolfsburg: I.-III. Partial results from 1999, 2000, 2001, *Research Report, 004.*

Orwat, D., Kirschner, W. & Meinlschmidt, G. (1996). Zur Notwendigkeit der Etablierung eines zielgruppenspezifischen Präventionsprogramms in Berlin [The necessity of a specific prevention programme in Berlin]. *Arbeits- und Sozialpolitik, 50*(7/8), 14-35.

Paul, K. & Moser, K. (2001). Negatives psychisches Befinden als Wirkung und als Ursache von Arbeitslosigkeit [Negative mental state as effect and cause of unemployment]. In J. Zempel, J. Bacher & K. Moser (Eds.), *Erwerbslosigkeit. Ursachen, Auswirkungen und Interventionen.* (pp. 83-110). Opladen: Leske and Budrich.

Porter, S. (1983). Reducing the pressure. Health care for the unemployed. *The Ohio State Medical Journal, 79*(11), 833-836.

Proudfoot, J., Guest, D., Carson, J., Dunn, G. & Gray, J. (1997). Effect of cognitive behavioural training on job finding among long-term unemployed people. *The Lancet, 350*, 96-100.

Radoschewski, M., Kirschner, W., Kirschner, R. & Heydt, C. (1994). Entwicklung eines Präventionskonzeptes für das Land Berlin [Development of a prevention concept for Berlin]. Senatsverwaltung für Gesundheit Berlin (Ed.), *Diskussionsbeiträge zur Gesundheitsforschung, 21*. Berlin.

Rosenbrock, R. (2003). Qualitätssicherung und Evidenzbasierung. Herausforderung und Chancen für die Gesundheitsförderung [Quality assurance and evidence base]. *Info-Service, Gesundheit Berlin e.V., 3,* 5.

Sammet, M. (1999). *Die gesundheitliche Belastung von Arbeitslosen: Empirische Belege auf der Basis von Auswertungen von Krankenkassendaten* [The health problems of unemployed]. Retrieved from http://www.gesundheitberlin.de/content/aktivitaeten/a_g/soziales/sammet.pdf.

Schwefel, D., Svensson, P.G. & Zöllner, H. (Eds.). (1987). *Unemployment, social vulnerability and health in Europe*. Berlin: Springer.

Schmidt, C.M. (2000). Arbeitsmarktpolitische Maßnahmen und ihre Evaluierung. Eine Bestandsaufnahme. [Labour market intervention and their evaluation] *Vierteljahreshefte zur Wirtschaftsforschung, 69*(3), 425-437.

Siegrist, J. & Joksimivic, M. (2001). *Tackling inequalities in health – ein Projekt des European Network of Health Promotion Agencies zur Gesundheitsförderung bei sozial Benachteiligten. Final report for the German branch.* Düsseldorf: Institut for Medical Sociology of the Heinrich Heine University of Düsseldorf, Bundeszentrale für gesundheitliche Aufklärung (BZgA).

Simon, W. (2003). *Arbeitslosigkeit und psychische Gesundheit. Eine Recherche nach evaluierten Gesundheitsförderungsprojekten für Arbeitslose* [Unemployment and mental health]. Unpublished bachelor thesis, University of Applied Sciences Neubrandenburg.

Trube, A. & Luschei, F. (2000). Entwicklungs- und Vermittlungsassistenz (EVA). Ein Instrument zur Wiedereingliederung Arbeitsloser. [Assistance in development and placement (EVA). An instrument for the reintegration of unemployed]. Final report. *Zentrum für Planung und Evaluation Sozialer Dienste/Universität - Gesamthochschule Siegen, 7.*

Trube, A. & Luschei, F. (2001). Entwicklungs- und Vermittlungsassistenz (EVA). Teilnehmerverbleib und Nachhaltigkeit von arbeitsmarktlichen Integrationseffekten (EVA II). [Assistance in development and placement (EVA II). Reintegration success follow-up of participants and sustainability of labour market integration effects.]. *Zentrum für Planung und Evaluation Sozialer Dienste/Universität - Gesamthochschule Siegen, 8.*

Vinokur, D.A. & Price, R.H. (1991). Long-term follow-up and benefit-cost analysis of the Jobs Program. A preventive intervention for the unemployed. *Journal of Applied Psychology, 76*(2), 213-219.

Vuori, J. & Silvonen, J. (2002). The Työhön Job Search Program in Finland. Benefits for the unemployed with risk of depression or disencouragement. *Journal of Occupational Health Psychology, 7*(1), 5-19.

Westcott, G., Svensson, P.G. & Zöllner, H. (Eds.). (1985). *Health policy implications of unemployment.* Copenhagen: WHO, Regional Office for Europe.

Preventive Group Intervention Promoting Quality of Employment and Mental Health Among Graduates of Vocational Schools

Jukka Vuori, Petri Koivisto & Elina Nykyri

1 Description of the intervention

The paper describes a Finnish intervention *From School to Work* among students entering working life, and assesses its feasibility in reducing the unemployment risks of young people.

The transition from school to work comprises two challenges: job-search and organisational socialization. Effective job-search requires career management skills related to e.g., searching a job by tips from one's social network, making connections with employers and presenting one's own strengths and abilities successfully in the job interview. Useful skills in organisational socialization are needed after finding the first job, in the learning process by which a newcomer develops relationships with co-workers and supervisors, learns skills needed in a new job, internalizes the young worker's role in the organisation and adjusts to the culture of the organisation (Feij, 1998).

For the graduates of vocational studies, the appraisal of competence to attain goals in job-search and organisational socialization depends on the psychological factor of self-efficacy. Strong self-efficacy beliefs in employment motivate commitment, planning and active behaviour in career goal construction and attainment (Lent et al., 1999). In addition to self-efficacy beliefs, the choice of the career management strategies is determined by personal attitudes (Ajzen, 1991; Caplan et al., 1997). It has been suggested that in addition to reinforcing self-efficacy, career education should strengthen participants′ action readiness, e.g., outcome expectancies and commitment to active career management strategies (Caplan et al., 1997; Lent et al., 1999).

Prolonged job-search due to setbacks and to barriers to employment, again, will increase stress and may weaken job seekers′ self-efficacy, outcome expectancies and commitment to work-related career goals. Inoculation against setbacks is an anticipatory stress management technique, which increases individuals′ ability to maintain active, goal directed behaviour and well-being when facing barriers or setbacks in the labour market (Caplan et al., 1997).

The aim of the preventive group intervention From School to Work is to promote career management and to prevent mental health problems among graduates of vocational schools (Koivisto et al., 2002). The group intervention deals with career management strategies related to achievement and with inoculation in the contexts of job-search, organisational socialization and life-long learning. The intervention is based on a general model (Caplan et al., 1997) developed at the Michigan Prevention Research Center (MPRC) and both on the preventive JOBS intervention for the unemployed workers (Vinokur et al., 1995) and on its Finnish adaptation, the To Work method (Vuori et al., 2002).

In addition to job-search skills training, similar to the typical training in the interventions for the unemployed, the From School to Work group deals also with skills of organisational socialization. The group model applies theories of social learning, and it focuses on boosting self-efficacy beliefs and attitudes at critical transition phases during the work career. The model comprises principles of supportive environment, active learning and anticipatory stress management. A central feature of the model is identifying the barriers associated with transition phases and inoculation against possible setbacks.

The impacts of the JOBS and the To Work methods have been examined in randomized field experimental studies and their contents will not be described here in detail (Caplan et al., 1989; Vinokur et al., 1995; Vuori et al., 2002). Results showed beneficial impacts on employment, income level and symptoms of depression among unemployed adult job-seekers. The beneficial impacts were stronger among unemployed, who were at risk of depression.

Based on earlier research we have hypothesised that the From School to Work group method would improve employment and its quality in terms of correspondence with education and personal career plans and that the intervention would support mental health among the young graduates, especially among those at risk of mental health problems.

2 Research methods

2.1. Participants and procedure

This field experimental study was carried out in five institutes of vocational education in western Finland during school years 2000-2001 and 2001-2002. The participants of this study were 17-24 year old graduates. Sixty-nine percent of them were women. After the baseline measurement (N=416) they were randomised to the experimental and the control groups. The members of the experimental group participated in the preventive groups that lasted for 5 days (20 hours) during their last school year. The controls performed homework on job-search training. The study had a 10-months follow-up (N=334), on average 7 months after participants graduated.

Each intervention group was trained by two co-trainers: one was a teacher from the vocational institute and the other an officer from the local employment office. The trainers used manuals and participant's workbooks made for supporting group work. Local employers were invited to groups for information interviews.

2.2. Indicators

Demographic characteristics were collected by standard survey questions for reporting age, gender, marital status, prior education and sector of education.

The employment status at the baseline was measured by asking participants who already have a job the following question: "Are you going to continue at this workplace after graduation?" The 4-point scale (1=not, ... 4=surely) was recoded to a dichotomous variable: 4=1, 3=1, and 2=0, 1=0. Participants who were not employed at baseline were coded 0.

The employment status during the follow-up was determined by asking two questions: 1. "What is your employment status now?" 2. "How many hours are you working weekly?" Respondents were classified as employed (coded "1") if they described themselves being "employed without subsidy from the state" or "self-employed" 20 hours or more a week.

The quality of employment in baseline measurement was defined as follows: Employed participants were asked to evaluate the extent to which their current job corresponded with their education (1=not at all, 2= to some extent, 3=perfectly). The measure was recoded to a dichotomous variable: 3=1, 2=0, 1=0. Unemployed participants were coded 0.

The quality of employment during follow-up was determined as follows: First, employed participants were asked to evaluate the extent to which their current job corresponded with their education (1=not at all, 2=to some extent, 3=perfectly). Second, participants also evaluated the extent to which their current job promotes their career plans (1=not at all, ... 4=perfectly). The correlation between these two scales of employment quality was .84 in follow-up. Both qualities of employment measures were recoded as a dichotomous variable as follows: 4=1, 3=1, 2=0, 1=0. Participants not employed were coded 0.

Psychological distress was measured by the GHQ-12 (Goldberg, 1972). The Cronbach alpha coefficients of the GHQ-12 were .88 at baseline and .90 at follow-up.

The employment self-efficacy measure used in this study consists of self-efficacy question-items related to two specific achievement contexts: job-search and organisational socialization. The measure includes five job-search self-efficacy items (e.g., “Do you believe you can give the best impression of yourself in a job interview?” van Ryn & Vinokur, 1992). In addition to job-search items the scale included four items concerning organisational socialization (e.g., Do you believe you can get along with fellow workers in a new job?") Participants gave their answers using a five-point scale (1=very badly, ... 5=very well). The Cronbach alpha coefficients were .85 at baseline and .87 at follow-up.

3 Results

Differences between the experimental group and the control group were studied using generalized linear models, controlling for age, gender, employment self-efficacy measurement, prior education and employment status at the baseline. The intervention explained employment statistically significantly (Table 1). Those participating in the group had a 1.65-fold probability to be employed during follow-up compared to the controls. The From School to Work method also had a beneficial impact on the quality of employment (Table 1).

Those who participated in the group that used the From School to Work method had a 2.08-fold chance to be employed in a job that corresponded to education and career goals compared to the control group. Regarding mental health effects, the decrease in symptoms of psychological distress, was not statistically significant. However, the interaction of the intervention with psychological distress at the baseline measurement had a significant impact on psychological distress at follow-up. Participation in the From School to Work group lowered psychological distress during the follow-up among those in risk of mental disorders at the baseline.

4 Discussion

The results showed that the From School to Work group intervention promotes employment and its quality among adolescents graduating from vocational education. The group intervention did not have a significant main effect on psychological distress, but did have an interaction effect on psychological distress for those at risk of mental health disorders.

On the basis of prior research (Vuori & Koivisto, 2003; Vuori & Silvonen, 2005) it can be assumed, that the positive effects of the From School to Work method are based on the strengthening of employment self-efficacies and inoculation against setbacks during the transition period. Strengthened career management strategies and stress management have probably promoted participants' employment in a job that corresponds with their education and career plans among the whole research group and prevented psychological distress due to set-backs and barriers met in employment among the risk group of mental disorders.

There are some limitations in this study. First, the results of the intervention may vary according to the labour market situation, which depends on occupation, economic trend and environment. Participants of this study came from western Finland and outside of economically growing areas, and most of them represent professions of social and health services, in which careers of young workers are often characterized by fixed-term jobs, part-time jobs and discontinuities. Second, more studies are needed to clarify what is the process mediating the positive impacts of the From School to Work intervention. If the strengthening of employment efficacies and inoculation against setbacks have a key role in this mediating process, the results of this study could be generalized to other methods aimed at strengthening employment related self-efficacy, attitudinal action readiness and inoculation against set-backs. Third, studying the long-term impacts of the intervention on career development and mental health would have needed a longer follow-up.

Table 1. **Results of logistic regression and linear regression analyses: Effects of the From School to Work intervention on employment, its quality, and on mental health. In the models age, gender, employment self-efficacy, prior education, employment status and the outcome variable at baseline were controlled.**

		Employment status T3[a]	Quality of employment T3[a]	Psychological distress T3[b]
Intervention	Experimental control group	**1.65***	**2.03***	.06
Control of outcome variable (T1)		non applicable	1.81 [e]	.33** [f]
Age (T1)		1.10	1.01	.003
Gender	Men vs. women	.89	.44 +	.003
Employment self-efficacy (T1)		1.16	1.55	.23**
Employment status (T1)	Employed vs. not employed	3.18**	1.68	-.09
Prior education (T1)	Higher than comprehensive school vs. not	.67	1.00	.03
N		330	296	281
R^2		.08 [c]	.05 [c]	.19
Proportion of R^2 explained by intervention		.016	.013	.022
Interactions: [d]				
Interaction of Intervention and psychological distress (T1)		.65	1.12	
- Experimental group				**.17 +**
- Control group				**.46****
N		326	293	281
R^2		.08 [c]	.06 [c]	.21
Proportion of R^2 explained by intervention		.012	.012	.022

Note. [a] Figures are standardised odds ratios (logistic regression, OR), [b] Figures are non-standardised parameter estimates (linear regression), [c] = Nagelkerke R^2-coefficient, [d] = Only coefficients of interaction terms and basic figures of the models are presented. + $p < .10$ * $p < .05$, ** $p < .01$, two-sided test,
[e] = employment to a job that corresponds to education in baseline measurement (T1),
[f] = psychological distress in baseline measurement (T1)

References

Ajzen, I. (1991). The theory of planned behavior. *Organisational Behavior and Human Decision Process, 50,* 179–211.

Caplan, R.D., Vinokur, A.D. & Price, R.H. (1997). From job loss to reemployment: Field experiments in prevention-focused coping. In G.W. Albee & T.P. Gullotta (Eds.), *Primary prevention works.* (pp. 341-379). Thousand Oaks: Sage.

Caplan, R.D., Vinokur, A., Price, R.H. & van Ryn, M. (1989). Job seeking, reemployment and mental health: A randomized field experiment in coping with job loss. *Journal of Applied Psychology, 74,* 759-769.

Feij, J.A. (1998). Work socialization of young people. In P.J.D. Drenth, H. Thierry & C.J. de Wolff (Eds.), *Handbook of work and organisational psychology III* (pp. 207-256). Hove: Psychology Press.

Goldberg, D.P. (1972). *The detection of psychiatric illness by questionnaire.* London: Oxford University Press.

Koivisto, P., Mäkitalo, M., Larvi, T., Silvonen, J. & Vuori, J. (2002). *Koulutuksesta työhön - menetelmä* [From School to Work method]. Helsinki: Työterveyslaitos.

Lent, R.W., Hackett, G. & Brown, S. (1999). A social cognitive view of school-to-work transition. *The Career Development Quarterly, 47,* 297-311.

van Ryn, M. & Vinokur, A.D. (1992). How did it work? An examination of the mechanism through which an intervention for the unemployed promoted job-search behaviour. *American Journal of Community Psychology, 20,* 577-597.

Vinokur A.D., Price, R.H. & Schul, Y. (1995) Impact of the JOBS intervention on unemployed workers' varying risk for depression. *American Journal of Community Psychology, 23,* 39-74.

Vuori, J. & Koivisto, P. (2003). Group intervention promoting quality of employment and mental-health among graduates of vocational schools. *Proceedings of the International Symposium on Youth and Work, 20-22 November 2002, Espoo, Finland.* Helsinki: Finnish Institute of Occupational Health.

Vuori, J. & Silvonen, J. (2005). The benefits of a preventive job search program on re-employment and mental health at two years follow-up. *Journal of Occupational and Organisational Psychology, 78,* 43-52.

Vuori, J., Silvonen, J., Vinokur, A.D. & Price, R.H. (2002). The Työhön job search program in Finland: Benefits for the unemployed with risk of depression or discouragement. *Journal of Occupational Health Psychology, 7(1),* 5-19.

Health-Oriented Counselling of Unemployed Workers and Employees on Short-Term Contracts

Peter Kuhnert, Türkan Ayan & Michael Kastner

1 Background of the study

This paper will give some first results of a pilot project dealing with health promotion for long-term and short-term unemployed, as well as for workers on short-term contracts. Studies show that these kinds of groups suffer intensely from physical, psychological and psychosomatic diseases (e.g. Kuhnert, 2004a; Paul & Moser, 2001). At the same time, measures of intervention for this target group, especially in terms of counselling, are missing in Germany (Kuhnert & Kastner, 2001, 2002; Kuhnert, 2004b).

Our project was financed by the Federal Association of the German health insurance fund BKK on the basis of statute § 20 SGB V. Its concept was developed and put into practice by the Institute of Organisational Psychology and Industrial Medicine (IAPAM) and the University of Dortmund as part of the developmental partnership N.A.G. (Network for job-market integrative health promotion). 70 participants were recruited in cooperation with BKK Hoesch and the transfer company PEAG in Dortmund out of a sample of 1,105 potential ones (aged 27 – 55). 45 participants were members of the health insurance company BKK who were unemployed, and 25 were employees of PEAG, a so-called transfer company that was instituted after a restructuring of another enterprise. The participants of PEAG were insured as members of BKK. Technically speaking the PEAG participants were short-term contracted with the transfer company but they did not hold a regular job. They were placed in the transfer company instead of becoming unemployed. Although they received professional counselling and better financial support than the unemployed they also were worried about their occupational future. They were chosen as a comparison group due to their similar occupational and social background. We expected differences, such as those between short- and long-term unemployed people, for example.

Our intervention – called BKK-Job-FIT – consisted of group counselling during a period of 5 weeks. The 49 male and 21 female participants were counselled in eight subgroups. Apart from health topics, we picked up every day issues like work or family problems. The idea was that especially those people occupied with serious everyday life problems should be motivated to gradually shift their focus from these problems to personal resources like health, quality of life and employability. A total of 20 hours of counselling per person was provided. In general, it is rather difficult to attract the target group of socially disadvantaged people to health promotion (Rosenbrock, 2001). Also, in our sample, people at first showed little interest in the subject or were sceptical concerning the cost-free programme. Especially those who claimed negative experience with job centres or social security offices showed much distrust. Another explanation for resistance might be that with increasing unemployment and shrinking job supply, health impairments are hidden and not thematized or played down (Hollederer, 2003).

A survey questionnaire was used to measure the health effects of our programme. It included several questionnaires: firstly, FABU (Questionnaire for unemployed and people on short-term contracts – German version; Kuhnert et al., 2004, Reick, 2005), which was developed by the researchers to examine potential health impairments in our target group.

The FABU questionnaire includes subscales related to the three categories (A) Physical Perception (body), (B) Psychosocial Health (mind) and (C) Emotional State and Social Support (life). The dimension A (body) helps to diagnose physical well-being and a general health state. Information about the client's state and personal feelings about client's state of health is of great importance. Besides somatic illness, psychosomatic symptoms play an equal role. The second dimension B (mind) collects data about psychosocial health. Questions are related to the following aspects:

- Expectation of self-efficacy
- Self-esteem
- Experience of a meaningful life
- Concentration
- Coping strategies
- Experience of control
- Mental impairment

The final dimension C (life) focuses on changes in emotions, vulnerability and social support during the unemployment phase. Unemployment is very strongly associated with decreased positive emotions like joy, satisfaction and pride, as well as increased negative emotions like anxiety, anger and embarrassment. The negative emotional states lead to changes in the social network of the client and enhance the vulnerability for illness. Unemployed people suffer more from physical, psychological and psychosomatic diseases.

Secondly, we also included the sense of coherence scale (SOC; Antonovsky, 1997) and thirdly, SF 36 scales (Bullinger & Kirchberger, 1998) in the questionnaire. The SOC scale reflects the person's general belief to live a predictable, meaningful and manageable life. It captures the subject's sense of coherence, which consists of the three components: sense of comprehensibility, sense of manageability and sense of meaningfulness.

The SF 36 scale deals with the health related quality of life and contains eight physical and mental dimensions in total.

The survey data from the target group were gathered at three times, T_1 baseline data, T_2 post intervention data, and T_3 4-8 weeks post-intervention follow-up data. The collected data were compared with and interpreted according to data of a normal population (Kuhnert et al., 2004). In addition, 18 case studies were described in the final report.

The underlying concept of the programme as a health promotion is based on the counselling concept SEBA (Self-Management Counselling for Unemployed, German version) by Kuhnert (1999, 2004b) and the Model of Work Life Balance by Kastner (2004). By focusing on individual resources and strengths rather than deficits and risks, the programme follows the salutogenic approach of Antonovsky (1997; Kuhnert & Kastner, 2002).

The programme aimed at:

- enhancing the motivation for health related behaviours and attitudes
- raising self-respect and self-control concerning one's own life-style
- becoming more aware of establishing trustful relations to others
- promoting skills to seek social support.

2 Social status and intervention effects

78% of the participants were older than 40 years, of this group 39% were over 50 years and 15% over 55 years old. People at this age have hardly any chance to gain a job on the German labour market. This proves to be a rather old sample compared to that of an earlier, rather similar study by Kuhnert and Kastner (2001, 2002) where 57% of 148 long-term unemployed in Dortmund were under the age of 40 and only 10% over 50.

Only 7% of the unemployed persons, but 16% of the short-term employees had graduated from a college. Three quarters had a secondary or just a basic education. More than half of the unemployed group (56%) consisted of single persons, only 40% had a partner and even less (11%) had children. In comparison, the vast majority of the employees on short-term contracts (80%) had a partner. Surprisingly, almost two thirds of the female subsample were single or single mothers (29%). Our results correspond to a longitudinal study by Bongarts and Gröhmke (2003) showing a very large share of single males among long-term unemployed. In another study (Bleck, 2003), one third of elderly male unemployed were socially isolated.

Taking into account that well-being and health are strongly dependent on social and emotional support by close partners, family or friends (Laireiter, 2002), male unemployed are at risk concerning this resource. Not surprisingly, in our counselling groups more men complained about the lack of a partner. Women seemed to be less troubled by this situation (see case studies) and showed a higher mean SOC score (women 4.91; men: 4.42). Compared to the male participants, women's problem solving strategies seemed more flexible, i.e., they were less dependent on employment as their focus of life. These results, however, stand in contrast to what is the gendered approach to unemployment in Germany, as described by Schuhmacher, Gunzelmann and Brähler (2000).

The majority of the unemployed participants (40%) had been unemployed for over 2 years, and 12% had been unemployed for over 5 years. It is remarkable that according to an IAB study (Cramer et al., 2002), only 39% of the subjects who were unemployed at least one year still expected to get a job. Looking at other German data only 19% of the long-term unemployed (>24 month) find a job in the so-called "first" job market[1] (Brixy et al., 2002).

The intervention effects measured by SOC and FABU scores were rather weak. The high number of elderly unemployed might be one reason for this (Figure 1). The intervention had the strongest effect on 11 persons with very low self-efficacy scores (see FABU) in the baseline-study. There is also an Australian study where participants with low self-efficacy scores and high stress scores showed progress after an intervention (Machin & Creed, 2003).

Figure 1. The average SOC and FABU scores in the target group before and after the intervention

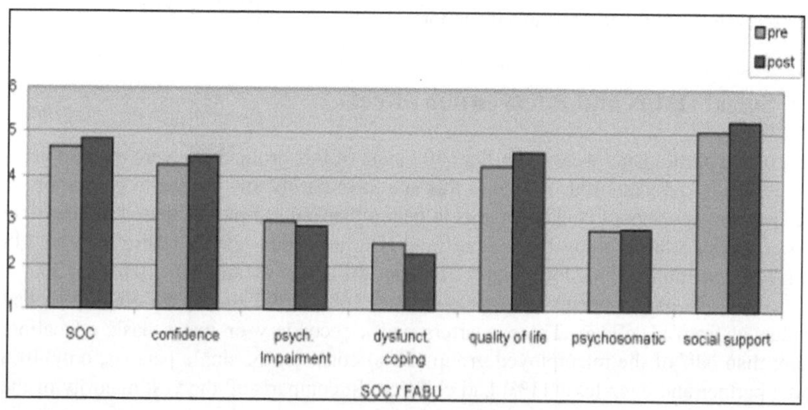

1 Employees in the primary labour market segment enter a regular employment relationship where they work full time and on a permanent-term basis (including tariff wage). In comparison the "secondary" job market offers work on temporary fixed-term basis (six month – two years). It is often subsidized from public funds and the employees rarely have a good career opportunity and do not belong to trade unions.

3 Social, occupational und health related behaviour

The vast majority of our sample (80%) stated that they were going to apply for a job after the intervention. This result is comparable with the one from an earlier study by Kuhnert (1999). Because of the higher age of the sample and their perspective of a transition into early retirement only 44% of them really applied for jobs during the follow-up period. Studies indicate that giving up the job search does not necessarily correlate with a relief in stress. It may even raise the risk of mental ill-health (Kieselbach et al., 1997). This is in compliance with our results that those participants with no applications showed the lowest SOC baseline scores (3,83). Neglect of health care was typical: only slightly more than one third of the unemployed participants (36%) attended medical check-ups during the follow-up period. By comparison, only 17% of the employees on short-term contracts failed to utilize health care provisions.

Three quarters of the employees on short-term contracts (74%) stated that they do have enough energy and vitality for taking action in everyday life issues (e.g. housekeeping, family, job), while almost half of the unemployed participants (47%) felt the opposite. With enduring unemployment, people engage less in physical exercise: about 40% of the unemployed clients did not exercise at all. Impairment due to back pain was rather common in both subgroups: 41% of the unemployed persons and 46% of the employees on short-term contracts mentioned some. The use of medication among the unemployed persons was 40%. This high share corresponds with the high prevalence of physical symptoms among the participants.

Looking at the emotional scales, a third of the unemployed (32%) suffered from anxiety, but this was also the case for 39% of the employees on short-term contracts. The result might be due to the fact that in the latter group many were about to lose their contracts. Guilt was a predominant emotion in the unemployment group (29%) but not in the group of employed (4%). Almost half of the employees on short-term contracts (46%) and one third of the unemployed felt unable to relax. It is remarkable that half of the unemployed participants claimed to be disappointed by people they trusted, while the corresponding share was less than a third in the group of employed. This disappointment, again, has a negative impact in establishing trustful relations as an important resource for the clients. High rumination scores among the unemployed respondents (38%) indicate a risk for mood disorders (employed on short-term contracts: 23%).

More than two thirds of the unemployed persons (68%) and half of those employed were dissatisfied with their living conditions. Almost half of all respondents (46%) anticipated negative future prospects. At least 71% of the unemployed participants (employed: 46%) said that they were desperate for changes in their life. This might be related to the perception that everyday activities seem senseless for a third of the unemployed subjects (for those employed: 22%). An important aim of group counselling was to foster enjoyment of life in various forms regardless of vocational status.

In general, the long-term unemployed (>24 months of unemployment) showed poorer scores compared to the short-term unemployed (<12 months) in all psycho-social scales (Figure 2).

Figure 2. SOC & FABU score of long-term vs. short-term unemployed persons in the intervention group

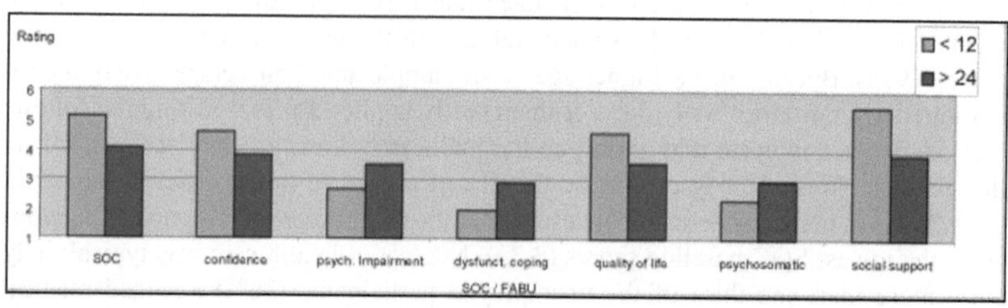

More than one third (35%) of the target group had problems getting by on their current income, and 44% stated they were in debt, of these 19% were highly indebted (Figure 3). Being in debt means high stress, which corresponds with the low SOC score of 3.38 for this subgroup.

Figure 3. Financial situation of the intervention group

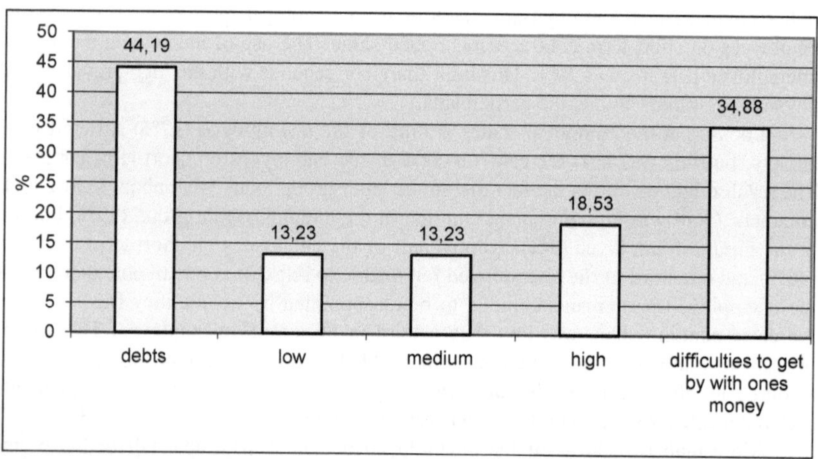

Social support is useful for coping with critical incidents. Critical incidents such as severe problems with family, marriage (divorce) or children (e.g. drug abuse) can have a negative influence on the capability to deal with problems. The Dortmund study on long-term unemployed persons (Kuhnert & Kastner, 2002) showed that almost two thirds (62%) of the subjects wished to have more social contacts. Therefore, we offered the participants peer-counselling and guidance through a self-help group in order to support sustainability of the behaviour changes potentially achieved by the counselling. Initially 20 participants showed an interest in self-help, and over an extended period of time, 8-10 people met every fortnight. In similar studies unemployed participants of a self-help group have showed significant progress in self-confidence, social behaviour and quality of life (Kager & Haslinger-Baumann, 2003). Our follow-up results 4-8 weeks after the intervention showed that the participants of the self-help group had statistically significantly better scores on psycho-social scales compared to non-participants. This emphasizes the importance of self-help groups within a counselling process.

By the end of the programme the mental health of most of participants had improved. The main challenge for clients seems to be how to transfer knowledge and techniques acquired during the counselling programme into everyday situations. It is not surprising that the positive effects of the programme diminished gradually in a problematic everyday life (e.g. due to unemployment, health problems, marriage problems etc.). In general, behavioural changes need time (Kanfer, Reinecker &Schmelzer, 1996; Kuhnert & Kastner, 2002). This finding corresponds well with our results from the self-help group. The poor baseline SOC score (58.17) in our target group is comparable to the one shown in a study of female alcoholics (see Table 1). This indicates the over-loaded situation of our clients and therefore the difficulty in achieving behaviour changes with a short-term counselling programme.

Table 1. SOC data in different studies (questionnaire life orientation, 13 items)

Authors	Year	Country	Sample	M-Age	SOC M	SOC SD
Kuhnert et al.	2003	Germany	65	45 (unemployed & un-steadily employed)	58.1	14.4
Callahan & Pincus	1995	USA	828	57 (rheumatism patients)	65.3	14.8
Carmel et al.	1991	Israel	230	45 (Kibbuzniks)	67.6	9.9
Ryland & Greenfield	1991	USA	302	38 (university assistants)	66.5	10.2
Franke et al.	2001	Germany	929	40,2 (alcohol-addicted women)	55.4	

The participants with a low SOC score were more troubled in different areas of life than participants with a high SOC score. Two thirds of all participants (67%) with a low SOC had limited work ability due to ill-health. More than half of the unemployed (55%) with a low SOC could hardly finance their cost of living, and they had serious difficulties managing with their available money. Almost one third of the participants (30%) with a low SOC felt highly stressed by conflicts with partners and family. Half of the participants (50%) with a poor SOC score stated that past conflicts (e.g. divorce, break up of a relationship or death of a partner) are still relevant at present. Participants with a low SOC score were also physically more impaired than those with a high SOC score (Table 2).

Table 2. Physical complaints depending on SOC score

Physical complaints	% of unemployed with low SOC	% of unemployed with high SOC
Shoulder-, neck and backache	60%	37.5%
Symptoms close to tinnitus/ hearing loss	40%	8.3%
Stomach trouble	35%	16.7%
Headaches	35%	29.2%
Cardiac problems	15%	8.3%
Tremor	15%	0%

4 Case studies with positive and negative prognoses

The following four cases were typical of all 18 case studies carried out in our survey. We interviewed the participants at the beginning of the counselling programme. The interview results and the results of the questionnaire were used to split the subjects into cases with ′good′ or ′poor′ coping strategies. In general our case studies indicate that (cf. Antonovsky, 1997):

a) Unemployed with high SOC choose useful strategies to cope with unemployment:
   - use existing resources
   - use healthy coping strategies in difficult situations

b) Unemployed with low SOC
   - feel stressed, are inconsistent in their behaviour, feel overtaxed or not challenged
   - show high vulnerability and critical life events
   - cope in a regressive, emotional, aggressive and forsaking way.

4.1 CASE 1. Long-term unemployed with multiple problems and negative prognosis

36 years, male, long-term unemployed (SOC: 1.85):

- Person: unemployed for 8 years, electrician, drop-out student, divorced, legal actions against ex-wife, lives in a new relationship, indebted

- State of health: occupational restrictions due to serious health impairments (30% handicapped); overweight, feet trembling, various musculo-skeletal problems, difficult mental state impairs work ability
- Denial: good state of perceived health
- During counselling: complaints, harms attributed to external causes (e.g. job centres), misses counselling sessions, future prospects vary from pension claim ("inability to work"), to television jobs and emigration

4.2 CASE 2. Employee with a short-term contract, with multiple problems and negative prognosis

47 years, male (SOC: 3.38):

- Person: lathe operator, married (one child), wife working, social support through family
- State of health: severe impairments (60% handicapped); major depression, nervous breakdown, hospital treatments, on medication, slipped disc, numbness in hands and legs, no physical exercise
- Denial: excellent state of perceived health, feels happy. In contrast: feels loss of energy, discouraged, nervous, sorrow, feels humiliated by his former employer
- During counselling: complaints, monologues concerning dissatisfaction with life and illness, no perceived job chances, no support from authorities, therapy is indicated

4.3 CASE 3. Drop-out, long-term unemployed

43 years, male, long-term unemployed (SOC: 3.62):

- Person: unemployed for 6 years, bricklayer and electrician, drop-out student, institutional child, single, no contact to women, financial difficulties, mother mentally ill (incapacitated)
- State of health: serious impairments (50% handicapped), depression, hospital treatment, on medication, crucial ligament problems, hip abnormality, damaged spinal column, chronic pain, stomach trouble, heart problems, headache, problems of hearing, irregular alcohol abuse, feels loss of energy, discouraged, hopeless and without orientation
- During counselling: anxious in groups, but involved; mental block about women, present critical incident (trouble with neighbours), crying fits, wrong address to follow up

4.4 CASE 4. Unemployed participant with a positive prognosis

45 years, single mother, unemployed (SOC: 5.00):

- Person: unemployed for 1.5 years, office clerk for 30 years, searches for work, sees her single life positive
- State of health: good, smoker, slightly overweight, emotionally stable, feels full of energy, strong will, "survivor"
- During counselling: realistic perception of the situation, follows career plans and health recommendations (walking, giving up smoking); wants to become self-employed, practically skilled

5 Changes in group dynamics

During the counselling process participants showed some positive changes in behaviour related to social, vocational and health aspects. They were:

a) Changes in the appearance of male unemployed persons, their clothing and hair-style. In one group three men went together to the hairdresser.

b) Changes in involvement. Participants gradually started to exchange private experiences and introduce materials or information concerning their situation (job offers, application, vocational training, etc.).

c) Changes in self-efficacy and job-related actions, leading to exchange and practical help in job search (e.g. taking care of application forms: self-employment, vocational training). In addition there was peer monitoring concerning appointments with authorities.

d) Changes in social behaviour and reactivation of relations. Participants started to meet again with old friends or tried to get in contact with new people (e.g. neighbours). They also tried to arrange some group activities (e. g., sports). This gave some of the isolated clients new optimism and enjoyment of life.

6 Implications for future programmes

The results of our survey offer some useful implications for future counselling programmes. For instance, 26% of our participants had suicidal tendencies in the past, and by the assessment of the project staff, 11% of the clients seemed to be in serious danger of committing suicide. In Germany the concepts for suicide prevention are still largely missing (Bergold & Schürmann, 2001), and there is a need to develop them in future programmes.

Approximately one fifth of the unemployed and those on short-term contracts were severely handicapped. These clients need special support by public services as guaranteed in the German legislation. For this purpose, improved co-operation with external services specialised for handicapped people should be set up also from side of such interventions schemes. Due to their particular problems that act as a barrier to re-employment they do need beyond the support from the employment services also assistance from social workers or as in some cases even psychotherapy.

The large number of singles in our groups indicates that group activities like trips, hiking, cycling or cooking would be useful to support closer relations, which will improve job chances and quality of life. This can be taken into account in future project work.

Future programmes should focus in particular on elderly jobseekers, they are more often impaired than the younger ones (cf. Hollederer, 2002, Adamy, 2003). In Germany a growing number of unemployed persons is over 50 years old (Bleck, 2003, Bellmann, Kistler & Hilpert, 2003), and taking health impairments into account, this group basically has no chances in the job market (Buck & Schletz, 2002). In our study the 40-50 year-old participants did not at all or only to a small proportion anticipate successful re-employment. Useful interventions for elderly jobseekers could be empowerment modules for coming to terms with the past, since these persons often show regressive behaviour. They seem to be less able to react flexibly to new circumstances (cf. Creed & Watson, 2003).

In general it is difficult to place any unemployed person with some disadvantage into work. Therefore, cross-administrative co-operation of employment authorities with social and health services is important. Vocational counselling and health promotion need to be more interlinked in Germany.

7 Summary

We can conclude that the BKK-Job-Fit Scheme was well received. In the final counselling session all participants evaluated the programme positively: 41% were "very pleased", 38% "pleased" and 21% were "half and half" pleased by the counselling programme. More than 80% stated that they would participate again in a similar project.

The participants' baseline scores for the psycho-social scales were below those of the standard population (Kuhnert et al., 2004). The employees on short-term contracts had a better state of health as compared to that of the long-term unemployed persons.

Except for five clients, all participants showed minor or significant progress in their state of health and social circumstances. The intervention effect tended to be more prominent for those employed. The best effects were achieved for those who joined the self-help group.

Our pilot scheme BKK-Job-Fit indicates that short-term counselling programmes help unemployed people with slight or average health impairments to improve their coping strategies. Some other short-term interventions have achieved even better results (Elkeles & Kirschner, 2003; Trube & Luschei, 2000). Unemployed with multiple problems and chronic diseases are in need of long-term support, which is not yet provided in

Germany. An analysis by the IAB indicates that structural unemployment will continue until 2015 (Lutz et al. 2002). Especially for people with multiple problems and dysfunctional coping strategies, the "Hartz IV"[2] reform results in additional pressure (e.g. financial sanctions). These aggravated circumstances might lead to increased stress for unemployed persons, thus increasing the risk of illness in the longer run.

It seems important that future projects should cushion the impact of long-term unemployment and support the maintenance and strengthening of existing skills as well as help to adopt new coping strategies. Multimodal programmes such as the scheme discussed above (cf. Kaluza, 2002) provide grounds for longer-term interventions: therefore group counselling concepts like the BKK-Job-Fit scheme will have future relevance.

References

Adamy, W. (2003). Ausgrenzung von Älteren statt Begrenzung der Frühverrentung. Wirkungen einer kürzeren Bezugsdauer beim Arbeitslosengeld [Exclusion of the elderly instead of reduced early retirement. Effects of a shorter employment benefit period]. *Soziale Sicherheit, 7*, 218-225.

Antonovsky, A. (1997). *Salutogenese. Zur Entmystifizierung der Gesundheit* [Salutogenesis, demystifying health]. Tübingen: DGVT.

Bellmann, L., Hilpert, M., Kistler, M. & Wahse, J. (2003). Herausforderungen des demografischen Wandels für den Arbeitsmarkt und die Betriebe [Challenges of demographic change for the labour market and companies]. *Mitteilungen aus der Arbeitsmarkt- und Berufsforschung, 36(2)*, 133-149.

Bergold, J. & Schürmann, I. (2001). Krisenintervention – Neue Entwicklungen? [Crisis intervention – New developments?] *Verhaltenstherapie und psychosoziale Praxis, 33(1)*, 5-15.

Bleck, C. (2003). *EQUAL – Entwicklungspartnerschaft „Offensive für Ältere"* [EQUAL – development partnership „Action for the elderly"]. Informationen für die Beratungs- und Vermittlungsdienste, 23. Juli 2003 (15/03). Bundesanstalt für Arbeit.

Bongarts, T. & Gröhmke, K. (1997). Soziale Isolation bei Arbeitslosen? Eine netzwerkanalytische Betrachtung. [Social isolation among unemployed – a network analytical consideration] In G. Klein & H. Strasser (Eds.), *Schwer vermittelbar* [Difficult to place]. (pp. 197-219). Opladen: Westdeutscher Verlag.

Brixy, U., Gilberg, R., Hess, D., & Schröder, H. (2002). Was beeinflusst den Übergang von der Arbeitslosigkeit in die Erwerbstätigkeit? [What influences the transition from unemployment to paid labour?] *IAB Kurzbericht, 1*, 1-3.

Buck, H. & Schletz, A. (2002). Sensibilisierungs- und Beratungskonzepte für eine altersgerechte Arbeits- und Personalpolitik [Concepts for sensitization and counselling for an employment policy and human resources policy appropriate for an aging workforce]. In Projektverbund Öffentlichkeits- und Marketingstrategie demographischer Wandel (Eds.), *Handlungsanleitungen für eine altersgerechte Arbeits- und Personalpolitik – Ergebnisse aus dem Transferprojekt* [Instructions for action for an age-appropriate work and human resources policy] (9-14). Stuttgart.

Bullinger, M. & Kirchberger, I (1998). *SF-36 Fragebogen zum Gesundheitszustand* [Health status questionnaire]. Göttingen: Hogrefe.

2 Hartz IV labour market reform: i.e. long-term unemployed (>12 months) receive an allowance of 345 € (plus reimbursement of the rent). Unemployed people with savings, private fortune or living together with an employed partner are excluded from receiving allowance (Kuhnert, 2005c).

Cramer, R., Gilberg, R., Hess, D., Marwinski, K., Schröder, H., & Smid, M. (2002). Suchintensität und Einstellungen Arbeitsloser [Job search efforts and attitudes of the unemployed]. *Beiträge zur Arbeitsmarkt – und Berufsforschung 261*. Nürnberg: Bundesanstalt für Arbeit.

Creed, P. & Watson, T. (2003). Age, gender, psychological wellbeing and the impact of losing the latent and manifest benefits of employment in unemployed people. *Australian Journal of Psychology, 55(2)*, 95-103.

Elkeles, T. & Kirschner, W. (2003). *Arbeitslosigkeit und Gesundheit - Interventionen durch Gesundheitsförderung und Gesundheitsmanagement- Befunde und Strategien (Ergebnisbericht)* [Unemployment and health – Interventions through health promotion and health management – Findings and strategies]. Expertise for the BKK-Bundesverband. Neubrandenburg/Berlin.

Hollederer, A. (2003). Arbeitslos – Gesundheit los – chancenlos? [Unemployed – ill-health – no chance?] *IAB-Kurzbericht*, 4/ 21.3.2003. Nürnberg: Bundesanstalt für Arbeit.

Kaluza, G. (2002). Förderung individueller Belastungsverarbeitung: Was leisten Stressbewältigungsprogramme? In B. Röhrle (Eds.), *Prävention und Gesundheitsförderung, vol.*2 [Prevention and health promotion] (pp. 195-218). Tübingen: DGVT.

Kanfer, F., Reinecker, H., & Schmelzer, D. (1996). *Selbstmanagement-Therapie. Ein Lehrbuch für die klinische Praxis* [Self management therapy: A handbook for clinical practice] (2nd rev ed.). Berlin: Springer.

Kieselbach, T., Scharf, G. & Klink, F. (1997). Interventionsmaßnahmen für Langzeitarbeitslose: Wiederbeschäftigung und psychosoziale Stabilisierung [Interventions schemes for long-term unemployed: Re-employment and psychosocial stabilization]. In G. Klein & H. Strasser (Eds.), *Schwer vermittelbar* [Difficult to place] (pp. 313-331). Opladen: Westdeutscher Verlag.

Kuhnert, P. (1999). *Bewältigungskompetenzen und Beratung von Langzeitarbeitslosen [Coping competencies and counselling of long-term unemployed]*. Dissertation, University of Dortmund.

Kuhnert, P. (2004a). Work Life Balance trotz Arbeitslosigkeit und instabiler Beschäftigung? – Paradoxie oder neue Chance? In M. Kastner (Eds.), *Die Zukunft der Work Life Balance* [The future of work life balance] (pp. 141-194). Heidelberg: Asanger.

Kuhnert, P. (2004b). Arbeitslosenberatung: Entwicklung und Perspektiven. In F. Nestmann, F. Engel & Sickendieck, U. (Eds.), *Das Handbuch der Beratung, vol.2* (pp. 959-975).Tübingen: DGVT.

Kuhnert, P. (2005c). Reformen der Arbeitsförderung – Irrwege oder Auswege für arbeitslose Menschen? [Reforms of work promotion – traps or way out for unemployed people?]. In M. Kastner, T. Hagemann & G. Kliesch (Eds.), *Arbeitslosigkeit und Gesundheit – „Arbeitsmarktintegrative Gesundheitsförderung"* [Unemployment and health – labour market integrative health promotion]. Lengerich: Pabst.

Kuhnert, P., Ayan, T., Hagemann, T., Iserloh, B., Reick, C., & Kastner, M. (2004). Gesundheitsorientierte Selbstmanagement-Beratung bei Arbeitslosigkeit und Beschäftigungsunsicherheit – Ergebnisse einer Evaluationsstudie [Health-orientated Self-Management Counselling in Unemployment and Job Insecurity - Results of a Longitudinal Study]. Final Report, University of Dortmund.

Kuhnert, P. & Kastner, M. (2001). Zusammenhänge zwischen sozialen und psychischen Faktoren bei der Bewältigung von Langzeitarbeitslosigkeit. In M. Kastner & J. Vogt (Eds.), *Strukturwandel in der Arbeitswelt und individuelle Bewältigung* [Structural changes at work and individual coping] (pp. 267-303). Lengerich: Pabst.

Kuhnert, P. & Kastner, M. (2002). Neue Wege in Beschäftigung – Gesundheitsförderung bei Arbeitslosigkeit. In B. Röhrle (Eds.), *Prävention und Gesundheitsförderung*, Band 2 [Prevention and health promotion, Vol. II] (pp. 373-406). DGVT.

Laireiter, A.-R. (2002). Internationale Tagung „Soziales Netzwerk und Soziale Unterstützung" in Salzburg [International conference "Social Network and Social Support" in Salzburg]. *Verhaltenstherapie und psychosoziale Praxis, 34 (4)*, 893-899.

Lutz, C., Meyer, B., Schnur, P. & Zika, G. (2002). Projektion des Arbeitskräftebedarfs bis 2015. Modellrechnungen auf Basis des IAB/INFORGE-Modells [Projection of the employee demands until 2015. Model calculations based on IAB/INFORGE models]. *Mitteilungen aus der Arbeitsmarkt- und Berufsforschung, 35 (3)*, 305-326.

Machin, A. & Creed, P. (2003). Understanding the differential benefits of training for the unemployed. *Australian Journal of Psychology, 55(2)*, 104-113.

Marstedt, G. (2002). Integration älterer und gesundheitlich beeinträchtigter Arbeitnehmer/innen des öffentlichen Sektors in die Erwerbstätigkeit [Integration of older and employees with health limitations working in the public sector]. *ForschungsInformationsDienst, 2,* 63-65.

Paul, K. & Moser, K. (2001). Negatives psychisches Befinden als Wirkung und als Ursache von Arbeitslosigkeit [Negative psychological mood as cause and effect of unemployment]. In J. Zempel, J. Bacher & K. Moser (Eds.), *Erwerbslosigkeit – Ursachen, Auswirkungen und Interventionen* [Unemployment – Causes, Consequences and Interventions] (pp. 81-110). Opladen: Leske + Budrich.

Reick, C. (2005). Möglichkeiten der Diagnostik arbeitslosigkeitsbedingter gesundheitlicher Beeinträchtigungen [Chances of diagnosing unemployment-related health impairments]. In M. Kastner, T. Hagemann & G. Kliesch (Eds.), *Arbeitslosigkeit und Gesundheit - Arbeitsmarktintegrative Gesundheitsförderung* (129-148). Lengerich: Pabst.

Rosenbrock, R. (2001). Primärprävention zur Verminderung sozial bedingter Unterschiede von Gesundheitschancen – was ist das und welche Rolle können die Krankenkassen dabei spielen? [Primary prevention to reduce the socially determined gap in health chances – what is it and what might be the potential role of health insurances]. In R. Geene, C. Gold & C. Hans (Eds.), *Armut macht krank! Teil 3. SGB V § 20 Gesundheitsförderung zum Abbau sozial ungleicher Gesundheitschancen* [Poverty makes sick! Part 3 Social Code § 20 Health Promotion for the reduction of social unequal health chances] (pp. 14-24). Berlin: b books.

Schach, E., Rister-Mende, S., Schach, S., Glimm, E. & Wille, L. (1994). *Die Gesundheit von Arbeitslosen und Erwerbstätigen im Vergleich* [Health of the unemployed and employed in comparison]. Bundesanstalt für Arbeitsschutz und Arbeitsmedizin. Dortmund.

Trube, A. & Luschei, F. (2000). *Entwicklungs- und Vermittlungs-Assistenz (EVA) – ein Instrument zur Wiedereingliederung Langzeitarbeitsloser (Abschlußbericht).* [Development and Placement Assistance (EVA) – An instrument for the reintegration of long-term unemployed. Final report]. ZPE-Schriftenreihe Nr. 7. Universität Gesamthochschule Siegen.

Changes in Mental Health: Young People in Labour Market Schemes in West and East Germany

Heike Behle

Introduction

This article deals with active labour market programmes (ALMPs), in particular, how they affect the mental health of young people in East and West Germany. The connection between unemployment and mental health as such has been well established in previous research (for reviews of the literature see Lakey et al., 2001; Kieselbach, 2000). Little work, however, has been done considering the changes of young people's mental health during a problematic school-to-work transition, which is characterised by the participation in an ALMP. This paper addresses this issue, and aims to contribute further towards a sociological understanding of the psychosocial conditions of underemployment (Prause & Dooley, 1997, 2001). The article draws on data gained from a longitudinal survey of participants in an ALMP.

The stabilisation of mental health is one important aim of active labour market policies within the employability discourse (Gazier, 1999). This article aims to discuss if and how active labour market programmes manage to stabilise young peoples' mental health. The underlying question of this analysis is whether and how changes in the mental health of participants in an active labour market programme can be traced. Which factors have a stabilising, which have a destabilising effect on mental health? Are there any differences between young people in East and West Germany?

To address these questions, the use of ALMPs and the social composition of participants in both parts of Germany will be described. The article will then draw on empirical results of an ALMP survey. Finally, the article discusses the possibilities and limitations of ALMPs, and concludes by explaining the different effects ALMPs have on young people in East and West Germany.

1 Active labour market schemes and mental health in both parts of Germany

Over the last decades, the participation in an active labour market scheme has become part of the school-to-work transition for many young people in Germany (Dietrich, 2001).

For approximately half a million young people each year, school-to-work is characterised by a movement between active labour market schemes, unemployment, vocational training, or (casual) jobs (Figure 1). The most important types of ALMP schemes are 'preparation for apprenticeship' and 'apprenticeship with a provider'. Young people who are not considered ready for vocational training get the chance to take part in a 'preparation for apprenticeship' scheme, where occupational orientation can be built up. In Germany, there were about 100 000 young people taking part in these schemes in 2004. Young people who are disadvantaged and capable of vocational training but fail to find a vacancy on the apprenticeship take part in an apprenticeship with a provider. About 125 000 young people were given this opportunity in 2004.

Figure 1. The participation in active labour market programmes in Germany

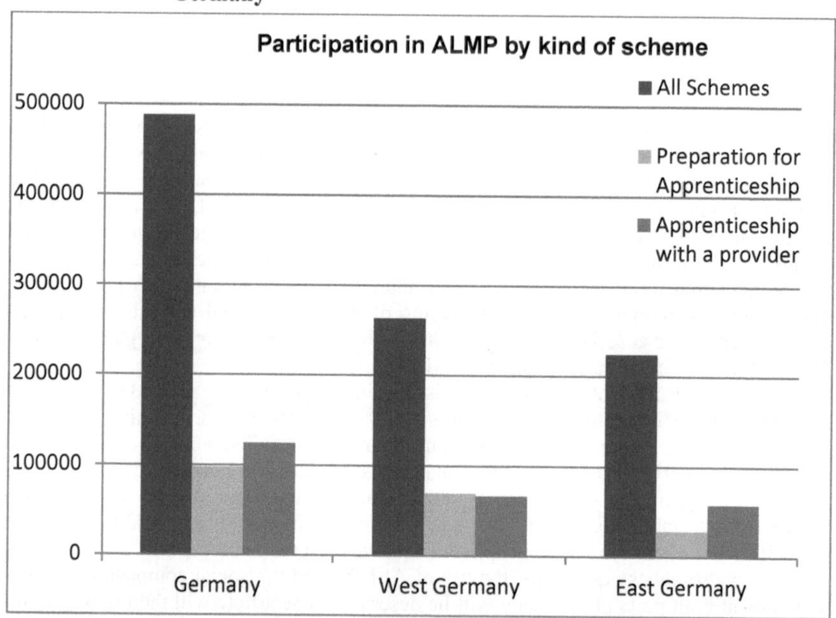

Source: Bundesagentur für Arbeit, 2005, pp. 107.

Differences between East and West Germany in the school-to-work transition of young people still exist more than fifteen years after reunification. In West Germany, in areas with persistent unemployment, many young people fail to find entry to the unsupported labour market. Especially young people with educational deficiencies were losers in the competition to find an apprenticeship. In East Germany, since the reunification in 1990, the vocational training market has been supported massively by ALMPs to help apprentices of bankrupt companies and school leavers. Thus, the East German vocational training structure was established as a publicly supported system after the reunification. It was expected that the need for apprenticeships with providers would fade away over time. This, however, has not yet been the case. Persistent unemployment and insecurity on the labour market have failed to encourage private employers to take on more apprentices.

In addition, there were demographic challenges. The population policy in the GDR was aimed to increase the population by financially rewarding people having children. As a result, there was an increase in the numbers of young people trying to enter the labour market during the 1990s. Also, directly after reunification, early retirement was facilitated. Therefore, the age structure of East German employers is very young. Both the GDR population policy and reunification measures created an oversupply of young people who did not manage to enter either company-based vocational training market or the labour market (Lutz et al., 1999).

Differences in the vocational training system together with economical and demographical circumstances have resulted in variations in the social composition of young people with problems in school-to-work transition in both parts of the country.

Young people with problems in school-to-work transition in West Germany can generally be described as low qualified due to employers' selection processes. Employers select and employ those candidates with the best prospects, which lead to a less qualified group of unemployed young people. Many lack a basic school degree and are not considered fit for vocational training. In the East, however, a different picture can be seen. Young people who experience exclusion from the labour market are often fully vocationally qualified and employable. In the denser East German labour market, there are fewer employers choosing from a larger pool of candidates, thus leaving behind a more heterogeneous group of young unemployed, containing many readily employable young people.

The differences are also represented in the choices of types of ALMP in both parts of the country (Figure 1). In East Germany, the ALMP scheme 'apprentice with a provider' was used twice as often than the 'preparation for apprenticeship' scheme. In West Germany, on the other hand, both types of schemes were used frequently. This can be explained by the 'trainability' of young people with problems in their school-to-work transition. The proportion of young people considered incapable of completing vocational training is smaller in East Germany. Those differences in the social composition can also be seen in the mental health of participants during the participation in the scheme (Korpi, 1994; Behle, 2005). Young people in East Germany who take part in ALMP were found to be mentally more stable than West Germans

The main aims of any ALMPs are to (re-)integrate young people to (company-based) vocational training and labour market. Previous research has shown that finding a way out of unemployment can have an impact on mental health (e.g., Tiggemann & Winefield, 1984; Hammarström & Janlert, 1997, Strandh, 2000). Changes in mental health are related to how the new status resolves the uncertainty faced in the unemployment situation. Therefore, changes in the labour market status after the scheme are expected to have an impact on mental health.

The participation in any ALMP schemes should directly help build up and maintain balanced mental health. (Further) occupational orientation, reliability, efficacy, motivation, teamwork, and creativity are abilities and skills that can be obtained during the participation. In addition, young people can increase their individual labour market chances by obtaining certified qualifications, work experience and contacts to employers.

Data gained for the evaluation of an active labour market programme for young people (JUMP) (Dietrich et al., 2001) were used to analyse changes in the mental health of participants. Depending on their previous labour market experience young people were allocated into different JUMP schemes such as 'Preparation for apprenticeship', 'Full vocational training (with a provider)' 'Employment (wage subsidies)'. 'Combined employment and qualification' and 'Special schemes for 'drop-outs from apprenticeships''

Participants' mental health were analysed to find out whether there were any changes in mental health after the participation in the scheme and if there were any differences between the groups of young people in West and East Germany.

2 Changes in mental health of JUMP participants

The JUMP survey
5,000 JUMP participants took part in a longitudinal survey with a non-proportional sample tiered according to the type of ALMP and personal characteristics drawn from the JUMP-participants database (Dietrich et al., 2001). Young people were interviewed three times 'during the participation' (t1), 'directly after the participation' (t2) and 'one year after participation in the ALMP' (t3). For this article, 1,832 interviews were used to analyse changes in mental health sustainable one year after the participation. The number of suitable interviews had to be reduced mainly due to panel mortality and the non-responses of some interviewees regarding their mental health.

Measures
Participants' mental health was operationalised using the Trier Mental Health Questionnaire, which contains a Likert scale consisting of 20 statements each with four possible answers (Becker, 1989). A Mental Health index (MH) can be calculated with values between 20 and 80 – the higher the score the more balanced is the observed mental health. Changes in mental health can now be described by a variable, which calculates the score difference between the mental health during the participation and one year after (t3-t1). The 'changes' variable can take values between -60 (extreme destabilisation) to +60 (extreme stabilisation). A value of 'nil' indicates that there were no changes in mental health. Becker (1989) confirmed for his calibrate sample that 'in general the distribution corresponds to a normal distribution' with a tendency towards left skewness (see also Table 1). Similar findings, which indicate high reliability of the sample, can be seen in the key data of the JUMP survey. Also, gender differences can be observed – both in JUMP as well as in the calibration sample. Although the sample underrepresents the number of East German participants it can generally be seen as representative for participants (Behle, 2007).

To evaluate differences in the group of young people, multiple regression models (Table 2 to Table 5) were estimated using 'changes in mental health' as dependent variable[1]. It was expected that the following variables showed an impact on changes in mental health:

1 The mental health score during the scheme will be used as an independent variable to take into account the level on which changes have taken place (analogous to Nordenmark & Strandh, 1999). This procedure results in the consideration of real net changes in mental health (for further discussion on the procedure see Jackson et al., 1983, p. 528; Frese, 1994, p. 195).

- *Type of labour market scheme.* Information provided by the job centre was used to distinguish between the different schemes 'wage subsidies', 'preparation for apprenticeship', 'special schemes for drop-outs', 'combined employment and qualification' and 'vocational training'.
- *Work Involvement.* Work involvement was operationalised using the Work Involvement Scale (WIS) by Warr et al. (1979). Due to the design of the study, both the initial work involvement as well as changes (t3-t1) in work involvement were included. Changes in work involvement are anticipated to affect the mental health positively if young people find an entrance to the labour market. In case young people are unemployed after JUMP, a negative connection of changes in work involvement and in mental health can be expected. Therefore, not only the changes of answers to the WIS scheme were used to evaluate differences in between the groups, also the interaction with further unemployment was looked at.
- *Current labour market status.* One year after finishing the scheme during the second interview, young people were either in vocational training, employed, unemployed, in a further scheme or had another status (e.g. maternity leave, long term illness). It can be expected that young people who were employed in an apprenticeship were more likely to have improved their mental health (Strandh 2000).
- *Social Support.* Social support has been operationalised using the Social Support Appraisal Scale (Vaux et al., 1986). In previous research it has been found that social support can moderate the effects of unemployment (Hammarstroem 1994).
- *Ethnic background.* The ethnic background of young people in Germany was broken down by their place of birth and nationality. German young people have both German nationality and were born in Germany. Foreign JUMP participants were defined as either not holding German nationality or having been born outside of Germany. The third category resettlers refers to a special group of migrants from East European states holding German nationality. Mental health of young participants with non-German ethnic background was found to be lower compared to their German counterparts during the scheme (Behle, 2005).
- *Labour market status between interviews.* Here, the months spent in full-time employment, apprenticeship/qualification and unemployment after the completion of the programme were measured.
- *Changes in relationships and financial changes* are expected to have impacts on changes in mental health which are unrelated to the ALMP or the problems in school-to-work transition.
- *Regional youth unemployment rate.* Regional unemployment in the highest tercile was operationalised as high. A high regional unemployment rate was expected to reduce the impact of the ALMP on the integration of young people to the labour market.
- *Expectation for future.* Young people were asked if they expected significant improvement by improved chances on the labour market due to the participation in current employment or training. It was expected that young people who had high expectations for the future were more likely to have increased their mental health.

3 Results

One year after participating in JUMP, sustainable changes in the mental health of both men and women existed (Table 1). Values varied from -29 to 36. The median for the whole sample for changes was 'nil', this value, however, was only found for 7% of young people. In the whole group, 47% showed stabilisation of mental health one year after the scheme. Destabilisation was observed in 45% of the young people. A high variance of approximately 46 indicated individual differences in changes of mental health.

A closer look revealed regional differences between participants in East and West German (Table 1). A higher mental health score among East Germans during the participation in the programme was observed, both for male and female participants. Also, significant differences in the changes of mental health existed. The proportion of East German men experiencing destabilisation after the scheme was higher than those of West Germans. Whilst in West Germany the median of mental health changes was 'zero', the corresponding figures in the East were -1. The group average means differed significantly (t-test $\alpha \leq 0.01$) from each other and within East German male participants' subgroup the average mean was significantly different to 0 ($\alpha \leq 0.05$). East Germans still, however, had a higher mental health score after the scheme, compared to

Table 1. Mental health during JUMP and sustainable changes of mental health by region and gender

Gender	Region	n	Min	Max	Average	Variance	Median	Skewness	Kurtosis
				Mental Health during JUMP (t1)					
Male	West Germany	814	32	80	64.45	52.88	65	-.408	.213
	East Germany	324	44	80	65.78	52.38	66	-.312	-.372
	Germany	1138	32	80	64,83	53,05	65	-0,38	0,06
Female	West Germany	503	37	80	62.73	59.25	63	-.394	.331
	East Germany	191	46	79	64.58	45.78	65	-.249	-.189
	Germany	694	37	80	63,23	56,16	63	-,393	0,302
				Sustainable Changes (t3-t1)					
Male	West Germany	814	-24	30	.523	44.74	0	.113	.960
	East Germany	324	-19	22	-.753	41.10	-1	.052	.742
	Germany	1138	-24	30	0.16	44.0	0	0.11	0.898
Female	West Germany	503	-29	36	.352	55.26	0	.244	2.767
	East Germany	191	-20	22	-.639	32.94	0	.098	.988
	Germany	694	-29	36	0.08	49.3	0	.263	2.773
Sample		1832	-29	36	0.13	46.07	0	.171	1.712

Source: JUMP participant survey (IAB-Project 486-1).

their West German counterparts. The same tendencies were observed with women although their average group change score did not differ significantly between the regions.

Four regression models for each of the subgroups 'West German Men', 'East German Men', 'West German Women' and 'East German Women' were estimated to gain further information on moderators of changes in mental health. For each subgroup, model 1 described the impact of JUMP related variables (type of labour market scheme, work involvement, and the current labour market status) on changes in mental health. Model 2 referred to general characteristics such as social support or ethnic background. In Model 3, changes between the interviews were researched. Finally, Model 4 referred to expectations for the future.

The following variables were found to show an impact on changes in mental health one year after the participation.

The *type of labour market scheme* only influenced the changes in mental health of West German young men. Compared to the preparation for apprenticeship, the participation in special schemes for young drop-outs increased their changes in mental health by 1.6. The type of ALMP scheme did not have any impact on men in East Germany or women in either West or East Germany.

Strengthening of *work involvement* was considered as one aim of the participation in JUMP. Nevertheless, in all groups a decrease of the WIS-Index was observed. West German young men scored on average the smallest decrease (-.47), East German women the largest (-1.5). West German men had lower scores than East German while on scheme, but East German women showed a greater deterioration (West German men 24.9, East German women 27). However, changes in work involvement only had an impact on changes in the mental health of participants in West Germany, there was no influence of the variable in the East. Changes in the difference-WIS scale of one unit resulted in an increase of .22 on the mental health scale of women in West Germany and in an increase of .13 on the mental health scale of men in West Germany.

The *current labour market status* which had a strong impact on men in West Germany did not have any impact on men in East Germany. In West Germany and for women in East Germany there was an increase in mental health for those in vocational training and employment (in reference to unemployment). Current unemployment in interaction with changes in mental health, however, resulted in a significant decrease of the mental health of West German women.

Social support only showed a significant positive impact on men in East and West Germany. It did not have any significant impact on changes in mental health of women. Young men with *foreign background* in West Germany were more likely to have lower mental health after the scheme. Also, the changes in mental health of men in East Germany were strongly influenced by the *number of months men spent unemployed* between the interviews. Financial changes only had a negative impact on the mental health of men in West Germany. Changes in relationships had a huge impact on changes in mental health of men in East Germany and also of women in West Germany.

The regional unemployment rate did not show any impact on changes in mental health. In all groups, the expectation of significant future improvement of labour market chances had a positive impact on changes in individual mental health. In addition to that, the power of the models to explain changes in mental health was significantly strengthened by the introduction of this variable, especially for women in East Germany.

4 Conclusion

The findings can be summarised as follows: Most young people in West Germany either improved or maintained their mental health one year after participating in JUMP. In the East, young people displayed a more stable mental health during the participation.

After the scheme, men were more likely to have lower mental health. East Germans, still had a higher score on the mental health scale than young people in West Germany one year after the scheme.

JUMP related variables (type of labour market scheme, work involvement, current labour market status), general characteristics (social support, ethnic background), changes between the interviews (experienced unemployment after the scheme, financial changes, changes in relationship) and expectations for the future had an impact on mental health changes. No influence was seen for the regional unemployment rate.

Active labour market schemes such as JUMP have different aims in East and West Germany. In West Germany, the main aim is to build up qualifications and work experience to facilitate employment. In East Germany, there are also some other aims. ALMPs are one possibility to bridge unemployment: instead of losing contact with the labour market and becoming more and more unstable due to a long period of unemployment, young people should spend their time building up and maintaining useful qualifications and skills.

There is a strong connection between expectations for the future and changes in mental health. If young people expect a significant improvement in future perspectives, they manage to improve their mental health. The more the achieved status after unemployment makes up for the the uncertainty faced in unemployment, the more people are able to improve their mental health (Strandh, 2000). Also, young people are in control of their lives and can (further) construct a life plan. The connection between expectations for the future and changes in mental health exist for both East and West German young people.

In the East, possibilities for ALMPs are limited as young people were already employable before participating in the scheme. JUMP was seen as 'yet another programme' aimed to provide qualifications and work experience for already qualified participants. It did not manage to change prospects for East German participants as the labour market was still very difficult and it continues to be so for young people. This, however, should not be seen as a failure of the programme as such. There is the possibility that the mental health of young people would have deteriorated further had they not participated in the programme.

The research poses questions about the future development of East and West Germans. It is possible that the mental health of young people in both parts of the country could be converging. However, two arguments contradict this interpretation: Firstly, there is no information available whether the downward trend of East German's mental health will stop at the same level as West Germans. It is possible, if not probable, that their mental health will continue to destabilise. Secondly, it could be seen that changes in mental health of young people in East Germany were influenced by different factors than the corresponding changes of young West Germans. Changes in the mental health of West German men and women were mainly influenced by labour market related variables: work involvement and the current labour market status had a strong impact on changes in the mental health. Neither impact factor presented (hardly) any significant impact on changes in mental health of East Germans, especially not for male participants. Changes in mental health of East Germans were mainly influenced by changes in their relationships and expectations for the future.

Schemes in JUMP were mainly developed in West Germany and then transferred to the East (Lutz et al., 1999). In conclusion, it seems necessary to develop appropriate schemes especially for East Germany to address relevant of labour market issues. A long-term solution, however, can only be achieved by an improved labour market for young people.

References

Becker, P. (1989). *Der Trierer Persönlichkeitsfragebogen (TPF).* [The Trier personality questionnaire] Göttingen: Hogrefe.

Bundesagentur für Arbeit (2005). *Arbeitsmarkt 2004. Amtliche Nachrichten der Bundesagentur für Arbeit, 53. Jahrgang, Sondernummer.* Nürnberg, 30. August 2005.

Behle, Heike (2010) The impact of active-labour market programmes on young people's mental health. possibilities and limitations. In: Evans, J, and Shen, W. (ed): Youth employment and the future of work. Strassbourg: Council of Europe publishing.

Behle, H. (2005). Moderators and mediators on the mental health of young participants in active labour market programmes: Evidence from East and West Germany. *International Review of Psychiatry, 17*(5), 337-345.

Behle, H. (1997). *Veränderung der seelischen Gesundheit durch arbeitsmarktpolitische Maßnahmen?* [Changes in mental health through active labour market schemes?]. IAB Bibliothek 308. Nuremberg.

Dietrich, H. (2001). Wege aus der Jugendarbeitslosigkeit - Von der Arbeitslosigkeit in die Maßnahme? [Ways out of unemployment – from unemployment into an employment or training measure?] *Mitteilungen für Arbeitsmarkt- und Berufsforschung, 34,* 419-439.

Dietrich, H., Behle, H., Böhm, R., Eigenhüller, L. & Rothe, Th. (2001). *School-to-Work-Transition und aktive Arbeitsmarktpolitik* [School-to-work transition and active labour market policies.] Nuremberg: IAB.

Frese, M. (1994). Psychische Folgen von Arbeitslosigkeit in den fünf neuen Bundesländern. [Psychological results of unemployment in East Germany]. In L. Montada (Ed.), *Arbeitslosigkeit und soziale Gerechtigkeit* (pp. 193-213). Frankfurt a.M.; New York: Campus.

Gazier, B. (1999). *Employability: Concepts and policies. Report 1998.* Berlin: Beschäftigungsobservatorium der Europäischen Komission DG V.

Hammarstroem, A (1994). Health Consequences of youth unemployment – review from a gender perspective. *Social Science and Medicine, 38:* 699-709.

Hammarström, A. & Janlert, U. (1997). Nervous and depressive symptoms in a longitudinal study of youth unemployment – selection or exposure. *Journal of Adolescence, 20,* 293-305.

Jackson, P.R., Stafford, E.M., Banks, M.H. & Warr, P.B. (1983). Unemployment and psychological distress in young people. *Journal of Applied Psychology, 68,* 525-535.

Kieselbach, T. (Ed.). (2000). *Youth unemployment and health.* Opladen: Leske & Budrich.

Korpi, T. (1994). *Escaping unemployment. Studies on the individual consequences of unemployment and labor market policy.* Stockholm: Universitet Stockholm.

Lakey, J., Mukherjee, A. & White, M. (2001). *Youth unemployment, labour market programmes and health.* Westminster: Policy Studies Institute.

Lutz, B., Ketzmerick, Th. & Wiener, B. (1999). Ausschlussrisiken und Grenzen herkömmlicher Arbeitsmarktpolitik. [Exclusion risks and the limits of traditional labour market politics]. In E. Wiedemann, C. Brinkmann, E. Spitznagel & U. Walwei (Eds.), *Die Arbeitsmarkt- und beschäftigungspolitische Herausforderung in Ostdeutschland* (pp. 267-284). Nuremberg: IAB.

Nordenmark, M. & Strandh, M. (1999). Towards a sociological understanding of mental well-being among the unemployed. *Sociology, 33,* 577-597.

Prause, J. & Dooley, D. (1997). Effect of underemployment on school-leavers' self-esteem. *Journal of Adolescence, 20,* 243-260.

Prause, J. & Dooley, D. (2001). Favourable employment status change and psychological depression. *Applied Psychology, 50*, 282-304.

Strandh, M. (2000). Different exit routes from unemployment and their impact on mental well-being. *Work, Employment & Society, 14*, 459-479.

Tiggemann, M. & Winefield, A.H. (1984). The effects of unemployment on the mood, self-esteem, locus of control, and depressive affect of school-leavers. *Journal of Occupational Psychology, 57*, 33-42.

Vaux, A., Phillips, J., Holly, L., Thomson, B., Williams, D. & Stewart, D. (1986). The social support appraisals (SS-A) scale. *American Journal of Community Psychology, 14*, 195-219.

Warr, P., Cook, J. & Wall, T. (1979). Scales for the measurement of some work attitudes and aspects of psychological well-being. *Journal of Occupational Psychology, 44*, 47-68.

Social Support and Organisational Integration of Long-Term Unemployed in the Volunteering Sector

Katrin Rothländer & Peter Richter

Introduction

Social integration and support of the unemployed are important factors to ensure their continued health and mental well-being. The following chapter introduces a special project which involves long-term unemployed in volunteering activities. Evaluation focuses on the potential of these activities to provide the unemployed with opportunities to seek and obtain more social support while also considering the organisational facets in such an integration programme.

Unemployment has increased dramatically in most European countries over the last two decades. Eastern Germany has faced particularly high unemployment rates since German reunification. Elderly people (aged 50 and above) have little prospect of returning to the first labour market and are most likely to face long-term unemployment. As a consequence, the government of the state of Saxony implemented the so-called TAURIS project in 1999. The TAURIS project aims to reintegrate long-term unemployed by offering employment in the volunteering sector. This project is also accompanied by an extensive evaluation of the participants' experiences over a period of three years. Members of the target group can choose between a variety of tasks from communities, churches, charitable or other public organisations. This non-profit work comprises 14 work hours per week and is compensated by a monthly allowance of 78 € in addition to the regular unemployment benefit. Compared to conventional unemployment measures, the TAURIS project is unique: First, the participation is voluntary in contrast to other unemployment measures where a refusal to take part can lead to the suspension of benefit payments. Second, this project is administered by local agents who work independently of the employment offices or job centres. Third, in principle the participation in TAURIS is indefinite and participants may continue as long as they wish. Terminations do occur, however, if the needs of the institutions change or if they are unable to pay the required share of the cost (20 € per month). However, TAURIS participants may still decide to continue their work for another non-profit organisation in the network. Until 2007, more than 10,000 long-term unemployed participated in the TAURIS project.

In this article, TAURIS serves as an example that describes how the volunteering experience of the long-term unemployed relates to their perception of social support and to their integration into the organisation they work for. As Mielck (2005) states, only little is known about the relationship between social support and social status, particularly in the case of the unemployed. Considering unemployment, it is taken for granted that the social network size shrinks due to the frequent loss of daily contacts with colleagues and supervisors. Yet, the degree of loss in perceived social support depends on the positive nature of those, now lost, former work-related social networks. If support is perceived as unsatisfactory, the termination of those relationships may evoke feelings of relief rather than of regret. Beyond such appraisal processes of one's social support network, the individual's behaviour in response to changes of the social network might be of importance as well. There is a need for longitudinal studies to investigate the extent to which the loss of work-related support could be avoided or compensated for by either maintaining contacts with former colleagues, by developing new relationships, or by intensifying existing relationships. Also, there is a lack of empirical work that measures the impact of unemployment, especially long-term unemployment, on the perceived personal social support (i.e., from a partner, from relatives and from friends). This may again influence the strategies applied by the unemployed in their search for and maintenance of social support.

Bongartz and Gröhnke (1997) investigated perceived social integration in a sample of 180 German long-term unemployed from Western Germany. One third of these participants were interviewed again at the end of the reintegration training. In a descriptive comparison with representative German data, only few differences between the two groups in terms of the size of their social support network were found. However, the long-term unemployed were almost three times more likely to be divorced than the general population. Accordingly, they were more likely to live without a partner (61%). Also, while 64% of the general representative sample reported a membership in associations and organisations (e.g., in unions), this was only the case for 19% of the long-term unemployed. In regard to the functions of different social support sources, Bongartz and Gröhnke found an obvious difference between relatives vs. friends. Relatives were the preferred source of advice if problems occurred, while friends were preferred as company for leisure activities. They also found that for most participants, the core social network size remained stable during the reintegration measure, though relatives have been assigned more central positions in the network than before. To interpret this finding, the authors drew on additional interview material and found that participants expected more support from their relatives than their friends in regard to their participation in the reintegration training. This may, in part, be due to the fact that it is often easier and less time-consuming to keep in contact with relatives than friends.

In a recent study from Australia, Creed and Moore (2006) compared the perceived social support between 94 unemployed and 77 underemployed persons. Underemployment is defined as a situation in which individuals are working less than is sufficient to earn a living as a result of which they receive additional unemployment benefits. In contrast to the above mentioned study by Bongartz and Gröhnke (1997) from Germany, the participants were younger and had been unemployed for shorter periods of time. Creed and Moore (2006) revealed significantly lower perceived support rates for the unemployed than for the underemployed. A second significant finding related to gender differences: the unemployed as well as underemployed women perceive significantly more social support than unemployed and underemployed men. This is consistent with other studies that show women are much more likely to seek social support as a coping strategy (e.g., Frydenberg & Lewis, 1991). Furthermore, it has been demonstrated that social support is a stronger predictor of health for women than it is for men (e.g. Denton & Walters, 1999).

In line with the presented comparisons between different subgroups of the unemployed (including underemployed), this inquiry is based on a comparison between long-term unemployed in the volunteering sector and long-term unemployed without such a voluntary engagement. The TAURIS project demonstrates that additional sources of social support and social integration beyond the family can be achieved through alternative engagement such as volunteering.

1 Research questions, sample and methods

Since this study on the success of the TAURIS project was of a rather exploratory nature, research questions were formulated instead of hypotheses. The main questions concerning social support were: (1) Did the extent of perceived social support from personal sources (partner, relatives and friends) differ for those who were volunteering compared to a similar group of non-volunteering long-term unemployed? (2) What was the nature of the relation between perceived social support from work-related sources (co-workers and supervisors) and perceived social support from private sources for the volunteering long-term unemployed? (3) Did perceived social support from personal and work-related sources change over time? (4) Which variables other than volunteering influenced participants' perception of social support? Further questions were posed regarding the organisational integration of long-term unemployed as a result of their volunteering work: (5) In regard to communication, cooperation and the participation, to what extent did the volunteering work meet objective demands of learning and health promotion? And finally: (6) To what extent did volunteering work meet the social needs of long-term unemployed?

1.1 Sample characteristics

The following analyses pertain to the second and third cohort of TAURIS participants from the years 2000 and 2002. In total, 122 surveys were sampled, 93 of which included the required information about participants' partnership status and their perceived social support. The characteristics of TAURIS participants were as follows: Two thirds (64%) of the investigated TAURIS group were female (this corresponds roughly to the overall gender distribution observed amongst all TAURIS participants). Similarly, 66% were in a relationship (60% were married) whereas 1 in 4 was divorced. Most had children (89%). In terms of their qualification, most TAURIS participants in the survey were skilled labourers (62% had a certificate as skilled worker). Prior to becoming unemployed, the TAURIS participants experienced periods of unemployment at least twice before (ranging from 1 to almost 4 years). In addition, 43% of the TAURIS participants reported that they were suffering from chronic illness.

A survey with another group of long-term unemployed (control group) provided a comparative baseline for the TAURIS survey results. Both groups were matched in terms of their socio-demographic characteristics with the following exceptions: With a mean age of 51 years compared to 48, TAURIS-participants were significantly older than members of the control group. However, the small age difference should not have an impact on the variables of interest. The control group also had slightly higher partnership rates: 75% were in a relationship. The members of the control group evaluated their financial situation as significantly less critical although the overall income per household, as an objective economic indicator, was similar across the two groups. This may be explained by the fact that the control group was financially more secure due to property ownership.

1.2 Measures

Perceived social support was assessed using a questionnaire by Rimann and Udris (SALSA, 1997), which differentiates between two work-related sources of support (coworkers, supervisor) and three functions of support (to rely on someone, to receive practical support, to have someone who listens, if one wants to talk about problems). Three private sources (partner, relatives, and friends) were added to the questionnaire. Perceived social support from their partner was also surveyed whether the person cohabited or not. Perceived support was measured by 4-point scales (from 1=*not at all* to 4=*totally*). Resulting values were aggregated. The reliability alpha coefficients ranged from .87 for the scale measuring perceived social support from colleagues to .90 for the scale assessing perceived social support from supervisors.

Trained work psychologists conducted an objective evaluation of the learning and health promoting potential of the volunteering tasks. They applied the computer-assisted analysis tool REBA (Richter, Hemmann & Pohlandt, 1999), which comprises 22 task characteristics, such as the requirements, forms and time-spans of cooperation, collective responsibilities, organisational the participation, information about organisation, organisational functions as well as information about work results and feedback. Based on the obtained results, the REBA instrument facilitates concrete recommendations to optimize the design of the evaluated tasks.

The subjective motivation potential of the tasks offered by TAURIS was determined through the implementation of the Job Diagnostic Survey by Hackman and Oldham (1975). Please note that this chapter focuses on the 7-point feedback subscale, whereby one half of it concerns feedback from supervisors or co-workers and the other half concerns feedback regarding the task. Schmidt, Kleinbeck, Ottmann and Seidel (1985) reported an internal consistency score of .68 for the feedback subscale.

Lastly, different scales by Trube (1995) were used to assess the motivation to take part in TAURIS, improvements and impairments of their personal situation since volunteering and wishes for change in regard to the conditions encountered in the TAURIS project. Motivation was measured using 10 dichotomous items, such as "The unemployment and inactivity got on my nerves", "I thought, without work one has always a bad reputation" or "I wanted to make new contacts to other people". One item is reserved for own additions. If someone confirms improvements or impairments since entry into the TAURIS project, the person is requested to choose for each with 8 concrete changes, whether they apply or not (e.g., "I have more contact with others", "I have difficulties with supervisors/ colleagues"). Finally, changes that TAURIS participants wish for their work were assessed using another 9 items on a 5-point scale.

The survey was conducted relatively similar across the two groups; both were surveyed twice at specific time intervals. Members of the control group were recruited through several job centres. They received 5 € each time they completed the questionnaire. In both groups, the first measurement was conducted in presence of the researchers, whereas the second measurement was mailed to participants. TAURIS participants were surveyed twice, once after they had started volunteering (around 4.5 months since the engagement commenced) and again 4 months later.

2 Results

No differences were found in terms of the perceived social support from partners for those who volunteered and those who did not. Members of both groups, when in a relationship, considered their partner to be the strongest source of support. Ranking the sources of support, perceived support from the partner is followed by support from relatives and lastly friends. Beyond the mere positions, that are congruent in both groups, the perceived support from relatives and friends differs significantly between volunteering and non-volunteering unemployed. As shown in Table 1, TAURIS participants in a relationship assigned significantly higher scores for relatives than corresponding members of the control group (t=3.61, p<.001). Furthermore, TAURIS participants with a partner reported higher perceived support from their friends compared to members of the control group, who are in a relationship as well (t=3.17, p<.01).

Table 1. Perceived social support for TAURIS vs. control group depending on partnership status

	TAURIS group (N=122)				**control group** (N=93)			
	with partner		without partner		with partner		without partner	
Sources of	(*N*=80)		(*N*=42)		(*N*=70)		(*N*=23)	
support (t1)	*M*	*SD*	*M*	*SD*	*M*	*SD*	*M*	*SD*
Partner	3.68	.57	-	-	3.64	.55	-	-
Relatives	3.41	.64	2.55	1.39	2.95	.89	2.89	1.31
Friends	2.85	.87	2.53	1.24	2.33	1.08	2.74	1.24
Co-workers	2.88	1.17	2.66	1.27	-	-	-	-
Supervisors	3.10	1.11	3.35	.77	-	-	-	-

The social support scores from friends within the control group are even higher for those without a partner than for those with a partner; however, no significant results were obtained. Finally, the TAURIS group reported higher perceived support from their friends compared to the control group (t=3.17, p<.01).

In an intra-group comparison of unemployed with and without partnerships, significant differences can only be determined within the TAURIS group. Volunteering long-term unemployed with a partner report significantly more social support from relatives than those without a partner (t=4.66, p<.001).

The support ascribed to friends by the total of the surveyed TAURIS participants has to be seen in perspective to the perceived social support from work-related sources (co-workers and supervisors). According to their reports, the TAURIS participants received significantly more support from the supervisors in the organisations they work for than from their friends (t=3.74; p<.001). The perceived support from colleagues corresponds to the perceived support from friends. In respect to work-related support, differences between TAURIS-participants with and without a partner remain below statistical significance. However, in a ranking of sources, for those volunteering without a partner, supervisors are by far the highest source of support relative to the largely identical but lower level of support received from co-workers, relatives and friends.

Table 2 demonstrates the development of perceived social support between the first measurement point (t1) and the follow-up four months later (t2). The perceived social support from relatives decreased significantly for the TAURIS-participants (t=2.17; p<.05), although the mean at t2 is still equal to that in the control group. All other evaluation of social support, including the perceived support from co-workers and supervisors, remained statistically stable. As for the control group, only the perceived support from the partner declined significantly over time (t=2.29; p<.05).

Table 2. Perceived social support for TAURIS vs. control group in a four-month follow-up

	TAURIS group (N=76)[a]				**control group** (N=64)[a]			
Sources of	**t1**		**t2**		**t1**		**t2**	
support	*M*	*SD*	*M*	*SD*	*M*	*SD*	*M*	*SD*
Partner[b]	3.68	.52	3.50	.80	3.62	.58	3.43	.80
Relatives	3.13	1.05	2.89	1.16	3.04	.89	2.96	1.04
Friends	2.84	1.06	2.73	1.00	2.42	1.17	2.21	1.34
Co-workers	2.74	1.36	2.75	1.36	-	-	-	-
Supervisors	3.22	1.03	3.04	1.19	-	-	-	-

[a] Here, only those are included, who responded at t1 and t2.
[b] Perceived social support from the partner is only reported for those long-term unemployed in a relationship (53 participants of the TAURIS group and 50 members of the control group).

Apart from the differences observed between the two groups, gender can be identified as a moderator of the reported support levels. Women overall perceived significantly higher support from relatives than men (t=4.29; p<.001). Amongst the TAURIS participants, this gender difference was also noted in terms of perceived support from colleagues: women perceived more support than men.

Shifting the focus from sources of support to the objective task conditions associated with the volunteer work at TAURIS, Table 3 lists selected variables and their positive and negative potential influence on organisational integration according to the REBA instrument.

The minimum standard for each variable was set by experts based on work tasks found primarily in the industrial setting. It is desirable to meet the minimum standard to prevent negative effects for the worker. The first three (requirements, forms and time-spans of cooperation) can be considered as potentials for volunteering tasks at TAURIS. The results suggested that the TAURIS group did not have to work isolated from others or merely shared workspace. Instead, they interacted with their co-workers to arrange organisational issues such as their work schedules. Further potential can be stated regarding the scores attained for 'collective responsibilities' which indicates transparency concerning the extent to which one's own performance contributes to the performance of the whole group.

For the fifth component, the scale 'organisational participation', the maximum score is attained if someone participates in processes of problem-defining and goal-setting whereas the lowest score is attained if someone is simply confronted with solutions elaborated by others. The obtained average score by TAURIS participants indicates that at work, they are not just presented with solutions but are also involved in the development and application of new solutions, which can be evaluated as a potential, too. The average participant score on the sixth scale 'information about the organisation' also slightly surpassed the minimum standard score. This shows that the received information reaches beyond the own workplace and concerns the work organisation of the whole department. TAURIS participants were also responsible for 'organisational functions' such as communicating with different departments and work groups.

Table 3. Objective task conditions of TAURIS work places

	Minimum standard	Means	SD	Potential
Requirements of cooperation	3	3.46	1.13	+
Forms of cooperation	4	4.41	2.29	+
Time-spans of cooperation	3	3.11	1.17	+
Collective responsibilities	2	2.36	.95	+
Organisational participation	3	3.55	1.73	+
Information about organisation	2	2.14	.82	+
Organisational functions	3	3.15	1.41	+
Information about work results	2	1.79	.74	-
Feedback	4	3.37	1.37	-

Deficits were found in terms of volunteers not receiving sufficient 'information about work results'. Furthermore, not obtaining sufficient feedback concerning one's own performance was identified by many participants. This implies delayed or sparse feedback which makes it not particularly effective.

Further data based on the third cohort underline the relevance of social aspects for the volunteering long-term unemployed. Asked about their level of motivation to take part, most TAURIS participants indicated that they sought new social contacts (73%). The second most important motive was their wish to contribute to something meaningful (71%). In retrospect, the majority of these participants (more than 90%) reported that their personal situation improved since they entered the TAURIS project, while the remaining 10% of the participants reported no change. Indeed, 70% of the third cohort (from 2002) stated that they had more social contacts since volunteering. Overall, the surveyed sample was satisfied with the social conditions they encountered in the non-profit sector.

Areas that need to be addressed in future research and initiatives of this kind were described by the participants as following: Volunteer work also has to offer a realistic and better occupational perspective (76%) and more opportunities for professional development (32%). In regard to social aspects, about one quarter would prefer more support, if personal problems occur.

3 Discussion

This chapter introduced the implications that volunteering can have for the long-term unemployed in terms of social support, demonstrated by the results of the TAURIS project from 2000 and 2002. Although volunteering does not solve the issue of unemployment, it can give individuals a sense of purpose and direction. This section discusses the main results and suggests some future research directions.

The main source of social support is often the private social network, starting with life partners. In the present sample, the majority of the volunteering and the non-volunteering long-term unemployed had such a source of support. Maintaining these despite the challenges and stress associated with unemployment is crucial. For those without a partner, new engagements might be essential for their social well-being, although new social contacts may be potentially difficult to acquire during unemployment.

Further important sources of support are relatives and friends. The results demonstrated that differences were found between the two investigated groups. Indeed, perceived social support from relatives and friends was higher for the volunteering group than indicated by the non-volunteering unemployed control group. Whether volunteering was a causal factor for these differences or whether this is due to a selection effect cannot be answered on the basis of the research design. However, retrospective data revealed a strong desire on part of the participants for new contacts before volunteering, which can be either interpreted as a deficit in the private network at that time or as an indication of higher social aspirations on behalf of the volunteering long-term unemployed.

It is unclear why social support from relatives and friends would increase with the uptake of the volunteering commitment. Retrospective data from the perspective of the volunteering unemployed do not support an observable enhancement of status since entry into the non-profit sector. An alternative explanation might be that this new engagement increased not only their level of activity, but that they became more self-confident, and used the extra money to spend time with others (e.g., they might spend the additional money to visit the cinema with friends or relatives, to join them for swimming or to make an appointment in a café). These findings support the idea that nobody is just a passive recipient of social support but that the unemployed also play an active role in mobilizing social support.

An additional dimension investigated was the social support relationship between the volunteers and their colleagues and supervisors. Results indicated that the partnership status of the volunteers played an important role here as well. For those volunteers in a partnership, the perceived private support from the partner and from relatives exceeded the level of perceived support by colleagues and supervisors. When comparing the overall support from colleagues and friends alike, the support levels seem comparable. It may help to consider working time as a form of shared time, because most TAURIS participants worked jointly with the remunerated staff at selected charitable and community institutions. This argument seems to mirror the suggestions outlined by

Bongartz and Gröhnke (1997) who described the participation in joint activities as an important basis for social support. The perceived support from the supervisor even surpassed these scores. To an even larger extent, this also applies for TAURIS participants without a partnership. Here, the perceived support from the supervisor exceeded support scores provided by all other sources. This underlines the fact that organisational leaders are important sources of support for the long-term unemployed, who perceive their supervisor as a reliable contact person if problems occur.

The time dimension also provides an interesting perspective to unemployment and social support work. In the follow-up, the TAURIS participants reported less social support from relatives than before. However, these support scores remained comparable to those reported for the control group. Considering that the perceived support from colleagues and supervisors remained stable across time, one can assume a potential compensation effect for TAURIS participants. In this light, it causes concern that social support from the partner declines over time in the control group, although no alternative sources of support are available apart from the personal network.

Another important factor to be considered in future initiatives as well as research is the observed gender difference. Its moderating effect mirrors results reported in previous studies. Reiterating earlier findings, it seems that female TAURIS participants also perceived more social support from co-workers than male respondents. This suggests that it is important to emphasize the significance of social support as a coping resource for the male long-term unemployed who may be particularly affected by the absence of social support. Social competence trainings could convey the necessary skills. At the same time, supervisors should be made aware of gender specific aspects of social support. This could be achieved by providing them with further information and potential competence training to recognize isolated individuals and those lacking social skills to effectively interact with others. Supervisors should also be aware that they are an important contact person and source of support even in time when the problems at hand are of a more personal nature.

An important consideration for similar projects should be strategy development to mobilize support from the personal network and to raise the unemployed individuals` awareness of social support available to them. It would also be desirable to publicize the positive contributions made by project members. Paid employment is not the only valuable employment. Therefore, efforts should be invested into improving the public's opinion and increasing their awareness about people's volunteering engagement outside the primary labour market. Especially since most volunteers receive little or no compensation for their efforts, society should at least acknowledge them for their work and commitment to charitable causes.

A special need for improvement in the voluntary setting concerns the lack of feedback as indicated by TAURIS participants. First, the tasks of the volunteering long-term unemployed should be better integrated and embedded into the work areas of their colleagues. This would allow for better organisational integration of all individuals at work, a prerequisite for successful work relationships and task completion. Second, supervisors should receive training on themes such as providing prompt, constructive feedback and conducting appraisals, which could be one component of quality management within the organisation.

There is clearly more room for improvement, also in some areas such as cooperation and organisational participation. For instance, one could think about the introduction of more group work, including collective self-organisation in regard to methods, functions and contents of work, potentially even performance assessments. Ideally, such an implementation process is prepared and accompanied by experts. Overall, the volunteer work of the participants seemed to fulfil the minimum standard for learning and health promotion in the workplace. The cooperation between the volunteering unemployed and their colleagues should be expanded beyond mere staff or organisational working arrangements. Volunteers should also be given the opportunity to contribute to more complex activities, such as problem definition, goal-setting and implementation of solutions. In addition, volunteers should be able to understand their role within the information and organisational framework. They should have insight into the functioning of the organisation and not be kept at the periphery as a result of their status as volunteers.

An interest in broadening one's social network is a powerful motivator for volunteers. Such engagement provides the long-term unemployed with a potentially structured environment, task and responsibility, as well as the chance to meet new persons and find new social support. Although this does not address the issue of employment as such, this engagement, experience of cohesion in groups, social support outside the home may be of important consequence for their social well-being. The majority of the TAURIS participants reported to be satisfied with the social support they perceive from their colleagues and supervisors. This trend and the findings of the TAURIS project warrant more longitudinal and cross-sectional studies involving alternative forms of work other than paid labour in relation to mental health, self-esteem, social support, and social skills maintenance. Any suggestions for interventions, such as social competence trainings for the volunteering unemployed, trainings in appraisal techniques for their supervisors or measures in task design, should be examined in a pre-post program-comparison group design in an effort to detect their effectiveness.

In summary, social support and social integration emerge as main quality characteristics of volunteering tasks for long-term unemployed. The TAURIS project demonstrated a new perspective which will hopefully be beneficial to other interested parties, particularly practitioners in work psychology. There is certainly a need for more research to consider roles and benefits of alternative engagements which carry both status and public respect in society – compared to the low social status often assigned to the unemployed. Optimizing the working conditions for volunteers and increasing their organisational participation may help to overcome the often described isolation of the unemployed from society. Getting the unemployed involved in a variety of community projects may represent an important step against further discrimination by society.

References

Bongartz, T. & Gröhnke, K. (1997). Soziale Isolation bei Langzeiterwerbslosen? Eine netzwerkanalytische Betrachtung. [Social isolation with long-term unemployed? A network-analytic examination]. In G. Klein und H. Strasser, *Schwer vermittelbar. Zur Theorie und Empirie der Langzeiterwerbslosigkeit* [Difficult to place. On theory and empirical research of long-term unemployment] (pp.197-219). Opladen: Westdeutscher Verlag.

Creed, P.A. & Moore, K. (2006). Social support, social undermining, and coping in underemployed and unemployed persons. *Journal of Applied Social Psychology, 36*(2), 321-339.

Denton, M. & Walters, V. (1999). Gender differences in structural and behavioral determinants of health: An analysis of the social production of health. *Social Science and Medicine, 48,* 1221-1235.

Frydenberg, E. & Lewis, E. (1991). Adolescent coping: The different ways in which boys and girls cope. *Journal of Adolescence, 14*(2), 119-133.

Hackman, J.R. & Oldham, G.R. (1975). Development of the Job Diagnostic Survey. *Journal of Applied Psychology, 60,* 159-170.

Mielck, A. (2005). *Soziale Ungleichheit und Gesundheit. Eine Einführung in die aktuelle Diskussion.* [Social inequality and health. An introduction into the current debate.] Bern: Huber.

Richter, P., Hemmann, E. & Pohlandt, A. (1999). Objective task analysis and the predicion of mental workload: Results of the application on an actionoriented software tool (REBA). In M. Wiethoff & F.R.H. Zijlstra (Eds.), *New psychological approaches for modern problems in Work Psychology* (pp. 67-76). Tilburg: University Press.

Rimann, M. & Udris, I. (1997). Subjektive Arbeitsanalyse: Der Fragebogen SALSA. [Subjective task analysis: The SALSA questionnaire.] In O. Strohm & E. Ulich, E. (Eds.), *Unternehmen arbeitspsychologisch bewerten. Ein Mehr-Ebenen-Ansatz unter besonderer Berücksichtigung von Mensch, Technik und Organisation* [Work-psychological assessments of organisations. A multilevel approach with special consideration of people, technology and organisation] (pp. 281-298). Zürich: vdf Hochschulverlag.

Schmidt, K.-H., Kleinbeck, U., Ottmann, W. & Seidel, B. (1985). Ein Verfahren zur Diagnose von Arbeitsinhalten: Der Job Diagnostic Survey (JDS). [A tool for the diagnosis of work contents: The Job Diagnostic Survey (JDS).] *Psychologie und Praxis. Zeitschrift für Arbeits- und Organisationspsychologie, 29,* 162-172.

Trube, A. (1995). Fiskalische und soziale Kosten-Nutzen-Analyse örtlicher Beschäftigungsförderung. Eine exemplarische Untersuchung. [Fiscal and social cost-benefit analysis of regional employment promotion]. *Beiträge zur Arbeitsmarkt- und Berufsforschung,* Nr. 189. Nuremberg: IAB.

Solution-Oriented Counselling of Unemployed under Perspectives of Hope

Sibylle Tobler

Introduction

"What can I hope for?" I was asked this question by a woman I was counselling in the context of my counselling practice as part of an unemployment project. She is from Bosnia, in her forties and recently divorced. If she doesn't find a job within three months, she will lose her residence permit. She is quite desperate.

Another situation: A young Tamil woman is looking for a job in accountancy. She has started a training course. She receives one rejection after another in her job search. She stays calm. She can even smile cheerfully. I ask her how she succeeds in remaining so positive. Her answer: "I just have hope. I know something will come up."

A third situation: A man. Swiss. 58 years old. For over twenty years he was committed to "his" enterprise. He was dismissed when production was curbed at work; now he is helpless. We begin to develop perspectives. It is hard for him, but he increasingly gets involved and engaged in the process.

Such situations motivated me to develop a counselling concept to support people trying to cope with unemployment. My goal was to integrate expertise regarding unemployment with a concept of counselling that stimulates counsellees to develop a solution-oriented attitude and to maintain a positive bias towards the future. This concept is based on ten years of counselling experience with unemployed people. To date, I have offered support in dealing with personal, social and professional challenges related to unemployment to around 400 people. In addition, the concept is based on theoretical research in this area[1].

In the following sections, I introduce three basic theses on which I built my theoretical research and my counselling concept. Secondly, I will outline the key aspects of this concept. Finally, I will summarize the results and conclusions.

---

1 I limit myself to the citation of only the most relevant references. Literature and detailed data about theoretical background and research results can be found in Tobler (2004).

1 Three basic theses

During my practical work as well as in my theoretical research, I developed three theses: 1. *The* unemployment does not exist. 2. Unemployed people depend on solutions. 3. Unemployed people also rely on motivation, courage and hope. What are the implications of these theses?

1.1 *The* unemployment does not exist

This thesis is not as obvious as it seems at first sight considering stereotypes about unemployment and unemployed people as well as efforts within research to find clear patterns of explanation.

Unemployment is influenced by various economic, structural, social and individual factors. It is a highly complex, dynamic phenomenon that cannot be reduced to simple explanations and generalizations[2]. The variability of individual situations, interpretations and manners of coping has finally been accepted as fact within unemployment research.

In the practice of counselling we are also confronted with a variety of situational and individual circumstances: Individual situations, attitudes, interpretations, manners of acting and coping responses differ across individuals. Simple concepts of counselling are therefore not adequate.

Nevertheless, counselling needs some orientation, a frame of reference. Otherwise, it might become meaningless or one might get lost in the variety of symptoms, interpretations, attitudes, and behavioural patterns. I developed a wider frame of reference which allows for this diversity, provides some orientation and facilitates focusing on certain concrete issues. This frame addresses four main challenges which unemployed people are confronted with.

The four main challenges for unemployed persons can be described as follows:

- Daily structure/attitude towards past and future: Unemployed people must find ways to create daily routines in the absence of occupational work. They have to come to terms with the past (that allows them to continue) without idealizing or regretting what was. And they must develop future perspectives in a situation where few or none of their preferred options seem to remain and the outcome is highly unpredictable.
- Personality development: When unemployed, individuals may be forced to question their attitudes, values, and patterns of behaviour in order to cope with their situation successfully. People may have to redefine themselves. They have to discover their resources and find ways to cope.
- Relations and social behaviour: Unemployment might lead to changes in relationships and roles. Conflicts might evolve. Unemployed people have to cope with various reactions and they might be confronted with negative social stereotypes about unemployment.

2 I have focused mainly on the more recent, differential subject- and coping-oriented research.

- Occupational development: Finally, unemployed people have to develop a realistic professional orientation and should be able to evaluate their experience, qualifications, resources, and chances. They have to develop and implement job search strategies.

1.2 Unemployed people depend on solutions

This notion is also not so obvious, considering the broad problem-oriented perspective in research as well as in working with unemployed people. To me, refocusing on solutions instead of problems constitutes the most central and significant part of my work.

For a long time, research focused on the distressing factors of unemployment. Results have shown that analyzing the stressful aspects of unemployment alone is not sufficient. While recognizing burdening factors may provide insight into people's experience of unemployment, it does not automatically provide keys to helping people cope with these factors. Bonss, Keupp and Koenen (1984) criticized the overemphasis on the "Belastungsdiskurs" (stress emphasis) already in the 1980s. Since then, research has shifted its emphasis more on what helps unemployed people and what keeps them healthy (instead of what burdens them and makes them ill). A similar change of perspective from a problem to a resource orientation can be observed within the field of psychotherapy and counselling.

Counselling experience furthered my own interest in the solution- and resource-oriented perspective: Unemployed people are often under pressure as they search for ways to cope with the numerous challenges posed by unemployment; they need to find solutions and a new job. It should further be considered that - as Grawe has also pointed out in his research on psychotherapy - the orientation toward resources, possibilities and solutions is more motivating for people to carry on in a challenging life situation rather than concentrating on what is difficult, does not work, goes wrong, etc. (Grawe, 1998).[3]

How can "solution orientation" be defined? I focus on concepts based on solution-oriented psychotherapy and counselling in the tradition of Steve de Shazer (de Shazer, 1999). They provide a profound basis for pragmatic, resource- and future-oriented counselling. Solutions consist of small (even the smallest) changes in perception and/or behaviour stimulated within counselling. A solution-oriented perspective must not be confused with perfectionism or naïve, simple optimism. It is a highly differentiated process that focuses on an individual's strengths instead of its weaknesses. With its focus on the search for solutions, this approach is resource and future oriented; it helps looking ahead and searching for what helps. The development of a "broad view" is encouraged; this gives access to alternate solutions and new options. The pragmatic development and implementation of solutions supports activity and strengthens self-confidence. With the counsellor's expressed esteem for the counsellee's success, positive expectation is stimulated.

3 The research of Grawe (1998) revealed that a resource orientation and active help to cope with problems belong to the most important elements of successful psychotherapy.

1.3 Unemployed people rely on motivation, courage and hope

It is evident from the unemployment research results as well as clinical research on coping that people in a challenging life-situation such as unemployment rely on motivation, courage and hope. To cope with such a life situation, a positive attitude of expectation is of central importance. Therefore, it is important to explore which elements avert despair and resignation amongst the unemployed, even in very severe situations. Nevertheless, relatively little research has been done in this area.

In my practical counselling work I have observed that counsellees spontaneously talked about hope. I am convinced that the topic of hope represents a stimulating frame of reference and a significant coping resource for the unemployed.

Hope is viewed differently across different disciplines, such as the practical counselling context, but also in philosophy, psychology and theology.[4] I prefer an open, dynamic, multidimensional definition. Hope includes temporal, individual, social and spiritual dimensions. Hope has to do with the attitude toward the future based on past life experience and this attitude is possibly challenged in the present. Hope is an attitude as well as a dimension of personality and a resource for coping. Hope has to do with emotions and cognitions, with the motivation to reach a goal as well as with trust, confidence, and faith. Hope is nurtured - or can be destroyed - in the context of relationships.

Hope, in its spiritual meaning, opens dimensions of existence that lead beyond intellect. Theologically, I define hope as a dynamic process that is oriented toward the promise of a fulfilled life and that leads back to active coping with present challenges without despair. Hope, in this sense, means to perceive future opportunities that might arise beyond what is currently possible and conceivable. Hope also means having the courage to go on, to become involved and not to give up or to become indifferent. This emphasizes the point that hope is quite different from escape and illusion; it is a resource that motivates people to cope faithfully, courageously and actively with life - just when it is most needed.

2 A concept of counselling

I have developed my counselling concept on the basis of these three theses. This means, this concept is built on the conviction that helpful counselling in the field of unemployment must be differentiated, solution-oriented and encouraging. What are the implications for counselling? Three principles can be described.

2.1 Begin with the individual situation

Considering the complexity, diversity and dynamics of individual situations faced by those in unemployment, counselling must provide a differentiated approach. At the same time, it is important to perceive individual situations in a greater context, to have a

---

4 In my dissertation you can find an overview of different concepts of hope in philosophy, psychology and theology, see Tobler (2004, pp. 175-195). I then focused on three theological concepts of hope: Moltmann ([13]1997); Lester (1995); Capps (1995).

wide and profound knowledge of unemployment and labour market conditions in order to avoid unrealistic, naïve or even damaging interventions.

Counselling must begin with an assessment of the individual situation in order to discover which possibilities are available to the counsellees. There has to be clarity about what unemployment means to this particular person. We have to explore how the counsellee perceives and interprets his/her situation, where he/she sees the biggest challenges, and how he/she handles the situation. The frame of reference described provides orientation during this process. Such an exploration finally enables us to focus on the concrete concern that the counsellee wants to work on.

2.2 Stimulate solution-oriented processes

It is important to explore what the counsellee considers to be the most significant concerns in his/her actual life-situation, but also to find out what concrete changes he/she wants for the future. Solutions must not be determined by the counsellor! Heterogeneous solutions might not only reduce motivation but could even evoke or intensify passivity and feelings of failure, helplessness, resignation and hopelessness. The question in which areas the counsellee would like to see changes supports self-responsibility and the experience of personal competence.

Having determined a focus to work on during counselling, exceptions, differences, resources, successes, and opportunities are explored: What helps? What "functions" "in spite of everything"? When does the problem feel a bit smaller? What are better moments? What is different then? What succeeds? What might be a motivating, realistic direction to follow? This exploration leads to the development, realization and evaluation of steps toward a solution. This process should not to be confounded by perfectionism; the goal is to stimulate a process. The recognition that even small changes and successes can make big difference is encouraging and energizing.

2.3 Stimulate processes of hope

Finally, counselling should stimulate processes of hope. This must always be adapted to the individual's situation. The main questions to focus on: What provides this person with motivation, courage and hope? What provides stability, support, and orientation? What gives them energy? What gives them the courage to go on "in spite of everything"? The counsellor supports the counsellee to recognize the motivating and hopeful elements in his/her life, encourages the development of perspectives that provide energy and helps the counsellee to develop strategies to give these elements more attention and room in everyday life.

3 Transfer into practical work

I would like to summarize the most relevant aspects regarding the context, structure and method of this counselling approach and point out the main objectives.

3.1 Context, structure and method of counselling

As solution-oriented counselling focuses on what is important to the counsellee, it can be adjusted to different groups of people and to different institutional contexts.

It is a characteristic of solution-oriented counselling that the number of sessions is limited (most common are about six to seven sessions). The goal is not to find an overall solution, but to stimulate a process and to provide the counsellee with experiences and skills that enable him/her to proceed by him-/herself. A session takes 45 to 60 minutes. The interval between two sessions ranges from a week to a month. It is common to have a follow-up evaluation about half a year after the regular sessions finish.

A session begins by organizing an agenda. This means, counsellees are asked to identify expectations, doubts, the purpose of the individual sessions and which themes they wish to focus on and should be worked on together. Referring to this focus, exceptions, differences, resources, previous strategies of coping, solutions, successes and possible steps toward a solution are considered next. The exploration of dimensions of hope takes place during the dialogue in a way that corresponds with the theme and the reaction of the counsellee. The focus may be on more concrete aspects of motivation or more on the extensive dimensions of hope as a spiritual resource of life that helps us go on in a challenging life situation. Finally, the main aspects of the session are summarized. The counsellor reinforces the process with realistic compliments regarding the counsellee's progress to date. The counsellee receives concrete tasks which will help him or her to deal with the issues at hand until the next session.

One important methodical instrument is solution-oriented questioning. There are questions that stimulate self-responsibility such as "What do you want to work on during this hour?" There are questions about a future without the problem such as "What will your life be like when you have reached your goal?"; they support motivation and let resources come into the picture. There are questions about exceptions and differences such as "Are there moments during which you feel better? What is different in these moments?" Questions about dimensions of hope are, for example, "What motivates you to go on in spite of everything?" or "What gives you hope that there will be a way out of this situation?" Another characteristic of solution-oriented counselling is the work with concrete tasks: Until the following session, the counsellee should think about or observe a specific issue concerning the problem, take notes or try to take a concrete step.

Nevertheless, it is important to remember that solution-oriented counselling must not be reduced to a way of interviewing, to a method. Such counselling will be superficial, or worse, enforces the counsellee to adopt an unrealistic positive view. This increases the likelihood that counsellees develop feelings of isolation and failure. Counselling with a solution orientation is more than a method; it is based on a positive and resource-oriented perception of the human being.

3.2 Objectives

The main objectives of solution-oriented counselling under perspectives of hope are the following:

- To stabilize, improve quality of life and well-being within unemployment;

- To develop positive, motivating perspectives and to maintain hope (regardless of current employment status);
- To improve the ability to cope with specific challenges posed by the individual circumstances while unemployed;
- To increase labour market opportunities.

4 Results and conclusions

Results from the exploratory transfer of the counselling concept to my professional practice (I tested the concept with seven people in the context of my dissertation) and further experience in counselling have led me to the following observations:

- Counsellees are able to engage in a solution-oriented process. Even if the actual situation might be very distressing and counsellees might initially perceive only few limited options for themselves, I have seen that they are able to investigate their situation, to experiment with different points of view, to define a focus and to train coping. I assume that a counsellor's resource- and solution-oriented attitude supports the counsellee's feeling of worthiness and stimulates the development of self-responsibility, activity and self-confidence. In my experience, counsellees are thereafter able to engage in a process during which they learn to cope better with their situation.
- Counsellees are interested in dimensions of motivation, courage and hope. They start to talk about spiritual attitudes without being explicitly asked. When they are asked what gives them the energy to cope with unemployment, I am surprised at how openly they start to talk about their hope, their faith, and their spiritual resources. Processes are different; some counsellees talk more about hope in terms of everyday life issues, such as finding a new job. Others start to talk about their values, their spirituality, or their religious faith. None of my counsellees have ever refused to talk about dimensions of hope.
- Counsellees improve their ability to define concrete steps for themselves and to transfer these into everyday life. It is my experience that the development of small steps leads to greater steps, to change and improvement, also in areas of the counsellees' life which are not directly related to the focus worked on in the counselling sessions. Even if counsellees initially feel that the only solution is finding a new job, they learn that small steps, even in an apparently unrelated area, lead to an improvement in the quality of life and therefore facilitate a more relaxed job search approach.
- Counsellees report improved self-confidence. To focus on what works instead of on what fails as well as the experience of developing and applying concrete steps and the subsequently observed successes are important factors in supporting self-confidence.
- Counsellees improve their chances in the labour market even if they have difficult starting positions. Five of the seven test persons in my dissertation found new jobs

during or shortly after the counselling process. In my professional practice, we reach an average placement rate of 70%.

I would like to summarize the most important conclusions:

- Results of unemployment research should be further integrated into multidisciplinary, multidimensional frames of reference. Bringing results of different research perspectives together will allow for a more differentiated view capable of depicting the complexity of unemployment effects. It will enable the development of multidimensional frames of reference, which facilitate orientation.
- Concepts of solution-oriented counselling should be further developed in the context of professional transition processes. My research and my practical experience have shown that solution-oriented counselling is a very helpful approach, especially in a problem-oriented context such as unemployment. It is worth exploring and developing this concept further.
- Concepts of hope should be further investigated as to their impact on coping with unemployment/professional transition processes. Spirituality as a resource for coping with unemployment should also be considered. From the results of my empirical and practical work, it is evident that the investigation of motivation and hope with regard to coping with unemployment should be advanced.
- Theoretical research and practical counselling should complement each other to support coping with unemployment. One of the most productive aspects of my work so far has been the combination of practical and theoretical work. I am convinced that an ongoing feedback cycle between the practice of counselling and research will lead to further indications of how we can best help people cope with unemployment.

References

Bonss, W., Keupp, h. & Koenen, E. (1984). Das Ende des Belastungsdiskurses? Zur subjektiven und gesellschaftlichen Bedeutung von Arbeit [The end of the "Belastungsdiskurs"? The subjective and social meaning of work]. In W. Bonss & R. Heinze (Eds.), *Arbeitslosigkeit in der Arbeitsgesellschaft* (pp. 143-188). Frankfurt a.M.: Suhrkamp.

Capps, D. (1995). *Agents of hope. A pastoral psychology.* Minneapolis, MN: Augsburg Fortress Press.

De Shazer, S. (1999). *Wege der erfolgreichen Kurztherapie* [Keys to solution in brief therapy] (7th ed.). Stuttgart: Klett-Cotta.

Grawe, K. (1998). *Psychologische Therapie* [Psychological therapy]. Göttingen: Hogrefe.

Lester, A.D. (1995). *Hope in pastoral care and counseling.* Louisville, KY: Westminster John Knox Press.

Moltmann, J. (1997). *Theologie der Hoffnung* [Theology of hope] (13th ed.). München: Kaiser.

Tobler, S. (2004). *Arbeitslose beraten unter Perspektiven der Hoffnung. Lösungsorientierte Kurzberatung in beruflichen Übergangsprozessen* [Counselling unemployed people with perspectives of hope. Solution-oriented short-term counselling in professional transition processes]. Stuttgart: Kohlhammer.

Solidarity Economy as a Solution for the Unemployment and Precariousness of Employment in Brazil: A Matter of Health Prevention

Maria Alice de Almeida & Ângela Patrícia Deiró Damasceno[1]

Introduction

The current crisis of the waged work force utterly exposes capitalism's promise of transforming everything and everyone into goods, ready to be offered and consumed in a market equalized by competitiveness. Millions of workers have been expelled from their jobs, therefore further increasing the precariousness of their working conditions and preventing them from exercising their labour rights. In this way, forms of employment once considered to be outdated and which should have been greatly diminished or even superseded, have spread and increased, thereby absorbing a great number of unemployed people, as observed in the constant increase of unemployment and jobs in the informal market, as presented by official data.

In Bahia, one of Brazil's oldest states, the precariousness and informality of the job market has reached astonishing numbers and there have even been news about unemployment in the casual employment (self-employment) market as well. The decrease in consumption caused by the large number of people without access to goods, as much in Bahia as well as the other states, involves two completely different socio-economic concepts, seen either as an "escape or refuge" from the capitalist market or as a survival strategy that rejects the centralisation of capitalism.

Some information, obtained by the Research of Employment and Unemployment in the Salvador Metropolitan Region (SMR), shows that in 2003 the total unemployment rate decreased in the capital of the state, for six successive months, falling from 26.8% of the Economically Active Population *(População Economicamente Ativa)* (EAP) in November to the current 26.0%. The contingent of unemployment was calculated in 440,000 people, while an estimated 1,694,000 individuals were present in the EAP that month (SEI, 2003).

---

1 Both authors work for the Mediterranean Institute of Environment, Health and Education (in Salvador de Bahia, Brasil), which uses agriculture, among other activities, as an axis of valuation of the human being.

The decrease in the unemployment rate continues to be explained by the reduction of the participation rate, which indicates the pressure of employed and unemployed persons on the job market, as well as by the increase in the occupational level. The participation rate was estimated at 62.9%. This rate decreased by 0.9%, coinciding with an exit of 12 thousand people from the job market (SEI, 2003).

As for occupation, there has been an increase of global level (0.4%) in SMR verified for six successive months. This occupational level increase reflects, however, different intersectional situations. During the month of analysis, positive variations in the aggregate "other sectors" (*Outros Setores*) had been registered that include civil architecture, domestic services, in addition to other activities (4.9%), in industry (2.8%) and commerce (1.0%), while the services sector presented a negative variation of 1.3% (SEI 2003).

Table 1. Participation and unemployment rates, December of 2003

Indicators	MRS	Salvador	Other cities in Bahia
Economically active population (in 1.000 people)	1,694	-	-
Rate of total unemployment (in %)	26.0	25.5	28.4
Unemployed and looking for a job	15.1	14.6	17.7
Unemployed from 1 to 5 years without expectation (due to lack of qualification)	13.9	11,0	10.8
Precarious job	7.9	-	-
Participation rate (EAP/PIA) (in %)	62.9	63.9	59.0

Source: PED RMS – SEI/SETRAS/UFBA/DIEESE/SEADE.

This dual concept has stimulated the emergence of successful experiences that have blossomed with promising hopes, based on simple solutions, regarding the issue of unemployment, precisely in the area that has been most affected in the last few decades: the rural zone.

In a reversal, tired of looking for better living conditions in urban centres characterized by slums, "palafitas[2]" and other degrading forms of housing, as well as a lack of medical and socio-economic assistance, and against the rejection of and disrespect towards rural cultural roots, peasants began to return to their native soil in search of the potentiality of the land. It is because of this quest that the joint action of others, also suffering, gave rise to economic support based on partnership principles of ancient primitive communal regimes.

Solidarity Economy, however, with data still not measured by official sources of the Union, has emerged today as a way to recover the workers' historical struggle in defence of their rights, as a strong effort against such issues as the exploitation of human labour and as an alternative to the way capitalism organises social relationships among human and between human beings and the environment.

In this scenario, under different titles, the Solidarity Economy or *Unifying Economy* has given rise to practices of social and economic relations, which, in a short period of time, have provided survival alternatives and a better quality of life to the people of

---

2 Shanty houses built in the sea using precarious material

the aforementioned community. And its horizon stretches as its solidarity practices are based on co-operative relations, inspired by cultural values that place the human being as the subject and purpose of the economic activity, rather than the private accumulation of general wealth and, in particular, capital.

Solidarity Economy makes up the foundation of a humanized globalization, of sustainable development, socially just and directed towards the rational satisfaction of the needs of the individual and of all citizens on Earth, following an inter-manageable path of sustainable growth in the quality of their lives and in environmental preservation.

As discussed in the "Fórum Nacional de Economia Solidária" (Solidarity Economy National Forum) held in Brasilia in June 2003, the main focus of the Solidarity Economy is work, knowledge and human creativity, not property or money. They represent an eagerness for integrated development that targets sustainability and economic, social, cultural and environmental justice through models of social management.

Thus the Solidarity Economy has become a new economic sector of society, distinct of the capitalist and monopolistic economy, strengthening the democratic state with the emergence of an autonomous social act, capable of creating new rules of rights and of regulating society on its own behalf, for it is based on the culture of co-operation, solidarity and sharing. Therefore, it rejects competitive, exploitative and lucrative practices. This way it responds to unemployment, giving people their dignity that was robbed by the exploiting capitalist system.

In this article readers will be able to learn about a very successful experience in Bahia, a north-eastern state of Brazil, where in this rural area the co-operatives and associations have brought self-esteem, quality of life, income and dignity to the workers.

Such experience could be adapted to other agricultural areas in developing and developed countries, culminating in the Solidarity Economy Model claimed by the Mediterranean Institute of Environment, Health and Education in Bahia as suitable for bringing a new global scenario to the region.

It is appropriate to stress that unemployment not only affects the physical and economic aspects of the person, but also his/her entire psychological and familiar structure. The reintegration of him/her into the job market promotes his/her value in society and with that comes a strengthening of social links, essential for the social welfare state.

1 Theoretical and practical model

The model in development by the Mediterranean Institute involved rural community awareness regarding the local potentialities developed by organised groups with the aim of implementing them at little cost and likely profitability. This is the first step, as such communities are unaware of their own productive capacity as well as being unaware of the organisations that can constitute social relationship models, capable of linking them all together, strengthening the group and its self-esteem.

After legal issues, the most important way to maintain production and group unity is the continuous learning of techniques and everything related to education and knowledge in general. That is the main aim of Mediterranean Institute, which only understands motivation allied with knowledge and the revelation of the "new".

Since 1993, professionals at Mediterranean Institute have been guiding and assisting the rural workers of a coastal community that used to live either on crops of coconut and roots, such as cassava and sweet potatoes or on very small incomes generated by the exchange of goods in local commerce.

However, one could easily see that this structure was not strong enough to survive capitalist demands, creating the necessity to identify alternative forms of production. With the technical team expanded and the participation in a number of popular and academic events discussing Solidarity Economy, the embryo of a productive model of popular economy began to take shape.

The first meetings showed that people from the rural area of Bahia work individually and do not know any sort of group action. They help one another during moments of emergency and disease, but not in daily production, which is strictly for the individual or family.

A change in their way of thinking was strongly needed and many meetings took place in the attempt to map out local potentialities, search for possible leaders, as well as familiarizing the group with the management of resources and the negotiable possibilities for a common economic position, which psychologically meant the fall of the barriers built in Brazil during the colonial period.

This possibility of thinking about oneself and about the "whole" reverts back to the movement of a solidarity character, close to the primitive model used by the Jesuits in the Brazilian Missions and the "Cone Sul" (Southern Cone) of Latin America. At that time they tried to add to the local culture a catholic model, civilized and class-based, where the aristocratic European church was the leader and its followers were the uncivilized "amerîndios" (American Indians), enslaved to be acculturated according to the European model. This movement was banned after the Pope's mediation.

Agriculture in the state of Bahia is very rich, but it is not scientific or technological and there is really no sense of leadership. The professionals of the Mediterranean Institute have adapted the common primitive system of the Jesuit Missions of Brazil into a co-operative partnership model.

Such an awakening occurred after the locals were given responsibility in economic and administrative co-participation for the upcoming activities. Of course there were some locals who gave up, frequently due to the difficulty of adjusting themselves to solidarity activities. However those who remained showed that they had learned a greater sense of unity and all public relations and representative activities were done on behalf of the group, based on collective deliberation, decided in meetings using a language that was understood by everyone and guaranteeing equal rights of speaking and voting. Through this "mixed democratic model", which gives priority to common-sense, research, content, and ethical and moral values rather than base decisions on simple voting, the professionals of Mediterranean Institute have attained great collective achievements.

Spreading the awareness of class and group work is probably one of the greatest challenges in the process of citizenship awareness, for the people of Brazil are shaped into a selfish and individualist American model, reinforced by a feeling of distrust, especially in the rural sector, due to past dictatorship and peasant struggles, still poorly researched nowadays.

Hopefully in the near future people will realize that the present Solidarity Economy, which will have reached and engaged most of the world's population, will be seen to be no more and no less than the new economic order of the Third Millennium.

2 Final considerations

The co-operative experience in "Vale do Açu da Capivara", in the district of Arembepe, Bahia, Brazil, demonstrates the viability of this economic model as an alternative in reducing the precariousness of employment in areas which generate social inequalities, mainly in the countryside. The first visits to the site revealed all sorts of needs, primarily regarding the legalization of the locals' land. Açu da Capivara was formed by uncultivated lands, where squatters had been living in social isolation for over forty years. None of them knew what to do, though they were all supposed to be granted an "udal", which is guaranteed by Brazilian law to all those cultivating or living on the same land for more than ten years.

With the help of professionals of the Mediterranean Institute of Environment, Health and Education, an NPO that supports new co-operatives, more than twenty squatters have organised and launched COPAÇU - Cooperativa de Agronegócios Fazenda Açu da Capivara (Açu da Capivara Farm Agro-business Co-operative). Today, starting with that first step, they have already managed to apply for a collective deed to their lands, succeeded in a small-shared business, as well as forecasted a profitable and productive future. Although this is only its first year of working co-operatively, the movement continues and the self-esteem among members of the group has improved considerably.

References

Cáritas Brasileira.(1999). *Solidariedade caminho para a paz.* [Solidarity way of peace] Brasília: Cáritas Brasileira.

Ferreira, H. & Buarque, A. de (1999). *Novo Dicionário da Língua Portuguesa.* Rio de Janeiro: Nova Fronteira.

Marx, K. (1867). *El Capital: Crítica de la Economia Política.* [The capital. Critique of political economy] México: Siglo XXI Editores.

Seoane, J. & e Taddei, E. (Eds.). (2001). *Resistências mundiais: de Seattle a Porto Alegre.* [Global resistency: From Seattle to Porto Alegre] Petrópolis: Vozes.

Critical Differences: The Development of a Community Critical Psychological Perspective on the Psychological Costs of Unemployment [1]

David Fryer

1 In two minds about psychology?

This year, 2009, it is twenty five years since I became involved, as an academic, with the field of unemployment and mental health. 1984 I saw the publication of my first journal article about unemployment and mental health (Fryer and Payne, 1984). Since then I have published scores of such papers, edited a book (Fryer and Ullah, 1987) and special issues of two different journals, had the privilege of travelling widely to give numerous conference keynote addresses or work with overseas research groups, played a role in the International Commission for Occupational Health, been elected a Fellow of the British Psychological Society and played the lead role in drafting, on behalf of the British Psychological Society, a report for a Select Committee of the British House of Commons, all the above in relation to unemployment and mental health. Despite, or perhaps because of, this long immersion in the field I have recently come to feel less and less satisfied with my understanding of the field. The more I have studied and thought, the less I have come to feel I understand or 'know' and the more I have come to feel there is 'something wrong' with the unemployment research literature. I have come to feel less and less comfortable with my own contributions, such as they have been, to the way the field has been constructed. What I had intended as progressive contributions, now seem increasingly problematic ideologically.

It is harder to say when I became involved, as an academic, with the field of community psychology. My first published community psychological paper to have little or nothing to do with unemployment appeared only fifteen years ago (Fryer, 1994). However, my publications referring to 'community' in the title first came out long before that (Darwin, Fitter, Fryer & Smith, 1987; Fryer, 1987), though with reference to unemployment. However, I now publish more papers and give more conference papers about community critical psychology than I do about unemployment and mental health,

1 An earlier version of this paper was originally presented at the *International Conference on Insiders and Outsiders*, May 14-16 2003 Graz, Austria

have served as Editor of the Journal of Community and Applied Social Psychology, am a Fellow of the Society for Community Research and Action (Division 27 of the American Psychological Society / Community Psychology) and am President Elect of the European Community Psychology Association from November 2009 for two years.

In public I kept these two spheres of activity pretty well separate for a long time. This was made easier to do (harder to do otherwise?) by the relatively little overlap between the people active in each sphere or between the different pre-occupations of members of each sphere. Increasingly, however, there has been 'leakage' between the two compartments. For example, I intended one of the Journal Special Issues I edited, primarily, to stimulate community critical psychologists to think and act differently in relation to unemployment and health issues. Conversely, I wrote a recent paper (Fryer and Fagan, 2003), published in a Journal Special Issue edited by two leaders in the unemployment and health field and whose special issue contributors were mostly not thinking or working in a community critical psychological way. I did this, primarily, to stimulate unemployment and mental health researchers to think and act more critically in relation to community critical psychology. In general, however, I have tended to keep my community critical psychology and my work in the unemployment field separate.

In this chapter, I try to bring together my community critical psychological thinking with my unemployment and health thinking and point out some of the contradictions. This is a personal perspective on the research literature but I hope it will prompt productive and progressive reflection by others.

After a brief summary of what I take community critical psychology to be, I continue with what takes the form of an autobiographical narrative (but is a text for which Foucauldian discourse analysis would appropriate) before engaging critically with a reading of the unemployment and health research literature.

2 Community critical psychology

Community critical psychology problematises exploitative, oppressive, unjust and pathogenic societal arrangements but also problematises reactionary psychologies and other social sciences which construct, maintain or collude with such societal arrangements.

It is an approach to understanding, reducing and preventing negative consequences for health and ill-being of the way society is organised (and the reactionary ways in which psycho-social sciences function) by working collaboratively at a variety of systemic levels with persons subject to those consequences, seeking to increase the power they have to control key factors affecting their lives and to facilitate their competence to bring about health-promoting social change through cooperative dismantling or circumvention of barriers, prevention of socially toxic situations and construction of salutogenic societal arrangements.

As a community critical psychologist, I assume that persons are characterised by subjective agency in the sense that they characteristically subjectively experience themselves as making real, if constrained, choices, bearing responsibilities, making sense of what is going on, formulating and carrying out plans in line with past memories, future expectations, imaginative hopes and transcendent values. I also assume that subjectivity is a defining feature of what it is to be a person, that this applies to all persons involved in research, action and praxis and certainly applies as much to researchers, social scientists, professionals and activists as it does to the people with whom they work.

However, crucial though the subjectivity of persons is I also assume that subjectivity not only impacts on and is impacted upon by phenomena which embrace, but go far beyond, the immediate environment to include material, environmental, family, peer-group, organisational, neighbourhood, policy, cultural, societal, multinational and discursive / ideological factors but is actually constituted by and constitutive of them. As a community critical psychologist, I believe that subjectivity is a means through which governmentality, in the Foucauldian sense, is achieved and that subjectivity is constituted out of dominant and counter discourses and dominant and subjugated knowledges. Agency, in this frame of reference, is a means through which people govern themselves in the interests of the status quo, a means through which dominant discourses are reproduced and dominant knowledges 'truthed', but also, potentially at least, a means through which dominant discourses can be problematised, counter discourses deployed, dominant knowledges problematised and subjugated knowledges surfaced and deployed.

It is important to realise that the point being made here is not the enlightened-positivist one that research on the person-in-community-context allows more ecologically valid generalisations than research on the person-in-laboratory-context but the radical point that persons are socially constituted and their very subjective experience is in part constituted by social discourses, some of which are constructed by, or with the collusion of, psychologists and other social scientists. Increasingly we understand our social worlds and our place in them in part through the dominant discourses and knowledges invented by social scientists.

I regard 'power' as central to understanding and intervention because I have observed that people with emotional and psychological problems are disproportionately likely to be located in disempowering contexts with little or no devolution to them of control over key factors affecting their lives, that inadequate power to collectively and individually determine events over time negatively affects mental health and that, over time, adequate power to do so affects individual and collective well-being positively. Increases in control over factors affecting their lives by some persons or groups, for example groups of unemployed benefit claimants, usually mean reduction in control for other persons or groups, for example state officials. Community critical psychology interventions therefore often involve attempts at redistribution of control, which of course would result in empowerment for some interest groups but de-powerment for others. Predictably those with most power do not like redistribution in the balance of power so usually they resist it. Moreover, I believe that "rather than seeking to engage with power as such we should be engaging with the way societal hierarchies are set up and maintained through wealth, class, labour market position, ethnic dominance (majority/minority status), gender etc., and the way societal structures impact on people both objectively and through their subjective understanding of them" (Fryer, 2008:242)

I am sceptical of the competence of professional so-called 'experts', and especially of 'psy-experts' and research experts, mindful that expertise is often wielded as power and that many disadvantaged community members only come across such experts as bearers of bad news and/or oppressive power: when their work is intensified in their jobs by time and motion experts, when they hear that they have not only lost their job but that they are also 'losing their mind' etc. On the other hand, I believe that the effectiveness of such experts in dealing with the problems brought to them by community members is often very limited and often problematic. On the other hand, as a community critical psychologist, I appreciate the competence and expertise possessed and exercised by 'ordinary' community members in relation to the issues of concern to them. They often have insightful understanding of the key phenomena and clear views about what is needed but inadequate resources to implement their solutions.

My ultimate aim is not to research, understand and document distress (including that associated with or caused by unemployment) but to make a difference through prevention or intervention.

3 Making sense of unemployment: A personal journey [2]

My childhood, adolescence and young adulthood were saturated by the protestant work ethic and by my emerging resistance to it. I below share some of what I have long understood of where I am positioned, and position myself, in my own and others' accounts in relation to employment and unemployment. It is a version that would be shared by some and contested by others, as are all subjectively constituted readings of 'reality'. Whatever its relationship with others' accounts of events, my understanding of it has shaped my work in this field, and continues to do so.

My maternal grandmother, Ruth, went 'into service' as a little more than a child, working as a live-in servant in one of the large houses which marked the division between rich and poor in the coal mining area of northern England where she lived. She was 'courted' by my grandfather, Joe, and in due course he made her a proposal of marriage but she refused to marry him until he changed his occupation from bare-knuckle fairground boxer, which she considered too risky an occupation for a husband and father, to a 'safer' and more respectable job. Joe became a coal miner, married Ruth and they settled down and brought up four children in a rented terraced house with two rooms upstairs and two downstairs in a mining village whose rhythms were regulated by 'work' and 'chapel'.

My grandparents slept in one bedroom and my mother, her sister and two brothers slept in the other. The toilet was outside at the end of a 'yard' and there was no bathroom. There were at that time no 'pit' baths so the whole family bathed in a zinc tub placed in front of the kitchen fire and filled up with buckets of hot water. The family was 'god fearing', hard working and saw the armed forces and education – for boys at least - as the ladders with the aid of which they could climb upwards out of working class drudgery to life with the more comfortable middle classes, whose conservative politics they adopted, perhaps as a badge of their aspirations. One of my maternal uncles went to University and became a teacher in the Royal Navy, the other became a self-employed shopkeeper. My maternal aunt became a nurse then worked her way up to being a senior health service manager. My own mother, who was 'good with figures', trained in book keeping courtesy of the local co-operative store and had a succession of jobs from store clerk to greyhound race-track clerk to box office manageress in a thea-

2 This is in the form of a biographical essay but as is noted above better seen as a text for Foucauldian discourse analysis to trace the presence of dominant and counter discourses, dominant and subjugated knowledges.

tre, whose users she regarded derisively as having 'more money than sense'. Of the five grandchildren of which I am one, all three males went to University, did PhDs and became academics. One of the females became a teacher and the other a librarian.

On the other side of my family, my paternal grandfather, Albert, became a live-in cabinet-maker apprentice at a 'Stately Home', married Florence, who had grown up as the daughter of a shop keeper, went off to serve in the 1st World War, survived to come home but died shortly after from septicaemia caused by war wounds. My paternal grandmother, a young mother of three when widowed, tried to make a living by running a tea-shop and brought up my own father and his two sisters as best she could. Eventually she met 'Will', a manual worker in a factory making wire; they married and had another daughter together.

My own father resented his stepfather and I grew up in a family climate that encouraged me to believe Will was lazy, parasitic and feckless. I grew up believing that Will had been 'unemployed for 50 years' having been made redundant in early middle age (he lived to the age of 96). I found out as an adult that this was actually an exaggeration but he had effectively been detached from employment as long as anyone could remember and certainly as long as necessary to be able to tar him with the stigma of being 'work-shy'. There were countless stories about Will, which illustrated his allegedly roguish qualities. For example, that he was once nearly thrown out of village fancy dress garden party, unrecognised by the local policeman, when he turned up dressed as a tramp having completed his preparation by an extended visit to the local pub followed by rolling under hedges and in cow pats. Will, a colourful character of whom, as a child I was full of awe and admiration in equal measure, served for me as an early icon of unemployed resistance to the employment ethic.

My own father started his employed life as a commercial artist designing wallpaper patterns but at the outbreak of the 1939-1945 war relocated to the city Birmingham to serve in the police force and met my mother in a hospital during the heavy bombing of that industrial city in the early part of the 1939-1945 war whilst she was working there as an emergency-trained nurse. He joined the Royal Navy during the later stages of the 1939-45 war but survived and ended up, after emergency post war training, working as a primary school teacher in Birmingham, England. In my family school teachers were regarded as having an easy number, short working days and long holidays. However, my father started work early (leaving the house before I was up), worked late getting home hours after me, falling asleep in his chair most evenings and he took other jobs during the school holidays[3].

My mother had a series of part time or full time jobs during my childhood and my parents appeared to work and save hard for the 'luxuries' - upon which they insisted - a daily cooked breakfast and an annual family holiday by the seaside. We moved house, as soon as my parents could afford it, from a council owned flat to their 'own' i.e. mortgaged, semi-detached home just as I was starting at secondary school.

I had failed my selective ('grammar') school entrance examination but after pressure on the Head teacher from my parents, and in particular from my mother, had been allowed to follow my older brother and take up a place at the grammar school. My brother was repeatedly held up to me at school and at home as a model pupil with an intelligence and industriousness I could never even approach. Whilst my brother was

3 Actually this aspect of his employment diligence proved later to be illusory.

positioned as the serious, studious and bookish son who spent hours and hours every night demonstrating his commitment to the work ethic on his homework, I was positioned – and positioned myself – as the laid back, 'practical' son who rushed through his homework – if at all – in half an hour. Whilst my brother was talked about as a natural candidate for Oxford or Cambridge University (though like me he actually went to London University), I was encouraged to join the armed forces. Actually, my schooling was financially subsidised by the Royal Navy with a commitment made that I would take up a position for training at Dartmouth Royal Naval College after leaving school. My parents were so proud of my passing the selection tests that the newspaper clipping recording the event were kept and revered like a holy relic. When as a 'revolting' teenager in the 1960s I grew my hair and rebelled at the prospect of joining the armed forces, my parents not only had to repay the Royal Navy money granted to keep me at school but were bitterly disappointed by me, hardly assuaged by my decision to go to University to study 'airy fairy' subjects (psychology and philosophy) incomprehensible to them.

Nevertheless, with the transition from being 'in service' as a maid or carpenter or coal miner to University graduates, my family had successfully made the transition, so all important to them, from working to middle class in the time it took for grandchildren and grandparents to become alienated from each other.

As a teenager I lived with my brother, father and mother in Birmingham in the then industrial heart of the West Midlands, close to the factory complex known locally as "the Austin", the volume car works at Longbridge. The plant was represented in local papers and national media as riven by industrial conflict, the message repeatedly drummed home that the British worker was bone idle, greedy, always looking for something for nothing, ready to go on strike at the drop at a hat, manipulated by trades unionists who were simultaneously represented as cunning and stupid, anarchist and communist and this was mirrored in my parents' positioning of our next door neighbour, an assembly line worker in the car factory. The message was loud and clear – British industrial workers as Longbridge deserved what was coming to them: unemployment, indeed their unemployment was the consequence of their militancy, stupidity, gullibility, laziness, anti-social attitudes and weak psychological grip on themselves. Yet with my own eyes I saw, morning after morning, men trooping compliantly and resignedly past my house on the way to the Austin factory in their brillcreemed hair and long grey gabardine coats with their canvas lunch bags over their shoulders, looking for all the world in my eyes as if they were trooping off for a day in purgatory and coming back an eternity later looking exhausted with their grey faces and stooped bodies.

I went to London University and read psychology with philosophy. I did well with psychology and finding philosophy difficult but fascinating stayed at London University to do a Master's degree in Philosophy. After what would now be called a gap year teaching English abroad in Athens, where I lived and taught during the fall of the fascist military junta and the restoration of democracy with the installation of the Karamanlis government, I decided to read for a PhD and went to Edinburgh to work on a language focused thesis. There was a ferment of interdisciplinary activity at Edinburgh in those days with linguists, logicians, psychologists, philosophers and computer scientists meeting in an innovative grouping of 'Epistemics' and others developing work in the sociology of knowledge. I got interested in these developments and I became fascinated by the historical and philosophical roots of the claim that intelligence could be understood by simulating it on a computer. The outcome was a thesis that focused on 18th century European mechanical automaton builders as an early artificial intelligentsia, which drew on and tried to contribute to the philosophy and history of psychology, sociology of knowledge and 'epistemic' sciences. Though the thesis led to the award of a Ph.D. it also left me pretty well unemployable in the conventional academic disciplines of psychology, philosophy and sociology with their rigid demarcation lines.

Lacking income and employment I signed on as unemployed, attended a job centre and when my partner obtained a (teaching) job in a rural area I relocated. 'Signing on' was a nightmare. When I attended the job centre I saw myself (and others in the queue) reflected in the eyes of the staff as a self-deluding, feckless, lazy, layabout looking for handouts at others' expense whilst maintaining unachievable aspirations to ensure remaining unemployed. I remember the derisive incredulity when asked what job I was looking for when I mentioned 'lecturer' or 'psychologist'. Because I lived in a rural area I was usually allowed to claim benefit by getting a neighbour to countersign a fortnightly form saying I had done no paid work in the reference period. This was a regular humiliation especially as the neighbours, farmers, seemed to work every daylight hour and kept suggesting where they had heard there was work going, as if my unemployment was just because I had not looked hard or well enough to find it. I applied for scores of jobs but when I went for interviews for labouring, factory or office work I had to hide the fact that I had three University degrees, thereby making my work record suspicious, or disclose it, which inevitably resulted in my being told I was over qualified. I also had little experience of anything but studying. Whatever the reasons, and there were plenty, I was offered nothing.

Life on the dole was poverty-stricken and I took any work in the black economy I could get: making hay, odd jobbing, pumping petrol etc. I felt ashamed to be seen in public in the day and when I needed to shop I would scurry around in other people's lunch hours. I was more and more frequently ill with one minor illness, or concern about illness, after another. After a few consultations, my middle class GP lost his patience with me and told me offensively – perhaps using the language with which he expected me as an unemployed man to be familiar and thus simultaneously positioning me in a derogatory way - to 'get off my arse and get a job'.

It would be a mistake to characterise my position as unemployed person totally in negative terms, however. Whilst in some ways fragile, I was also fit and strong from physical work. I had discovered a resourcefulness within myself I had not suspected. I acquired, as a matter of necessity, skills in fixing car engines for myself and for friends, for example, and once I undertook and completed, totally alone, a job refurbishing the entire interior of a small ocean going yacht. I had no previous do-it-yourself experience and had to try out lots of the tools I was given on spare timber to see what they did before I could use them. I laid the foundations for a local film club, which actually became established after I had moved on. I was active in a local centre left political party and was even approached about being put forward by the branch at constituency level as a potential parliamentary candidate.

I do not want to individualise this period. My partner was employed, bringing in income and she and her local family were very supportive. We had a good circle of friends centred on the local pub, a thriving social life and a very extensive extended network of contacts, friends of friends. Benefits were low, but I was sometimes able to supplement them on the black economy and self-provisioning was possible e.g. for firewood. There was a subculture of surfers, artists and greens, with whom we rubbed along, who had honed post-industrial survival skills to an individual and collective art form and regarded employment as a form of exploiting subservience to others' will in the service of alien values, rather than seeing themselves as 'amoral drop outs' most saw becoming employed as dropping out of higher order ethical and ideological obligations into a form of sub-human slavery and a noxious foolishness - and to be avoided at all costs. Moving in these circles provided me with a form of massively influential conscientization and with critical discourses counter to the dominant discourses of my upbringing.

I alternated between these worlds, applying for jobs - academic and other - signing on with the state bureaucracies and jumping through their hoops when necessary, approaching hard working neighbours to legitimise me as one of the 'deserving poor', doing work 'on the side' for worthy citizens who despised the unemployed but were happy enough to employ one if it reduced the wages bill, mixing with political activists, film buffs and all manner of members of the counter culture. My experience of unemployment was thus complicated, multifaceted, ambivalent and contradictory. Whilst my experience was unique to me, I would expect others' experience to be equally complex.

It was at this time that I saw a job advertised at Sheffield University to join a team investigating the psychological consequences of unemployment. I applied for the job less in the hope of being offered the job, as I had no experience of social, applied, health or clinical psychology, than in the hope of getting interviewed and being able to make money on interview expenses. To my surprise and consternation I was offered the job and - after a wobble of confidence and commitment – relocated to Sheffield to take up a position as a post-doctoral research fellow to do research into unemployment and mental health at the Social and Applied Psychology Unit.

In brief summary, as a boy and young man, my positioning in relation to employment and unemployment was ideologically speaking enormously complicated and contradictory – just like, I submit, that of most other people. I was immersed in a dominant discourse in which employment was positioned as a moral duty, the highest calling, unemployment positioned as a source of profound shame associated with fecklessness, laziness and trouble-making and labouring and working class status positioned as best left behind as soon as possible with the armed forces and education as the escape routes. But I was also immersed, increasingly so as I grew up, in a counter discourse in which employment was positioned as a form of oppressive drudgery, physically and spiritually crushing, to be avoided at all costs whenever possible and unemployment was positioned as a glorious, self-determining, life-affirming, romantic slap in the face of bourgeois convention. Both, I absorbed without ever being explicitly taught. My own experience as an unemployed person confirmed for me the humiliating power of others' gaze to create me as a lazy, parasitic, third-rate layabout looking for handouts at others' expense and to consign me to financial hardship, illness and shamed isolation yet also that a fulfilling and fulfilled socially valuable life is not the same as an employed life.

4 From unemployed person to unemployment researcher

The Social and Applied Psychology Unit (SAPU) at Sheffield University, which was internationally influential at the time because of its major programme of research on unemployment and mental health, was a very privileged place to work. Although there was a hierarchical team structure, the Director, Professor Peter Warr, who was located at the apex, managed virtually all the necessary bureaucratic work and allowed staff considerable autonomy, even to depart on occasions from the characteristic preferred house style which was widely referred to as 'Medical Research Council (MRC) shape and size' research, the MRC being the Unit's main funding body. 'MRC shape and size' research, at least in relation to unemployment, was at that time typically quantitative, survey design research using packages of measures, with fieldwork carried out by contract survey organisations, a statistician to advise on data analysis, data preparation people to input data into computer programmes from questionnaires and secretaries to type up the papers. The role of researchers was largely confined to choosing the samples to be targeted, choosing the measures and writing the papers for publication at the end.

Typical SAPU MRC shape and size unemployment studies assessed the mental health of large groups of people in school, employment or unemployment and using measures of proven reliability and validity like the General Health Questionnaire, then followed them through labour market transitions, periodically measuring the mental health of those who got jobs and those who did not and comparing group mean scores cross-sectionally and longitudinally. Again and again the evidence suggested poor mental health was the consequence rather than the cause of the labour market transition.

I had lots of reservations about the way SAPU operated but articulated few of them at the time. Putting aside any qualms about the overwhelming privileging of the interests of unemployed white males, it seemed to me that SAPU bureaucracy and MRC methodological conformity strait-jacketed innovation leading to repetitive and unimaginative work which stripped away the context which would have made the experience of the research participants meaningful and understandable and produced accounts which were naïve and lacking in authenticity so that unemployed people, as represented in SAPU research, seemed to me pale two-dimensional shadows of the unemployed people I knew and one of whom I had recently been. This also made the work subtly alienating.

I soon learned at SAPU that an extraordinary international interdisciplinary consensus had developed in the field that unemployment *caused* mental health problems. I found that evidence had been presented as consistent with that claim over a very long period of time. Scholars had traced the roots of the claim back for about two hundred years and there had been a steady stream of publications to that effect ever since with floods of publications in the 1930s, the 1980s and the 1990s. The effect of unemployment had come to be taken to be a powerful, fundamental, process transcending relatively superficial historical differences in the nature of employment and unemployment. Evidence consistent with the claim that unemployment caused mental health problems had also been reported from studies carried out across Europe, North America and Australia so had come to be taken to be not just a phenomenon unique to a particular country or culture (though, perhaps tellingly, studies from Africa, China, the Far East, India, Russia and South America were rather hard to find). Evidence consistent with the claim that unemployment caused mental health problems had also been reported by researchers using a wide range of research methods (from epidemiology to longitudinal quantitative surveys to qualitative techniques) and had also been reported by studies using a wide range of ways of measuring mental health (so had been widely taken to be not just an artefact of a particular research method or technique). However most emphasis tended to be placed upon the results of longitudinal research following very large groups of people through various employment status transitions and showing, time and time again and almost without exception, changes in group mean scores on measures of mental health consistent with employment causing improvement and unemployment causing deterioration in mental health. SAPU was, as I would later put it, constructing and legitimating ('truthing') claims of dominant knowledge and at the same time subjugating the counter knowledges to which I had had access and which I still found authentic and authoritative.

When I started unemployment research, politicians and commentators were still suggesting that the relationship between unemployment and poor mental health was better explained by people with poorer mental health being more likely to become and remain unemployed (`individual drift` or `selection`). Whilst I accepted at that time that, in some cases, poor mental health did lead to unemployment, I also regarded the notion of 'social cause' as problematic if one was trying to steer a course between the inexorable socio-structural determinism and the naïve psychologism of the self determining individual. I was aware that many people's experience of the labour market was dynamic rather than static, involving iterative cycles of both cause and apparent selection (one could lose one's job because of a mental health problem and have other mental health problems caused by that unemployment; one could be perfectly healthy, lose one's job, have mental health problems caused and then be selected out of re-employment because of that problem or at least because of employer prejudice against those 'with mental health problems' etc). I was also convinced that the individual drift or selection interpretation of the relationship between unemployment and poor mental health was oppressive and the social causation interpretation progressive. So, to engage in unemployment research which demonstrated the process through which unemployment negatively impacted on well being was almost a moral imperative.

At that time I assumed, naively, that research demonstrating beyond reasonable doubt that certain economic policies damaged public mental health, whilst an irritant to the status quo, perhaps even leading to the 'shooting' of the messenger, but would lead inexorably to those policies being changed. To my surprise I found, on the contrary, that when at conferences, in press releases, reports to politicians etc., I (re) asserted that unemployment put mental health at risk, there was little if any reaction and far from being uncomfortable with that message, the establishment seemed very happy with it. For example, following a press conference organised by the British Psychological Society, at which I addressed journalists, related headlines appeared in at least 17 major newspapers including "Death, Disease and Despair on the Dole" (The Daily Mirror), "Unemployment linked to high suicide rates" (The Independent), "Middle aged men react worse to redundancy" (The Daily Telegraph) and "Redundancy – fate men cannot handle" (Yorkshire Post women's page). On the other hand whenever I reported work suggesting that unemployment could be 'good for' the unemployed person and others in society (Fryer and Payne, 1984; Fryer, 1985), the left right and centre all seemed to come down on me like a ton of bricks!

It dawned upon me that whilst it was obvious whose interests would be served by the drift / selection argument that poor psychological health causes unemployment which blamed unemployed people for their own unemployment (and thus for their own distress) and simultaneously absolved governments and businesses of responsibility for the health consequences of unemployment, it was not necessarily the case that the alternative, social causation, argument that unemployment caused poor psychological health was in the interests of unemployed people and detrimental to the interests of the status quo.

This was reinforced for me when I realised that mass unemployment served the interests of the status quo in a range of ways *but most ways also required unemployment to be a condition so undesirable that no-one wished to become unemployed and which all unemployed people wished to leave as soon as possible.* Mass involuntary unemployment guarantees there are potential workers willing to do the most boring, dead end, underpaid, temporary, insecure, unpleasant jobs (i.e. the ones being created in the so-called flexible labour market) and mass involuntary unemployment functions effectively as an incomes policy because it guarantees that there are unemployed people competing for the jobs of the employed thus facilitating employers in reducing wages and working conditions.

What are the conditions under which involuntary mass unemployment would be in the interests of the status quo? The existence of far fewer jobs than potential workers seeking them would, of course, create mass unemployment but that would not in itself necessarily ensure that unemployment was involuntary, after all large numbers of rich people seem to manage quite happily without paid employment or disastrous psychological consequences of not being in paid employment. Ensuring that unemployed people are poverty stricken and that they had to go through intrusive and degrading rituals to get the pittance they get to keep them healthy enough to compete for work but not comfortable enough to have a viable alternative life style would help. Ensuring that unemployment was a stigmatised condition with orchestrated campaigns by the media and politicians reinforcing the view that unemployed people are feckless, anti-social idlers living a life of luxury at taxpayers' expense, fraudulently claiming income and two-timing the system would also help. Making a point in press reports of mentioning whenever criminals were unemployed would help taint unemployment with criminality. A psychological literature demonstrating that unemployment caused mental ill health whilst mental illness was simultaneously socially constructed as frightening, dangerous and deviant would also help. A psychological profession also implicated in promoting 'employability', active labour market policies and individualistic cognitive interventions would virtually guarantee unemployment was involuntary.

Moreover, I observed that when unemployment increased, the value of the stock exchange tended also to increase and vice versa. I also noticed that The Governor of the Bank of England caused outrage in 1998 when he unguardedly disclosed that he thought job losses in the north of England were an acceptable price to pay for curbing inflation (which would disproportionately benefit the employed south), that many economists agreed that a rise in unemployment usefully reduced pressure for wage increases and that there was even an acronym (NAIRU) coined for the "Non-accelerating inflation rate of unemployment", the level of unemployment (about 4-6%) regarded as necessary to keep inflation from rising. Moreover, the unanimity that unemployment caused mental health problems was not just due to an accumulation of empirical 'findings'. By far the most dominant explanation of the relationship between unemployment and mental health, that offered by Marie Jahoda, also reinforced the causal claim. Jahoda argued that whilst most people sought and took a job in order to earn a living (what she called the manifest function), employment as a social institution had inevitable unintended consequences (what she called latent functions) for people's psychological state – irrespective of individual differences in their feelings, thoughts, motivations and aims. This was, according to her account, because of the employment oriented time-structure imposed upon employees' working days, weeks and years; the social contact and shared experiences with non family members which employees are compelled to have; the required participation of employees in collective goals and purposes beyond their scope as individuals; the social identity imposed upon employees by their employed status and the level and nature of activity enforced upon employees as employees. According to Jahoda the psychological consequences of unemployment are simply the absence of the psychological consequences of employment.

I long ago offered both a theoretical and methodological critique of Jahoda's account on a variety of orthodox grounds (Fryer, 1986) and a study (Fryer and Payne, 1984) of unemployed people who appeared to be doing just fine without Jahoda's latent functions but I want, here, to ask whose interests Jahoda's account serves? Because it suggests a mechanism through which unemployment negatively affects mental health, Jahoda's account supports claims that unemployment causes mental ill-health rather than the reverse. Moreover, it does this by positioning employment as not only psychologically benevolent but also perhaps necessary for psychological well-being and positioning people as dependent upon employment for the structures which keep them psychologically healthy.

In case it seems far fetched to consider Jahoda's theory as manna for the status quo, consider the following. On 27th September 1993 The Times, the so-called 'top people's paper', carried on page 19 a 'Leading article' based on a report by Nigel Hawkes, the paper's science editor. The leader is worth quoting extensively (the italics are mine):

"WORKING TO LIVE"

Having a job seems to be necessary for self-esteem

Buried in the almost impenetrable jargon of a British Psychological Society paper, is an observation about modern attitudes that has real political significance. What it suggests is that the kind of self-esteem and satisfaction that most feel to be necessary for a fulfilled life can rarely be achieved outside of paid work. In other words, no matter how worthwhile or energetic a leisure pastime may be, it cannot impart the same sense of purpose and self-respect as a "proper" job.

The research paper, by Dr John Haworth of Manchester University, does not simply equate this unique kind of satisfaction with the fact of being paid. There seems to be a more subtle psychological difference between working for a livelihood and even the most assiduous hobby. What is suggested by the responses of those who filled in Dr Haworth's questionnaires is *that the very constraints of working life are what make it satisfying*. The findings refer to factors like "time structure, social contact, collective purpose, social identity or status, and regular activity". In plain English, these amount to having to be at a given place at a particular time with actual deadlines for completed tasks, working toward some larger goal with a team of people in which everyone has a specified role and having all of this take place in some customary, habitual way.

Being compelled to take part by some force outside of personal whim – what Dr Haworth calls "extrinsic motivation" as opposed to "intrinsic motivation" – *seems to be the key factor in making paid work a more valuable source of psychological well-being*. Having objectives and structures imposed by others lends credibility to an enterprise. In leisure activities – even ones that are socially useful – the freedom to create personal goals and time limits often degenerates into an open-ended activity in which people find it difficult to maintain a sense of purpose.

People specifically mentioned that *being at work often involved doing things which were initially disliked. Overcoming their own resistance to complete the task gave a form of gratification that was peculiarly difficult to match outside of the workplace.* (. . . continued . . .)"

Note that Jahoda's account was here being used in one of the Establishment's favourite newspapers to bestow scientific legitimacy upon claims that employment is not only good but necessary for psychological well-being and that it is the constraining and compelling aspects of employment, requiring employees to do things they do not like doing, that are particularly beneficial psychologically.

5 Reflections from a community critical psychological perspective

From a community critical psychological perspective the unemployment and health literature is problematic in many ways. In general, it refrains from engaging with contextually constituted subjectivity as an irreducible phenomenon with emergent properties; it treats unemployed people as 'respondents' rather than agents (albeit agents of their own governmentality); it steers clear of 'power' issues in theory and frequently disempowers participants in process and in practice; it is deferential to so-called 'experts', especially of 'psy' and research experts yet fails to appreciate the competence and expertise possessed and exercised by unemployed people; and it is concerned with documenting distress rather than preventing or reducing it.

However, it is arguably most problematic in that it is fundamentally acritical. I started out by describing community critical psychology as a 'critical' approach and this is, for me, the key point. By a 'critical perspective', in this chapter, I do not mean a perspective which is 'disparaging', 'disapproving', 'nit-picking', 'judgmental', 'unsympathetic', 'derogatory', 'fault-finding', to refer to the terms listed in Microsoft Word's Thesaurus in the 'critical - unfavorable' sub-category. Rather, I mean it is a critical social science in the sense of Pihama (1993): "exposing underlying assumptions that serve to conceal the power relations that exist within society and the ways in which dominant groups construct concepts of 'common sense' and 'facts' to provide *ad hoc* justification for the maintenance of inequalities and . . . continued oppression" (Pihama, L., 1993:186). Community critical psychology problematises both the way neo-liberal societies are structured, what the consequences are of that, for whom, in which ways and simultaneously problematises 'psy' and related sciences, which thrive in neo-liberal societies: asking what systems of socially constructed and maintained meanings are 'truthed' as 'realities' by such sciences , how exactly do they 'knowledge' such 'realities' to give them the social status of 'truth', for what purposes are they deployed, whose interests do they serve when this is done and what political and ideological positions are promoted through them?

In particular in this paper, I mean a perspective which asks awkward questions about whose interests are served by what is thought, written and done by those doing and publishing research concerning the psychological costs of unemployment, what the implications are of various positions for the empowerment of some and disempowerment of other interest groups and whether or not problematic assumptions underlie various aspects of the relevant literature.

Thus, in reflecting *critically* on claims in the research literature, I am not asking whether they are theoretically coherent, whether they are methodologically sophisticated, whether they are well supported or whether they are ethically viable. Rather I am asking whether the claims are ideologically problematic and to whose benefit and to whose detriment it would be if these claims were widely believed.

To give an example, some work by clinical psychologists has evaluated attempts to reduce the negative mental health consequences of unemployment by giving cognitive behaviour therapy to unemployed people. The underlying logic is that by increasing the effectiveness of unemployed people in job search, one can increase the likelihood of them becoming re-employed and so remove the risk of the noxious psychological effects of unemployment upon them. Orthodox questions can be asked about whether such interventions 'work' i.e. whether they do indeed increase the effectiveness of the job search of the unemployed people involved in the studies, whether the theoretical basis of cognitive behaviour therapy is coherent, how faithfully such interventions implement this theory in its practice, how representative the participants are of the wider population of unemployed people, how suitable the research methods used are, how appropriate the techniques are which are used to analyse the data etc.

Whether or not the above are appropriate questions to ask depends on one's epistemological and ontological frame of reference but they are not *critical* questions in the sense being used here.

In reflecting *critically* on this work I would be interested in whose interests it would serve for it to be widely believed that the mental health problems of unemployed people are caused not by external socio-economic events like recession but by internal psychological cognitions.

I am interested in what the implications of this work are for the attribution of blame for the psychological damage wrought by unemployment and thus responsibility for reparation?

I am interested in how the interests of the various parties would be served if it were believed that mental ill health caused by unemployment was reversible through a few sessions of talking therapy.

I am interested in whose interests would be served if it were believed that mass unemployment could be tackled by *individual cognitive treatment* of people after they become unemployed (by cognitive behaviour therapy) rather than *collective socio-economic prevention* before people become unemployed (through job creation or redistribution).

I am interested in whose interests would be served by interventions that create and maintain an excess of potential employees over vacancies and coach such people to compete with each other for success in getting one of the inadequate number of vacancies compared with interventions that create and maintain an excess of vacancies over potential employees and coach employers to compete with each other for success in getting one of the inadequate number of potential employees.

I am interested in whose interests would be served by the creation of an illusion of effective intervention of increasing unemployed people's chances of reemployment which, fundamentally, merely redistributes the misery of unemployment from one subgroup to another whilst maintaining an unchanged total of unemployed people, and thus an unchanged number of people at risk of psychological damage by unemployment.

Within mainstream psychology, the notion of 'ideology' is very often used in a derogatory way to dismiss the work of others by (re)presenting it as more a product of a pre-existing political 'bias' than of any systematic process of scholarship or research. However, here I am using the term "ideology" as Stainton Rogers (2002) uses it i.e. to mean "the use of knowledge to promote the power of certain groups". For me, all human activity, including all research, is ideological. The issue is not whether or not it has implications for the distribution of power but what those implications are, for whom and with what consequences. I am committed to ideologically progressive praxis which constructs and deploys knowledge to promote the power of the most exploited, powerless and psychologically oppressed: people who, under current societal arrangements, are unemployed, impoverished, psychiatrically abused, disabled etc. In particular I am interested in to what extent the construction and use of certain 'knowledges' about unemployment, employment and mental health are used to promote the power of the status quo rather than that of unemployed people.

The suggestion that there are a variety of 'knowledges', each of which promotes the interests of some as opposed to other interest groups, may seem an odd one to those who operate on the modernist assumption that knowledge is fundamentally a cognitive representation of 'what is the case' in the 'real world' arrived at through a combination of rationality and empiricism. However, according to a critical perspective, there are many 'reality-versions', in each case 'reality' being constituted, in an emergent process, through subjective sense-making of a unique intersection of societal structures and then socially manufactured into 'knowledges'. Within this frame of reference, the dominant version of 'knowledge' is just the 'reality-version' that serves the interests of the most powerful group. The societal structures of which subjective sense is made include systems of ideas within which one is immersed and through which one is, at least in part, constituted. There are, however, not just one set of systems of ideas but many alternative systems, which serve different interests in different ways. Permeating and constituting much of psychology as a discipline, and as it is used by the status quo, are systems of ideas which imply that psychological distress and illness are caused and maintained at the individual level by intra-psychic forces and processes and that it could not be any other way. Such systems of ideas suit the status quo for, not only, do they *not* require the status quo to change in practice in order to prevent or reduce distress and illness but also they imply that psychological distress and illness could not be prevented in principle. From a *critical* perspective, it does not make sense to ask what is 'really' the case. Rather one can only ask how 'realities' are constituted for each contextualised person and why some realities come to dominate as received 'knowledge'.

From the above I hope it is now clear why I think it important to engage critically with the unemployment research field, to bring to the surface the subjectivity of all those involved in constructing 'knowledges', dominant and subjugated, try to surface the discourses and ideologies which constitute them, grapple with power issues in process and outcome, contest the disempowerment of unemployed people and the collusion of psychologists and other social scientists in its construction and maintenance and go beyond documenting distress associated with or caused by unemployment or the flexible labour market to make a difference by preventing or reducing it.

Of course, the status quo is the status quo because it resists challenges to its power, exerts control reactively when challenged and exerts control proactively as an everyday function of it existence. Some of this control is obvious but much is far less obvious because built into our multi-level social context and taken for granted.

The behaviour of unemployed people is controlled in some obvious ways by the status quo. For example, unemployed people are required to search diligently for jobs (which often don't exist) in return for subsistence income. Less obviously, unemployed people are 'governed' ideologically through their internalisation of dominant discursive representations of unemployed people as inadequate social pariahs individually responsible for their own unemployment through their psychological dysfunction. Less obviously still, unemployed people are governed by interconnected assemblages of apparatuses in the Foucauldian sense, of practices, procedures, policies and discourses. The 'surfacing' and contesting of such governmentality - and the collusion of psychologists and other social scientists in its construction and maintenance - is a central task of community critical psychology in my view.

The remarkable consistency of claims across time, culture, method and technique, generally taken as constituting *replication* and *convergent validity*, i.e. as particularly strong grounds for confidence in the empirical claim that unemployment causes mental ill health and the theoretical accounts like Jahoda's which reinforce it, may be just another manifestation of that control. Could the extraordinary consistency just provide grounds for confidence that 'knowledge' about unemployment and mental health has been consistently constructed, amongst others by social scientists, and used ideologically by the status quo to promote the power of certain powerful interest groups. Is the consistency of 'knowledge claims' across time, culture, method and technique an indication of the ideological saturation of social science research in this field which reflects the consistent interests of the status quo and in particular employers, company owners and shareholders in neo-liberal economies. If unemployment were not psychologically and physically destructive, wouldn't it be necessary for social scientists and others to construct it as such in order for it to function effectively as an instrument of social control in the interests of the status quo?

Community critical psychology is not an individualistic or psychologistic approach. It recognises that oppressive systems are generally enacted by hard-working, principled, people of good will. If the knowledges constructed by unemployment researchers function oppressively, that does not mean that the individual researchers are oppressive in a simple sense and this paper is not meant to be read as a simplistic personal reprimand, rebuke to or slur on colleagues. To emphasise that I have positioned myself unambiguously as a contributor to a reactionary knowledge whilst emphasising my progressive intentions. However, I believe that we each can and should not evade our responsibility for the ideological consequences of what we construct and enact ("the use of knowledge to promote the power of certain groups" (Stainton Rogers, 2002) by appealing to the 'inquisitiveness' of the scientist or 'contributions to science'. Nor do I believe can we devolve our moral and ideological responsibilities to others, ethicists and philosophers, to do on our behalf. As a white male I am aware that my male and white privileges are difficult for me but easy for women and minority group members to see, that patriarchy and racism are wired into practices, procedures, policies and discourses in my society and can not be 'refused' by individual acts of will but I also believe that I have an obligation to do my best to avoid enacting patriarchy and racism in my thinking and acting and to surface and contest it in my society and my discipline. I believe the same is true in relation to the oppression of unemployed people and the social sciences which play their parts in constructing and maintaining it.

References

Cassell, C., Fitter, M., Fryer, D. & Smith, L. (1988). The development of computer applications by unemployed people in community settings. *Journal of Occupational Psychology, 61,* 89-102.

Darwin, J., Fitter, M., Fryer, D. & Smith, L. (1987). Developing information technology in the community with unwaged groups. In P. Bjerknes, P. Ehn, M. Kyng. *Computers and Democracy*. Aldershot: Avebury/ Gower.

Fryer, D. (1985). The positive functions of unemployment. *Radical Community Medicine, 21,* 3-10.

Fryer, D. (1986). Employment deprivation and personal agency during unemployment. *Social Behaviour, 1,* 3, 3-23.

Fryer, D. (1987). Monmouthshire and Marienthal: Sociographies of two unemployed communities. In D. Fryer & P. Ullah, *Unemployed people: Social and psychological perspectives.* Milton Keynes: Open University Press, 74-93.

Fryer, D. (1994). Commentary: "Community Psychology and Politics" by David Smail. *Journal of Community and Applied Social Psychology, 4,* 11-14.

Fryer, D. (2008). Power from the people? Critical reflection on a conceptualization of power. *Journal of Community Psychology, 36(2),* 238-245.

Fryer, D., & Fagan, R. (2003). Towards a community psychological perspective on unemployment and mental health research, *American Journal of Community Psychology, 32, 1/2,* 89-96.

Fryer, D., & Payne, R.L. (1984). Pro-activity in unemployment: Findings and implications. *Leisure Studies, 3,* 273-295.

Fryer, D., & Ullah, P. (Eds.). (1987). *Unemployed people: Social and psychological perspectives.* Milton Keynes: Open University Press.

Pihama, L. (1993). Tungia te Ururua, Kia Tupu Whakaritorito Te Tupo o te Harakeke: A critical analysis of parents as first teachers. MA thesis, University of Auckland. In L.T. Smith (1999). *Decolonising methodologies: Research and indigenous peoples.* Dunedin: University of Otago Press.

Stainton Rogers, W. (2002). Critical approaches to health psychology. In D. F. Marks (Ed.). *The health psychology reader* (Ch. 21, 286-303). London: Sage.

4. ORGANISATIONAL RESTRUCTURING AND HEALTH: SOCIAL CONVOY AND OCCUPATIONAL TRANSITIONS

Effects of Socially Sensitive Enterprise Restructuring on Unemployment and Health: The ILO Perspective

Nikolai Rogovsky

1 ILO and Socially Sensitive Enterprise Restructuring (SSER)

Although our programme "Socially Sensitive Enterprise Restructuring" (SSER) of the International Labour Organisation (ILO) is primarily aimed at enterprises, we operate in the context of the overall ILO strategy aimed at the promotion of social dialogue and decent work for all. This means that we view ourselves as a part of the ILO team. In practice it means that when we at the ILO deal with a particular issue, such as, for example, health and unemployment, we try to make sure that such an issue is addressed from different angles. In this particular case the issue addressed by the ILO programmes is the one dealing with job creation, employment promotion, social protection and occupational safety and health. Our programme approaches this issue from the enterprise point of view in the context of economic restructuring. The growing importance of this approach is explained by the fact that in today's world the role of business is growing day by day.

The world is constantly changing. It is not the same as it was 10-15 years ago. Nowadays, society, in general, and enterprises in particular have to face the following trends that cannot be ignored:

- Globalization
- Technological change
- Changes in corporate ownership (we now see a growing number of worldwide mergers and acquisitions, management buyouts, privatization programmes, the increasingly important role played by institutional investors, such as pension funds).
- Industrial society is being replaced by an information society
- Demographic changes
- Growth of Foreign Direct Investments

But perhaps most importantly, we are witnessing serious changes in the demands and expectations that civil society has made on the role of business in society.

In general, it is hard to deny that nowadays business plays a much more important role in society than it did 10-15 years ago. In the situation when in many parts of the world, society's trust in political parties, religious organisations, governments and other institutions of civil society is diminishing, business becomes a dominant institution in society. Needless to say, a dominant institution in any society should accept certain responsibilities and obligations. However, business does not have such a tradition, and still has to realize its new role in the society.

The International Labour Organisation (ILO) is trying to help business to realize this role.

2 What is the International Labour Organisation?

The ILO is a tripartite organisation. This means that all decisions are made by representatives of governments, employers and workers. Consequently, a "reality check" is built into international labour standards. From its inception, the ILO recognized the importance of creating win-win strategies for workers and employers, as well as for society as a whole.

The ILO works to promote equality of treatment, social justice, and improved working and living conditions for all.

This, it does through:

- international labour standards (ILS) which are benchmarks for good governance in both the public and private sectors;
- technical assistance to help governments implement ILS; and
- education and promotional activities to raise awareness and understanding of ILS.

The ILO covers a broad range of topics. Today my focus is on one of such topics – social aspects and consequences of enterprise restructuring.[1]

3 Why is it important to look at restructuring?

Enterprise restructuring is one of the key topics covered by the activities of the ILO Employment Sector. In general, this sector looks at what we define as Global Employment Agenda, which highlights a number of the major challenges current labour markets present all over the world. It emphasizes that such a problem as unemployment could no longer be solved only by what we call government active labour market policy. Globalization expands the labour market beyond national boundaries; government funds are no longer sufficient to cover the programmes of vocational training and re-training. Due to these and other factors, it is becoming increasingly difficult not to look at what the enterprises could do voluntarily, beyond the legal requirements, in order to help the societies address a number of social and economic problems, unemployment being one of those.

A lot of companies now are marketing themselves as socially responsible enterprises. It has become a commonplace for a company to state: "employees for us are not costs, they are our assets". However, whether or not the company is indeed a socially responsible one, can only be seen at times of economic slowdown.

1 More information on the ILO can be found at www.ilo.org

When everything is going well it is not so difficult to be socially responsible, especially if it does not cost too much. The real test is how a company behaves with regards to its employees and with regards to society at large when the things are not going well. This brings us to the issue of socially sensitive enterprise restructuring.

4 What is restructuring?

What is restructuring? This term is used in many different ways. First of all, restructuring is a profound change in the ways company operates. This involves changes in company's strategy, structure, etc. This term is also used when it comes to downsizing. However, from our point of view, restructuring is a much broader issue than downsizing.

As we investigated the approaches that various companies, large and small, public and private, adopted in their efforts to restructure, what became obvious to us was that companies differed in terms of how they viewed their employees. Indeed, they almost seemed to separate themselves logically into two groups. One group, *by far the larger of the two,* saw employees as costs to be cut. The other, much smaller group saw employees as assets to be developed. Therein lays a major difference in the approaches they took to restructure their organisations:

- *Employees as costs to be cut.* These are the "downsizers". They constantly ask themselves, "What is the minimum number of employees we need to run this company? What is the irreducible core number of employees the business requires?"
- *Employees as assets to be developed.* These are the responsible restructurers. They constantly ask themselves, "How can we change the way we do business, so that we can use the people we currently employ more effectively?".

The downsizers see employees as commodities – like microchips or light bulbs, interchangeable, substitutable, and disposable, if necessary. In contrast, responsible restructurers see employees as sources of innovation and renewal. They see in employees the potential to grow their business.

The legitimacy of the latter position seems to be grounded in the recent global market trends, such as easier access to finance, the inability of many companies to protect their technological advantage in a long-term perspective, etc. In such a situation human and social capital, or, to put it simply, the way company treats workers and community, become important sources of competitive advantage.

Of course, even the most socially sensitive companies go through restructuring when they have to. Very often, the purpose of restructuring is not only financial and economic improvement of enterprise performance, but the very survival of the enterprise itself.

We are far from saying that the companies could and should not go through restructuring. In many cases, restructuring is the only solution. However, we believe that restructuring could be carried out in a socially sensitive way. In other words, companies could try to maximize economic benefits through restructuring and, at the same time, address the needs of employees and communities.

Such a position is grounded in international standards adopted at the European level and in other parts of the world, and also in the ILO International Labour Standards

(ILS), known as Conventions and Recommendations. In particular, one should mention here Convention No. 158 on Termination of Employment, which is accompanied by the Recommendation No. 166. Both documents were adopted by the International Labour Conference in 1982, but are still very relevant. These documents emphasized that the ILO recognized that termination of employment could take place for economic reasons. They also stressed that this decision could be made by the enterprise management. At the same time, the ILO emphasizes the necessity of a long-term approach to HR planning and the importance of creating and maintaining a multifunctional workforce, and its continuous training and development.[2]

The ILO also emphasizes importance of:

- Consultations between workers and employers before, during and after the period of restructuring
- Creation of the most preferential conditions for the workers affected, so that they can continue their professional career
- Non-discriminatory policies and practices in restructuring, based on such characteristics as age, gender, union membership, etc.

However, in practice a socially sensitive approach to restructuring is not always the case. In contrast, we see the following trends very often:

- Downsizing is often the first thing that company does when the economic situation deteriorates
- Downsizing is often the first thing that the new owner does when he/she acquires another company
- Downsizing is often made without social dialogue and making any consideration of the interest of the employees affected.

These trends exist all over the world, and, unfortunately, the West European countries are not the exceptions.

At the same time, some good examples of socially sensitive enterprise restructuring do exist, and the goal of our programme is to make sure that these examples are advocated and promoted.

Before going into specifics, please allow me, however, to share with you some data:

- Loss of job is more than just a loss of income. This means a loss of self-respect, respect of others, structure of the day, often – loss of *raison d'etre*. For example, the international study "The International Social Survey Programme" shows that while for 50 percent of the unemployed in Europe and North America a loss of income remains the major problem, for 25 percent a loss of self-respect and respect from others are more important than a loss of income.

2 See Nikolai Rogovsky et al., "Restructuring for Corporate Success: A socially sensitive approach", ILO; Geneva, 2005.

- It is often believed that downsizing has a positive impact on the performance of the company and the so-called "survivors". This is not always true. For example, data from the USA shows that for some time (one year or more) after downsizing[3]:
  - 70 percent of companies that underwent downsizing reported a decline in work-related morale among survivors
  - only 35 percent of companies increased the quality of their products
  - only 34 percent of these companies reported an increase in productivity.

There is also overwhelming evidence that suggests that downsizing might have very negative health effects not only in affected workers, but also in so-called survivors".[4]

The ILO is involved in a number of training, promotional, research and policy-related activities on socially sensitive enterprise restructuring.

Let me share with you some of the lessons that we, in the ILO, have learned by carrying out our SSER activities in more than 30 countries throughout the world:

- In order for restructuring to be successful, it should be linked to the long-term development strategy of the company, country, region
- Company management should always know what to do if restructuring is inevitable. This concerns, first of all, the company's human resources
- Restructuring should be based on the joint agreement between employers and workers, and, in some cases, the government
- Estimate not only costs, but also benefits of SSER
- Consider ALL the options before downsizing, try to use less painful options
- If restructuring is inevitable, execute it in a socially sensitive way (suggested tools)

Let me focus on some of these tools. It is important to note here that these tools should not be viewed as some sort of an "emergency kit". On the opposite, we believe that these tools should be a part of the company's long-term HRM strategy.

5 Psychological help

Being made redundant is a trauma. People's reactions to a broken work relationship are comparable to the breaking up of a private relationship – common states of mind are: denial, fear of the future, anger, and depression. While management's decision must be binding and irreversible, members of the management team should always be ready to listen to those workers who are affected, to let them address and express their concerns.

Additionally, management needs to remember that it is not just the employees being dismissed that are affected by this; survivors also experience emotional distress during this time either because friends were let go or because they fear for the future of their own job. Communicating with both of these groups is critical to the success of the

3 Wane F. Cascio, "Responsible Restructuring: Creative and Profitable Alternatives to layoffs", Berrett-Koehler Publishers, Inc. San Francisco, CA , 2002, pp. 28-31.

4 Ibid.

restructuring effort, as well as putting in place counselling sessions to help the employees get through restructuring.

The counselling sessions should be a tool to help redundant employees with the emotions coinciding with the breaking of the work relationship, and help them focus on their strengths, regaining self confidence and a positive outlook.

For surviving employees, the counselling session should focus on the feelings they have regarding the company, their future, and any other issues that may arise. It is important to reinforce security and comfort for the employees who will continue to be responsible for the functioning of the company. This counselling effort could even be extended to the families of affected workers.

6 Skills assessment

A change in job, whether within the company or externally, will often require a shift in skill requirements. A company can help employees realize both their strengths as well as their areas for improvement by providing a skill assessment previous to the change.

Gone are the days of the life-long career with a large company and a steady advance up the corporate ladder. More now than ever in the expanding European Union, as well as other parts of the world, people are realizing that job security has to be replaced in many cases with employment security. And only high level of skills, multiplied by the government active labour market policy, can provide people with stable employment.

7 Training/Employability

Faced by this uncertain job market, employees have realized that the more skills they have, the more valuable they are to the company and, in general, the more marketable they are. In other words, employees are increasingly looking at how "employable" they are. As a result, many people seek employment with the companies that can provide them with training and skill development. Needless to say, this is still the government system of vocational training and re-training that needs to be the most important provider of skills. Governments should not abstain from this task, and should work further to strengthen such systems. However, employers, as well as trade unions' training centres in many countries have to provide additional training to workers. And this is the reality that, perhaps, shows that in some countries a governmental system of vocational training no longer reflects the needs of the market.

The question that employers may ask is as follows: "If we agree that our relationship with employees is changing, and that he/she might wish to leave our company in 3-5 years, why should I invest in this person and, in fact, train an employee for my competitors?" To this I usually reply: "If you invest in your current employee now, there is a good chance that your competitors are investing in your future employees at the same time. The rules of the game have changed and this is the reality of today!"

Training should not be limited to just job skill training however, but should also include job search skill training which includes training on how to write a CV, to pass an interview, to analyze one's own strengths and weaknesses, how to handle stress, etc.

8 Internal job search

At the time of restructuring, some jobs might be available within the company itself, in another plant or service of the same plant, or in a subsidiary. These are the first to consider. Within the human resource department, it must be checked which jobs are available, and what skills are needed. Depending on the size of the company, there may already be tools in place to aid this process, such as a company intranet site that lists vacancies, or bulletin boards that post upcoming openings.

9 External job search

If employees have to leave the enterprise, they should be prepared both psychologically and technically through counselling and skills assessment/training, and be able to get a job according to their skills, either locally or in another region. Collective search of job ads is normally more efficient if done by a specialized team. In preparing the employee, there are a number of important points to be considered at the very beginning:

- Prepare the employee as described above (psychological support, assessment, job search, etc.)
- Compile and classify job ads (newspapers, magazines, journals, internet, intranet, etc.) every day, prepare an overview in an easy-to-use format.
- Contact other companies to learn about their vacancies. Usually, this is common practice between large enterprises dealing in different domains. Do not limit it to big companies. Hundreds of contacts are generally necessary.
- Contact professional organisations to identify corporate needs.
- Keep in permanent contact with employment organisations (public and private).
- Contact and propose help to recruitment agencies.

Another practice used to encourage employees to be proactive in their job search used by some European multinational corporations is the 'speed bonus' which is a monetary sum awarded to employees that find a job within, let's say, 4 months, in addition to the severance package, with or without the help of the job search unit.

Another trend is to convince an employee to use the facilities and the help of the job search unit, and for that, give a higher severance package if he/she accepts help. This prevents employees from taking the money and waiting until the last moment to look for a new job without any help or advice. Some companies give only the legal amount of severance pay if an employee refuses help, and much more in case that he/she accepts.

Finally, some companies maintain previous work contract until the end of the new job trial period. This helps the employee to feel safe and calm, and helps him/her to test a new job with little risk.

10 Mobility help

Job mobility can be handled mostly through skills assessment and training; however geographical mobility is more complex as it often requires uprooting the entire family, making the emotional and financial impact even more significant.

Geographic mobility could be encouraged through offering attractive relocation packages, which could include:

- Financial resources to cover moving expenses
- Paid days off to conduct the move
- Home search assistance and potentially a housing allowance
- An augmentation of salary to bridge potential gaps in terms of cost of living of the areas
- An orientation to the new area, which would include information on local schools, facilities and other areas of interest of the family. In the case of relocation to an area where the language is different, financial allowances for language training for the entire family could be offered.
- Administrative support in terms of registrations, service installations, understanding local regulations, etc

11 Conclusion

The International Labour Organisation views restructuring as an on-going challenge that needs to be addressed by the mutual efforts of all the social actors.

References

Cascio, W.F. (2002). *Responsible restructuring: Creative and profitable alternatives to layoffs*. San Francisco: Berrett-Koehler.

Rogovsky, N., Ozoux, P., Essex, D., Marpe, T. & Broughton, A. (2005). *Restructuring for corporate success: A socially sensitive approach*. Geneva: ILO.

Health Promotion and Health Initiatives in Restructuring Organisations: Potentials, Barriers and Recommendations

Debora Jeske & Thomas Kieselbach

1 Innovative health initiatives in restructuring organisations

The following chapter outlines innovative approaches to better manage the impact of restructuring on health. The examples of the following four initiatives[1] show how health can become a central issue prior or during restructuring which needs to be addressed by the concerned organisation, consultant or other institutional bodies responsible for managing, supporting, or financing the restructuring. Health initiatives, such as the examples included in this report, allow a company, consultant or other institutional body to directly confront the issue of unemployment and health prior - but also during - the restructuring. We put special emphasis on describing the development of new initiatives so that the process of how new ideas were conceived and implemented can be better understood.

Measures typically used to assess the effectiveness of downsizing from a corporate perspective are clearly inadequate as a means to understand and manage the impact of this process on all stakeholders such as employees and the local community (Shaw & Barrett-Power, 1997). Such measures usually relate to economic performance indicators, as profitability, productivity, investment returns, consumer satisfaction, etc. But the fallout effects for those employees being dismissed and affected by the restructuring produce a new cost factor to be considered at work. Restructuring acquires a new dimension beyond the obvious financial and political connotations – the consequences and implications of restructuring for the health of those being directly or indirectly being affected by this process.

Restructuring changes not just the actual work of the individual, but also the social fabric at work and as a result the culture and organisational climate. The workload increases and thus does the individual stress just as employees are also experiencing an

1 The report is based on some of the results of the European Social Funds project „Monitoring Innovative Restructuring in Europe – MIRE". Coordinator: Frederic Bruggeman, Syndex, France. In addition to MIRE, this report also draws on a secondary analysis of case studies produced by MIRE partners from other European countries. These case studies were further supplemented by additional interviews. More information available at: www.mire-restructuring.eu and in Gazier & Bruggeman (2008).

increasing lack of control and greater role ambiguity (Mohr & Udris, 1996). The reduction of skill discretion and lower job security moreover contribute to the individual stress levels (Kivimäki, Vahtera, Ferrie, Hemingway & Pentti, 2001).

Even if employees are not directly affected by downsizing, the experience nevertheless results in adverse changes of psychosocial factors and thus a great amount of psychological stress. Survivors report role overload, role ambiguity and role conflict (Tombaugh & White, 1990). The reduction of skill discretion and lower job security moreover contribute to the individual stress levels (Kivimäki et al., 2001). Emotions such as feelings of anger, anxiety, guilt, and uncertainty are common (Thornhill & Saunders, 1998). These are furthermore accompanied by increased work stress, perception of job insecurity, lack of job satisfaction and decreased organisational commitment. Other warning signs are increased absence rates, reduced sense of loyalty towards the organisation, risk avoidance, resistance to further change and reduced work commitment. These forms of behaviour often vary according to the stage of the occupational transition a person feels he or she is at, such as the point of preparation, confrontation, adjustment and stabilisation (Nicholson & West, 1988). Modified jobs may also lead to more musculoskeletal problems, particularly in the case of an increased physical workload following workforce reductions (Kivimäki et al., 2001). The result of these demands and work changes can also be observed in lower productivity rates and lead to potentially chronic dissatisfaction, decreased performance and health issues (Kieselbach, 2006; Mackay & Cooper, 1987; Maes, Vingerhoets & van Heck, 1987).

According to Leka, Griffiths and Cox (2003), stressful work is characterised by an excessive demand and pressure at work in association with a potential gap between the demand and the employee's knowledge, abilities and skills to complete the set tasks, particularly in a situation where the employee has neither control over his or her work nor any support from others. Not only increased workload, but also adverse psychosocial environments such as reduced job control and job insecurity can produce more musculoskeletal morbidity after downsizing (Kivimäki et al., 2001). It is not surprising that as the nature of work and the organisations themselves are undergoing change, both mental and emotional risk factors increase (Paoli, 1992, 1997). According to some estimates, 30-50% of absenteeism is caused by psychosocial problems at work (Schreurs, Winnubst & Cooper, 1996). These circumstances in turn further increase time constraints sharply which lead to more performance and time pressures at work. It is not all surprising to note that absenteeism and accident rates as well as unsafe working practices increase in such a situation (Leka, Griffiths & Cox, 2003). Not only increased workload can lead to health problems amongst employees, but also adverse psychosocial environments such as reduced job control and job insecurity can lead to health problems and result in more musculoskeletal morbidity after downsizing (Kivimäki et al., 2001). It is not surprising to hear that particularly likely to have health problems are those employee groups who have only a primary education, do physical work and have a limited social network (Schreurs, Winnubst & Cooper, 1996).

Once the stress and coping resources of an employee are exceeded, chronic fatigue, depression, withdrawal and auto-aggression as well as the adoption of unhealthy behaviours go hand in hand with increased morbidity and mortality (for a brief overview, see Froneberg, 2003; Kieselbach, 2000; Eliason & Storrie, 2003). The increased uncertainty also leads in some cases to vulnerable individuals resuming drug habits or using unprescribed medication which are often readily available (Thiel, 1999). The likelihood that vulnerable employees with former drug addictions relapse seems to be increased by the circumstances usually produced in restructuring, such as the threat of dismissals. In particular, those aspects usually found in restructuring organisations such as high occupational demands, lack of control and recognition of work, as well as little assistance from supervisors, are factors which have consequences for employees vulnerable to addiction (i.e. Polli, 1998).

The unhealthy behavioural changes are often associated with employees who are being dismissed or already entered (potentially long-term or permanent) unemployment. As outlined, the individual's behaviour may change for the worse with an increasing tendency to smoke, to consume alcohol, to having unhealthy sleep patterns and a lack of physical exercise. All these risk factors affect the long-term health of the individual as stipulated above, as the incidence of ill-health related to heart and circulatory systems which are often found to be on the increase amongst the unemployed (Bormann, 1992). If these are the effects of work stress on individuals, it is very realistic to assume that a stressful restructuring will have far-reaching consequences for large swaths of an organisation's workforce. Appelbaum, Lopes, Audet et al. (2003) showed on the example of a downsizing telecommunications company that the effectiveness with which a restructuring programme is managed has significant impacts on the surviving employees' attitudes and behaviour. The authors were able to show that the problems associated with the programme management and downsizing resulted in decreased production, attendance, motivation, emotional health, job satisfaction, and confidence in management. Frequent restructurings increasingly force organisations to acknowledge new causes for stress: uncertainty, employability, an increasingly individualised assumption of responsibility on part of the employee in maintaining and getting new work. Taking this into account, all these stressors are more likely to be reproduced in new combinations and to impact negatively on the workforce when the job descriptions, working groups and entire departments change or disappear in the wake of a restructuring.

The following examples of health promotion approaches in the MIRE project demonstrate how different circumstances can create health problems, but also how organisations can address these with health promotion initiatives. Example of initiatives outlined here include: rehabilitation initiatives to improve the reintegration chances of sick employees; group workshops to increase health awareness so as to reduce stress and (restructuring related) ill-health following dismissal; social support and counselling in redeployment situations; and staff health monitoring and internal health initiatives to reduce stress and increase health awareness following restructuring; as well as new anticipatory methods, such as accessible health and stress tools for employees to monitor health while giving the company better means to monitor stress during normal operations, but also reorganisations. Even though innovation always depends on the national and organisational context these examples of how companies can address health in various ways show how to become aware, potentially improve and monitor the health status of the workforce. This first step may also lead to a wider, more appropriate concept of health including the responsibility of all social actors within a country, including employers.

1.1 Health rehabilitation during the notice period

Like most telecommunications companies, Ericsson in Sweden started to develop and implement restructuring programmes in order to refocus its business activities in the mid-1990s to adjust to the increasing competition and technology developments[2] (see also Armgarth, in this volume). In 2001 the company made the decision to reduce the workforce by 12.000 staff of the 40.000 Swedish employees[3]. In return for the concession used when selecting employees for dismissals, unions negotiated various options for the employees, including the extensive re-employment package in form of a twelve-month career change programme called "Forum for the Future". The aim of this new initiative was to find all employees new solutions: a new job, assist employees with starting a new business, new studies and education. A total of 9.500 employees chose to participate in this career change programme. The programme was supported by the Swedish Job Security Foundation and a number of consultancies such as Manpower, Right and Antenn. Dismissals were taking place every six months and each dismissal wave marked the beginning of a new 'Forum'.

The administrators, managers as well as the career coaches soon became aware of the number of unexpected health problems amongst the participants, which meant that career coaching alone was no help to the participants. The fact that all employees entered the programme without their personnel files with their health history having been forwarded to the Human Resources (HR) manager of the Forum represented an unexpected problem for the consultants and the administration. They found there were significant differences between the various subgroups, some of them reacting very strongly and visibly to the challenges of impending unemployment, whereas others withdrew. A careful estimate from the programme manager suggested that about 10% of the participants had a longer health history with physical as well as mental health problems (i.e. alcoholism, depression). Another 10-15% of the people were not performing well in interviews, were not coping with the new situation or participating at all.

The problems these employees were struggling with included gambling, long-term sickness absences from work, but also alcoholism and drugs, physical and other disabilities (i.e. dyslexia, limited language skills, blindness and deafness), but also cultural discrepancies. Some employees lacked the social skills and competences necessary for managing interviews and conversations, suffered from personality disorders, and self-esteem problems. Therefore another approach had to be found to gain a better overview of the health and additional needs of the employees.

A number of new experts were brought in to assist – the Prästbyrån organisation which consists of priests and also included qualified therapists and psychologists, to provide psychological counselling to assist with the rehabilitation and health status assessment. The Prästbyrån consultants had already been involved in the pilot where they took over the supervision and support of those participants with health issues. The manager of Prästbyrån conducted interviews to identify general themes amongst those

2 Case study reference: Diedrich, A. & Bergström, O. (2005). The process of restructuring at Ericsson. School of Business, Economics, and Law, Göteborg University and Institute of Management of Innovation and Technology (IMIT).

3 Employees had four options available to them: the career change programme, early retirement, severance pay or to accept the ordinary notice to quit. For more details please see Armgarth & Hvarfner (2009).

employees and was responsible for identifying and organizing support for those participants who would benefit from professional help. He also trained the general Forum manager to help her and through her the rest of the staff to manage the difficult interviews and situations with the employees. As a result of the experience and first trials using counsellors, certain procedures were introduced to better monitor the situation, access employee information in advance and gather feedback from participants. Attendance was regularly monitored to make sure that all participants were cooperating within the programme.

The initial success rate (new solutions) envisaged for all 9.500 participants of the career programme was set at 80%. About 7.500 participants were successful in finding a new solution, such as starting a new job, starting a new business, or deciding to continue education. However, this initial target had to be corrected downward due to the observed health and social competence issues mentioned above. Ericsson released no data about their employees' health statistics. However, some estimates from the programme's HR manager suggest that overall 10-15% of the concerned employees benefited from rehabilitation services (about 900-1.400 people of the total of 9.500 people) between 1998 and 2005. The priest at the site of the Forum reported that 500-1.000 employees visited her office over the course of the years. Ericsson also requested additional statistics from the various consultancies for about 1.200 employees who were supported by their consultants.

The consultancy Right released some results for about 15% of Ericsson's employees which were supervised by Right. The consultancy reported that about 25% went off to work elsewhere, whereas about 15% left the programme to start studying or to join further long-term education programmes. 20% remained on continuous sick leave due to cancer, multiple sclerosis or similar illnesses. 10% went to the public employment service's rehabilitation provisions. For the remaining 31%, no information was available. The consultancy Antenn reported slightly different results for their participants. They stated that amongst those, 55% returned to work, 25% went on to study, 10% remained on sick leave, and 10% went to the public employment service's rehabilitation provisions. It needs to be remembered, however, that the dissimilar results are due to the special characteristics of the various Ericsson business units as well as the allocation of employees to different consultancies according to their needs. Of the 400 participants supported by the consultancy Manpower, around 10% were on continued sick leave or received disability pensions whereas the large majority were successful in finding new employment (Armgarth & Hvarfner, 2009).

Including provisions to rehabilitate employees was the end-product of a variety of problem-solving approaches, with the help of the Prästbyrån network, the company came up with a tool that allows them to actively participate in the way that employee health issues are dealt with by including a referral system into their re-employment programme. With their help, the company found a way to assist those employees in need of specialist help rather than referring them on to the general Swedish health agencies where this assistance might not be as quickly available. The programme was thus a learning process for the employees and for the management board which realised the complexity of health issues faced by its workforce. Since the restructuring, the company has introduced a certain protocol and procedure for dismissals on how to communicate and support employees upon their dismissal (see Armgarth, 2009).

The introduction of resource coaches to aid rehabilitation also resulted in more permanent, long-term successes. The cooperation between various consultancies led to the development of a new consultancy agency which focuses particularly on health concerns in restructuring organisations. The consultancy Manpower decided to form subsidiaries for special services, creating Empower in 2001 to handle career coaching on a bigger scale. Empower merged with the company Right Management in 2003, thus leading to the founding of a new company called Hälsopartner that provides specialised health consulting services to companies who are restructuring.

1.2 Redeployment challenges and support mechanisms

Like most state companies which were privatised in the 1980s and 1990s, the Swedish telecommunications network Televerket, later known as Telia, underwent significant changes in the last 15 years[4]. Following major restructuring, dismissals and redeployment in the 1990s, Telia and the Finish company Sonera (formerly known as Telecom Finland) merged operations in 2002 and became TeliaSonera in 2003. In 1995, the redeployment programme, which later became known as Division P programme, was set up by the management to reduce the workforce by up to 5.000 employees over a period of three years without redundancy. The work of the redeployment unit between 1996-1998 first highlighted the need to address health within the organisation. In total, 23.000 people were involved in this first, company-wide redeployment initiative in 1996. The purpose of the programme was to give employees sufficient time to apply and find new work (outside or potentially also in Telia), to consider other solutions such as studies or start off a new business. In addition, the new division had to take care of existing expertise and protect future know-how within the company, develop employees in new areas of work and towards future professions, distribute work assignments and efficiently deploy staff within the Telia Business Group.

An important characteristic of the programme relates to the working conditions of all participants. The unit had eliminated many aspects commonly shared by other temporary programmes: staff no longer had any real work, any working area such as a desk. They shared IT resources and were expected to come to the redeployment unit every day just like a normal employee. In order to manage the redeployment effectively, the programme relied on the sole expertise of its trainers and HR managers who had entered the programme just like their general staff. No external consultants were employed to support internal staff. The only provision made related to services being offered and accepted by the company's health insurance agency and external providers such as job brokers and job placement agencies.

The advantage of remaining employed when applying for new jobs could not stop motivation to decline rapidly without any meaningful work. The link between job position and status was for many an important aspect of their identity. The effects quickly became noticeable: some staff as well as managers were increasingly stressed and found it difficult to cope with this situation. The impact that the redeployment situation had on

4 Case study reference: Diedrich, A. & Bergström, O. (2006). Developing Restructuring Practice – Workforce Reduction at a Large Swedish Telecommunications Company. School of Business, Economics, and Law, Göteborg University and Institute of Management of Innovation and Technology (IMIT).

some of the staff also led them to question their life's work, in some cases they returned to drugs which was totally unanticipated. It quickly became apparent during the first few months that the designers of the programme had underestimated some of the issues that arose for all staff. The managers needed special skills to manage the challenging environment. A number of provisions were made as a result: TeliaSonera introduced provisions so that all staff nation-wide could contact counsellors by phone or in their offices. In addition, special training and a mentorship scheme were organised for the managers to help them manage the challenges of supervising people in such difficult situations.

During the three years of the Division P programme (1996-1998) solutions could be found for a total of 6.500 employees and only 2% of 6.500 staff had to be dismissed because no solutions had been found. Overall, since 1996 about 7.000 staff left the company Telia and its predecessor and a further 11.000 individuals were outsourced. The programme thus successfully exceeded all expectations from 1995. However, similar to Ericsson, the managers at Division P found that a small number of employees relied solely on the company to find them work and remained passive throughout the entire length of the programme. In this case, the long-term nature of the programme did not work to their advantage. Since the psychological counselling and also the health care were all externally provided, there are no data available regarding the use of counselling provisions. However, careful estimates suggest that about 5% of staff in the redeployment unit used the offer to contact the counsellors at the health insurance agency.

The experiences gained during the first redeployment period in 1996 to 1998 had several positive repercussions for the application of such programmes and the general managerial approach at TeliaSonera in terms of the importance of health provisions in the case of future restructurings. The know-how also played an important role in the later conception of redeployment services, particularly in terms of the number of counsellors and managers needed to support the large number of employees.

1.3 Staff health monitoring and online health tools

British Telecom was repeatedly restructured in the past fifteen years, similar to all public companies being privatized in the UK[5]. In the course of these organisational changes, employees were redeployed and many locations closed, moved and changed. BT had continually monitored its sickness absence rates and was aware that around 20 per cent were due to mental health issues. In response to these data, BT introduced two different tools to monitor and improve health at work, one of which is STREAM – an online stress tool, and Work Fit – a programme to increase the physical health of employees. The launch also related to the UK's statutory body, the Health and Safety Executive (HSE) and its initiative to improve mental health issues in the workplace (see Wallington, Jefferys & Moore, 2009).

STREAM was launched in 2004 and aims at identifying and subsequently addressing stress throughout the workforce. This tool allows individual employees voluntarily and confidentially to report stress scores which are then summarized and analysed

---

5 Case study reference: Moore, S. (2006). BT – A case study of health initiatives in the context of continual restructuring. Working Lives Research Institute, London Metropolitan University.

for the entire workplace pool. The online questionnaire has been developed by a clinical psychiatrist and evaluated in workshops. STREAM picks up on "excessive or intolerable pressure leading to physical or psychological effects on the human body". The individual employee also benefits by receiving a colour-coded assessment (going from red to amber or green) and suggestions on how to lower stress levels. In addition to the employee receiving this feedback, the system also forward the confidential report to the line manager (the report will only be released to the employee and his or her managers). In case of high stress result, the employee will be offered a chance to discuss the concerns on a one-to-one basis with the manager so as to identify possible solutions. If the employee prefers, he or she may also have this discussion with another immediate supervisor, such as the second line manager - recognising that the line-manager could be the issue.

STREAM may identify a number of problems and available solutions could include workload, childcare provision, work scheduling within flexible working policies, or debt counselling. Furthermore, the employee may also opt to talk to a counsellor in the company's own Employee Assistance Programme, contact the free confidential telephone support service or request a free face-to-face counselling meeting (subcontracted to a counselling service through a third party). Managers are also able to request assistance to accurately address issues that their employees may have raised in their reports. According to a union representative, 20-25% of the workforce had used the tool by summer 2006. The tool also includes a 'Mental Health dashboard', which monitors sick days related to mental health. It also lists referral information for the occupational health service and the outcomes of the STREAM process.

The second programme, Work Fit, addresses the physical aspects of health, particularly obesity, high blood pressure and diabetes. This voluntary, sixteen-week programme has been developed in conjunction with a number of public and private organisations interested in health and focuses on promoting a healthy diet and exercise. The problems of the increasingly ageing workforce are also considered as these employees are likely to be less physically active and thus more susceptible to cardiovascular diseases. The programme is delivered almost entirely over the BT intranet and by e-mail. Work Fit sets participants weekly tasks to help them to eat healthier and become more physically active, often resulting in weight loss as well. Participants also collaborate and join teams. This enables groups to compete with one another. This further increases their motivation to participate and, for example, enables them to raise money for charity in the process. Work Fit is also free and confidential.

The STREAM results can be aggregated to monitor health issues more generally without identifying individuals. Such assessments are supported by the union as well as the Health and Safety Committee, particularly given the stressful working environments which some of the employees at BT – such as their call centre employees – face on a daily basis. Although STREAM is not used in the context of restructuring, higher stress ratings have been found during company unit reorganisations. The reports available via the STREAM tool allows the management to regularly address problems on the shop floor and to potentially intervene or assure staff in times of increased uncertainty. Just like STREAM, Work Fit enables employees to regularly monitor their own, but in this case also their team's progress. At the end of the programme, employees are invited by the company to celebrate their achievement. The participation in this programme has so far exceeded all expectations. Whereas only 5.000 employees were estimated to sign up, at total of 16.500 employees registered in the first twelve months of the programme with a fifth (3.500) completing the programme successfully by summer 2006.

In addition to these two projects, BT has also invested into two new health initiatives: 'Positive Mentality', another 16-week campaign which encourages employees to take better care of their mental health by focussing on the maintenance of mental health and a smoking cessation program which tries to help employees to reduce and potentially quit smoking altogether (following the company wide smoking ban). This last campaign is organised in cooperation with the local primary care trusts and further supported by a company agreement which allows employees to take time-off during work hours to attend counselling on giving up smoking. It is expected that these initiatives will enable the company to monitor health during reorganisations and to assist their employees in a more insecure, competitive and volatile working environment. This means, rather than relying on problem-solving approaches alone, these new initiatives help to address the repercussions of heightened job insecurity and recurring company restructuring for the workforce, which, in turn, have led to more preventive and proactive health initiatives being employed.

2 General trends and barriers identified in case studies

The organisations were motivated to invest into such health initiatives either because they had to address problems encountered during restructuring (the most common reason), or because they had gained greater insight into the health costs following restructuring. The forms and extent of health promotion initiatives seems to vary across organisations. Whereas some focus on seminars and group meetings alone, others included access and encouraged the participation in confidential psychological counselling (i.e. "Resource coaches" in the ‚Forum for the Future' programme set up by Ericsson) combined with ecclesiastical and medical counselling (i.e. addiction counselling and publication of guidelines for employees as in the case of Ericsson). In some cases, the organisations also tried to implement long-term health related provisions in form of professional and qualified contact persons for staff (i.e. psychological counsellors at TeliaSonera; confidential telephone help line and access to counsellors at BT). So whereas some approaches started from a basic level - disseminating health information and improving health understanding (behaviour in relation to nutrition, sport, sleep, addiction), other companies went beyond this and tried to change or influence employee behaviour. These initiatives aimed to improve individual management of situations and employees' coping skills, help them to objectively assess and improve their labour market chances and opportunities, and – hand in hand with most general transfer measures – supported employees to analyse and improve upon their skills and qualification gaps.

Whereas the first form of support in terms of information is the easiest to install, the second does require the expertise of appropriate consultants and managers – as well as the appropriate budget to support such often long-term initiatives. However, there is also a third level which can be classified as the most challenging for most managers: helping employees to stay motivated, giving them courage and confidence, and supporting them emotionally during their transition despite the threat or reality of unemployment. This means that health initiatives need the commitment of the organisation and its supervisory staff and should not be a means to avoid conflicts and thus abdicating responsibility for health by making the individual employee alone responsible for maintaining physical and mental health at work. Particularly important here seems to be the awareness of many social actors to what extent they should and are able to support health initiatives during restructuring financially as well as organisationally.

There are first signs that health initiatives are increasingly successful and sustainable. One such an example includes the setting up of new health observatories and the foundation of new service providers to provide specific health services during restructuring (i.e. Hälsøpartner consultancy following Ericsson initiative). First indications also reveal that some health initiatives have improved longevity (increasing popularity of BT's sixteen-week programme Work Fit to improve health). In addition, consultants increasingly realize the need to recruit psychological counsellors for transfer agencies and to offer these services in combination with outplacement transfer services, and continuous professional development. These first signs promise success, however, the reality still is that the majority of innovative initiatives are still passive, problem-oriented approaches which are initiated due to higher sickness rates, longer absences, and conflict situations during restructuring.

Despite the increasing attention paid to health at work and during unemployment, health initiatives during restructuring are still very rare. The cases presented are still the exception rather than the norm. The barriers and hurdles preventing innovative measures to take hold are diverse. On the one hand, the legal framework and the health insurance systems for the unemployed simply do not address employability and its relationship to health. Furthermore, whereas some health insurance funds will retain unemployed members, employees in other European countries or also the US will not have the same insurance access while being unemployed (see also Prüßmann, this issue). Depending on the professional sector and the relationship between companies and health insurance funds, there is certainly a need for some insurance funds to consider their services to companies and their role as information source and health promoter. Another difference exists between legal frameworks for disabled workers. There are significant differences between countries and companies to what extent they make it easier or more difficult for disabled or health-impaired employees to register for early retirement.

The positive examples do not reflect the usually limited view which is shared by many managers: restructuring is not part of an organisation's development but conceived of as an interfering process to be completed as quickly as possible. This negative association prevents many beneficial long-term projects or their employment beyond the active restructuring - that is dismissal - stage. This situation is further complicated by HR policies that do not include provisions for sudden or repeated staff cuts, managing the dismissal process, often ignoring health altogether. Without procedures to monitor and address health company-wide as an organisational responsibility, ad hoc conflict management is the likely result. This is often compounded by management's insufficient knowledge and information about the needs, background and health status of those employed and to be dismissed (as aptly demonstrated in the case of Ericsson and TeliaSonera which were both partly unaware of the problem until the restructuring process had started). Furthermore, there seems to be a common lack of knowledge amongst organisational representatives which public and private bodies and consultancies can be contacted for information, funding, and expertise exceeding – usually simply physical - health and safety risks at work.

Just as this emphasis on physical health is pre-eminent still in most organisations, the concept 'employability' is often a difficult concept for Human Resource departments to tackle. It is here that social key players such as the Public Employment Service would be well advised to offer training and help to organisations to address this issue themselves. The lack of provisions in terms of health and employability are compounded even more by the following circumstances: Since health initiatives are often not considered a valuable investment of time by most organisations, individuals who are about to leave or those who are not in employment will generally have even less access to health advice and general health promotion initiatives – and, as pointed out above, measures to increase their employability in terms of health promotion. Many counselling offers at job centres do not address health and many companies are very defensive about making health an issue at work, since many companies fear that these initiatives will increase employee concerns to whether their health will be an assessment criterion for their job performance and future job cuts. Therefore, initiatives are rare due to a lack of infrastructure as well as misunderstandings on both sides of employers and employees.

At present, many exemplary pilot projects are suffering from lack of expertise in terms of the systematic evaluation of these projects and dissemination and implementation complicated by diminishing participant numbers and social conditions (job placement successes; the participation due to shift work, mobility, family obligations, motivation etc). Only if innovations and approaches are known, misunderstandings are put aside, and social actors lead a real dialogue will health initiatives become more commonplace and accepted.

An important prerequisite relates to restructuring and health promotion. Both need to be integrated into the organisational development plan of a company: requiring organisations to address potential production and employee issues into their strategic business plans to allow more flexibility in times of change restructuring may trigger already existing health problems or may lead to chronification of previous health problems. Organisational instability and change also have an effect on employee trust, and perception of justice. Employers assume that employees will be able to manage the process of occupational transition themselves. Health promotion at the worksite before downsizing or dismissal creates positive sustainability which enables employees and organisations to better cope with change by developing new employment perspectives. Despite the potential benefits, most Occupational Health Services do not include preventive health promotion initiatives. Whereas the lack of specific knowledge regarding the effects of restructuring on the dismissed and the surviving employees are explanatory factors, the lack of interest on the side of many social actors – public and private – is certainly also to blame for the current situation. Incorporating employee concerns into the organisational development, particularly focusing on sustainability of expertise and production efficiency by addressing stress and employability, increase the organisation's as well as employees' options in cases of restructuring.

3 Recommendations

The following factors have thus been identified as essential environmental requirements to encourage organisations to employ innovative health initiatives at work: Organisations, particularly management, need to actively engage with the debate on health and participate in pioneering activities. Passive information provision is not leading to health gains without role models and active encouragement. A second factor concerns the lack or reluctance regarding the funding of such initiatives. Legal frameworks can only go so far, but organisational policies and initiatives (such as the preparation of managers in Ericsson prior to the arrival of employees in the 'Forum') need to be in place to actively consider which forms of initiatives and which groups would benefit most from their implementation. A third concern regards the goals and inadequate description of success criteria for health initiatives, which in turn lead to no or a very flawed evaluation of the results. Quite often, the understanding and concept of what constitutes health and illness varies and carries hidden messages that employees get classified according to health, social class, and behaviour. These unclear and mistaken notions impair critical analyses and can lead to data misinterpretations. In addition, the majority of case studies in this MIRE project introduced health initiatives to solve a problem – thus focusing less on the proactive prevention of health issues that may arise in the future. Another concern touches on the lack of expertise among public and private service providers able who may be willing to provide assistance to organisations, but who are not knowledgeable enough to implement as well as evaluate pilot projects.

The following recommendation can be made based on the case study analyses, expert interviews, and health workshops during the MIRE project to increase the acceptability and availability of health initiatives at work and primarily, their application during restructuring.

Modifying current practices in organisations represents one first step towards integrating more health initiatives in the full process of restructuring. This will require multi-stakeholder approaches such as public and private bodies (i.e. Public Employment Service, OHS, professional associations) in the wider environment, and the responsible parties for health and safety, HRM, and organisational development within the company. However, employees must also be able to voice their concerns, thus becoming empowered participants in the social dialogue regarding health at work. Accessibility of health benefits and initiatives need to be increased during working hours, just as SMEs need to have better access to expertise and opportunities to participate. A further step relates to official definition of health at work which needs to encompass not just health and safety, workplace accident management, or occupational disease diagnostics. A number of new pioneers as well as existing actors could also play a crucial support role for managers responsible for managing the restructuring, provide training to vulnerable employees (i.e. those with previous health problems), and give access to information about health promotion to small and medium sized organisations. These actors – particularly the much needed pioneers - should be supported on a national as well as European level. Fostering local initiatives in association with local commerce associations and others is another step forward towards modifying organisational practices.

Certification and standardisation of currently available health-related training programmes are another step beyond simply modifying organisational practice and encouraging existing and new actors to implement, support and promote health initiatives.

Health promotion tools need to be modified in order to be more compatible with the requirements of the changing organisational environment, thus exceeding the traditional risk approach. In addition, certificates of excellence could further smooth the introduction of health initiatives and increase health awareness in organisations. The participation of works councils and unions in the social dialogue is a positive step forward in terms of the employee trust, health and sense of justice in uncertain times: their commitment to the agreed goals of the restructuring process leads to smoother transitions.

Legislation is closely tied to standardisation and certification. For example, the currently existing and very limited criteria regarding the involvement of company physicians at work need to be revised to go well beyond annual visits to inspect hazardous working areas. Company physicians should – in addition to the health and safety representatives – get involved into risk assessments and health initiatives. Another proactive step forward would be new legislation that required all healthcare providers (national services as well as company health insurance funds) to provide annual statistics to enable companies as well as national bodies to monitor health more closely. In addition, changing current health regulation to include a clause that requires these actors to introduce, fund, and monitor preventive health concepts as part of their services would furthermore redefine and redistribute the responsibility for prevention and health promotion beyond the predominantly clinical approach limited to physical ill-health at work.

In order for the above recommendations to be put in place, new research efforts must be focused on assessing the health impact of restructuring, the importance of employability, and the introduction of new policy recommendations beyond those outlined above. Health must become a central aspect of employment as well as of Corporate Social Responsibility. In addition, health should be put on the agenda of international bodies and associations such as internationally operating works councils and publicly funded company initiatives so as to furthermore encourage a more positive public discourse and a more specific focus on improving HRM practices (in terms of reducing negative stressors, maintaining or broadening support systems, and equally taking on the concerns of the survivors-of-layoffs).

4 Concluding remarks

The above case studies, discussion of trends and barriers and recommendations reveal the need for further work and persistence in regard to promoting health at work, particularly in the case of restructuring. The vicious cycle of restructuring is resulting in a counterproductive loss of productivity (evidence which is supported by the ILO's programme SSER - "Socially Sensitive Enterprise Restructuring", see Rogovsky, Ozoux, Esser, Marpe & Broughton, 2005). The ILO results and the MIRE case studies[6] confirm that indiscriminate self-interest of companies to restructure are far more costly than some of the potential gains. Companies restructure in order to become more competitive, productive and efficient. Organisations must get away from actual crisis management.

6 Overall project results can be found in Gazier and Bruggeman (2008). The country-specific results for Germany are available in Kieselbach, Knuth, Jeske and Mühge (2009).

Restructuring has been shown to trigger already existing health problems or may lead to chronic health problems. At the same time, restructuring may also reveal underlying health problems, and even lead to a potential increase of ill-health in the future by increasing job insecurity and generally uncertainty over a longer period of time resulting in higher rates of depression, work-related burnout, a deterioration of concentration and performance after the restructuring officially concluded. Health is a competition factor just like the location of a production site. Adapting existing good health practice and adapting them to address the restructuring-related increase in ill-health is not sufficient: worksite health promotion needs to become a fixed component of employee relations and adequate HR Management.

References

Armgarth, E. (2009). Human resources protocol on restructuring. In T. Kieselbach et al. *Health in restructuring: Innovative approaches and policy recommendations* (pp. 187-191). München: Hampp.

Armgarth, E. & Hvarfner, A. (2009). Restructuring and individual health: Ericsson and Manpower Health Partner/Sweden. In T. Kieselbach et al. (Eds.), *Health in restructuring: Innovative approaches and policy recommendations* (pp. 180-186). München: Hampp.

Appelbaum, S.H., Lopes, R., Audet, L., Steed, A., Jacob, M., Augustinas, T. & Manopoulos, D. (2003). Communication during downsizing of a telecommunications company. *Corporate Communications: An International Journal, 8*(2), 73-96.

Bormann, C. (1992). Arbeitslosigkeit und Gesundheit. Empirische Analysen auf der Basis der Daten aus dem 1. Nationalen Gesundheitssurvey der Bundesrepublik Deutschland aus den Jahren 1984 bis 1986 [Unemployment and Health. Empirical analysis based on the first national health survey of the Federal Republic of Germany between 1984-1986]. *Sozialer Fortschritt, 41*(3), 63-66.

Eliason, M. & Storrie, D. (2003). *Displaced workers and mortality. A fourteen year follow-up of all plant closures in Sweden 1987 and 1988 with a matched control group.* Unpublished manuscript, 23 April 2003.

Froneberg, B. (2003). Psychological stress and well-being at work. *Asian-Pacific Newsletter on Occupational Health and Safety, 10,* 28-30.

Gazier, B. & Bruggeman, F. (Eds.). (2008). *Innovative Restructuring in Europe.* Cheltenham: Edgar Elgar Publishers.

Kieselbach, T. (Ed.). (2000). *Youth unemployment and health. A comparison of six European countries. Psychologie sozialer Ungleichheit,* Vol. 9. Opladen: Leske + Budrich.

Kieselbach, T. (Ed.). (2006). *Social convoy in enterprise restructuring in Europe. Concepts, instruments and views of social actors in Europe.* München: Hampp.

Kieselbach, T., Knuth, M., Jeske, D. & Mühge, G. (2009). *Innovative Restrukturierung von Unternehmen: Fallstudien und Analysen [Innovative restructuring of organisations: Case studies and analyses].* München: Hampp.

Kivimäki, M., Vahtera, J., Ferrie, J.E., Hemingway, H. & Pentti, J. (2001). Organisational downsizing and musculoskeletal problems in employees: A prospective study. *Occupational and Environmental Medicine, 58,* 811-817.

Leka, S., Griffiths, A., & Cox, T. (2003). *Work organisation & stress. Systematic problem approaches for employers, managers and trade union representatives.* Protecting Workers' Health, Series No. 3. Geneva: World Health Organisation.

Mackay, C.J. & Cooper, C.L. (1987) Occupational stress and health: Some current issues. In M.J. Schabracq, J.A.M. Winnubst & C.L. Cooper (Eds.), *Handbook of work and health psychology.* New York: Wiley.

Maes, S., Vingerhoets, S. & van Heck, G. (1987). The study of stress and disease: Some developments and requirements. In M.J. Schabracq, J.A.M. Winnubst & C.L. Cooper (Eds.), *Handbook of work and health psychology*. New York: Wiley.

Mohr, G. & Udris, I. (1996). Gesundheit und Gesundheitsförderung in der Arbeitswelt. [Health and health promotion in working life]. In R. Schwarzer (Ed.), *Gesundheitspsychologie [Health Psychology]* (2nd ed.). Göttingen: Hogrefe.

Nicholson, N. & West, M.A. (1988). *Managerial job change: Men and women in transition.* Cambridge: Cambridge University Press.

Paoli, P. (1992). First European survey on the work environment 1991-1992. In M. Kompier & C. Cooper (Eds.), *Preventing stress, improving productivity. European case studies in the workplace*. London & New York: Routledge.

Paoli, P. (1997). Second European survey on working conditions 1996. In M. Kompier & C. Cooper (Ed.) *Preventing stress, improving productivity. European case studies in the workplace.* London & New York: Routledge.

Polli, E. (1988). Betriebliche Sekundäranalyse [Company-based secondary analysis]. In C.G. Hoyos & D. Frey (Eds.), *Arbeits- und Organisationspsychologie. Ein Lehrbuch.* Weinheim: Psychologieverlag Union/Beltz.

Rogovsky, N., Ozoux, P., Esser, D., Marpe, T. & Broughton, A. (2005). *Restructuring for corporate success. A socially sensitive approach.* Geneva: International Labour Organisation.

Schreurs, P.J.G., Winnubst, J.A.M. & Cooper, C.L. (1996). Workplace health programmes. In M.J. Schabracq, J.A.M. Winnubst & C.L. Cooper (Eds.), *Handbook of work and health psychology.* New York: Wiley.

Shaw, J.B. & Barrett-Power, E. (1997). A conceptual framework for assessing organisation, work groups and individual effectiveness during and after downsizing. *Human Relations, 50*(2), 109-127.

Thiel, B. (1999). Alkohol, Medikamente und Drogen am Arbeitsplatz. [Alcohol, medication and drugs in the workplace]. In C.G. Hoyos & D. Frey (Eds.), *Arbeits- und Organisationspsychologie. Ein Lehrbuch.* Weinheim: Psychologie Verlags Union/Beltz.

Thornhill, A. & Saunders, M.N.K. (1998). The meanings, consequences and implications of the management of downsizing and redundancy: A review. *Personnel Review, 27*(4), 271-295.

Tombaugh, J.R. & White, L.P. (1990). Downsizing: An empirical assessment of survivors' perceptions in a post layoff environment. *Organisation Development Journal,* Summer, 32-43.

Wallington, D., Jefferys, S. & Moore, S. (2009). Health policy in BT under continuous restructuring. In T. Kieselbach et al. *Health in restructuring: Innovative approaches and policy recommendations* (pp. 154-160). München: Hampp.

Restructuring and Individual Health: The ERICSSON Case of Restructuring

Elisabeth Armgarth

Introduction

My name is Elisabeth Armgarth. I am on early retirement since January 2006 from the telecommunication company Ericsson. Ericsson is a world-leading provider of telecommunication equipment and related services for both mobile and fixed network operators. The parent company is Telefonaktiebolaget LM Ericsson and dates back to 1876. The headquarters are located in Stockholm, Sweden. Over 1,000 networks in 175 countries operate with Ericsson network equipment and 40 percent of all mobile calls are made through its systems. Before I left the company, I was Program Manager for the Swedish part of the restructuring from 2001 until the end of 2005.

The following chapter is a summary describing how past restructuring experience was used in subsequent restructuring approaches. The description is divided into two parts: the first part describes the events between period 2001 and 2005 (before I left the company) and the second part is a follow-up from March 2008.

1 Part I: Do we ever learn? Modelling the restructuring

1.1 Background

In the period 2001 to 2005, Ericsson reduced the number of employees from 107,000 to 48,000 worldwide. This personnel reduction was partly due to the reduced orders from telecom operators as well as the technology shift. In 2001, Ericsson had more than 40,000 employees in Sweden alone. From 2001 to 2005, we reduced the number of employees to around 22,000. Of the original 40,000 employees, 10,000 individuals were outsourced to other companies (mainly hardware manufacturing, but also IT support/development, health care, etc). This meant that around 12,000 Ericsson employees received their dismissal notice. The dismissal process proceeded in waves, two to three notice periods every year with about 1,000 people given notice each time.

Ericsson offered an extensive support package to the people given notice and they were given three choices:

- To participate in a career change program (up to 12 months with salary),

- To select early retirement (from 58 years of age) or
- To choose a severance payment (up to 11 months' salary).

The contents of the support package and its details were negotiated with the labour unions (see below) before every new notice period. During the notice period, local representatives from Ericsson and the unions were also negotiating which employees had to leave the company and who could stay on. However, this aspect of the restructuring process is not dealt with in this report.

Close to 10,000 people entered the career change programs. These programs were managed by six external companies: Right Management Consultants (former Empower and Manpower), Antenn (former Proffice), TRR (Trygghetsrådet), Agora, Hudson and AS/3. We formed a steering group for each of these companies. Steering group members included representatives from the external company, representatives from the unions, me as the program manager and an assistant from Ericsson. There were four labour unions involved in the negotiations and the steering group meetings: SIF (today Unionen), the Swedish Union of Clerical and Technical Employees in Industry; CF (today Sveriges Ingenjörer), the Swedish Association of Graduate Engineers; Ledarna, the Swedish Service Organisation for Leaders; and IF Metall, the Swedish Metalworkers' Union (for further details regarding the restructuring at Ericsson, please see Diedrich and Bergström, 2006). The union representatives were often the same individuals who had participated in the original negotiations of the support package. The steering groups met on a regular basis, typically every six weeks.

The purpose of the steering group meetings was to assess what kind of progress people made in terms of obtaining new employment, returning to education, or starting their own businesses (these three options were called New Solutions). If the progress was too slow, the steering group took actions to ensure progress was made by considering new measures for either a group of people or for specific individuals. At the end of each career change program, more than 90% of the people reported that they were satisfied or very satisfied with the assistance provided by the external companies. Approximately 80% found new employment, started their own businesses or continued with long term studies.

1.2 What to do with gained experience?

In line with the personnel reduction in Sweden, Ericsson decided to reduce the human resource (HR) staff by half by the end of 2003. The company offered HR staff the option of voluntary resignations plus a support package. I took the offer and applied for early retirement.

At this point, many of those individuals (i.e., Ericsson and the union representatives) involved in the prior restructuring initiatives realized that in order to preserve the knowledge gained over the years, some documentation was needed. We at Ericsson and the union representatives felt we had done something good for a large number of people during the difficult notice periods. As a matter of fact, a couple of times every year since then, union representatives continued to ask me to admit an employee to the career change program (although he or she had not been made redundant) in order to help that person find a new, more suitable work situation outside Ericsson. In response to these considerations, an agreement was drawn up between Ericsson and the unions in which I was appointed to the task to document the "good parts" from five years of restructuring work.

After several interviews with HR managers, line managers and union representatives, I learnt that there were different problems to be solved. Being an engineer by training I followed the principles of good design (Nam P. Suh, 1990) and proposed one distinct solution for every problem to Ericsson and the unions. It was rather tricky to label the different solutions, since every person interprets a situation differently depending on his or her own experience. In order to reduce misunderstandings, I decided to call the solutions Model 1 (M1), Model 2 (M2), etc. We defined seven models in total. The four unions then appointed representatives to participate in a working group to develop the proposed models. The work started in autumn of 2004. Not all models were of such nature that they had to be negotiated with the unions. In those cases, the models were merely presented and comments and/or clarifications were added.

M1 – Restructuring by Notice to quit

This model contains seniority principles along with support packages. All notices given between 2001 and 2005 (as described above) fall into this category.

M2 – Competence Shift

In 2003, a competence shift program was introduced within the subsidiary Ericsson Microwave System which affected many jobs. A total of 450 people (out of 2000) left the subsidiary voluntarily with a support package. An important component of this program was the introduction of mandatory training to develop communication skills for the managers at this subsidiary. This step was taken to ensure that the people leaving the subsidiary did that on their own free will. The managerial training course was later further developed and incorporated into Model 4 (see Part II).

M3 – Partial Restructuring

Model 3 was agreed by the working group in the spring of 2005, a smaller version of Model 1. So far, this model has not been used. Ericsson has been able to use Model 7 instead which is based on voluntary principles.

M4 – Individual Career Change

While running the large career change programs during 2001-2005, we found that roughly 10-15% of all people entering the programs were, for various reasons, not ready to look for or enter new employment. The majority of these people were not on any sick leave. Some had difficulties coping following the dismissal notice but many of these individuals had been underperforming for years. The goal was to find satisfactory New Solutions for 80% of all program participants. With so many people not being ready to take on new employment, it was impossible for me to reach this target. Thus, after discussions of the steering groups with Right and Antenn, the consultancies recruited so-called resource coaches with special skills in psychology and therapy – who also had experience in the Swedish medical service system. This was a very successful solution and is further described in Kieselbach and Jeske (2008).

The use of resource coaches represents a core feature of Model 4. This model was already introduced in late 2004 on an individual case-by-case basis. It is suited to assist those individuals who, for some reason, are not functioning properly in his or her work. The model requires an individual agreement and can be used without the consent of the unions. The official agreement with the unions regarding this model was reached in January 2008 (see Part II).

In 2005, a larger unit organised a M4-project approved by the local unions. Around 20 people within the unit were identified by Ericsson to be eligible for Model 4. By December 2006 the result, 8 of these 20 people were back on track at work, 8 people had joined a career program and left Ericsson. This project was extended in 2006 and is still ongoing. The number of eligible participants is now 84 (see Table 1 in Part II).

M5 – Rehabilitation

The fifth model is the rehabilitation program that all Swedish companies are obliged to follow according to Swedish law. I just added it as a model to illustrate that it is different from Model 4. Very often, people are not on sick leave even though they might be entitled to. In such cases, the company cannot apply the rehabilitation program (Model 5) and, consequently, Model 4 would be applicable.

M6 – Individual Career Planning

Model 6 is a career program for someone who is functioning well in his or her job, but needs a push for further development. Ericsson has developed very good relations with several external companies (Right, Antenn, etc) during the restructuring years, all of which also have a good reputation among the employees due to the successful career programs. The structure of the M6 program, as well as the coaches, is very much the same as used within M4.

M7 – Increase of Flexibility

Due to the drastically reduced number of staff during the period between 2001 and 2005, three important organisational consequences need to be mentioned briefly here. These consequences are probably not unique to Ericsson. Firstly, people are afraid to apply for other vacancies, both within the company and externally. Secondly, after 2005 the age distribution within Ericsson was skewed, with most employees between 35 and 58 years old. Thus, fewer people will retire in the near future. This means that Ericsson cannot, for example, recruit engineers with new skills for research and development without increasing the number of employees. Thirdly, as Ericsson often recruits the managers internally, the seniority principle means that many managers will not be laid off. Thus, managers stay on, very often with too few subordinates, or they step down a level or two. This is, of course, not an optimal situation, neither for the organisation nor for the manager. Model 7 was thus formed to create a more flexible work force. The lay-out of this model will be adapted on a case by case basis. The unions will always be involved when forming the structure of M7.

2 PART II: Lesson learned! Implementing the models

To date, Ericsson has used Model 4 continuously and Model 7 twice. In addition, a new module was added to the regular Ericsson leadership program in 2008. These two models are therefore described in more details next.

2.1 Implementing model 4

Model 4, approved by the unions in January 2008, applied regularly by the managers. There are employees in all organisations that, for various reasons, do not perform as expected. The starting point of the model is any employee who does not meet the expected performance level in his or her current position. Examples are inadequate work performance, absence from work, problems in cooperating with others, illness, does not follow rules, etc. Ericsson learnt during the restructuring process that it is of utmost importance for these employees to be identified as early as possible. This ensures that possible physical or psychological issues will be resolved quickly. Clear feedback from the manager, appropriate support activities and action plans will assist the person to perform well. The fundamental concept (Figure 1) can be described as follows:

Figure 1. Model 4 concept

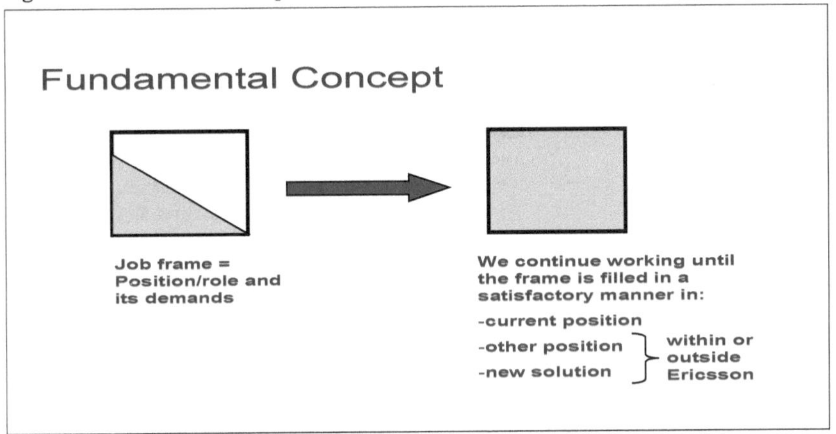

The process of Model 4 is defined as follows (see Figure 2). Note: CPx means check point number x. At the check point all necessary actions are checked by the manager, the HRM and the union. The next phase of the process can start as soon as approval has been given by all relevant parties. This method follows the Ericsson General Project Model.

During Step One, the manager and the HRM gather facts about the employee and his or her current situation to understand as much as possible about the actual problem. The manager draws up a plan including demands of the job position and goals as well as job activities, all in accordance with labour laws. The union is informed, and CP0 can be passed. Next, the dialogue between the manager and the employee starts, and the employee states his or her perception of the situation. Based on this dialogue, an action plan with a time constraint (normally 9-12 months) will be drawn up together. Several coaching sessions with an external coach are often part of the plan. The manager and the employee meet on a regular basis to ensure that the plan is followed. The union is also regularly informed during the whole process. When passing CP1, the action plan has been followed, the targets have been assured, and all parties - manager, employee, HRM and union – have agreed upon a common route forward.

Figure 2. Model 4 process

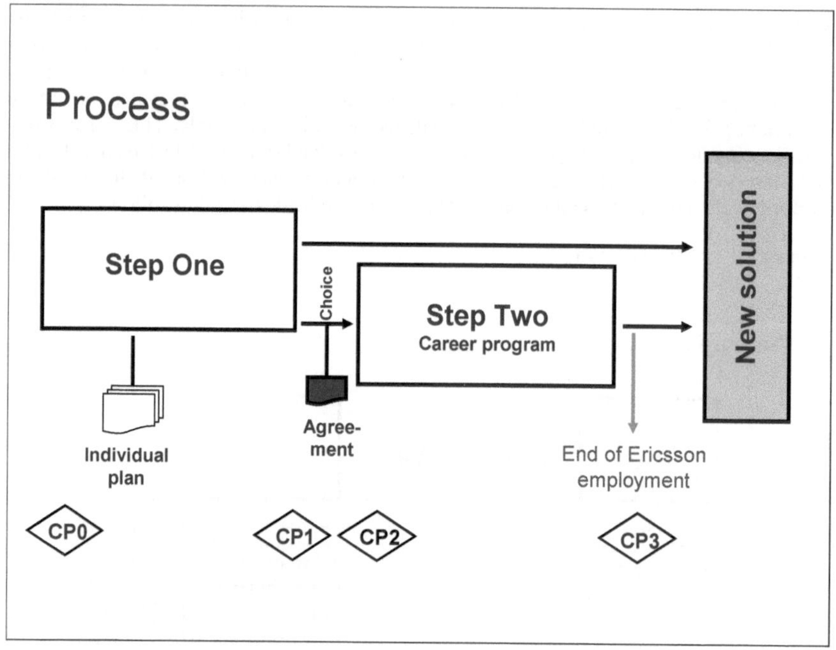

As Step One is completed one of three things will take place: (1) The employee is back on track in his or her current position, has found another job within or outside Ericsson or another solution. (2) The employee agrees to enter a career program provided by Ericsson. (3) The employee does not agree to participate in a career program.

In the second case, when employees enter the career program, CP2 involves the manager and the employee coming to a written agreement regarding the six months' career program and set a date when the employment will come to an end. The full-time career program at Step Two takes place outside Ericsson premises and is largely identical with the original career change program used during the earlier restructuring period. The usual outcome, following Step 2, is that the employee has found a New Solution (that is, employment, continuing education, or starting up a business). However, if this is not the case, the program ends and the employment will also come to an end (CP3) as stated in the written agreement.

In the third case, where the employee does not agree to enter the career program, the action plan from Step One will be reconsidered. A possible transfer to another position will also be analyzed. If the problem still exists after about six months, the career program will be offered again. In most cases, if the employee still refuses, there will be sufficient legal grounds for termination of employment. During the whole process, it is very important that all involved parties know their roles and responsibilities in the process. These parties are the individual employee, his or her line manager, HRM, and the trade unions. The responsibilities are outlined in detail in Figure 3.

The goal is that all managers will work in accordance with the concept of Model 4 on a daily basis. Still, there is some backlog. When people were given notice during the reduction period between 2001 and 2005, they were often surprised to hear that their performance was perceived as being too low. In the future, this should never come as a surprise. All employees should be aware of the expected level of work performance required of them at work. To date, around 150 people have entered the process that is Model 4. About half have left Ericsson for new jobs or other solutions (also see Table 1 for results achieved based on project mentioned in Part I with the Model 4).

Figure 3. Model 4 roles and responsibilities of the various parties

Manager	Responsible for creating an individual action plan and driving the process onwards Clear feedback to/dialogue with the employee Clarify choices in the process for the employee Document all activities and plans Inform the trade union together with the HRM
Employee	Responsible for fulfilling his/her part of the employment agreement and contributing to the organisation Responsible for following agreed action plan Responsible for his/her own actions and choices
HRM	Responsible for providing professional support and functioning as a "sounding board" for managers prior to discussions with employees Labour legislation pertaining to employer-related issues Proposals for solutions/alternatives in action plan Quality assurance of the process Responsible for the terms of the support package Can take part in three-party meetings when necessary
Trade Unions	Responsible for supporting their members

Table 1. Model 4 results, status September 2007

	Initiated cases	Concluded – internal solution	Concluded – external solution	On-going
No. of people	84	26	22	36
Clarification		Back to existing job – 13 New job within Ericsson – 13	Career program – 4 (18 people chose to transfer to Model 7 in April 2006, see below)	

2.2 Implementing model 7

Model 7 has been another model applied at Ericsson. In April 2006, the management decided to offer this new flexibility program to deal with the issues associated with age structure as well as low personnel turn-over. The offer was directed to specific working parties in Sweden. At that time, almost 640 people (70%) were aged between 35 and 50 (average 40) and around 360 (40%) managerial staff. Thus, this program addressed one of the restructuring consequences from 2001 to 2005: the surplus of managers staying in the organisation due to the seniority principle (cf. last section of Part I). As acceptance of the offer was voluntary, the unions were merely informed about the process. The goal was to reduce staff numbers by a 1,000 people. The offer was made to people from 35 years of age with at least six years employment and consisted of a financial package worth 12-18 months' salary plus EUR 5 000 as a capital investment, as well as the participation in a career program. The offer was accepted by 910 individuals.

A second such offer was made to specific work groups as well as geographical areas in 2007. The goal was to reduce staff numbers on a voluntary basis by 300 to 400 people, especially within marketing, sales and product management. This offer was negotiated with the unions as always and was accepted by 271 employees in the target group who left Ericsson only three months after the offer was first made. The average age of this employee pool was 43 years and their tenure at Ericsson 15 years. This allowed Ericsson to quickly recruit younger engineers for their research and development divisions without increasing the number of personnel.

2.3 Leadership training

As mentioned in the section on the Model 2 Competence Shift in Part I, the mandatory management training first employed in 2003 was further developed and is now used for line managers supervising employees participating in Model 4 initiatives. It all started with the cooperation between Ericsson and one of the suppliers, Right Management. By listening to the people entering the career change programs during the period 2001-2005, the coaches at Right obtained an extensive knowledge about Ericsson managers. This experience, combined with their coaching knowledge, provided the basis for the development of a new important leadership program at Ericsson. The purpose of the training program is to strengthen leadership for increased performance by enhancing the managers' ability to be concise, confident, and better communicators.

The training program consists of three modules, each covering specific aspects: Module 1 trains managers to act with confidence as an employer and handle difficult leadership situations, all in accordance with the labour laws (1 day). Module 2 focuses on coaching leadership (5 days), including role playing based on cases and own situations. Module 3 emphasizes communicative leadership skills (½ day).

The implementation plan for 2008 is to engage 400 managers in Sweden and to initiate pilot programs globally. All managers within the Ericsson group are expected to have successfully completed the training program by 2010.

3 Conclusion

By formulating the presented models, people in the organisation, including union representatives, have gained a more common understanding for what has to be done and in which order. By using a common vocabulary, misinterpretations can be reduced and qualms can be diminished. Also, the cooperation between different parties has improved as they can gather around a mutual strategy. Following the strategies in the models, the managers can feel certain that they are treating their employees with respect and understanding. Also, the employees feel reassured that the process is subject to rules and the routines, adhered to by all parties. This reduces the risk of harassment and/or neglect at work in connection with restructuring. At the same time, Ericsson managers will improve their ability to be concise, confident and good communicators via the new leadership program. In addition, the overall restructuring process is faster than before.

Another benefit of having this structured approach is that with the given models, the effect of the restructuring on the individual's health is to some extent minimized by the clear methodology of the established models. Most importantly - there is a much greater understanding that we, as employees, go through changes and develop during our job careers and that it is important to realize that the job careers change and develop with us.

References

Diedrich, A. & Bergström, O. (2006). *The process of restructuring at Ericsson.* Göteborg University, School of Business Economics and Law, MIRE.

Kieselbach, T., & Jeske, D. (2008). Health impacts and innovative approaches (pp. 308-338). In B. Gazier, & F. Bruggeman, F. (Eds.), Restructuring work and employment in Europe. Managing change in an era of globalisation. Cheltenham: Edward Elgar.

Suh, N.P. (1990) Massachusetts Institute of Technology, *The principles of design,* Oxford University Press.

Strategies and Efforts of the Statutory Health Insurance in Germany: The Example of the Federal Association of Health Insurance Funds (BKK BV)

Jan-Frederik Prüßmann

Introduction

The Federal Association of Company Health Insurance Funds (in the following called BKK BV) is engaged in a campaign for "More Health to All" to reduce the socially determined health inequalities. In contrast to many other countries in the European Union, the German health system is organised on a subsidiary basis. The prevention and the medical treatment of diseases are regulated by German state laws. The individual services in the health system are not carried out by national organisations, but by institutions such as hospitals or medical doctors, which are more or less private enterprises. The services are financed in most cases by an insurance system. Health insurance coverage for the individual citizen is governed by German legislation. The German health insurance companies are called the "statutory health insurance (system)", and the executive bodies are the individual statutory health insurances. German legislation stipulates that health insurance funds should provide primary prevention and health promotion services according to the § 20 of the Social Security Code V (SGB V), the purpose of which is to improve the general health status of the population and in particular to reduce socially determined health inequalities. The BKK BV – the head organisation of the individual Company Health Insurance Funds considered how to meet the legal obligations. How this was achieved is described in this article.

1 First phase: The first measures

In order to provide a starting basis how to proceed, the BKK BV appointed several experts to provide information on how and where prevention and health promotion should be applied for socially disadvantaged people. One of these leading experts was Rolf Rosenbrock from the Science Centre for Social Research in Berlin (WZB), Working Group for Public Health, who produced a report on the conditions, the need and opportunities for prevention and health promotion for socially disadvantaged people (Rosenbrock, 2004a, 2004b, 2004c; Rosenbrock, Bellwinkel & Schröer, 2004).

The BKK BV identified and chose socially disadvantaged children and adolescents, immigrants and unemployed as specific targets groups (Geene & Philippi, 2004; BKK BV 2004, 2005, 2006). This article will focus on the unemployed and discuss services and projects which were developed, financed and initiated by the BKK BV as well as other measures the BKK BV is engaged in.

1.1 Unemployment as a chronic problem: Novel measures of the statutory health insurance providers

Unemployment has become a chronic problem in today's societies in the European Union. In this context we should speak of structural unemployment, because unemployment is no longer just a seasonal phenomenon, but moreover a permanent feature of modern society caused by structural changes. People are becoming more and more aware of the fact that the workplace itself is continually developing into a socially sought-after and appreciated commodity. In spite of the social security systems in European societies, to have or not to have a job is a major criterion in the development of social inequalities. Technological development and globalization are causing new competition between the former more nationally oriented labour markets. Economic upheavals and restructuring result in more people facing unemployment during their working life: Job insecurity and threat of unemployment are becoming commonplace in society. For people affected by these changes, dealing with job insecurity or unemployment and the resulting problems is a challenge. No one has prepared them for dealing with this situation.

Because of its insight into the working experience of those insured by the BKK providers, the issue of "unemployment" is viewed as highly relevant in dealing with health inequalities. The BKK BV identified the main questions for its future initiatives as follows: What can the statutory Health Insurance Funds in Germany do to help people better cope with unemployment and to protect their health? How can statutory health insurance providers in Germany address this task? What are the possibilities available to them, but also, what are the limitations facing them

Work is the most effective method for overcoming health problems caused by unemployment. But creating new jobs is beyond the means of the health insurance funds. It is impossible for the German statutory Health Insurance Funds to change the structural conditions. Services by the Health Insurance Funds can only focus on individuals. Responsibility for their fate should not be put just on the individuals affected – in the sense of "blaming the victim". Instead, measures should focus on areas in their life that are under the influence and control of the same individuals. In the following, some of the methods being used by the BKK BV will be outlined.

1.2 Expert report on unemployment and health

First the research literature on the issue of unemployment and health was reviewed next by the BKK BV staff. This overview gave them a qualified basis for further planning. Thomas Elkeles of the University of Applied Sciences, Neubrandenburg, was also appointed to produce an expert report on the health situation of the unemployed, which outlined interventions and how these can be implemented (Elkeles & Kirschner, 2004; Elkeles, 2004).

On the basis of these findings and to acquire yet more information, different model projects for prevention and health promotion for the unemployed have been developed and put into practice. The specific projects were the FIT Consulting approach, BKK Job Fit, a health module in the Job-related Integration and Working Initiative (BEAM), and health-oriented counselling. In the following these examples will be introduced.

1.3 FIT Consulting

FIT Consulting is an approach based on the motivational interviewing technique for individual counselling. This approach basically consists of health-related counselling for the unemployed. This activity was carried out in cooperation with the Institute for Prevention and Health Promotion at the University of Duisburg-Essen (IPG) and the Institute for Therapy Research (IFT-Nord), Kiel, at different places in Germany in different settings, for example labour exchanges, adult colleges etc. The counselling was given on an individual basis (Hanewinkel & Stephan, 2004; Wiborg et al., 2005).

1.4 BKK-Job-Fit Project

This BKK-Job-Fit-project was carried out in co-operation with the Company Health Insurance Fund of Hoesch in Dortmund. Members of the Hoesch health insurance fund are traditionally steel workers of the former Hoesch AG. The issue of unemployment is almost an everyday problem in the steel industry which has been in crisis for a long time. In the BKK-Job-Fit-project, unemployed individuals and those only temporarily employed in transfer organisations (that means, people imminently threatened by unemployment) received counselling on health-related issues in groups (Kuhnert, 2004). The counselling was provided by qualified staff of the University of Dortmund and the Institute for Psychology of Work and Medicine of Work (IAPAM) in Herdecke under the direction of Michael Kastner. The concept of the counselling is based on the self-management method by Kanfer (Kastner, 2005).

1.5 Health module in the project "Job-related Integration and Working Initiative" (BEAM)

This health module in the project Job-related Integration and Working Initiative (BEAM) is an initiative for unemployed people with mental illnesses and/or a drug addiction. The aim of this project is to stabilize the participants and reintegrate them or introduce them into the labour market for the first time. Work is seen as a major psychologically stabilizing factor in this project. The participants therefore receive training in the "housekeeping/hotel/catering trade", in "horticulture" and in "office or administration". The emphasis in this model is not just on individuals getting qualifications but also on therapeutic training.

Because of their situation, health has only very little priority for the target group and/or they are unaware of the linkage between their health and their employability. At the same time, they are highly exposed to mental and physical stress factors, which can then lead to problematic health behaviour. The project has therefore been extended by introducing a health module supported by the BKK BV. The curriculum for this health module incorporates several aspects such as self-management, time-management, learning to cope with everyday life, stress, learning about nutrition and health behaviour. The aim of the 10-month module is for the participants to develop an understanding of health as relevant to their own situation, to recognize their own opportunities, assets and limitations, and to learn adequate behavioural strategies for increasing their chances of reintegration into the labour market (Schulze, 2004).

1.6 Health-oriented placement-counselling

The literature analyses revealed a need for action to address the concerns of those groups imminently threatened by unemployment, because the risks associated with impending unemployment or job insecurity and insecure working conditions are very similar to the health risks found amongst the long-term unemployed. Moreover, the group of people threatened by unemployment as the target group have two important advantages compared to the group of unemployed:

(1) The group of unemployed is difficult to reach, as they generally have no common setting. The only place where the unemployed can be easily identified is at the Labour Exchange, because they have to report there regularly. But for unemployed people this place is very often associated with negative feelings which often impact their attitude in a negative way. People who are acutely threatened by unemployment (for example, due to restructuring measures) or who are in insecure jobs can, by comparison, be reached easily at the workplace.

(2) Different studies have shown that the earlier the health prevention/promotion measures for the unemployed commence the more effective they are. When people are seriously threatened by unemployment or job insecurity, specific initiatives for prevention and health promotion can be put into practice before unemployment occurs. In comparison to the long-term unemployed, the chances of success are therefore considerably higher.

Before the literature analysis was completed, one of the BKK insurers of a big steel works in Bremen announced their interest in a project for promoting the health of people threatened by unemployment. Jobs of around 1,700 steelworkers in the company were due to be cut in a restructuring process. According to the staff at the BKK, uncertainty among the staff and the first health problems were already becoming apparent. Michael Kastner from the University of Dortmund was appointed to develop a concept based on the self-management approach, focusing on health promotion counselling for people in the process of occupational transitions (Kastner, 2005). Because the steelworks in Bremen had already planned extensive placement measures, the focus of this initiative was on conveying information about health stabilising aspects. In addition, two more institutions were involved. Thomas Kieselbach from the University of Bremen was appointed to research the possibilities of implementing "health promotion counselling for people threatened by unemployment" (see Kieselbach and Beelmann in this volume). The Institute for Prevention and Health Promotion at the University of Duisburg-Essen was appointed to analyze both reports and to develop a practical overall plan (Prüßmann, 2005).

After the pilot was implemented at the steel company in Bremen, plans were made to extend the project throughout the region in Bremen which has been badly affected by unemployment. However due to internal conflicts of interests in the steel company, this project was not extended. Instead, the BKK BV offered the project to the other BKK providers. It has been implemented at several places all over Germany (Prüßmann, Friedrichs & Bellwinkel, 2009; Prüßmann, 2007, 2005).

1.7 Dissemination of concepts and measures

The individual projects have been evaluated und the results and findings have become part of a consultation concept, which is tailored to the individual BKK. This consulting concept describes ways in which BKK providers can get involved in prevention activities focusing on the field of unemployment and health.

1.8 Further activities

The BKK BV is engaged in many activities at the national as well as European level. As the initiator of different networks, the BKK BV has tried to put the issue of unemployment and health on the agenda of each of these organisations and networks. A new forum on Labour Market Integration and Health Promotion was integrated into the existing German Network for Workplace Health Promotion (DNBGF) where the issue of unemployment and health was established as a core concern.

In this network, practitioners from companies, agencies or health insurance funds meet up with scientific experts. This set-up provides a platform for experts from different fields involved in unemployment issues to exchange their experiences and search for common and comprehensive (i.e. holistic) solutions.

The BKK also focuses on unemployment and health in their own networks and is trying to recruit network members that are interested in the issue. Two initiatives, carried out by the BKK BV, deal with this issue on a European level. The European network Social Insurance for Health (SifH) is made up of social experts from different European countries (Meggeneder, Breucker & Järvisalo, 2006). The SifH has placed unemployment and health on their programme. Also the European Network for Workplace Health Promotion (ENWHP) is addressing problems of unemployment and health. Moreover, the BKK BV acts as a WHO Collaborating Centre and addresses the theme of unemployment and health in research studies carried out in collaboration with WHO (cf. Prüßmann, Wehmhöner & Stephan, 2003).

2 Second phase: Continuing measures and outlook

The experiences gained from the different projects were gathered and fed into new activities. It became clear that it would be effective to combine measures of prevention and health promotion (for the unemployed or those threatened with unemployment) with other measures, such as counselling or qualification measures. The advantage of such a combination is that different actors in the field of unemployed and health promotion can work together more closely. As a consequence, they can adapt the measures to fit the needs of the people concerned.

This method is both more efficient and effective. It has to be acknowledged that the primary interest of most unemployed individuals is to obtain a new job. Therefore a combination of services focussing on getting a new job with services for health promotion seems to be a good solution. Moreover it seems to be more efficient to combine different methods like individual counselling, group counselling and self-help. These considerations were integrated in some new projects presented in the following.

2.1 Promotion of health competence for job hunters in North Rhine Westphalia (NRW)

The forum for Labour Market Integration and Health Promotion of the German Network for Workplace Health Promotion (DNBGF) put forward a proposal for the cooperation of different institutions in a joint project. The company for Innovative Employment Promotion (Gesellschaft für Innovative Beschäftigungsförderung, G.I.B.) is leading this project called Promotion of Health Competence for Job Hunters in NRW. The project is supported by the Institute for Prevention and Health Promotion (IPG) at the University of Duisburg-Essen. The BKK BV and the state of North Rhine Westphalia are responsible for the strategic management, the controlling and the finances. The aim of this project is to develop and implement an everyday practical and future oriented model to combine labour market integration and health promotion. The different services are to be linked, so that – suited to their individual needs – unemployed people can be offered qualification and health promotion measures at one central location.

The aim of this project is also, to combine the interests and tasks of politicians, labour market integration bodies and the BKKs. The result should be a model with variable opportunities for the different structures and needs. The expertise of the DNBGF forum for Labour Market Integration and Health Promotion should be involved in developing this model.

The next major step in this project will be the production of an overview. The plan is to develop a matrix that includes variables to point out locations and identify relevant institutions in different regions that are running suitable projects. On the basis of this overview, a map of external suppliers and courses will be produced – information which can then be offered to the unemployed when necessary (Bellwinkel, Faryn-Wewel & Roesler, 2009a, 2009b; Bellwinkel et al., 2008; Bellwinkel, 2007).

2.2 Health-promoting self-management counselling for those threatened by unemployment and in insecure jobs

The concept originally designed for the Steel Works Bremen has now been implemented by a number of BKKs in different places throughout Germany. The concept, a combination of group counselling and individual counselling, was adapted to the conditions at each place. The plan is to gradually develop the counselled groups to self-help groups. The group counselling will gradually be replaced by self-managed groups. The self-help groups will be intermittently supervised by the counsellors. A method was deliberately chosen that allowed the development of a self-help group gradually over time, because of unemployed people's prejudices against self-help activities. Those prejudices are reflected in the generally low participation rates of unemployed in self-help groups. The activities of the self-managed groups are due to run for 20 months and the self-help activities will continue indefinitely. If the counselling takes place during regular working hours, the participants are released from work by the company for the duration of the individual counselling sessions. This helped to increase the acceptance of the measures (Prüßmann, Friedrichs & Bellwinkel, 2009; Prüßmann, 2007, 2005).

3 Conclusion

Besides the projects described above, there are further activities planned. It is intended to take part in a comprehensive regional project, in which health promotion for unemployed people will be inter-linked through case-management. Qualification measures, health promotion, prevention and assistance in finding employment will be provided by one source to unemployed people. Furthermore, there are plans to develop health promotion counselling for the unemployed which consistently links elements of both individual and group counselling and self-help. The most effective elements of these methods and model projects will be incorporated into a new concept.

The BKK BV also cooperates with a variety of other institutions, as the German Federal Centre for Health Education (BZgA) or the Regional Association for Health in Berlin e.V., in order to build a network for health promotion projects for the socially disadvantaged. In addition to the BZgA, associated offices will be set up in the different federal states. At least two of these offices will be financed and staffed by the BKK BV.

These examples demonstrate the possibilities available to health insurance funds in Germany and hopefully inspire similar projects across Europe to jointly tackle the issue of unemployment and health – prevention, health promotion and risk awareness are all key for a healthy future of Europe's workforce as individuals face recurrent transitions between jobs but also face unemployment and job insecurity.

In the meantime, the two previously described examples (see chapter 2.1 and 2.2) have in particular become standard instruments for health promotion and primary prevention targeting the issue of unemployment and insecure jobs. They have been evaluated and extended to further regions of Germany (e.g., Bellwinkel, 2007; Prüßmann, Friedrichs & Bellwinkel, 2009). The next step will be to combine both approaches and extend them to the setting of transfer organisations (i.e. organisations supporting people at risk of unemployment getting a new job).

References

Bellwinkel, M. (Ed.) (2007). *JobFit Regional. Ein Modellprojekt zur Verbesserung der Beschäftigungsfähigkeit von Arbeitslosen durch Gesundheitsförderung* [JobFit Regional. A model project for the improvement of the employability of the unemployed via health promotion]. Gesundheitsförderung und Selbsthilfe, 20. Essen: BV BKK; Bremerhaven: Wirtschaftsverlag NW.

Bellwinkel, M., Faryn-Wewel, M. & Roesler, J. (2009a). *Verknüpfung von Gesundheits- und Arbeitsmarktförderung* [Linking health and labour market promotions]. JobFit-Leitfaden. Ratgeber. Essen: BV BKK.

Bellwinkel, M., Faryn-Wewel, M. & Roesler, J. (2009b). *Linking health and employment promotion. JobFit Guide.* Guideline. Essen: BV BKK.

Bellwinkel, M., Faryn-Wewel, M., Busch, G. & Schupp, C. (2008). *Und keiner kann's glauben – Stressfaktor Arbeitslosigkeit. Trainermanual für den Präventionskurs [And nobody can believe it – stress factor unemployment. Training manual for preventive measures]*. Praxishilfe. Essen Bundesverband der Betriebskrankenkassen.

BKK Bundesverband (Ed.) (2004). *Mehr Gesundheit für alle. Praxisbeispiele* [More health for all. Practical examples]. Essen: BV BKK.

BKK Bundesverband (Ed.) (2005). *Mehr Gesundheit für alle. Praxisbeispiele 2005* [More health for all. Practical examples 2005]. Essen: BV BKK.

BKK Bundesverband (Ed.) (2006). *Mehr Gesundheit für alle. Zwischenbilanz 2006* [More health for all. Current status 2006]. Essen: BV BKK.

Elkeles, T. (2004). Arbeitslosigkeit und Gesundheit – Einführung in gesundheitsförderliche Interventionsstrategien. In R. Geene & T. Philippi (Eds.). *Mehr Gesundheit für alle. Die BKK-Initiative als ein Modell für soziallagenbezogene Gesundheitsförderung* [More health for all. The BKK initiative as a model for socially focused health promotion]. Gesundheitsförderung und Selbsthilfe, vol. 6 (pp. 61-70) Bremerhaven: Wirtschaftsverlag NW.

Elkeles, T. & Kirschner, W. (2004). *Arbeitslosigkeit und Gesundheit. Intervention durch Gesundheitsförderung und Gesundheitsmanagement – Befunde und Strategien.* [Unemployment and health. Intervention via health promotion and health management – Findings and strategies]. Gesundheitsförderung und Selbsthilfe, vol. 3. Bremerhaven: Wirtschaftsverlag NW.

Geene, R. & Philippi, T. (Eds.) (2005). *Strategien und Erfahrungen. Mehr Gesundheit für alle. Die BKK-Initiative als ein Modell für soziallagenbezogene Gesundheitsförderung [Strategies and experiences. More health for all. The BKK initiative as a model for socially focused health promotion]*. Gesundheitsförderung und Selbsthilfe, vol. 14. Bremerhaven: Wirtschaftsverlag NW.

Geene, R. & Philippi, T. (Eds.) (2004). *Mehr Gesundheit für alle. Die BKK-Initiative als ein Modell für soziallagenbezogene Gesundheitsförderung* [More health for all. The BKK initiative as a model for socially focused health promotion]. Gesundheitsförderung und Selbsthilfe, vol. 6. Bremerhaven: Wirtschaftsverlag NW.

Hanewinkel, R. & Stephan, C. (2004). Motivierende Gesundheitsgespräche zur Förderung der Selbstverantwortung. In R. Geene & T. Philippi (Eds.). *Mehr Gesundheit für alle. Die BKK-Initiative als ein Modell für soziallagenbezogene Gesundheitsförderung* [More health for all. The BKK initiative as a model for socially focused health promotion]. Gesundheitsförderung und Selbsthilfe, vol. 6 (pp. 79-83). Essen: BV BKK; Bremerhaven: Wirtschaftsverlag NW.

Kastner, M. (2004). *Selbstmanagement für unsicher Beschäftigte und Arbeitslose* [Self-management for those in insecure employment and the unemployed]. Gesundheitsförderung und Selbsthilfe, vol. 9. Bremerhaven: Wirtschaftsverlag NW.

Kuhnert, P. (2004). Selbstmanagement-Beratung für Arbeitslose: Befähigung zur Nutzung gesundheitsförderlicher Aktivitäten. In R. Geene & T. Philippi (Eds.). *Mehr Gesundheit für alle. Die BKK-Initiative als ein Modell für soziallagenbezogene Gesundheitsförderung* [More health for all. The BKK initiative as a model for socially focused health promotion]. Gesundheitsförderung und Selbsthilfe, vol.. 6 (pp. 71-74). Essen: BV BKK; Bremerhaven: Wirtschaftsverlag NW.

Meggeneder, O., Breucker, G. & Järvisalo, J. (Eds.) (2006). Social insurance for health. The role of health promotion and prevention within social insurance in Europe. Frankfurt am Main: Mabuse.

Prüßmann, J.-F. (2007). Gesundheitsorientierte Selbstmanagementberatung bei Veränderungsprozessen. In *Prävention und Gesundheitsförderung* [Prevention and health promotion]. Band 2, Supplement 1, Oktober 20076. Deutscher Kongress für Versorgungsforschung und 2. Nationaler Präventionskongress (pp. 50-51). Heidelberg: Springer Medizin Verlag.

Prüßmann, J.-F. (2005). Gesundheitsorientierte Selbstmanagementberatung in Veränderungsprozessen. In R. Geene & T. Philippi (Eds.). *Strategien und Erfahrungen. Mehr Gesundheit für alle. Die BKK-Initiative als ein Modell für soziallagenbezogene Gesundheitsförderung* [Strategies and experiences. More health for all. The BKK initiative as a model for socially focused health promotion]. Gesundheitsförderung und Selbsthilfe, vol.14 (pp. 229-235). Bremerhaven: Wirtschaftsverlag NW.

Prüßmann, J.-F., Friedrichs, M. & Bellwinkel, M. (2009). *Gesundheitsförderung in Veränderungsprozessen. Gesundheitsorientierte Selbstmanagementberatung bei drohender Arbeitslosigkeit und prekärer Beschäftigung* [Health promotion in change processes. Health oriented self-management counselling in the case of threatened unemployment and precarious employment]. Gesundheitsförderung und Selbsthilfe, vol. 22. Bremerhaven: Wirtschaftsverlag NW.

Prüßmann, J.-F., Wehmhöner, M. & Stephan, C. (2003). Veränderte Arbeitsbedingungen beeinflussen die Gesundheit. Studie im Rahmen der Kooperation mit der WHO [Changed working conditions influence health. Study in cooperation with WHO]. *Die BKK, 91*(11), 563-569.

Rosenbrock, R. (2004a). *Primäre Prävention zur Verminderung sozial bedingter Ungleichheit von Gesundheitschancen* [Primary prevention for the reduction of socially caused inequality of health chances]. 13 Befunde und Empfehlungen zur Umsetzung des § 20 Abs. 1 SGB V durch die GKV. Essen: BV BKK.

Rosenbrock, R. (2004b). Primäre Prävention zur Verminderung sozial bedingter Ungleichheit von Gesundheitschancen – Problemskizze und ein Politikvorschlag zur Umsetzung des § 20 Abs. 1 SGB V durch die GKV. In R. Rosenbrock, M. Bellwinkel & A. Schröer (Eds.). *Primärprävention im Kontext sozialer Ungleichheit. Wissenschaftliches Gutachten zum BKK-Programm "Mehr Gesundheit für alle"* [Primary prevention in context of social inequality. Scientific expertise to the BKK program „More health for all"]. Gesundheitsförderung und Selbsthilfe, vol.. 8 (pp.7-149). Bremerhaven: Wirtschaftsverlag NW.

Rosenbrock, R. (2004c). Sozial bedingte Ungleichheit von Gesundheitschancen. In R. Geene & T. Philippi (Eds.). *Mehr Gesundheit für alle. Die BKK-Initiative als ein Modell für soziallagenbezogene Gesundheitsförderung* [More health for all. The BKK initiative as a model for socially focused health promotion]. Gesundheitsförderung und Selbsthilfe, vol.. 6 (pp. 19-36). Bremerhaven: Wirtschaftsverlag NW.

Rosenbrock, R., Bellwinkel, M. & Schröer, A. (Eds.) (2004). *Primärprävention im Kontext sozialer Ungleichheit. Wissenschaftliches Gutachten zum BKK-Programm "Mehr Gesundheit für alle"* [Primary prevention in the context of social inequality. Scientific expertise to the BKK program „More health for all"]. Gesundheitsförderung und Selbsthilfe, vol.. 8. Bremerhaven: Wirtschaftsverlag NW.

Schulze, D. (2004). Ein Bewusstsein schaffen für die eigene Gesundheit: das Gesundheitsmodul im Projekt "Berufliche Eingliederung und Arbeitsmaßnahme (BEAM) für Sozialhilfeempfänger/innen. In R. Geene & T. Philippi (Eds.). *Mehr Gesundheit für alle. Die BKK-Initiative als ein Modell für soziallagenbezogene Gesundheitsförderung* [More health for all. The BKK initiative as a model for socially focused health promotion]. Gesundheitsförderung und Selbsthilfe, vol.6 (pp. 75-77). Bremerhaven: Wirtschaftsverlag NW.

Wiborg, G., Stephan, C., Wewel, M., Hanewinkel, R., Brouwer, M., Isensee, B., Stickan-Verfürth, M., Bellwinkel, M. & Niestrath, H. (2005). *Die FIT-Beratung. Motivierende Gesundheitsgespräche für Arbeitslose* [The FIT counselling. Motivating health talks with the unemployed]. Praxishilfe. Essen: BV BKK.

Health-Promoting Transitional Counselling in a Steel Work Company Undergoing Restructuring

Thomas Kieselbach & Gert Beelmann

Introduction

The Institute for Psychology of Work, Unemployment and Health (IPG) had the opportunity to investigate the potential for integrating health-promoting measures to personnel cuts at the steelworks Bremen in association with the Federal Association of Health Insurance Companies (BKK BV) as a part of an expert report. The following specific questions are discussed in this chapter:

- Inspection of planned measures (health-promoting measures to overcome the individual and social consequences of personnel cuts) in terms of their appropriateness and effectiveness during the restructuring process at the Bremen steelworks
- Review for the implementation of health-promoting measures during the outplacement process at the Bremen steelworks
- An examination of the potential opportunities to motivate concerned individuals to participate in the measures.

1 Restructuring of the Bremen steel works

The regional press in Bremen announced in September 2002 that the steel works in Bremen belonging to the ARCELOR group were planning to invest heavily into the company and continue the restructuring process which had started already at the end of the 1990s so as to reduce costs. There were no official details available regarding the extent of planned personnel cuts, but according to rumours, the numbers lay somewhere between 1.000 to 1.500 jobs (Weser Kurier, 23/09/02). Three months later, the Weser Kurier reported that the management board and work council had agreed on a number of measures which were aimed at reducing the impact of the personnel cuts in the most socially responsible manner possible. At this point, the first concrete figures were provided about how many people were to be affected by the job cuts, namely 1.700 employees (of a total of 4.800 employees). The regional press first reported about the framework of the social plan called FIT in July 2003, which was meant to enable a more socially cushioned dismissal process for the concerned employees. The FIT programme included a number of different options, such as part-time arrangements for older workers, training as well as financial compensation for those who left the organisation right away (Weser-Kurier, 17/07/03).

The FIT programme at the Bremen steelworks included various offers which should enable the affected employees to make use of a large number of support measures when leaving the organisation. In particular, employees affected by the restructuring were offered the following alternatives (see also Stahlwerke Bremen, 2003):

1. *Direct exit:* The employees could choose to leave the organisation voluntarily. This was only an option for those employees whose jobs had been considered redundant. In exchange for this decision to leave, the concerned people received a compensatory payment between € 7.000 and 51.200 according to their current income, age and their tenure.
2. *Transfer company (BreTraG):* The BreTraG was a transfer company which was founded in May 2003, which operated as an independent company, using labour market instruments such as structural short time compensation (till the end of 2003 under article §175 SGB III), transfer short time compensation (from January 2004 onwards under Article (§216b SGB III) for those who voluntarily transferred to the BreTraG so as to achieve a successful occupational transition via training, counselling and support.
3. *Part-time work for older employees:* This offer was open to all employees over 55 year olds who wished to go into early retirement. The offer entailed employees working 2.5 years working full-time and subsequently receiving 85% of their last net salary for another five years.
4. *Transfer within the organisation*: Due to the restructuring at the Bremen steelworks, some positions became redundant, whereas new job positions became available in other areas requiring retraining of existing staff. These employees received tailored training so that they could be placed accordingly into different jobs within the organisation.
5. *Reserve employment:* Some employees could no longer be employed in their actual workplace nor was any employment available at their specific work sites. These employees were offered the chance to transfer to so-called reserve employment sites. They stayed within the company for the moment and could be placed into temporarily available positions so as to extend their employment within the company for as long as possible. In addition, it was possible for these employees to be lent to outside companies via an employee loan system.

2 Unemployment and health-promoting interventions

Using health promotion initiatives during occupational transitions is an approach that is being discussed in a number of countries. Some organisations have already employed this approach. Usually, those involved are more traditional players in the field and include outplacement companies, educational and training institutions, transfer agencies and labour administration. These actors have past experience and can contribute important insight to this new approach.

Combining transitional counselling and health prevention represents a new trend in society to acknowledge the psychosocial costs associated with loss of work and transitional change, thus shifting the emphasis to preventive action rather than initiatives following dismissal. This trend assumes that integrating people into the labour market is significantly more difficult, costly and demanding following long-term unemployment and a history of ill-health during unemployment.

The effects of interventions for the unemployed are discussed in detail below. These findings support an approach of early interventions being integrated into the early phases of transition counselling rather than using post-hoc initiatives once employees have been dismissed (for more information see Kieselbach, Klink, Scharf & Schulz, 1998). The lack of reemployment perspectives in the German labour market is an important factor which explains partly the high rate of long-term unemployment.

In the case of the "Bildungswerkstatt 90" in Bremen (Germany), the reintegration concept of an intervention scheme for the Public Employment Services (PES) was led by a group of committed professionals and the initiative itself was aimed at a particularly challenging group of male, long-term unemployed with health problems. However, this carefully constructed one-year reintegration scheme only resulted in a reintegration rate of 26% (stable at 22% six months later).

Therefore, it is important to ask which alternative future approaches might be more successful, efficient and effective. The actors involved in the "Bildungswerkstatt 90" conducted an evaluation of the initiative on behalf of the PES in Bremen and came to the following conclusions (Kieselbach, Klink, Scharf & Schulz, 1998, p. 231):

> *"The reintegration quota of 26% (and 22% respectively after a six month period) naturally raises the question of efficiency of such integration measure. The reintegration ratio is normal for as well as similar to the results associated with other interventions in the local region as well as across German-speaking territory. Nevertheless, it is essential to confront the problem of the remaining 78% of participants who did not manage to get back into the labour market in the following six months after the programme despite considerable monetary means that have been dedicated to such holistic measures as the one mentioned above.*
>
> *A number of psychosocial stabilisation effects have been observed amongst the remaining unemployed participants, which can be interpreted primarily as a result of their participation in the initiative. However, the primary goals of such a measure should be the reintegration of the unemployed into the market and this is where the attention should be focused.*
>
> *The relative high rate of participants who were not successful reflects, in our view, mainly the extent of the psychosocial challenges faced by the participants due to the loss of employment and the subsequent length of the unemployment. These findings have also been supported by social science research. The psychosocial elements can themselves constitute a barrier for later reintegration into the labour market (i.e. in the case of an economical upturn). If the individual exclusion from the labour market continues to be tolerated by society for such long periods of time, as was found in the current participant group, the expenditure for such delayed reintegration measures is not at all surprising. The high financial costs for occupational reintegration measures resulted in a primary success for a*

> *minority of participants. This expenditure must be considered part of the substantial costs of mass unemployment for society and should further the attempts of our society to intervene more vehemently and earlier (therefore, also using preventive means)."*

However, the low reintegration rate of the long-term unemployed populations taking part in such measures does not mean that such initiatives should be limited to other groups because of the assumption that this will allow for more efficiency and reflect a more purposeful investment. If our society retains the aim of a holistic approach it must go beyond debating the social exclusion as a result of long-term unemployment. Instead, society must address the failings for the insufficient labour market politics and compensate for these decisions. This means, systematic approaches are required which counteract the "creaming-off effect" often associated with such measures (which tend to be more successful and effective for those participants that have fewer problems preventing a re-entry into the labour market). This effect can be primarily attributed to the often quite narrow success criteria used in these measures and the demand for visible and quick placement of the unemployed into jobs.

Preventive measures imply a different understanding of individual versus social responsibility, which is an approach inherent in many of planned measures put forward by the company-based health insurance companies (BKK) and represents their main argumentation for such measures. Psychosocial support and counselling in occupational transitions should be placed within the context of current employment, so that all social actors get involved and take social responsibility (at least in the interest of a networked approach so as to benefit from joint-actor synergy).

Such an approach has the following advantages:

1. This approach moves the treatment of psychosocial problems experienced by the unemployed away from the context of social work with problem groups, a context which by itself produces stigmatisation effects (not intentional on the part of the assisting social workers, but perceived as such by the recipients of such help).

2. This leads to the normalisation of the helper-recipient relationship away from the concept of benefits, which provides a better basis for the accessibility, acceptability and effectiveness of offers to help.

3. In this context, asking for help becomes a legitimate claim in contrast to being dependent on assistance due to long-term unemployment. This will also foster the self-confidence and self-esteem of the concerned individuals (which tends to be affected negatively if a person is perceived and considers him- or herself as being in need of help).

4. It combines prevention and occupational transition in the context of the actual employment situation from which people are being dismissed. The restructuring employer therefore becomes a responsible party and has to accept partial responsibility for the situation of the employee. This demonstrates the corporate social responsibility to other employees and the society more widely and shows that the responsibility of the company extends beyond the actual employment to assisting employees during their occupational transitions as well.

When judging the various possibilities associated with preventive health measures and other health-related interventions targeted at the unemployed, considerable conceptual research and work are still required in the following areas: the goals, targeted groups and how they can be reached; the methods and measures used; the relevant needs, acceptance and demands of different health measures; the type of intervention and the institutional connection; the use of individual health promotion activities or joint initiatives (linked also to employment-related measures), as well as the evaluation concepts being used (see Elkeles & Kirschner, 2003).

The following factors are of great importance to the success of measures aiming at health promotion, particularly when these are long-term process, namely which type of health promotion measures are used in a company restructuring, which company activities are used to promote employment initiatives, and what other activities are available from other social actors.

Labour market policies meant to aid employment promotion are not generally limited to labour market concerns alone. These programs of employment promotion often include other elements which can be described as health promotion measures in a broader sense, or implicitly contain health promotion aspects (i.e. psychosocial problems are often discussed and appropriate strategies are outlined during the initial counselling interviews in labour market interventions). This also becomes apparent when reviewing the research on labour market interventions. The quantity of programs as well as the heterogeneity of approaches is astonishing. Therefore, there is no sign of a systematic concept of employment and health promotion measures at this point (see also Kuhnert & Kastner, 2002).

3 Preventive health behaviour and the problem of reaching the unemployed

Unemployment or even the threat of unemployment are life events which require considerable personal resources which means there is usually little time and energy available for preventive activities. In addition, the experience of job insecurity reduces the time span available to make plans, which means that any actions that produce outcomes only further down the line are less likely to be engaged into. This is the only explanation why unemployed individuals are much more likely to seek medical assistance in the case of acute health problems in contrast to those in employment. This trend would also explain why preventive offers are much less likely to be used and accessed by those in employment as well.

According to Kieselbach (1998), this relationship was already pointed out in the chapter on unemployment and health in the National Health Report of the Federal Republic of Germany. This was replicated in a study by Catalano and colleagues (see Catalano, Satariano & Ciemins, 2002). They could prove using register data of regular mammography screenings performed between 1985 and 1997 in Detroit and Atlanta that the participation of women in this preventive health program varied according to regional unemployment rates. The authors explained these findings by pointing out that the overstretched labour market prevented women with symptoms to seek appropriate medical assistance, alternatively, this might have also diverted their attention away from discovering and attending to the symptoms. African-American women were less likely to seek preventive health care attention given the situation on the labour market. This group generally had lower socioeconomic status but also held the greatest risk to develop an undetected tumour.

The researchers came to the conclusion that screening programs should go beyond monitoring regular screens being performed. In particular, they recommended that they should pay particular attention to the regional economic development. Phases of economical change, which also harbours the potential for increased health problems, should be accompanied by additional health promotion programmes.

The stress resulting from unemployment can also change people's attitude towards institutions offering assistance and thus affect utilization. Unemployed individuals with psychosocial problems often avoid seeking professional help so as to avoid further diminishing their already low self-esteem due to the loss of work. They often tend to experience self-doubt, shame, and stigmatization, yet hide these from others. This can lead to the protraction of the health issues and the future development of chronic health problems. In addition, reemployment changes are closely linked to people's ability to work and perform. This is likely to result in people describing as well as dismissing health problems as harmless. On the other hand, experiencing ill-health can, in the case of a continuous failure to find reemployment, result in the subjective perception of relief.

4 Description of planned measures

Based on projects such as the FIT program pioneered by the BKK BV and the BKK (the health insurance fund of the company), a variety of health promotion initiatives were considered to be implemented into the restructuring and personnel reduction process at the Steelworks Bremen. These measures were intended to be pilot projects and can be considered to be models for the region and other companies.

The following initiatives were proposed:

- The implementation of a module-based self-management training for the unemployed over the time frame of two years (Michael Kastner, Universität Dortmund)
- Accompanying course offers for the unemployed, remaining employees, and other stakeholders in the context of the Steelworks Bremen (Thomas Kieselbach, Universität Bremen)
- Development of a model project for other health insurance companies in the North German region based on the company's experiences and implementation (Thomas Kieselbach, Universität Bremen).

5 Self-management for the unemployed and employees in insecure employment

The offer of a self-management training for the unemployed is a counselling concept developed for those who are unemployed or in uncertain unemployment. The main purpose is to build up their competences, enabling them to analyse and manage critical life situations independently. The aim is to relearn and try out new behaviours and strategies so that these individuals are able to develop but also use the newly acquired learning and coping skills. This concept was based on past experience made by the developer in the Ruhr Area, where this counselling initiative was first run.

The programme includes several specific stages:

- The formation of a trustful relationship between the counsellor and the employed (or unemployed): Clarification of counselling process and achievement of a joint consensus regarding the cooperation between the counsellor and the counselled individual (length: at least three meetings, duration 3-4 hours)
- Focus on problematic behavioural patterns and adjustment: Here urgent problems are tacked and identified. The main problems of the counselled person are reconstructed and named.
- Motivation: Increase the person's motivation regarding the areas in question, where adjustments need to be made.
- Development of an orientation schemata of 'investment and consumption activities': The focus here is on critically reflecting on one's own options in terms of the benefit or costs associated with changing behaviour.
- In the next step, diagnostic tools are used (such as tests, interviews, and observation) to analyze actual behavioural patterns.
- Each difference between the ideal behaviour and the actual behavioural patterns is discussed with the counselled individual to set specific and achievable goals.
- Agreement of counselling goals: Agreement on intervention strategy and steps.
- Interventions: Specific aspects are addressed one by one.
- Preventions: Not only the interventions are central, but so are preventive strategies aimed at achieving the permanent continuation of these behaviours.
- Behavioural controlling and evaluation: All previous steps are subject to continuous evaluations.
- Completion of the process: Counselling hours are reduced continuously depending on the level of the individual's autonomy from the counsellor. If more hours are required, additional counselling was made available.

In preparation for the self-management training (Kastner, 2004), the BKK organised an information event led by Kastner in January 2004. There were 40 participants (50% women), half of which were employees of the Steelworks Bremen as well as candidates who had been specifically recruited by various doctor's offices. The remaining participants were members of the works council, the BKK, the BKK BV and further experts.

Following the event, another 20 minute discussion followed which focused on the current company situation at the steelworks and essential communication line improvements in relation to the planned programme of the BKK. In particular, participants were consulted on which communication and motivation strategies were considered most useful so as to motivate more individuals to join the counselling process:

- Presentation at the next works council meeting,
- Information presentation,
- Forwarding of concrete information about the programme,
- Systematic marketing,
- Offer of introductory courses to recruit target groups to join the programme.

Despite a costly brochure action prior to the event, some of the participants complained about not having received information about the programme until shortly before the actual information event and the BKK's programme.

6 Accompanying course offers for the unemployed, for the remaining employees in the company and other social actors from the local community in the vicinity of the Bremen steelworks

In addition to the intensive and time-consuming counselling initiative, other offers were developed by the BKK and the University Bremen (Thomas Kieselbach) which represented either an alternative to the proposed self-management training or could potentially be used in addition to the training. These courses should, even though they were conceived of as independent initiatives by the BKK, enable a strategic alliance with the FIT programme in cooperation with the BreTraG.

As such, the following courses were offered:

(1) *New life perspectives beyond the working life:* This course focused on which new life perspectives were available for each individual once they have lost their work or retired. Assistance and alternatives are discussed as well as the achievement of personal wishes, ideas and aims.

(2) *Healthy balance:* This course was aimed at both the unemployed and those in employment by identifying new ways how a balance can be achieved between professional and private life goals, in particular focusing at challenges and stressful situations - be it at work or home – and counterbalancing these with compensatory and coping strategies.

(3) *Time for two:* Leaving working life has important implications for the home life in terms of stress, disagreements in the home and getting accustomed to each other in the home. This initiative was aimed at supporting couples to manage everyday life, meeting each others' needs and creating a healthy daily life.

(4) *Health competence:* This seminar offered a mix of various preventive medical checks, physical activities, nutrition suggestions as well as possibilities to reduce stress.

These courses were considered supplementary to the self-management training and targeted particularly those who showed interest in the health promotion measures, but were less interested in the intensive and longer training options.

7 Development of a transferable model

The development of a transferable model based on the Steelworks Bremen experience was of particular interest to the BV BKK at the time as the partners were contemplating to integrate health-promoting transition counselling into the service provided by individual health insurance funds. In particular, the implementation of the intervention would enable a dialogue and discussion with other social, particularly regional, actors – possibly enabling a regional network which, in turn, would serve as a means to evaluate the intervention and assess its general transferability.

The benefit of this project certainly carried over to the European Social Funds project called MIRE („Monitoring Innovative Enterprise Restructuring") which started at the end of 2004 in which the IPG of Bremen University was one national partner for Germany. This project entailed the construction of national networks including various social stakeholders - employers, unions, work council representatives, consultancies, as well as the BKK BV - in order to produce case studies outlining innovative approaches in restructuring companies.

8 Recommendations

The subsequent recommendations are based on participants' experience of and feedback to the restructuring process:

8.1. Review of the planned measures (health promotion to manage the individual and social consequences of dismissals) during the Steelworks Bremen restructuring in terms of their relevance and adequacy

> Increased transparency regarding the developments and decisions made in the dismissal process

The BKK was requested to organise regular restructuring information updates in association with the company management. These information events could also be used to simultaneously publicise health promotion offers of the BKK.

> *Participation of the health insurance funds in health promotion measures in the context of personnel cuts*

Advantages

- BKKs have the required experience in the implementation of health promoting measures and have access to companies
- BKKs can function as emotional buffers for the workforce since employees often feel left alone in the course of a restructuring process. The trust basis between employees and management is usually disrupted because of the restructuring. Health promotion as a preventive measure assumes that employees are able to confide and entrust their concerns, which usually requires a trusting rather than a strained context.

Disadvantages

- In the context of restructuring, measures of company-based health promotion can be considered a job for Human Resource Management – rather than something that should be outsourced - as emphasized by the ILO in relation to the increasing externalization of outplacement-consulting.
- BKKs should avoid trying to correct for the failings of company managements by offering health insurance initiatives.

a. Screening for opportunities - the implementation of health promotion measures in the outplacement process of the Steelworks Bremen.

> *Conception of strategies and measures by the BKK independently of the company's management*

Advantages of independent BKK initiatives:

- Developments in the company's dismissal process, if evaluated negatively by employees, are not connected to the activities of the BKK. Therefore, employees will be more likely to consider offers as being independent from the company's activities. Existing psychosocial offers provided by the company might benefit as a result, particularly because these company measures are often viewed negatively by employees and thus not used very often.

Disadvantage of independent BKK initiatives:

- The independent nature of measures does not allow for the placement of this health initiative in the appropriate labour market context which would certainly be beneficial
- The danger of duplicating offers: This should certainly be avoided for those employee groups who are very difficult to recruit to participate in health promotion initiatives in the first place.

- The lacking use of synergy opportunities between personnel departments as well as institutions working collaboratively on the FIT-Programme (i.e. the specific offers such as "Work and Life" provided for those in early retirement by the health insurance funds).

> Planning of current and future interventions of the BKKs
The integration of accompanying interventions supporting outplacement processes (which are part of the FIT programme) must be comprehensible. In addition, employees should be able to use these interventions independently of the overall programme. The beneficiaries should have the feeling that they have control over and influence on their personal situation.

> Dealing with the remaining employees following organisational restructuring („survivors")

- Health promotion offers are certainly appropriate for remaining survivors, particularly since this group is often neglected in the dialogue on health at work.
- "Survivors-of-layoffs" can be assumed to experience subsequent and long-term uncertainty at work, which can lead to a loss of productivity, lower identification with the company and also a reduced tolerance for stress given new job challenges and the exhaustion phenomenon. Once the main restructuring stages are completed, initiatives such as health circles might enable an easier start for these employees. These could be conceived of in cooperation with the BKK-BV, and would certainly be received positively by the works council which would increase the initiative's attractiveness. Following such health circles, additional intervention programmes could be initiated that are based on the work accomplished by these first measures.

> Using the referral function of the various institutions involved in the FIT programme
As can be seen in other research on help-seeking behaviour of people in unemployment or unstable employment, these individuals often face significant barriers in terms of searching for information on unemployment and locating professional help with psychosocial problems.

An unemployed person or an employee in insecure employment faces the problem of having to focus on obtaining a new job position, often considering any change-related health or psychosocial problems as distracting from the main issue. This is certainly the case at the early stage following loss of employment. Only when proof regarding the relationship between the inadequate management of this situation and the reemployment chances is provided will many individuals open up to more health-focused examination of the occupational transition challenges. The inadequate management is often described as the hysteresis effect which entails increasing psychosocial problems the longer the unemployment lasts (it is often these psychosocial factors that may, in turn, represent additional significant reemployment barriers).

Due to this complex interaction of factors, it becomes imperative to build up good cooperation structures between all actors involved in the FIT programme so as to allow employees to access help when facing occupational transitions. If the various actors are willing to cooperate rather than compete for clients, a certain referral process could be developed which could reduce people's initial apprehensions about participating in BKK initiatives and help them to overcome aforementioned barriers to seeking help.

b. Review of potential options to motivate and encourage the participation in planned initiatives

> Concerning opportunities and problems when motivating employees to participate in health promotion initiatives

- Strategic arrangements between the management and the BKK are essential to this process so as to make bilateral offers transparent and credible.
- Information events at work should be organised as joint initiatives supported by the management and the BKK. At the same time, individual actor responsibilities need to be clearly clarified and differentiated.
- The BKK should remain autonomous regarding the health promotion measure implementation, however the strategic planning and the direct access to employees requires close cooperation with the management.
- The BKK should have a trusted contact person in place that accompanies the process in the company and serves as a permanent, accessible and identifiable contact person for employees.
- The attractiveness of a health promotion measure could be increased by installing a bonus system that offers discounted rates to employees.

> Central intervention and accompanying offers
In addition to the main intervention which was led by Kastner and colleagues, a number of additional seminars and workshops were planned. These initiatives had to meet the following requirements:

- The topics of the additional seminars should dovetail the existing curriculum for the main intervention, each covering one of six main themes.
- Many of the concerned employees were left with one option only, early retirement. Quite understandably, this group reacted very negatively to this prognosis. As a result, the works council proposed that a specific seminar for this group of individuals should be drawn up in order to assist them in their psychosocial transition from being a full-time worker entering retirement. This proposal certainly indicated the difficulties faced by employees in the absence of appropriate counselling.
- Any related seminars must align with the offers by the various providers which are already in place. Related workshops give provides additional opportunities to publicise offers.

The evaluation of these initiatives in the case Steelworks Bremen led, according to the works council representatives, to a number of improvements in the workforce's working conditions – improvements that had been discussed but failed to be implemented long before to the instalment of health circles.

The main purpose should be to motivate potential participants to attend regular events offered by provider. Examples of such motivational events include the fortnightly information events run by the Bremen transfer company BreTraG. The close cooperation with professionals or specifically designated work council representatives is an essential requirement for the success of independent initiatives.

Finally, it is important that those in insecure employment understand that the BKK is neither involved in the negotiations nor responsible for any decisions made related to the dismissals. BKKs should emphasize their independent position in such initiatives, having developed a high profile for developing and running company quality circles and health promotion activities. Simultaneously, the BKK should be prepared to agree on a joint concept in a strategic alliance with the management and the works council. This balancing act between autonomy and cooperation is required in order to attain effective solutions and increase transparency of the process to the recipients of the offers.

References

Catalano, R.A., Satariano, W.A. & Ciemins, E.L. (2002). Unemployment and the detection of early stage breast tumors among Africans and non-Hispanic whites. *Annals of Epidemiology, 13*, 8-15.

Elkeles, T. & Kirschner, W. (2003). *Gesundheit und Arbeitslosigkeit. Gutachten für den BKK-Bundesverband, Kurzfassung. [Health and unemployment. Report for the BKK-BV].* Neubrandenburg, Berlin.

Kastner, M. (2004). *Selbstmanagement für unsicher Beschäftigte und Arbeitslose. [Self-management for employees in insecure employment and for the unemployed].* Expertise für den BKK-Bundesverband. Herdecke.

Kieselbach, T. (1998). Arbeitslosigkeit [Unemployment] (Kap. 4.10). In Bundesregierung (Ed.), *Gesundheitsberichterstattung des Bundes* (pp. 117-122). Wiesbaden: Statistisches Bundesamt.

Kieselbach, T., Klink, F., Scharf, G. & Schulz, S. (1998). *„Ich wäre ja sonst nie mehr an Arbeit rangekommen". Evaluation einer Reintegrationsmaßnahme für Langzeitarbeitslose.* [„I would never have gotten work otherwise". Evaluation of a reintegration measure for the long-term unemployed]. Weinheim: Deutscher Studien Verlag.

Kuhnert, P. & Kastner, M. (2002). Neue Wege in Beschäftigung - Gesundheitsförderung bei Arbeitslosen [New ways into employment – health promotion for the unemployed]. In B. Röhrle (Ed.), *Prävention und Gesundheitsförderung Bd. II* (pp. 373-406). Tübingen: DGVT Verlag.

Stahlwerke Bremen/Arcelor Gruppe (Ed..). (2003). *Wegweiser *Wir tun was! Informationsbroschüre zu den Restrukturierungsmaßnahmen [Guide * We are doing something! Informational brochure regarding restructuring measures].* Bremen 2003.

Weser-Kurier (2002). *Personalabbau steht fest. [Staff dismissal certain].* (23.09.2002).

Weser-Kurier (2002). *Es geht auch ohne Kündigung. [It can be done without dismissals].* (13.12.2002).

Weser-Kurier (2003). *Die Älteren sind zuerst dran. [Older workers are the first to go].* (17.07.2003).

About the Authors

Ala-Kauhaluoma, Mika, Dr., (1972), works as a special researcher at the Centre for Rehabilitation Research and Development of the Rehabilitation Foundation, Helsinki. She completed her Master's Degree in Social Sciences in 1998 and her Ph.D. in Social Sciences in 2007. Her main research areas focus on evaluation studies, vocational rehabilitation, unemployment and the connection between work and illness.

Armgarth, Elisabeth, B.Sc., (1946), is currently self-employed working as a consultant within restructuring. She holds a Bachelor of Science in Electronic Engineering. Amongst others, she was a program manager for the Swedish part of the restructuring at Ericsson between 2001 and 2005. Her activities included the design of the restructuring process, negotiations with unions, as well as passing information to all parties concerned. Furthermore, she participated in the ESF project on Monitoring Innovative Restructuring in Europe (MIRE) and the EU project (DG Employment of the EU Commission) on Health in Restructuring (HIRES).

Beelmann, Gert, Dr., (1971), completed his doctoral thesis in 2003 in Psychology at the Institute for Psychology of Work, Unemployment and Health (IPG) at the University of Bremen. His main research interest focuses on the effects of unemployment and concepts for creating occupational transitions. He has been working in two European projects on youth unemployment and outplacement counseling. Since 2005, he is managing director of an outplacement agency. Selected publications: Beelmann, G. (2003). Long-term unemployed youth in Germany. An action related analysis of personal und situative factors. Hamburg: Dr. Kovac Verlag; Kieselbach, T., Beelmann, G. & Wagner, O. (2009). Job insecurity and successful reemployment: Examples from Germany. In T. Kieselbach, S. Bagnara, H. De Witte, L. Lemkow & W. Schaufeli (Eds.), Coping with occupational transitions: An empirical study with employees facing job loss in five European countries (pp. 115-167). Wiesbaden: VS.

Behle, Heike, Dr., (1970), is a quantitative sociologist and holds a Ph.D. from the University of Konstanz, Germany. Currently she works as a Research Fellow at the Warwick Institute for Employment Research (IER) at the University of Warwick. Previously, she has worked as a Research Fellow for the Institut für Arbeitsmarkt- und Berufsforschung (IAB) in Nuremberg, Germany. Her research interests include the school-to-work transition of young people and vocational training systems. She was involved in various projects dealing with education, employment and careers, especially with career guidance. Selected publications: Behle, H. (2006). Veränderungen der seelischen Gesundheit durch Teilnahme an arbeitsmarktpolitischen Maßnahmen. Evaluationsergebnisse zum Sofortprogramm zum Abbau von Jugendarbeitslosigkeit (JUMP)

[Changes of mental health by the participation in active labour market programmes. Results of an evaluation]. In A. Hollederer & H. Brandt (Eds.), Arbeitslosigkeit, Gesundheit und Krankheit [Unemployment, health, and illness] (pp. 113-122). Bern: Huber; Behle, H. (2005). Moderators and mediators on the mental health of young participants in active labour market programmes: Evidence from East and West Germany. International Review of Psychiatry, 17(5), 337-345.

Berth, Hendrik, PD Dr., (1970), is since 2010 acting director at the University Hospital Carl Gustav Carus, Department of Medical Psychology and Medical Sociology, Technical University Dresden, Germany. His main research areas are unemployment and health, content analysis (Gottschalk-Gleser-method), psychosocial aspects of human genetics and coping with cancer. Selected publications: Berth, H., Balck, F. & Brähler, E. (Eds.). (2008). *Medizinische Psychologie und Medizinische Soziologie von A bis Z* [Medical Psychology and Medical Sociology from A to Z]. Göttingen: Hogrefe; Berth, H. (Ed.) (2010). *Psychologie und Medizin – Traumpaar oder Vernunftehe?* [Psychology and Medicine – partners or antagonists?] Lengerich: Pabst.

Borges, Livia de Oliveira, Prof. Dr., (1960), is Professor of Psychology at the Federal University of Minas Gerais in Belo Horizonte (Brazil). From 1990 to 2008 she was Professor at the Federal University of Rio Grande do Norte (Brazil). She is past-president (2003–05) of the Brazilian Society of Organisational and Work Psychology (http://www.sbpot.org.br). Her main research areas focus on the relationship between mental health and work conditions, the meaning of work, organisational socialization and values. She has also conducted research on workers with low formal education. Selected publications: Borges, L. O. & Yamamoto, O. (2004). O mundo do trabalho. In J. C. Zanelli, J. E. Borges-Andrade & A. V. B. Bastos (Eds.), *Psicologia, Organizações e Trabalho no Brasil* (pp. 24-62). Porto Alegre: Artmed; Borges, L. O. (2008). Valores humanos en trabajadores de bajo nivel educativo en Natal (Brasil). *Revista de Psicología Social, 23*(3), 377-394.

Borghi, Vando, Prof. Dr., (1965), is Associate Professor in the field of sociology of economic processes, work and organisation at the Department of Sociology at the University of Bologna, Italy. He is member of the scientific board of the International Centre of Sociological Studies on Labour Problems (U Bologna), vice-director of the journal "Sociologia del lavoro" and member of the Scientific Committee of the journal "Partecipazione e conflitto". He is also member of the research centre "Sui Generis – Laboratorio di ricerca sociologica sull'azione pubblica" at the University of Milano-Bicocca in Italy and of the Active Social Policy European Network (A.S.P.E.N.). Furthermore, he is member of the Board of the section "Economy, labour and organisation" of the "Italian Sociological Association", scientific coordinator of the regional Institute of Social and Economic Research (Ires Emilia Romagna, a trade union's research institute) and member of the Scientific Committee of the Foundation Mario Del Monte in Modena. His research focuses primarily on transformations of work, with particular attention to the quality of work, and to the relationship between transformations of work, vulnerability and institutions. Selected publication: Borghi, V. (2007). Do we know where are we going? Active policies and individualisation in the Italian context. In R. van Berkel & B. Valkenburg (Eds.), *Making it Personal. Individualising Activation Services in EU* (pp. 163-192). Bristol: Policy Press.

Bormann, Cornelia, Prof. Dr., (1954), is Professor for Public Health at the Institute for Economy and Health at the University of Applied Sciences in Bielefeld, Germany. Previously, she was employed at the University of Bremen, the University of Marburg as well as the German Aerospace Center – Project Management Agency (DLR-PT), Cologne. Her research areas are Public Health, Health Care Research, Social Inequality and Health, Gender and Health. Selected publications: Bormann, C. (2005). *Geschlechtsspezifische Aspekte zum Zusammenhang zwischen Erkrankungen und Erwerbstätigkeit mit besonderer Fokussierung auf die Arbeitslosigkeit in den alten und neuen Ländern Deutschlands* [Gender-related aspects of the link between diseases and employment with a special focus on unemployment in the old and new Federal States in Germany]. Regensburg: Roderer; Bormann, C. (2007). Theoretische Aspekte der Versorgungsforschung [Theoretical aspects of health services research]. In C. Janssen, B. Borgetto & G. Heller (Eds.), *Medizinsoziologische Beiträge zur Versorgungsforschung: Theoretische Ansätze, Methoden und Instrumente sowie ausgewählte Ergebnisse* [Medical sociological contributions to health services research: Theoretical approaches, methods and instruments as well as selected findings] (pp. 13-24). Weinheim: Juventa.

Brähler, Elmar, Prof. Dr., (1946), is since 1994 Head of the Department of Medical Psychology and Medical Sociology and also Vice Dean of the Medical Faculty, both at University of Leipzig, Germany. He is member of the Reviewer's Board of the German Research Foundation (Deutsche Forschungsgemeinschaft - DFG) in the field of Clinical Psychology, Differential Psychology and Diagnostics and Medical Psychology. After completing his Ph.D. studies in Mathematics and Physics in 1976, he habilitated in Medical Psychology in 1980. His main research areas involve East-West-research, migrants, psychooncology, aging and gender specific aspects of health and disease. Selected publications: Berth, H., Förster, P., Brähler, E. & Stöbel-Richter, Y. (2007). *Einheitslust und Einheitsfrust. Junge Ostdeutsche auf dem Weg vom DDR- zum Bundesbürger. Eine sozialwissenschaftliche Längsschnittstudie von 1987 – 2006* [Enjoyment and frustration about the German Unity. Young East-Germans on the way from GDR to Federal Republic of Germany citizens. A socialscientific longitudinal study from 1987 – 2007]. Gießen: Psychosozial-Verlag; Decker, O. & Brähler, E. (2008): *Bewegung in der Mitte. Rechtsextreme Einstellungen in Deutschland 2008 - mit einem Vergleich von 2002 bis 2008 und der Bundesländer* [Motions in the centre. Attitudes of rightist extremists in Germany in 2008 – a comparison of 2002 and 2008 and of the German States]. Berlin: Friedrich-Ebert-Stiftung, Forum Berlin.

Brenner, M. Harvey, Prof. Dr., (1939), is Professor of Epidemiology at the University of Technology in Berlin since 1996, Germany. He is currently Director of the Institute for Social Medicine and Professor of Social and Behavioral Sciences at the University of North Texas, Health Science Center, USA. Furthermore, he is Professor Emeritus at the Department of Health Policy and Management at The Johns Hopkins University in Bloomberg. He works as a consultant for the European Commission, the International Labour Organisation, the United Nations Social Defense Research Institute as well as the World Health Organisation. His main research interests involve the impact of employment and economic growth on mortality, life expectancy and health services utilization. Selected publications: Brenner, M. H. (1973). *Mental Illness and the Economy.* Cambridge: Harvard University Press; Brenner, M. H. (1979). Mortality and the National Economy: A Review and the Experience of England and Wales, 1936-1976. *The Lancet, 2,* 568-573.

Buchtová, Božena Šmajsová, Prof. Dr., (1947), is since 1992 Associate Professor of Psychology at the Philosophical Faculty of Masaryk University in Brno/Czech Republic. She was involved in several international activities in Poland, Austria, Russia and the Netherlands. Furthermore, she has been a member of the departmental committee of the Grant Agency of the Academy of Sciences in the Czech Republic as well a member of the evaluation committee of the Faculty of Economics and Administration at Masaryk University. Her current research interest involves the influence of new eco-technologies on employment and requalification in companies. Selected publications: Buchtová, B. (Ed.). (2007). *Unemployment – Technological and Social Changes of Labour.* Brno: Masaryk University; Buchtová, B. (Ed.). (2004). *Psychology and Unemployment. Experience and Practice.* Brno: Faculty of Economics and Administration.

Catalano, Ralph Anthony, Prof. Dr., (1946), is since 1989 Professor of Public Health at the University of California at Berkeley, USA. He is both Director of the UC Berkeley Robert Wood Johnson Health and Society Scholars Program and the UC Berkeley Center for Health Research. He is also involved in the UC Berkeley Fogarty Center Training Program for Central Europe. From 1972 – 1989, he used to work as a Professor of Social Ecology and Management at the University of California at Irvine. His main research area spans the health effects of population stressors. Selected publications: Catalano, R.A., Bruckner, T. & Smith, K. (2008). Ambient temperature predicts sex ratios and male longevity. *Proceedings of the National Academy of Sciences of the United States of America, 105,* 2244-2247; Catalano, R.A. (2007), Economic Factors and Stress. In G. Fink (Ed.), *Encyclopedia of Stress* (2nd ed., Vol. 1). (pp. 884-888). Oxford: Academic Press.

Coimbra, Joaquim Luís, Prof. Dr., (1955), is Associate Professor at the Faculty of Psychology and Education Sciences, Porto University, Portugal. His most relevant research areas are career development and intervention, vocational education, lifelong learning, psychosocial consequences of unemployment, youth personal and social development. Selected publications: Coimbra, J.L. (2005). Subjective perceptions of uncertainty and risk in contemporary societies: Affectives-educational implications. In I. Menezes, J.L. Coimbra & B.P. Campos (Eds.), *The Affective Dimension of Education: European Perspectives.* Porto: Fundação para a Ciência e Tecnologia – Ministério da Ciência, Inovação e do Ensino Superior; Santos, B.J. & Coimbra, J.L. (2000). Psychological separation and dimensions of career indecision in secondary school students. *Journal of Vocational Behaviour, 56,* 346-362.

Dalbert, Claudia, Prof. Dr., (1954), is Professor of Psychology at Martin Luther University of Halle-Wittenberg, Germany. She is Editor-in-Chief of the International Journal of Psychology and former President of the International Society for Justice Research (ISJR). Her research interest involves Justice Psychology as well as Educational Psychology. Selected publications: Dalbert, C. & Sallay, H. (Eds.). (2004). The justice motive in adolescence and young adult-hood: Origins and consequences. London: Routledge; Dalbert, C. (2001). The justice motive as a personal resource: Dealing with challenges and critical life events. New York: Kluwer Academic/Plenum Publishers.

De Almeida, Maria Alice, M.Sc., (1955), holds a Master of Science of the Family. Currently she works as a consultant and is involved in international activities in the areas of coaching and mentoring. Her main research areas are family and the quality of life in work.

Demiral, Yucel, Prof. Dr., (1964), is Associate Professor at the Department of Public Health at Dokuz Eylul University in Izmir, Turkey. Since 2002, he is member of the International Commission on Occupational Health (ICOH) and in 2003, he became National Secretary for Turkey for ICOH. Furthermore, he is since 2004 member of the Steering Committee of the SC Unemployment, Job Insecurity and Health of ICOH. Selected publications: Demiral, Y., Soysal, A., Bilgin, A.C., Kılıc, B., Unal, B., Ucku, R. & Theorell, T. (2006). The association of job strain with coronary hearth disease and metabolic syndrome in municipality workers in Turkey. Journal of Occupational Health, 48(5), 332-338; Ergör, A., Demiral, Y. & Piyal, Y.B. (2003). A significant outcome of work life: Occupational accidents in a developing country, Turkey. Journal of Occupational Health, 45, 74-80.

Dornbusch, Peer, Dipl. Psych., (1967), holds a Diploma in Psychology and works as a research assistant at Fachklinik Wilhelmsheim in Oppenweiler, Germany. Selected publications: Henkel, D. & Zemlin, U. (Eds.). (2008). *Arbeitslosigkeit und Sucht: Ein Handbuch für Wissenschaft und Praxis* [Unemployment and addiction. A manual for research and practice]. Frankfurt: FH; Henkel, D., Zemlin, U. & Dornbusch, P. (2005). Prädiktoren der Alkoholrückfälligkeit bei Arbeitslosen 6 Monate nach Behandlung: Empirische Ergebnisse und Schlussfolgerungen für die Praxis [Predictors of alcohol abuse relapses of unemployed 6 months after treatment: Empirical results and conclusions for the practice]. *Suchttherapie, 6*(4), 165-175.

Dragano, Nico, Dr., (1972), is senior researcher in the Department of Medical Sociology at the University of Düsseldorf, Germany. His main fields of research are occupational health, cardiovascular epidemiology, urban health and social inequalities in health. In particular, he studies micro-macro links between the socio-political context and health. Selected publications: Dragano, N., He, Y., Moebus, S., Jöckel, K.H., Erbel, R. & Siegrist, J. (2008). Two models of job stress and depressive symptoms: Results from a population based study. *Social Psychiatry and Psychiatric Epidemiology, 43,* 72-78; Dragano, N., Verde, P.E. & Siegrist, J. (2005). Organisational downsizing and work stress: Testing synergistic health effects in employed men and women. *Journal of Epidemiology and Community Health, 59*, 694-699.

Elkeles, Thomas, Prof. Dr., (1952), is Professor at the Department of Health, Nursing, Management at the University of Applied Sciences Neubrandenburg, Germany. He is member of the European Society of Health and Medical Sociology (ESHMP) and the International Association of Health Policy, Europe (IAHP). In addition, he is also member of the Berlin School of Public Health (BSPH). His main research interest involves health and social policy, unemployment and health, health inequality, social epidemiology, health reporting, evaluation research as well as health, organisational and rural sociology. Selected publications: Elkeles, T. & Kirschner, W. (2005): Unemployment as a determinant of health. In L. Georgieva & G. Burazeri (Eds.), *Health determinants in the scope of new public health* (pp. 96-131). Lage: Hans Jacobs Publishing Company; Elkeles, T., Heinze, S. & Eifel, R. (2007): Healthcare by a DMP for Diabetes mellitus

Type 2 -Results of a survey of participating insurance customers of a health insurance company in Germany. *Journal of Public Health 15*(6), 473 - 480.

Fryer, David, Prof. Dr. B.A., M.A., Ph.D., F.B.Ps.S., F.S.C.R.A, (1949), is since 2009 Professor of Community Critical Psychology at Charles Sturt University in New South Wales, Australia; Professor Extraordinarius, University of South Africa and Honorary Senior Research Fellow, University of Stirling, Scotland. He is also President Elect of the European Community Psychology Association. After completing his Ph.D. in Psychology at the University of Edinburgh, he was Reader in Psychology at the University of Stirling in Scotland. His research area lies in Community Critical Psychology. Selected publications: Fryer, D. (2008). Power from the people? Critical reflection on a conceptualization of power. *Journal of Community Psychology, 36*(2), 238-245; Fryer, D. (2008). Some questions about "The History of Community Psychology". *Journal of Community Psychology, 36*(5), 572-586. Email: dafryer@csu.edu.au.

Hammarström, Anne, Prof. Dr., (1951), is Professor of Public Health at the Department of Public Health and Clinical Medicine at Umeå University, Sweden. She is currently member of the Steering Committee of the Scientific Committee Unemployment, Job Insecurity and Health of ICOH, the Swedish Research Councils′ priority committee, the steering committee of Centre for Gender Excellence at Umeå University and member of the board of the Journal of Social Medicine. Furthermore, she used to work at Karolinska Institute at Uppsala University and has been co-ordinator of a European project on unemployment and health. Her main research areas are social epidemiology and gender studies in public health. Selected publications: Reine, I., Novo, M. & Hammarström, A. (2008). Does transition from an unstable labour market position to permanent employment protect mental health? Results from a 14-year follow-up of school-leavers. *BMC Public Health 8: 159* ; Hammarström, A. & Janlert, U. (2002). Early unemployment can contribute to adult health problems – results from a longitudinal study of school-leavers. *International Journal of Epidemiology and Community Health, 56*(8), 624-630.

Henkel, Dieter, Prof. Dr., (1944), is Professor Emeritus of Psychology and Addiction Research at the Institute of Addiction Research at the University of Applied Sciences in Frankfurt a.M., Germany. His research focuses on unemployment and addiction, addiction and social inequality. Selected publications: Henkel, D. & Zemlin, U. (Eds.). (2008). *Arbeitslosigkeit und Sucht: Ein Handbuch für Wissenschaft und Praxis [Unemployment and addiction: A manual for research and practice].* Frankfurt: FH; Henkel, D., Zemlin, U. & Dornbusch, P. (2005). Prädiktoren der Alkoholrückfälligkeit bei Arbeitslosen 6 Monate nach Behandlung: Empirische Ergebnisse und Schlussfolgerungen für die Praxis [Predictors of alcohol abuse relapses of unemployed 6 months after treatment: Empirical results and conclusions for the practice]. *Suchttherapie, 6*(4), 165-175.

Ishitake, Tatsuya, M.D., Ph.D., (1960), is Professor at the Department of Environmental Medicine at the Kurume University School of Medicine in Japan. He is a member of ICOH Scientific Committee Unemployment, Job Insecurity and Health. His research areas involve occupational and social epidemiology as well as occupational and environmental medicine. Selected publications: Ishitake, T. & Matoba, T. (2006). Health and lifestyle of re-employed and unemployed people following the Japanese *corporate reorganization law. In T. Kieselbach, A.H. Winefield, C. Boyd & S. Anderson*

(Eds.), Unemployment and health. International and interdisciplinary perspectives (pp. 127-133). Bowen Hills: Australian Academic Press; Tamaki, H., Kohshi, K., Ishitake, T., et al. (2010). A survey of neurological decompression illness in commercial breath-hold divers (Ama) of Japan. *Undersea Hyperb Med, 37(4), 209-17.* Nagatomi, K., Ishitake, T., Hara, K., et al. (2010). Association between the transition from unemployment to re-employment after abrupt bankruptcy and the depressive symptoms. *Kurume Med J, 57, 59-66.*

Jeske, Debora, M.Sc., is a doctoral student in the Industrial-Organizational Psychology program at Northern Illinois University/USA. She holds a B.Sc. in Psychology from Westminster University, a M.Sc. in Organisational Psychology from City University, London, UK, and a M.A. in Psychology from Northern Illinois University, USA. Prior to coming to the US, she worked as a Junior Researcher at the Institute for Psychology of Work, Unemployment and Health (IPG) at University of Bremen, Germany. She has also held training/program development and project-related appointments at Middlesex University, UK, and Jacobs University Bremen, Germany. Her main research interest is in employee monitoring, organizational development and change. Selected publication: Kieselbach, T., Knuth, M., Jeske, D. & Mühge, G. (2009). *Innovative Restrukturierung von Unternehmen: Fallstudien und Analysen* [Innovative restructuring of enterprises: Case studies and analyses]. Mehring/München: Hampp.

Kieselbach, Thomas, Prof. Dr., (1944), is Professor Emeritus of Work and Health Psychology and was head of the Institute for Psychology of Work, Unemployment and Health (IPG) at University of Bremen, Germany, now at Förderwerk Bremen. Between 1993-1998 he was Professor for Health Psychology at the University of Hannover. He published 33 books and more than 250 journal articles on work and health psychology, unemployment, evaluation of interventions, enterprise restructuring. He is coordinator and partner of several EU-funded research and policy projects on unemployment, social exclusion and restructuring and Editor of *Psychology of Social Inequality* (VS-Verlag für Sozialwissenschaften). Between 2000-2009 he chaired the Scientific Committee Unemployment, Job Insecurity and Health of the International Commission on Occupational Health (ICOH); since 2009, he is member of the board of ICOH. Selected publications: Kieselbach, T., Heeringen, K. van, Lemkow, L., Sokou, K., Starrin, B. (Eds.). (2001). *Living on the edge - A comparative study on long-term youth unemployment and social exclusion in Europe.* (Psychology of Social Inequality, vol.11). Opladen: Leske + Budrich; Kieselbach, T., Winefield, A., Boyd, C. & Anderson, S. (Eds.). (2006). *Unemployment and health. International and interdisciplinary perspectives.* Bowen Hills: Australian Academic Press.

Kirschner, Wolf, Dr., (1951), is since 1996 head of the Department of Evaluation Research at the private Institute Research, Consultancy + Evaluation in Berlin. From 1987 to 1996 he was Managing Director of the Institute Epidemiological Research Berlin. Since 1979 he is working on epidemiological research especially in the field of infectious diseases. In the last decade he has focused his studies on prevention and evaluative and interventive research. In the field of intervention he has developed two programs for the reduction of preterm deliveries in Germany. Evaluative research is concentrated on the evaluation of prevention programs especially dealing with health promotion of the unemployed. Selected publications: Elkeles, T., Kirschner, W., Graf, C. & Kellermann-Mühlhoff, P. (2009): Health care in and outside a DMP for type 2 diabetes mellitus in Germany – results of an insurance customer survey focussing on differences in general education status. *Journal of Public Health, 17*(3), 205-216; Kirschner, W., Halle, H. & Pogonke, M.-A. (2009). Kosten der Früh- und Nichtfrühgeburten und die Effektivität und Effizienz von Präventionsprogrammen am Beispiel von BabyCare - Eine Schätzung auf der Grundlage der Diagnose Related Groups (DRG) unter Berücksichtigung der Primäraufnahmen in der Neonatologie [Costs of preterm delivery and non-preterm delivery and the efficacy and efficiency of prevention programs, exemplified on BabyCare – An estimation on basis of the Diagnose Related Groups (DRG) in consideration of the primary admission to neonatology]. *Prävention und Gesundheitsförderung, 4,* 41-50.

Kuhnert, Peter, Dr., (1955), is since 1999 research assistant at University of Dortmund, Germany and board member of the Institute for Psychology at the Chair of Organisational Psychology at Technical University of Dortmund. Furthermore, he is senior researcher of the Department „Unemployment, Counselling and Health Promotion" and since 2007 he is university lecturer. From 2002 to 2008, he was project manager of two EQUAL-development partnerships (European Social Fund) and managing several research projects improving health equity of long-term unemployed and other excluded groups. His main research area spans health promotion, prevention of mental illness and coping strategies in unemployment and precarious work. Selected publications: Kuhnert, P. (2007). *Arbeitslosigkeit bewältigen und Lebensmut erhalten – Beratung von Langzeitarbeitslosen* [Overcoming unemployment and obtaining courage to face life – counselling of the permanently unemployed]. Saarbrücken: VDM Verlag Dr. Müller; Kuhnert, P., Deutschmann, A. & Kastner, M. (2008). Gesundheitsförderung für Arbeitslose mit Suchtproblemen: Kritische Bestandsaufnahme und Perspektiven [Health promotion for the unemployed with addiction problems: Critical assessment and perspectives]. In D. Henkel & U. Zemlin (Eds.), *Arbeitslosigkeit und Sucht. Ein Handbuch für Wissenschaft und Praxis* [Unemployment and addiction. A manual for research and practice] (pp.127-162). Frankfurt a. M.: Fachhochschulverlag/Verlag für angewandte Wissenschaften.

Lantz, Annika, Prof. Dr., (1955), is Associate Professor at Uppsala University Sweden and responsible for the research, supervision and the teaching of work and organisational psychology at the Department of Psychology. She is a senior advisor at Fritz Change Company and works part time as a consultant to companies and organisations. Her main research interest has been learning at work, and specifically how the design of work impacts on individuals' and groups' possibilities to go beyond the stipulated tasks, take initiative and be active in change and developmental activities. She has been a partner in several European projects and is cooperating with researchers at German universities regarding methods development. Selected publications: Lantz, A. & Brav, A. (2007). Job design for learning in work groups. *Journal of Workplace Learning, 19*(2), 269-285; Brav, A., Andersson, K. & Lantz, A. (in press). Group initiative and self-organisational activities in industrial work groups. *Journal of Work and Organisational Psychology.*

Levi, Lennart, Prof. Dr., (1930), is Professor Emeritus of Psychosocial Medicine and Member of the Swedish Parliament. He is Temporary Advisor to WHO, ILO and EU. He used to work as Professor of Psychosocial Medicine at the Karolinska Institute in Stockholm/Sweden and has been Director of the National Institute for Psychosocial

Factors and Health. His research involves public and occupational health, stress medicine and health promotion. Selected publications: Levi, L. & Levi, I. (2000). *Guidance on work-related stress - Spice of life or kiss of death?* Luxembourg: European Commission; Levi, L. (1972). Stress and distress in response to psychosocial stimuli. Laboratory and real life studies on sympathoadrenomedullary and related reactions. *Acta Medica Scandinavica, 528*(191),1-166.

Mahendran, Kesi, Dr., (1967), is lecturer in social psychology at the Open University, UK. She is member of the International Commission on Occupational Health – Scientific Committee on Unemployment, Job Insecurity and Health. She completed her Ph.D. at the University of Stirling in 2002 and worked as an analyst within the Scottish Government's analytical services from 2002 to 2006 working on labour market transitions and afterwards as a senior analyst on international labour mobility. She is currently pursuing research in dialogical psychology, labour mobility, migration, integration and citizenship. Selected publications: Mahendran, K. (2003). The transition of a Scottish Young Persons Centre - a dialogical analysis. In C.B. Grant (Ed.), Rethinking communicative interaction: New interdisciplinary horizons. Pragmatics and Beyond New Series (pp. 235-256). Amsterdam/Philadelphia: John Benjamins.

Maignan, Carole Juliette Marie-Anne, Dr., (1972), is researcher and consultant at the University Iuav of Venice in Italy, Meyer University Children's Hospital and UNESCO. Until 2007, she worked at the WHO Venice, where she participated in the elaboration of regional development strategies for the health sector. In particular, she was responsible for a project looking at the economic role of the health sector. She worked for WHO as liaison with global WHO and United Nations initiatives. These included the Millennium Development Goals, the Commission on Macroeconomics and Health and the Commission on Social Determinants of Health. She has also participated actively in international conferences, seminars and organised workshops in the field of sustainable development. Selected publications: Pinelli, D., Maignan, C., Ottaviano, G.I.P. (2003). ICT, clusters and regional cohesion: A summary of theoretical and empirical research. FEEM Working Paper N° 58.03; Maignan, C., Pinelli, D., Francesco Rullani, F. & Ottaviano, G.I.P. (2003). Measuring diversity in Economics: Insights from Biology and Ecology. FEEM Working Paper N°13.

Mannila, Simo, Prof. Dr., (1951), works with the Department of International Affairs of the National Institute for Health and Welfare and is Adjunct Professor of Sociology at the University of Helsinki as well as Adjunct Professor of Social Policy at the University of Turku (Finland). He is a member of several Nordic and EU research networks and has a long career as a consultant of social policy in Ukraine, Russian Federation, Romania and Mongolia. He has carried out extensive research into unemployment and social exclusion, and his present research interest focuses on immigration. Selected publication: Mannila, S. & Reuter, A. (2009). Social exclusion risks and their accumulation among Russian-speaking, ethnically Finnish and Estonian immigrants to Finland. *Journal of Ethnic and Migration Studies, 35*(6), 939-956.

Matoba, Tsunetaka, Prof. Dr., (1935), is Professor Emeritus at Kurume University School of Medicine, Japan. He was Professor at the Department of Environmental Medicine at Kurume University School of Medicine. Since 1999, he is chairing the "Meeting of Unemployment and Health in Japan". His research focuses on industrial medicine, unemployment and bioethics. Selected publications: Matoba, T., Ishitake, T. & Noguchi, R. (2003). A 2-year follow-up survey of health and life style in Japanese unemployed persons. *International Archives on Occupational and Environmental Health, 76*, 302-308; Ishitake, T. & Matoba, T. (2006). Health and lifestyle of reemployed and unemployed people following the Japanese corporate reorganisation law. In: T. Kieselbach, A.H. Winefield, C. Boyd & S. Anderson (Eds.), *Unemployment and health. International and interdisciplinary perspectives* (pp. 127-133). Bowen Hills: Australian Academic Press.

Nyman, Juha Tapio, Dr., (1958), is since 2007 Senior Planning Officer at the City of Helsinki, Health Centre, Strategy Department, Finland. He is vice-chairman of the board at the registered social firm, limited company PosiVire, in Helsinki. From 1989 – 2007 he used to work as a lecturer at the Helsinki Polytechnic. His main interest is supported employment of the disabled and long-term unemployed. Selected publications: Nyman, J.T. (2006). Metropoliluotain, sosiaali- ja terveyspalvelut pääkaupunkiseudulla vuonna 2015 [The metropolis sounder, Finland]. Helsingin kaupungin tietokeskus,Tutkimuksia, 1; Nyman, J.T. (2009). Sosiaalinen yritys hoiva-alalla [Social enterprise in the caring sector]. Helsingin kaupungin tietokeskus, Tutkimuskatsauksia, 2.

Otsuka, Yasumasa, Prof. Dr., (1975), is Associate Professor at the Department of Psychology at Hiroshima University Graduate School of Education, Japan. Previously, he worked at the National Institute of Occupational Safety and Health in Japan. His research focuses on job stress, Occupational Health Psychology and Clinical Psychology. Selected publications: Otsuka, Y., Takahashi, M., Nakata, A., Haratani, T., Kaida, K., Fukasawa, K., Hanada, T., & Ito, A. (2007). Sickness absence in relation to psychosocial work factors among daytime workers in an electric equipment manufacturing company. Industrial Health, 45, 224-231; Kaida, K., Takahashi, M., Haratani, T., Otsuka, Y., Fukasawa, K., & Nakata, A. (2006). Indoor exposure to natural bright light prevents afternoon sleepiness. Sleep, 29, 462-469.

Otto, Kathleen, Prof. Dr., (1975), works as an Assistant Professor at the Work and Organisational Psychology Unit at the University of Leipzig, Germany. She previously worked for Martin Luther University of Halle-Wittenberg, Germany. Her main research areas are organisational justice, mobility readiness, job insecurity and career development. Selected publications: Otto, K. & Schmidt, S. (2007). Dealing with stress in the workplace: Compensatory effects of belief in a just world. *European Psychologist, 12*, 272-282; Otto, K., Glaser, D. & Dalbert, C. (in press). Mental health, occupational trust, and the quality of working life: Does the belief in a just world matter? *Journal of Applied Social Psychology*.

Pensola, Tiina Helena, Dr., (1966), is currently Research and Development Manager at the Centre for Rehabilitation Research and Development of the Rehabilitation Foundation, Finland. She is member of the steering committee of the SeniorForce project that belongs to the ENEA project of the EU. Previously, she worked at the Population Research Unit at University of Helsinki. Her main research interests involve social differences in health and work ability (social and lifecourse epidemiology), ageing and workability and unemployment and health. Selected publications: Pensola, T., Järvikoski, A. & Järvisalo, J. (2008). Unemployment and work ability. In R. Gould, J.

Ilmarinen, J. Järvisalo & S. Koskinen (Eds.), *Dimensions of work ability* (pp. 123-130). Helsinki: Finnish Centre for Pensions, The Social Insurance Institution, National Public Health Institute, Finnish Institute of Occupational Health; Pensola, T. & Martikainen, P. (2004). Life-course experiences and mortality by adult social class among young men. *Social Science & Medicine, 58*, 2149-2170.

Prüßmann, Jan-Frederik, Dipl. Sociologist, (1966), is since 2002 researcher at the Institute for Prevention and Health Promotion at the University of Duisburg/Essen (IPG). His main working fields are social inequality/inequity and health, health promotion for disadvantaged people, unemployment and health, health promotion and prevention by health insurance as well as health systems. He works for the WHO Collaborating Centre on Health Inequities in Insurance-based Health Systems and works closely together with the WHO Regional Office for Europe, Office for Investment for Health and Development (Venice). Selected publications: Prüßmann, J.-F., Friedrichs, M. & Bellwinkel, M. (2009). Gesundheitsförderung in Veränderungsprozessen. Gesundheitsorientierte Selbstmanagementberatung bei drohender Arbeitslosigkeit und prekärer Beschäftigung [Health promotion in change processes. Health focused self management counselling when facing the risk of unemployment and precarious work]. *Gesundheitsförderung und Selbsthilfe, 22*. Bremerhaven: Wirtschaftsverlag NW; Prüßmann, J.-F. (2008). *Social determinants of health and socially conditioned health inequalities – Activities of the statutory health insurance sector in Germany.* Essen: WHO Collaborating Centre for Health Promotion at the Workplace and Institute for Prevention and Health Promotion at the University of Duisburg/Essen.

Rogge, Benedikt G., Dipl. Psych., MA, (1979), is currently Research Assistant at the Department of Sociology at the University of Bremen and Ph.D. candidate at Bremen International Graduate School of Social Sciences, Germany. He completed his Diploma in Psychology in 2006 and also holds a Master's Degree in Sociology. His research focuses on mental health, emotions, stress and the sociological identity theory. Selected publications: Rogge, B.G. (2009): Entwertete Zeit? Erwerbslosenalltag in Paarbeziehung und Familie [Devaluated time? Everyday life of unemployed people living as couples and in families]. In M. Heitkötter, K. Jurczyk, A. Lange & U. Meier-Gräwe (Eds.), *Zeit für Beziehungen? Zeit in und Zeitpolitik für Familien* (pp. 67-90). Opladen: Barbara Budrich; Rogge, B. G., Kuhnert, P. & Kastner, M. (2007): Zeitstruktur, Zeitverwendung und psychisches Wohlbefinden in der Langzeitarbeitslosigkeit [Timestructure, time management and mental well-being in longterm unemployement]. *Psychosozial, 109*, 85-103.

Rogovsky, Nikolai G., Dr., (1965), is currently Senior Specialist at the International Labour Office in Geneva, Switzerland. From 1995-1996, he was Associate Professor at California State University in Hayward, USA. He also worked as consultant at Multinational Research Advisory Group at Wharton Business School in Philadelphia and as Senior Research Officer at the Economic Research Institute of the State Planning Agency of the USSR (GOSPLAN) in Moscow. His research focuses on international human resource management, cross-cultural management, international industrial relations, Corporate Social Responsibility and Socially Responsible Restructuring (IOL-SSER). Selected publications: Rogovsky, N. (Ed.). (2005). Restructuring for corporate success: A socially sensitive approach. Geneva: ILO; Rogovsky, N. & Sims, E. (2002). People-oriented management: Social dimension of corporate success. Geneva: ILO.

Rothländer, Kathrin, Dipl. Psych., (1977), works as a Research Assistant at the Institute of Work, Organisational and Social Psychology at Technical University of Dresden, Germany. Her research areas include health promotion, long-term unemployment and temporary agency work. Selected publications: Rothländer, K. (2009). Training psychosozialer Kompetenzen für Arbeitslose am Beispiel des Gesundheitsförderungsprogramms AktivA [Training of psycho-social skills for the unemployed, using the example of the health promotion program AktivA]. In A. Hollederer (Ed.), Gesundheit von Arbeitslosen fördern! Frankfurt a.M.: Fachhochschulverlag; Mühlpfordt, S. & Rothländer, K. (2008). Gesundheitsförderung für Arbeitslose bei Bildungs- und Beschäftigungsträgern [Health promotion for unemployed at training providers and employment agencies]. In B. Bergmann, U. Pietrzyk & J. Klose (Eds.), Beschäftigungsfähigkeit entwickeln, Innovationsfähigkeit und Kompetenz fördern (pp. 85-94). Dresden: Technische Universität Dresden.

Siegrist, Johannes, Prof. Dr., (1943), is since 1992 Director of the Department of Medical Sociology and the Postgraduate Training Programme Master of Science in Public Health at Heinrich Heine-University Duesseldorf, Germany. His major research interests involve the influence of employment conditions on health, social determinants of health in midlife and early old life, with a special focus on psychosocial stress at work. He is author of the work stress model effort-reward imbalance. Selected publication: Siegrist, J. & Marmot, M. (Eds.) (2006). *Social inequalities in health: New evidence and policy implications.* Oxford: Oxford University Press.

Sousa Ribeiro, Marta, (1976), is a Ph.D. student at the Faculty of Psychology and Education Sciences at Porto University, Portugal. Her main research areas include psychological consequences of unemployment, career development, vocational education, and lifelong learning. Selected publication: Sousa Ribeiro, M. & Coimbra, J.L. (2008). Os Trabalhadores Seniores Face ao Mercado de Trabalho [Senior workers and the labour market]. *Formar, Revista para Formadores, 62*, 28-32.

Tobler, Sibylle, Dr., (1962), works currently at her own practice in the field of counselling individuals and organisations in processes of change both in the Netherlands and Switzerland. Her international activities include lectures and trainings. From 1995 till 2004 she was general manager of two praticefirms in Bern, Switzerland and has worked as an assistant at the Institute of Practical Theology at the University of Bern, Switzerland. Her main research interests involve psychological research on unemployment, psychological aspects of individual transition processes, helpful aspects of individual coping with change, counselling with a focus on solution-oriented short-term concepts in the tradition of Steve de Shazer and self-empowerment. Selected publications: Tobler, S. (2004). *Arbeitslose beraten unter Perspektiven der Hoffnung. Lösungsorientierte Kurzberatung in beruflichen Übergangsprozessen* [Counselling unemployed people with perspectives of hope. Solution-oriented short-term counselling in professional transition processes]. Stuttgart: Kohlhammer; Tobler, S. (2010). *Neuanfänge – Veränderung wagen und gewinnen* [Making a new start - Dare to change and win] (2nd ed.). Stuttgart: Klett-Clotta. E-Mail: sibylle.tobler@sibylletobler.com. Homepage: www.sibylletobler.com

Vuori, Jukka, Prof. Dr., (1954), is Research Professor at the Finnish Institute of Occupational Health and Director of the Life Course and Work theme. He is Visiting Scholar at the Institute for Social Research at the Prevention Research Centre at the University of Michigan and was scientific expert for WHO preparing the agenda for the First European Ministerial Congress on Mental Health in 2005. He is reviewer in several international journals. Furthermore, he is member of organizing committees, keynote speaker and chair of symposia of international conferences and symposia. Since 2009 he is chairperson of the Scientific Committee Unemployment, Job Insecurity and Health of the International Commission on Occupational Health (ICOH). His research areas involve sources and consequences of occupational stressors, work life transitions and coping and prevention during the work life course. Selected publications: Vuori, J., Koivisto, P., Mutanen, P., Jokisaari, M., & Salmela-Aro, K. (2008). Towards working life: Effects of an intervention on mental health and transition to post-basic education. *Journal of Vocational Behavior, 72*, 67-80; Vuori, J. & Vinokur, A. (2005). Job-search preparedness as a mediator of the effects of the Työhön job-search intervention on re-employment and mental health. *Journal of Organisational Behavior, 26*, 275-291.

Wacker, Alois, Prof. Dr., (1942), is Professor Emeritus at Leibniz University of Hannover, Germany where he has formerly been Professor of social psychology. His research focuses on research methods, individual and social meaning of work and psychosocial impact of unemployment. Selected publications: Wacker, A. & Kolobkova, A. (2000). Arbeitslosigkeit und Selbstkonzept - ein Beitrag zu einer kontroversen Diskussion [Unemployment and self concept – a contribution to a controversial discussion]. *Zeitschrift für Arbeits- und Organisationspsychologie, 44*, 69–82; Wacker, A. (2003). Zur Wirksamkeit der Total Design Method (TDM) nach Dillman - am Beispiel der Hannoverschen Absolventenstudie [The efficacy of the Total Design Method (TDM) of Dillman – an example of the Hannover alumni study]. In J. Allmendinger (Ed.), *Entstaatlichung und soziale Sicherheit. Verhandlungen des 31. Kongresses der Deutschen Gesellschaft für Soziologie in Leipzig 2002* [Denationalization and social security. Negotiations at the 31. Congress of the German Society for Sociology in Leipzig 2002]. Opladen: Leske + Budrich.

Zemlin, Uwe, Dr., (1951), is Head of Clinical Psychology and Psychotherapy at Fachklinik Wilhelmsheim in Oppenweiler, Germany. His research interests are unemployment and addiction therapy. Selected publications: Henkel, D. & Zemlin, U. (Eds.). (2008). *Arbeitslosigkeit und Sucht: Ein Handbuch für Wissenschaft und Praxis* [Unemployment and addiction. A manual for research and practice]. Frankfurt: FH. Henkel, D., Zemlin, U. & Dornbusch, P. (2005). Prädiktoren der Alkoholrückfälligkeit bei Arbeitslosen 6 Monate nach Behandlung: Empirische Ergebnisse und Schlussfolgerungen für die Praxis [Predictors of alcohol abuse relapses of unemployed 6 months after treatment: Empirical results and conclusions for the practice]. *Suchttherapie, 6*(4), 165-175.

Zhou, Zhi-jun, Prof. Dr., (1964), is Professor and Chair of the Department of Occupational Health and Toxicology / WHO Collaborating Center for Occupational Health, School of Public Health of Fudan University in Shanghai. He got his bachelor (1985) and master (1988) degree in Medicine from Shanghai Medical University, and his Ph.D. in (1996) from University of Erlangen, Germany. His main research area span health effects of occupational exposure to industrial chemicals, such as pesticides, heavy metals and endocrine disruptors, the industrial toxicology and experimental therapy of pesticides and the management and control of occupational hazards at workplaces. Selected

publications: He Y., Miao M., Herrinton L.J., Wu C., Yuan W., Zhou Z. & Li D.-K. (2009). Bisphenol A levels in blood and urine in a Chinese population and the personal factors affecting the levels. *Environmental Research, 109,* 629-633; Chang X., Shao C., Wu Q., Wu Q., Huang M. & Zhou Z. (2009). Pyrrolidine dithiocarbamate attenuates paraquat-induced lung injury in rats. *Journal of Biomedicine and Biotechnology, Article ID 619487, 8 pages; doi:10.1155/2009/619487;* Wu C., Liu P., Zheng L., Chen J., Zhou Z.(2010). Urinary dialkylphosphate metabolites concentrations of organophosphorous pesticides among occupationally exposed workers and general population in Shanghai of China. Journal of Chromatography B.; 878: 2575-2581; Li W., Shibata E., Zhou Z., Ichihara S., Wang H.,Wang Q., Li J., Zhang L.,Wakai K., Takeuchi Y.,Ding X., Ichihara G. (2010) Dose-dependent neurologic abnormalities in workers exposed to 1-bromopropane. Journal of Occupational and Environmental Medicine;52(8):769-77; Wu Q., Ban T., Chang X., Wu Q., Zhou Z.(2010) Effects of Acute and Subchronic Exposures to Dimethoate on Rat Cerebral Cortex GABAergic System. Journal of Health Sciences 56(3): 267-274.